THE NORTH-WEST EUROPEAN SHELF SEAS: THE SEA BED AND THE SEA IN MOTION
I. GEOLOGY AND SEDIMENTOLOGY

FURTHER TITLES IN THIS SERIES

1 J.L. MERO
THE MINERAL RESOURCES OF THE SEA
2 L.M. FOMIN
THE DYNAMIC METHOD IN OCEANOGRAPHY
3 E.J.F. WOOD
MICROBIOLOGY OF OCEANS AND ESTUARIES
4 G. NEUMANN
OCEAN CURRENTS
5 N.G. JERLOV
OPTICAL OCEANOGRAPHY
6 V. VACQUIER
GEOMAGNETISM IN MARINE GEOLOGY
7 W.J. WALLACE
THE DEVELOPMENT OF THE CHLORINITY/SALINITY CONCEPT IN OCEANOGRAPHY
8 E. LISITZIN
SEA-LEVEL CHANGES
9 R.H. PARKER
THE STUDY OF BENTHIC COMMUNITIES
10 J.C.J. NIHOUL (Editor)
MODELLING OF MARINE SYSTEMS
11 O.I. MAMAYEV
TEMPERATURE—SALINITY ANALYSIS OF WORLD OCEAN WATERS
12 E.J. FERGUSON WOOD and R.E. JOHANNES
TROPICAL MARINE POLLUTION
13 E. STEEMANN NIELSEN
MARINE PHOTOSYNTHESIS
14 N.G. JERLOV
MARINE OPTICS
15 G.P. GLASBY
MARINE MANGANESE DEPOSITS
16 V.M. KAMENKOVICH
FUNDAMENTALS OF OCEAN DYNAMICS
17 R.A. GEYER
SUBMERSIBLES AND THEIR USE IN OCEANOGRAPHY AND OCEAN ENGINEERING
18 J.W. CARUTHERS
FUNDAMENTALS OF MARINE ACOUSTICS
19 J.C.J. NIHOUL (Editor)
BOTTOM TURBULENCE
20 P.H. LEBLOND and L.A. MYSAK
WAVES IN THE OCEAN
21 C.C. VON DER BORCH (Editor)
SYNTHESIS OF DEEP-SEA DRILLING RESULTS IN THE INDIAN OCEAN
22 P. DEHLINGER
MARINE GRAVITY
23 J.C.J. NIHOUL
HYDRODYNAMICS OF ESTUARIES AND FJORDS

Elsevier Oceanography Series 24A

THE NORTH-WEST EUROPEAN SHELF SEAS: THE SEA BED AND THE SEA IN MOTION
I. GEOLOGY AND SEDIMENTOLOGY

Edited by

F.T. BANNER, M.B. COLLINS and K.S. MASSIE

Department of Oceanography, University College of Swansea, Swansea, Great Britain

ELSEVIER SCIENTIFIC PUBLISHING COMPANY
Amsterdam - Oxford - New York 1979

4N

ELSEVIER SCIENTIFIC PUBLISHING COMPANY
335 Jan van Galenstraat
P.O. Box 211, 1000 AE Amsterdam, The Netherlands

Distributors for the United States and Canada:

ELSEVIER NORTH-HOLLAND INC.
52, Vanderbilt Avenue
New York, N.Y. 10017

Library of Congress Cataloging in Publication Data

Main entry under title:

The North-West European shelf seas.

(Elsevier oceanography series ; 24 A)
Includes bibliographies and index.
CONTENTS: v. 1. Geology and sedimentology.
1. Continental margins--Europe, Northern. 2. Oceanography--North Atlantic Ocean. 3. Geology--Europe, Northern. I. Banner, Frederick Thomas, 1930-
II. Collins, M. B. III. Massie, K. S.
GC84.2.N67N67 554 78-14524
ISBN 0-444-41734-6 (v. 1)

ISBN: 0-444-41734-6 (Vol. 24A)
ISBN: 0-444-41623-4 (Series)
ISBN: 0-444-41740-0 (Set)

Printed in The Netherlands

Contributing Authors

R.J. BAILEY	Exploration and Production Dept., BP Trading Ltd., Britannic House, Moor Lane, London, EC2Y 9BU, U.K. (formerly: UCNW Marine Science Laboratories, Menai Bridge, North Wales)
F.T. BANNER	Department of Oceanography, University College, Swansea, West Glamorgan, SA2 8PP, U.K. and Department of Geology, University College London, Gower Street, London WC1E 6BT, U.K.
P. BISHOP	Department of Geology, University College, Gower Street, London, WC1E 6BT, U.K.
K.F. BOWDEN	Department of Oceanography, The University of Liverpool, Bedford St. North, P.O. Box 147, Liverpool, L69 3BX, U.K.
D.J. BLUNDELL	Department of Geology, Chelsea College, University of London, 271, King Street, London, W6 9LZ, U.K.
J.D. BURTON	Department of Oceanography, The University, Southampton, S09 5NH, U.K.
V.N.D. CASTON	British Petroleum Co. Ltd., Britannic House, Moor Lane, London, EC2Y 9BU, U.K.
M.B. COLLINS	Department of Oceanography, University College, Swansea SA2 8PP, West Glamorgan, U.K.
S.J. CULVER	Department of Paleobiology, Smithsonian Institution, Washington, D.C. 20560, U.S.A. and Department of Oceanography, University College, Swansea SA28PP, U.K.
M.R. DOBSON	Department of Geology, University College of Wales, Aberystwyth, Cards, SY23 3DB, U.K.
L.R. DRAPER	Marine Information and Advisory Service, Institute of Oceanographic Sciences, Brook Road, Wormley, Godalming, Surrey, GU8 5UB, U.K.
D. HAMILTON	Department of Geology, Queen's Buildings, University Walk, University of Bristol, Bristol, BS8 1TR, U.K.
D.M. HARDING	Department of Forestry and Wood Science, University College of North Wales, Bangor, Gwynedd, LL57 2UW, U.K.
D.A. HUNTLEY	Department of Oceanography, Dalhousie University, Halifax, Nova Scotia, CANADA. (formerly at the Institute of Coastal Oceanography and Tides, Bidston Observatory, Birkenhead, Cheshire, L43 7RA, U.K.).
E.J.W. JONES	Department of Geology, University College London, Gower Street, London, WC1E 6BT, U.K.; now at Bliksem Geophysical Services Ltd., South Quay, Gt. Yarmouth, U.K.
A.J. LEE	Ministry of Agriculture, Fisheries and Food, Fisheries Laboratory, Lowestoft, Suffolk, NR33 0HT, U.K.
K.S. MASSIE	Department of Oceanography, University College, Swansea SA2 8PP, West Glamorgan, U.K.
J. MAUCHLINE	Scottish Marine Biological Association, Dunstaffnage Marine Research Laboratory, P.O. Box 3, Oban, Argyll, Scotland, U.K.
R.D. PINGREE	Institute of Oceanographic Sciences, Brook Road, Wormley, Godalming, Surrey, GU8 5UB, U.K.
E.M. WILSON	Department of Civil Engineering, The University, Salford, M5 4WT, U.K.
M.L. YOUNG	Department of Oceanography, The University, Southampton, S09 5NH, U.K.

Preface

These invited reviews seek to summarise present knowledge of the physical nature of the sea bed and of the movement of the sea waters and sediments on the north-west European continental shelf. In recent years, national and international economic needs have accelerated exploration by governmental organisations, industrial companies and universities, paralleling the growth of the marine sciences as a whole. Probably, more quantitative data on the north-west European shelf has been accumulated during the last few decades than in all of the preceding centuries of human activity on and in these seas. Therefore, although the time to begin a synthesis has clearly come, the reviews are limited to the region north of Ushant (Ile de l'Ouessant) and south of the northerly limits of the Norwegian Trough; the Baltic Sea has been omitted entirely, as it would require separate studies of its own. We have sought to focus attention on those areas and aspects which particularly merited synoptic study, ranging from the geological origins, structure and history of the shelf, with the geomorphology and sediments of the sea bed, to the nature and effects of tides and waves, and the movement of water masses. Mineral resources (e.g., coal) from below the sea bed, which are unaffected by present oceanographic processes, are not dealt with; the petroleum geology and hydrocarbon reserves of the North Sea have been already reviewed extensively (e.g Pegrum et al., 1975; Woodland, 1975; Odell and Rosing, 1976; Selley, 1977), but this still represents only a small part of what is known: most of the results of commercial sub-sea-bed exploration remain inaccessible. Consequently, we concern ourselves with the physical resources found at and above the sea bed, whose distribution, as well as development, is affected by the oceanographic processes of the present day; for the pre-Quaternary geology of the North Sea, the reader must look elsewhere.

The geotectonic origin of the shelf, the development and geological nature of the continental margin and the geological history of the western edge of the shelf itself are discussed by Bailey and by Dobson (Chapters 2 and 3), who concentrate their attention upon areas immediately west of the British Isles. The deep structure of the south-western part of the shelf is interpreted by Blundell (Chapter 4), and the sea bed of this region, the "Celtic Sea", is described by Hamilton (Chapter 5) who also is concerned with the canyons of the continental slope to the west, and the palaeovalleys and their geological contexts in the English Channel to the east. The Sea of the Minches has been chosen by Bishop and Jones (Chapter 6) to exemplify the interrelationships between geology, geomorphology, bathymetry, sedimentology, hydrography and dynamical oceanography, in a partially enclosed, relatively deep, sea. In contrast, Caston describes the sea bed of the broader, relatively shallower North Sea, especially in relation to its Quaternary history and modern sedimentology (Chapter 7). The nature of the sediments themselves is reviewed in Chapter 8. The tidal regimes and wave climates of the north-western shelf as a whole are described and discussed by Huntley (Vol. II, Chapter 9) and by Draper (Vol. II, Chapter 10), and the effects of the interplay of tidal and wave-induced currents on the sea bed are introduced in Vol. II, Chapter 11. The physical and dynamical oceanography of the Irish Sea is described by Bowden (Vol. II, Chapter 12), of the Celtic Sea and English Channel by Pingree (Vol. II, Chapter 13) and of the North Sea by Lee (Vol. II, Chapter 14). Burton and Young (Vol. II, Chapter 15) describe the input of trace metals to the sea and their distribution and fate therein, and Mauchline (Vol. II, Chapter 16) follows this with an account of radio-isotopes and their significance in the interpretation of the movements of sea waters and sediments. The natural, inorganic resources of the seabed are described in Vol. II, Chapter 18; the economic potential of the sea water itself is dealt with in Vol. II, Chapter

19, and Wilson (Vol. II, Chapter 20) discusses the development of real and possible schemes for harnessing the power of the tides. Finally, the socioeconomic implications of the development of marine resources, including reclaimed land and other coastal developments, are discussed by Harding (Vol. II, Chapter 21).

It is hoped that the references to previously published work, given at the end of each chapter, will be a guide both to sources and to further reading. In the text and figures, metric units are preferred but it has occasionally been necessary to retain imperial units where they have been used by earlier authors and their direct conversion would be misleading or confusing (metric equivalents are then noted).

We are very grateful, not only to the authors whose work is presented here, but also to our colleagues for their help and forebearance during the preparation of this book. We wish to thank Mrs. J. Humphreys, Mr. Keith Naylor and Mr. J. Uzzell Edwards for their help in the preparation of many of the figures; especially, we must than Ms. J. Greengo for her assiduous and conscientious work in manuscript and index preparation.

References

Odell, P. and Rosing, K., 1976. The North Sea Oil Province. Kogan Page, London, 60 pp.

Pegrum, R., Rees, G. and Naylor, D., 1975. Geology of the North-West European Continental Shelf. 2. The North Sea. Graham Trotman Dudley, London, 225 pp.

Selley, R.C., 1977. The habitat of North Sea oil. Proc. Geol. Assoc., 87: 359–388.

Woodland, A.W. (Editor), 1975. Petroleum and the Continental Shelf of North-West Europe. 1. Geology, Applied Science Publishers, London, 501 pp.

F.T. BANNER
M.B. COLLINS
K.S. MASSIE
January, 1978

Contents

Volume II. Physical and Chemical Oceanography, and Physical Resources

Contents

Chapter 1

Introduction

F.T. BANNER

The post-Pleistocene flooding of some 942,000 km² of the continental shelf of north-west Europe, west of the Baltic Sea, has created some of the most important epicontinental shelf seas of the world. Not only have the English Channel and the southern North Sea, in particular, some of the busiest shipping clearways in the world (see, e.g., Lee and Ramster, 1976, sheet 16), serving as they do much of the European Economic Community, but the seas are also sources of its industrial wealth (fisheries, petroleum, minerals, power) and sinks for disposal of the refuse from its intensively urbanised and industrialised coasts. The total area of the north-western European shelf seas comprises only about 3¼% of that of all of the shallow seas which fringe the world's continents, but the significance of these seas is far greater than this proportion might suggest.

It could even be claimed that continental-shelf oceanography began here. Although the rise and fall of the tide had been noted earlier by Herodotus and Aristotle, it seems to have been the spectacular tidal ranges of the English Channel which led Pytheas the Greek, in the fourth century B.C., to measure them, to recognise the phase inequality of springs and neaps and to observe some practical, economic use to which they could be put (MacMillan, 1966). Although these observations, as recorded by Strabo and Diodorus, did not prevent Julius Caesar from accidentally stranding his ships, three centuries later, high on a British beach during the ebb of a springs tide, the phenomenon was well known to the seafaring, roving Vikings and Danes, as the Old Teutonic "tîdiz" (iterative events, in this case repeating with a lunar period) confirms.

By the early part of the eighth century A.D., Bede was able to compile the first quantitative observations which would now be called "cotidal", and to deduce the north to south progression of the tidal crest down the North Sea coast of England (See Vol.II, Chapter 1). By the thirteenth century, the relationship between the times of new moon and of the tidal flood at London Bridge enabled the Bishop of St. Albans to produce the first tidal predictions, which were later to become generalised in the equilibrium theories of Sir Isaac Newton and Bernoulli (1740) and particularised in Whewell's cotidal charts (1833) and Kelvin's tide-prediction machines (1882). Even before tidal predictions had become a matter of routine determination and prediction, Nils Gissler had statistically shown the effects of atmospheric pressure changes on sea level, leading to the understanding of storm surges, and Sir Henry de la Beche, geologist, had turned his attention to the effects of tidal currents on the transport and sorting of sea bed sediments (Chapter 8), thus beginning the science of dynamic sedimentology. The seventeenth century had already seen the first comparative salinity determinations (Boyle, 1673), and the eighteenth was one in which there was speculation on the significance of rates of dilution of sea water by river discharges (Halley, 1715); by the early nineteenth century, marine chemistry established its foundations in the recognition, by Albert Marcet of London, of the constancy of composition of sea salt, later confirmed by Forchhammer. By the late nineteenth centry, the "Knight Errant", "Goldseeker", "Porcupine", "Lightning", "Shearwater", "Ingolf" and many other small ships (later, notably, the "Michael Sars") were surveying systematically the north-western European continental shelf: their observations added data to the pool which was to form the basis of modern physical oceanography (e.g., Sandström and Helland-Hansen, 1902; Knudsen, 1903), and to our knowledge of the nature of the sea bed. The latter was improved beyond measure by the invention of asdic (Richardson, 1912), later to be developed as precise depth-recorders, sub-bottom profilers and side-scan sonar recorders. It was here that GLORIA (the geological long-range inclined asdic) first produced its revelatory acoustic images of a shelf sea-bed (Belderson

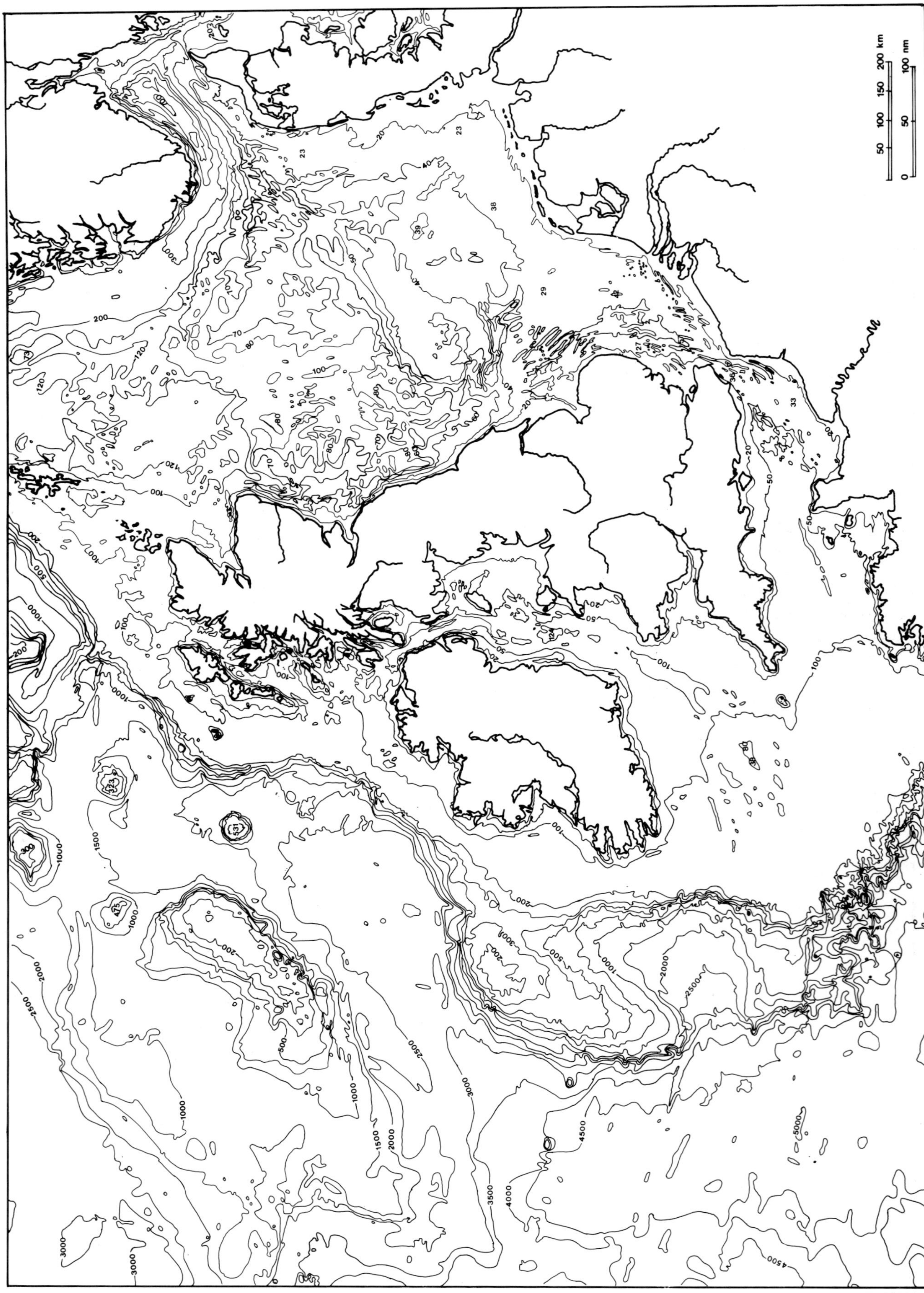

Fig. 1.1 Bathymetry of the north-western European shelf.

Fig. 1.2. Topography of the north-western European shelf.

and Stride, 1969; Rusby and Revie, 1975). Up to the end of the era closed by World War II, there existed practically no quantitative knowledge of the physical processes of wind-wave generation or propagation; there was a wide gulf between the mathematical studies of idealised wave forms and the largely visual records made by mariners and by devoted amateurs such as Vaughan Cornish. The first serious application of modern techniques to the recording and analysis of real waves was made in the early 1940's by Dr. (now Sir George) Deacon and his group, which later formed the nucleus of the U.K. National Institute of Oceanography (now the Institute of Oceanographic Sciences, Wormley, U.K.) (Cartwright, 1967). With the development of sea-bed wave recorders and spectrum analysers (Barber and Ursell, 1948, and their classical analyses of waves and swell arriving on the coast of Cornwall), the processes of wave generation, propagation, growth and decay came closer to quantitative comprehension, and forecasting became possible.

The bathymetry of the north-western European shelf is broadly depicted in Fig. 1.1. The low relief, which follows from planation during the late Tertiary and the Quarternary (Chapters 3, 5–8), is typical of a trailing edge (of the Eurasian crustal plate, see, e.g., Laughton, 1975) which has reached geomorphological maturity ("Amero-type" of Inman and Nordstrom, 1971). To the south-west of the Celtic Sea (Fig. 1.2), its margin is dissected by active canyons (Belderson and Stride, 1969, cit. supra; Hamilton, Chapter 5; Francis, 1962; Whittard, 1962; Hadley, 1964; Curray et al., 1966a,b) while, to the west and north-west of the Malin Sea and the Sea of the Hebrides, the margin is apparently more stable sedimentologically, and relict canyons are buried (Curray et al., 1966a,b). To the west and north-west, especially, the slope is steep and topographically betrays its origin in the plate separation, rifting and faulting which shows itself so spectacularly in the separation of the Rockall Bank microcontinent (Roberts, 1975a,b), with its slumped eastern margin (Roberts, 1972), and the thin crusts of the Rockall Trough (Scrutton, 1972; Roberts, 1975c). Much further to the north (strictly, beyond the scope of this book), off Norway, the boundary between oceanic and continental crust – i.e., the geological definition of the margin of the continent – is recognisable by the buried scarp of the Vøring Plateau and the Faeroe–Shetland Escarpment, with a segment of the Jan Mayen Fracture Zone between them; they define the landward boundary of buried basement highs on the continental slope, which are a direct continuation of typical oceanic basement (Talwani and Eldholm, 1973; Sundvor and Nysaether, 1975). The Escarpment continues south-west from the gentle, Norwegian continental slope to the Faeroes–Shetland Channel and may be continuous with the ocean-wards limits of the West Shetland Shelf to mark the continental margin of Scotland (Laughton, 1975). Further to the south, the Porcupine Bank, and its separation from the continental mass to form the Porcupine Seabight (Fig. 1.2), are described in Chapters 2, 3 and 4.

The shelf itself is, essentially, a flooded section of north-western Europe, and has suffered the imprint of European orogenies, especially those which are called "Caledonian" (middle Palaeozoic, which created the Caledonides mountains of Scotland and the Isles) and "Armorican " (or "Hercynian" or "Variscan", late Palaeozoic, which were primarily responsible for the existence of Armorica and Cornubia). Their lineations are still impressed upon the outlines of the western coasts (Fig. 1.3). However, both deep and shallow geological structures, below the sea bed, have left little direct regional impact upon the present-day bathymetry (Fig. 1.4, cf. Fig. 1.1), although the nature of the bed-rocks which outcrop at the seabed (or which exist below a cover of Quaternary deposits) have directly or indirectly influenced both topography and sedimentation (Chapters 3–8) and cannot be ignored. Consequently, the geological origins and structure of the western parts of the shelf are described in ensuing chapters by Bailey, Blundell, Hamilton and Dobson. The geology of the North Sea has already been extensively described in publications by Kent (1975a), Pegrum et al. (1975), Selley (1977), Whiteman et al. (1975), Ziegler (1975), and various authors in Woodland (1975), i.a. (see Brooks and Thomas, 1974), although much information must remain oil-company property and is still unpublished. The tectonic development of the shelf as a whole has been admirably outlined by Kent (1975b), who, with Naylor and Mounteney (1975), has provided concise descriptions of the nature of the major sedimentary basins and highs which form its structure (see also Delany, 1970).

The numberless eustatically and isostatically induced transgressions and regressions of the sea which have occurred since the end of Miocene time (i.e., during the last 5 million years, following the chronology of Berggren, 1977), aided during the Quaternary by successive advances and retreats of the icesheets and glaciers (Chapters 7 and 8), have sculpted the shelf into the basins and channels now occupied by the north-western European shelf seas (Fig. 1.2). None is wholly isolated from any other, although their entrances may suffer restriction (especially those of the Sea of the Minches, notable for its relative narrowness and depth, see

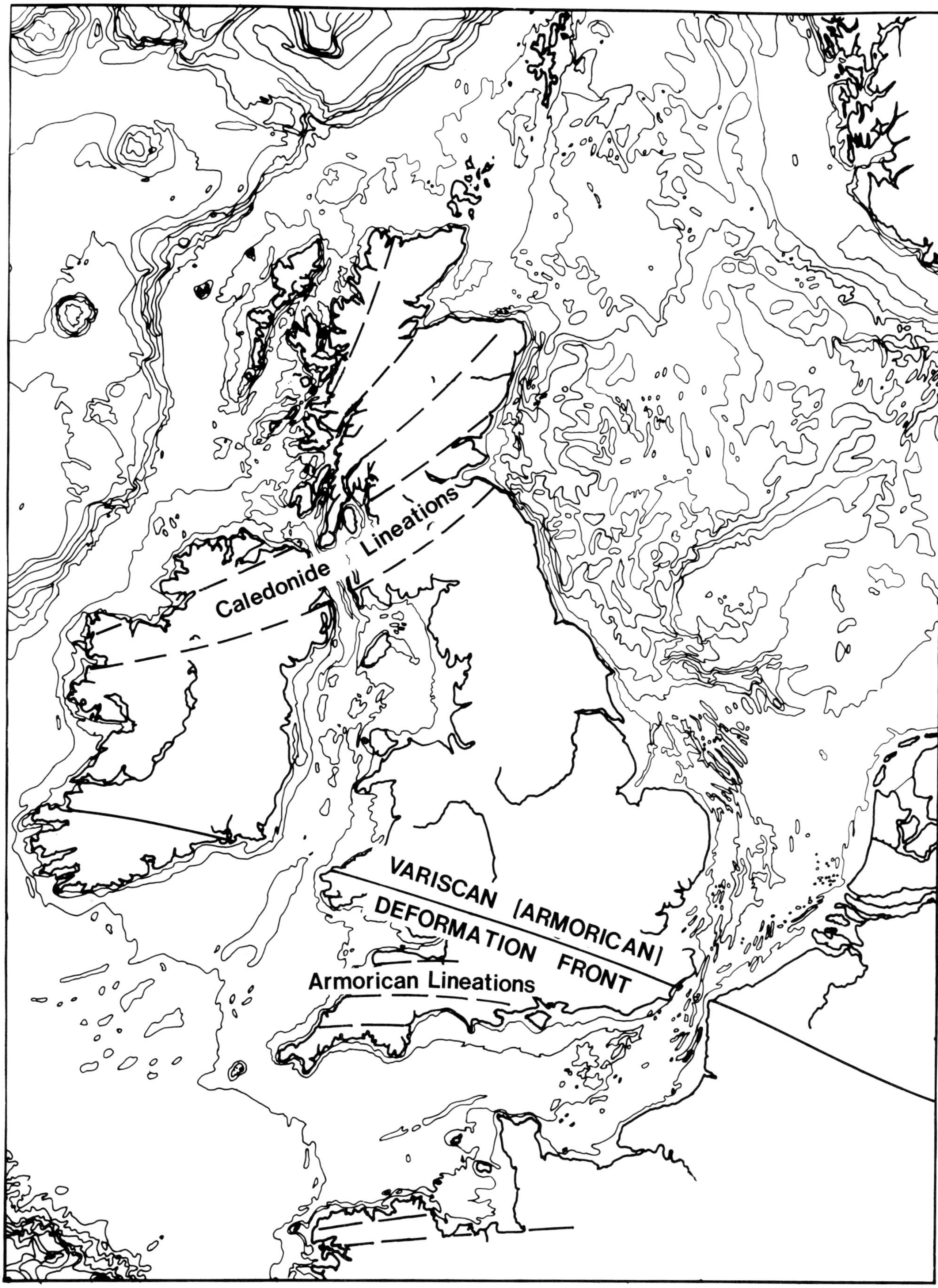

Fig. 1.3. Bathymetry and orogenic lineations.

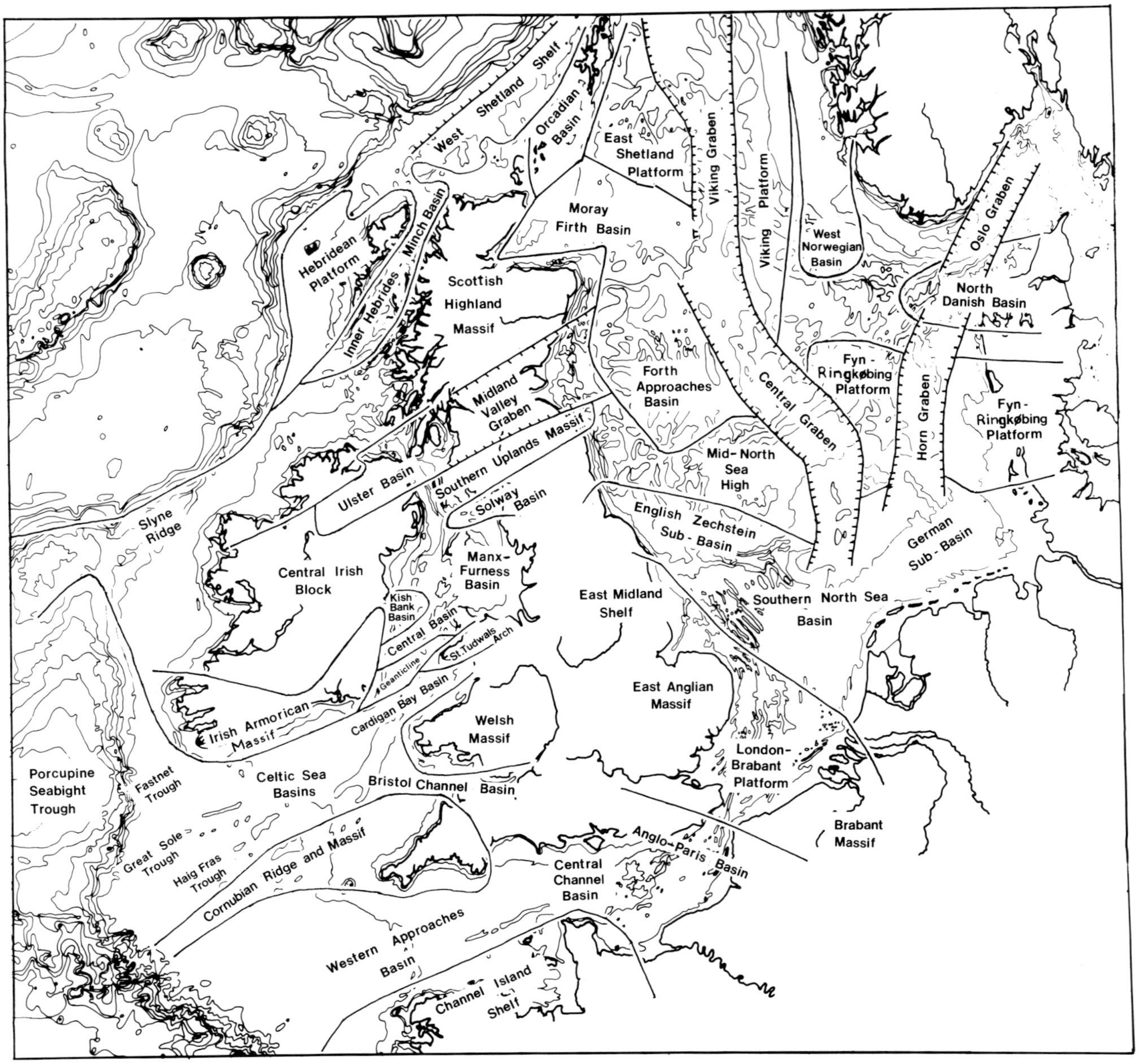

Fig. 1.4. Major geological structural units (modified after Kent, 1975a; Pegrum et al., 1975, et auctt.).

Chapter 6). The Hebridean, Malin and Celtic Seas openly fringe the north-eastern Atlantic Ocean, and are subjected directly to the northwards-progressing eastern Atlantic tide, which enters the Irish Sea and the North Sea by their northern entrances (Doodson and Warburg, 1941, pp. 211–217), there to develop particular tidal regimes (Vol.II, Chapter 9). The shelf is northerly of mid-latitudes; the Coriolis effect operates on the tides of the English Channel, so that eastwardly a progressive Kelvin wave induces tidal ranges on the southern side which can be used as a source of power (Vol.II, Chapter 12). Shallow-water effects on the tides of some inlets make their exploitation possible there also (Vol.II, Chapter 20). Tidally-induced currents on this shelf can be among the most powerful of any in the world, and they have been held responsible for much of the sculpture, scour and sediment transport of the modern sea-bed (Chapters 5, 7 and 8) as well as for the transport, dispersion and mixing of water masses (Vol.II, Chapters 12–16).

Exposed as it is to the northern Atlantic, with possible directions of fetch in excess of 5000 km, it is not surprising that the westerly facing shelf has been classed as one of the "storm wave" coastal zones of the world (Davies, 1964). Although Emery (1970) did not class the north-western European shelf as one of those dominantly eroded by waves (preferring to emphasise the role of ice), there can be no doubt that wave action has contributed, and does contribute, greatly to the movement of all particles, organic and inorganic, on the sea bed, and has played a major part in its erosion and in the redistribution of its sediments (Chapter 8; Vol.II, Chapters 10 and 11). The combined forces of tides and waves create turbulence which produces vertically homohaline waters in all seas except the northern North Sea (Vol.II, Chapter 14), and which are homothermal except for the fair weather of the height of summer (see, e.g. Vol.II, Chapter 13); this has immediate effects upon the transportation, mixing and dispersal of both natural and artificial additives to the sea waters (Vol.II, Chapters 14–16), making it possible to use them as waste-disposal sinks without the excessively cumulative pollution which has damaged so many other seas.

In its oceanic context, this westward facing shelf is situated peripherally not only to the North Atlantic and north-west Atlantic tidal amphidromes (see Doodson and Warburg, 1941, p.212, fig. 24.15) but also to the mid-latitudinal, anticyclonic current system which

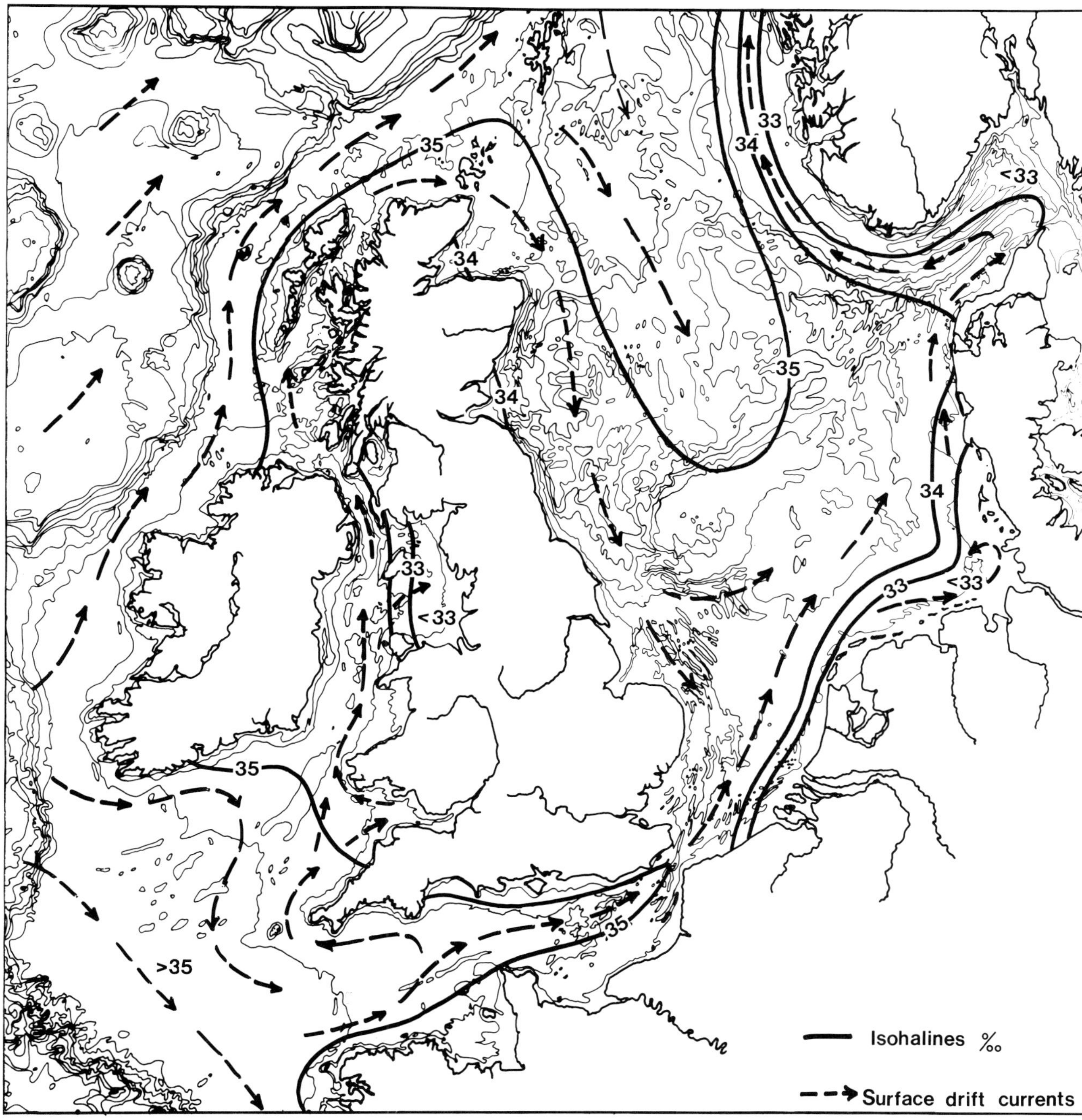

Fig. 1.5. The surface isohalines (‰) (monthly means for 1905–1954, ICES data, 1962) for February, adapted from Lee and Ramster (1976), with average surface drift-current and residual current directions of transport.

is dominated by the Gulf Stream in its western boundary, transforming into the North Atlantic current system beyond the Grand Banks (see, e.g., Stommel, 1965), apparently ending in great, slow moving, ill-defined eddies off the margins of the European shelf. Energy transport by these currents greatly affects the north-eastern Atlantic climate, directly by influencing the paths and intensities of atmospheric cyclones and anticyclones across the North Atlantic towards north-west Europe, and indirectly by warming the north-east Atlantic surface waters to temperatures above the global mean for their latitude. Even though the North Atlantic current is subjected to long-term variations (Maximov et al., 1972) its transport of heat largely determines the

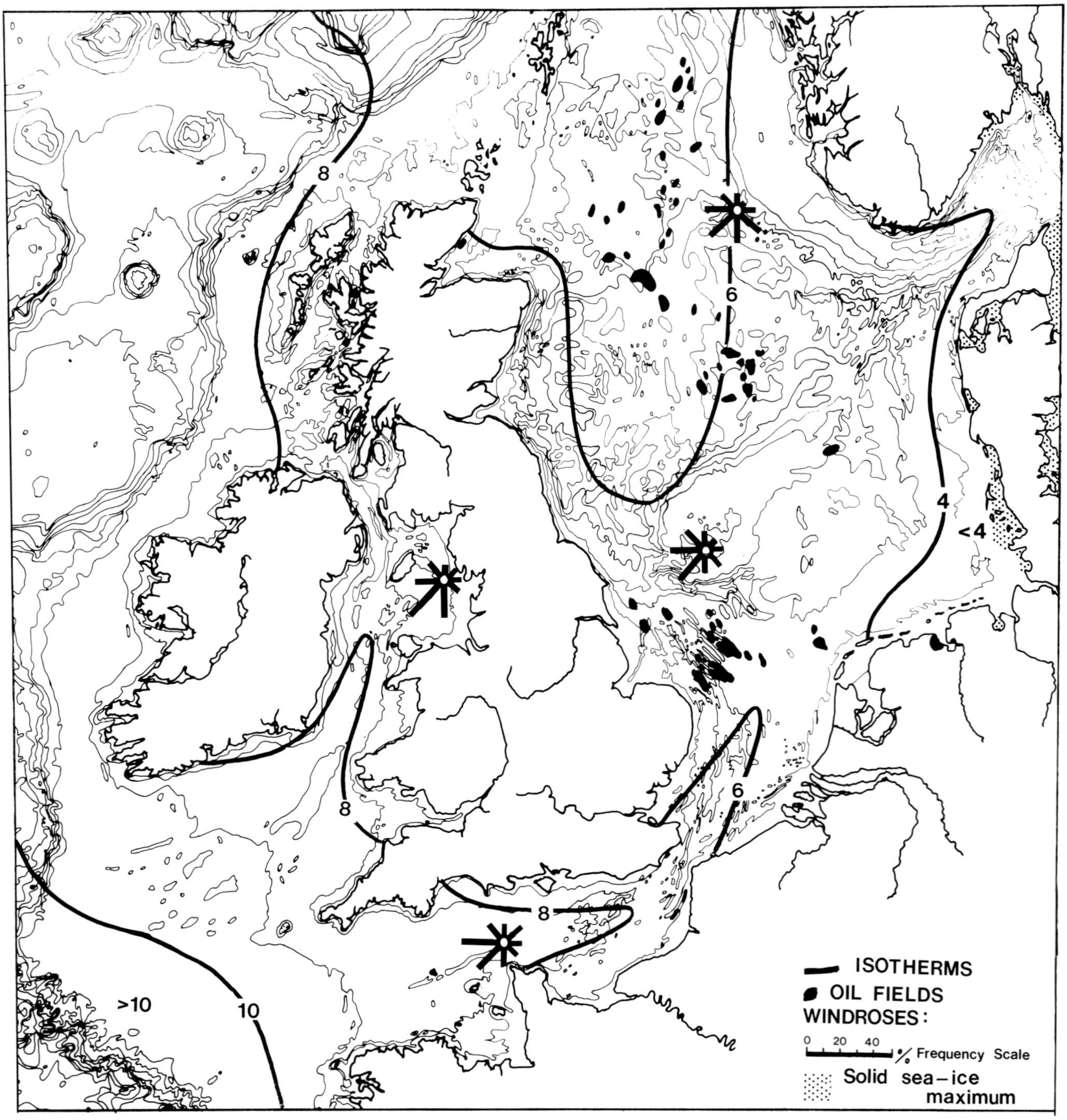

Fig. 1.6. The surface temperatures (°C, monthly means for 1905–1954, ICES data, 1962) for February and wind-roses (means for monthly data up to 1956) for January, adapted from Lee and Ramster (1976), with heaviest solid sea-ice developments (1896–1955, General Ice Chart of the German Hydrographic Office, as adapted from Fairbridge, Encycl. Oceanogr., Reinhold, 1966). Oilfield and gasfield development sites are marked for comparison with Figure 18.1 of Vol. II.

North Atlantic temperature regime, and produces positive surface temperature anomalies (mapped by Böhnecke, 1936) in the oceanic approaches to northwest Europe: the surface waters of the north-western European shelf are, on average, 2°C warmer than the mean for their latitude. Severe sea-ice formation is rare, and is to be expected (and, even then, only in rare winters) on and near to the coasts of Germany and Denmark (Fig. 1.5), where the continental, more extreme climate has its greatest influence on a vertically homogenous water mass (Vol.II, Chapter 14).

The surface drift of oceanic waters (as "residual currents") onto the continental shelf is physically recognisable especially by the distribution of isohalines (Fig. 1.5), which immediately indicate the intrusions of oceanic water from the south-west, into the Celtic Sea and thence to the Irish Sea and the English Channel, and from the north-west (bringing water from the Norwegian current as well as from the North Atlantic drift), into the North Sea. The rates and directions of surface drift are greatly affected by wind stress, as the relationship between drift directions and wind-rose dominance immediately suggests (Fig. 1.6). This is discussed fully in Vol.II, Chapter 13. Each intruding water mass brings with it, as well as its heat and characteristic proportions of solutes, hydrobiological index-species of the plankton (see, e.g., Pingree, Vol.II, Chapter 13 and Newell and Newell, 1966), which both identify the waters and indicate their sources. The progress and transformation of these waters, as they mix with each other and with the coastal waters of the shelf, are described in Vol.II, Chapters 12–16.

The gamut of physical and dynamical phenomena and processes which can occur on continental shelf seas (summarised neatly by Inman and Brush, 1973) exists, or has existed, on the north-western European shelf. We hope that this book will have some value as an introduction to some, at least, of them, as a synopsis of present knowledge for others, and as a source which will be of some use to readers for the advancement of their own interests in marine science, whatever they may be.

References

Barber, N.F., and Ursell, F., 1948. The generation & propagation of ocean waves and swell. Philos. Trans. R. Soc. London, Ser. A., 240: 527–560.

Belderson, R.H., and Stride, A.H., 1969. The shape of submarine canyon heads revealed by Asdic. Deep-Sea Res., 16: 103–104.

Berggren, W.A., 1977. Late Neogene planktonic foraminiferal biostratigraphy of the Rio Grande Rise. Mar. Micropaleontol., 2: 265–313.

Bernoulli, D., 1740. Traité sur le flux et reflux de la mer. Pièces qui ont remporté le prix de l'Academie en 1740. Paris, 55.

Böhnecke, G., 1936. Temperature, Salzegehalt und Dichte an der Oberfläche des Atlantic Ozeans. Meteor-Werk, 5 (1), Atlas, Berlin.

Boyle, R., 1673. Tracts Consisting of Observations about the Saltness of the Sea. R. Davis, Oxford.

Brooks, J.R.V. and Thomas, P., 1974. A selected bibliography of North Sea geology (up to 1974). Proc. Geol. Assoc., 85 (1): 103–112.

Cartwright, D.E., 1967. Modern studies of wind-generated ocean waves. Contemp. Phys., 8 (2): 171–183.

Curray, J.R., Moore, D.G., Belderson, R.H. and Stride, A.H., 1966a. Continental margin of western Europe; slope progradation and erosion. Science, 154 (3746): 265–266.

Curray, J.R., Moore, D.G., Stride, A.H. and Belderson, R.H., 1966b. Slope progradation and subsequent erosion, continental margin, western Europe. Proc. Int. Oceanogr. Congr., 2nd., Moscow, 90.

Davies, J.L., 1964. A morphogenetic approach to world shorelines. Geomorphol., 8 (Spec. No.): 127*–142*.

Delany, F.M. (Editor), 1970. The Geology of the East Atlantic Continental Margin. 2. Europe. ICSU/SCOR Working Party Symp., Cambridge, 1970; Rep. Inst. Geol. Sci., 70/14.

Doodson, A.T. and Warburg, H.D., 1941. Admiralty Manual of Tides. HMSO, London.

Emery, K.O., 1970. Continental margins of the world. In: F.M. Delany, (Editor), The Geology of the East Atlantic Continental Margins. Inst. Geol. Sci. Rep., 70/13 (1).

Francis, T.J.G., 1962. Black Mud Canyon. Deep-Sea Res., 4: 457–464.

Hadley, M.L., 1964. Continental margin southwest of the English Channel. Deep-Sea Res., 11: 767–779.

Halley, E., 1715. A short account of the cause of the saltness of the ocean. Philos. Trans. R. Soc. London, 29: 296–300.

Inman, D.L. and Brush, B.M., 1973. The coastal challenge. Science, 181 (4094): 20–32.

Inman, D.L. and Nordstrom, C.E., 1971. On the tectonic and morphological classification of coasts. J. Geol., 79: 1–21.

Kelvin, Lord (Thompson, Sir W.) 1882. The tide gauge, tidal harmonic analyser and tide predictor. Min. Proc. Inst. Civ. Eng. (1/3/82).

Kent, P.E., 1975a. Review of North Sea Basin development. J. Geol. Soc. London, 131: 435–468.

Kent, P.E., 1975b. The tectonic development of Great Britain and the surrounding seas. In: A.W. Woodland (Editor), Petroleum and the Continental shelf of North West Europe. 1. Geology. Applied Science Publishers, London, Chap. 1, pp. 3–28.

Knudsen, M., 1903. Gefrierpunkttabelle fuer Meerwasser. Publ. Circ. Cons. Explor. Mer, 5: 11–13.

Laughton, A.S., 1975. Tectonic evolution of the Northeast Atlantic Ocean: a Review. Norg. Geol. Unders. (Publ.), 316: 169–193.

Lee, A.J. and Ramster, J.W., 1976. Atlas of the seas around the British Isles. Min. Agric. Fish. Food, Fish. Res. Tech. Rep., 20.

MacMillan, D.H., 1966. Tides. C.R. Books, London.

Maximov, I.V., Sarukhanyan, E.I. and Smirnov, N.P., 1972. Long term variations in the North Atlantic current and their possible causes. In: A.J. Lee and H. Charnock (Editors), Physical Variability in the North Atlantic. Rapp. Proc.-Verb. Reun., Cons. Int. Explor. Mer, 162: 159–166.

Naylor, D. and Mounteney, S.N., 1975. Geology of the North-West European Continental shelf, 1, Graham Trotman Dudley, London.

Newell, G.E. and Newell, R.C., 1966. Marine Plankton: a Practical Guide. Hutchinson Educational, London, pp. 130–140.

Pegrum, R.M., Rees, G. and Naylor, D., 1975. Geology of the North-West European Continental Shelf. 2. The North Sea. Graham Trotman Dudley, London, pp. 1–125.

Richardson, L.F., 1912. Apparatus for warning a ship at sea of its nearness to large objects wholly or partly under water. British Patent No. 11,125, Patent Office, London.

Roberts, D.G., 1972. Slumping on the eastern margin of the Rockall Bank, North Atlantic Ocean. Mar. Geol., 13: 225–237.

Roberts, D.G., 1975a. Tectonic and stratigraphic evolution of the Rockall Plateau and Trough. In: A.W. Woodland (Editor), Petroleum and the Continental Shelf of North West Europe. 1. Geology. Applied Science Publishers, London, pp. 77–89.

Roberts, D.G., 1975b. Geology and tectonics of the area beyond the shelf west of the British Isles. Offshore Europe '75 Conf., Aberdeen, 1975, Spearhead Publ., Pap. OE-75.202, 8 pp.

Roberts, D.G., 1975c. Marine geology of the Rockall Plateau and Trough. Philos. Trans. R. Soc. London, 278 (1285): 447–506.

Rusby, J.S.M. and Revie, J., 1975. Long-range sonar mapping of the continental shelf. Mar. Geol., 19: M41–M52.

Sandström, J.W. and Helland-Hansen, B., 1902. Ueber die Berechnung von Meereströmungen. Rep. Norw. Fish. Invest., 2 (4).

Scrutton, R.A., 1972. The crustal structure of the Rockall Plateau microcontinent. Geophys. J. R. Astron. Soc., 27 (3): 259–275.

Selley, R.C., 1977. The habitat of North Sea oil. Proc. Geol. Assoc., 87: 359–388.

Stommel, H., 1965. The Gulf Stream: a Physical and Dynamical Description. Cambridge University Press, London, 2nd ed., pp. 1–248.

Sundvor, E. and Nysaether, E., 1975. Geological outline of the Norwegian continental margin between 60° and 80°N. In: C.J. Yorath, E.R. Parker and D.J. Glass (Editors), Canada's Continental Margins and Petroleum Exploration. Can. Soc. Pet. Geol. Mem., 4: 267–281.

Talwani, M. and Eldholm, O., 1973. Boundary between continental and oceanic crust at the margin of rifted continents. Nature, 241 (5388): 325–330.

Whewell, Rev. W., 1833. Essay towards a first approximation to a map of cotidal lines. Philos. Trans. R. Soc. London, 1833: 147–236.

Whiteman, A., Roberts, D.G. and Sellevoll, M.A. (Editors), 1975. Petroleum geology and geology of the North Sea and north-east Atlantic continental margin. Norg. Geol. Unders. (Publ.), 316: 376 pp.

Whittard, W.F., 1962. Geology of the western approaches to the English Channel. Proc. R. Soc. London, Ser. A., 265: 395–406.

Woodland, A.W. (Editors), 1975. Petroleum and the Continental Shelf of North West Europe. 1. Geology. Applied Science Publishers, London, Chaps. 9–38. (various authors).

Ziegler, P.A., 1975. Geological evolution of the North Sea and its tectonic framework. Bull., Am. Assoc. Pet. Geol., 59 (7): 1073–1097.

Chapter 2

The Continental Margin from 50 N to 57 N: its Geology and Development

R.J. BAILEY

Introduction

The British Isles lie within a gentle salient in the European continental margin, their essential outlines reflecting patterns of crustal subsidence established in earliest Mesozoic times. The North Sea, the Irish Sea, the Celtic Sea and the approaches to the English Channel feature sediment-filled grabens and basins which provide a record of this protracted subsidence and of a predominantly tensional tectonic regime (Kent, 1975; Ziegler, 1975). This regional pattern of tensional rifting and crustal subsidence extends to the continental shelves west of the British Isles (Fig. 2.1, cf. Fig. 1.3), often involving the reactivation of old, Caledonoid or Hercynoid, lines of weakness (Binns et al., 1974; Bailey, 1975a). It is widely regarded as a manifestation of plate-tectonic processes which culminated in lithosphere plate separation and the creation of a new arm of the Mesozoic Atlantic Ocean – the Rockall Trough. The geology of the continental margin between 50° and 57°N is concerned with the great submarine escarpment forming the eastern flank of this trough and the shelf seas lying to the landward.

The continental margin – 50°–53°N

Deeper crustal structure

This sector of the continental margin is deeply embayed by the Porcupine Seabight (Fig. 2.2). Marine geophysical studies (Gray and Stacey, 1970; Scrutton et al., 1971; Whitmarsh et al., 1974; Young and Bailey, 1974; Buckley and Bailey, 1975; Blundell, 1975) show that the shelf, to the east, Porcupine Ridge, to the west, and Slyne Ridge, to the north, are all characterised by crust which is continental in type, some 30 km in thickness, and overlain by a generally thin Mesozoic–Recent sedimentary sequence. The Seabight, however, contains around 5 km of such sediments, resting upon unusually thin crust. In the north, there are indications of crustal thinning towards a narrow axial zone in which the crust–mantle interface lies at the more typically oceanic depth of 15 km (Fig. 2.3; Buckley and Bailey, 1975). To the south, where the Seabight broadens and takes a more south-westerly trend prior to merging with the Atlantic Ocean floor, it appears to be underlain by a uniformly thin quasi-oceanic crust.

Steep gravity gradients, culminating in a linear gravity "high", locate the eastern limit of the deeply-buried and thinned crust of what is hereinafter termed the Seabight Trough. The form of this "high", with its associated positive magnetic anomalies, suggests that continental basement lies at relatively shallow depth, perhaps in the form of a marginal basement ridge flanked to the landward by a thickening Palaeozoic succession (Young and Bailey, 1974; Buckley and Bailey, 1975). A more pronounced gravity and magnetic anomaly present over this supposed basement ridge at lat. 53°N is interpreted as indicating the presence of a basic intrusive centre, comparable in scale, and perhaps in age, with those of the Hebridean Tertiary volcanic province (also see Riddihough and Max, 1976; Fig. 2.3).

The western flank of the Seabight Trough, i.e. Porcupine Ridge, shows a similar pattern of gravity and magnetic anomalies, and seismic refraction (Whitmarsh et al., 1974) suggests that along the ridge axis the 6.0 km/s "continental basement" refractor lies no more than 2 km below sea level. Hence, both flanks of the Seabight Trough show a relatively thin sedimentary succession – Upper Palaeozoic and younger on the east, probably similar, though thinner on the west (Figs. 2.3 and 2.4) – overlying elevated continental basement.

The gross crustal thinning, the thick sedimentary succession and the buried marginal basement ridges are characteristics that the Seabight Trough shares with

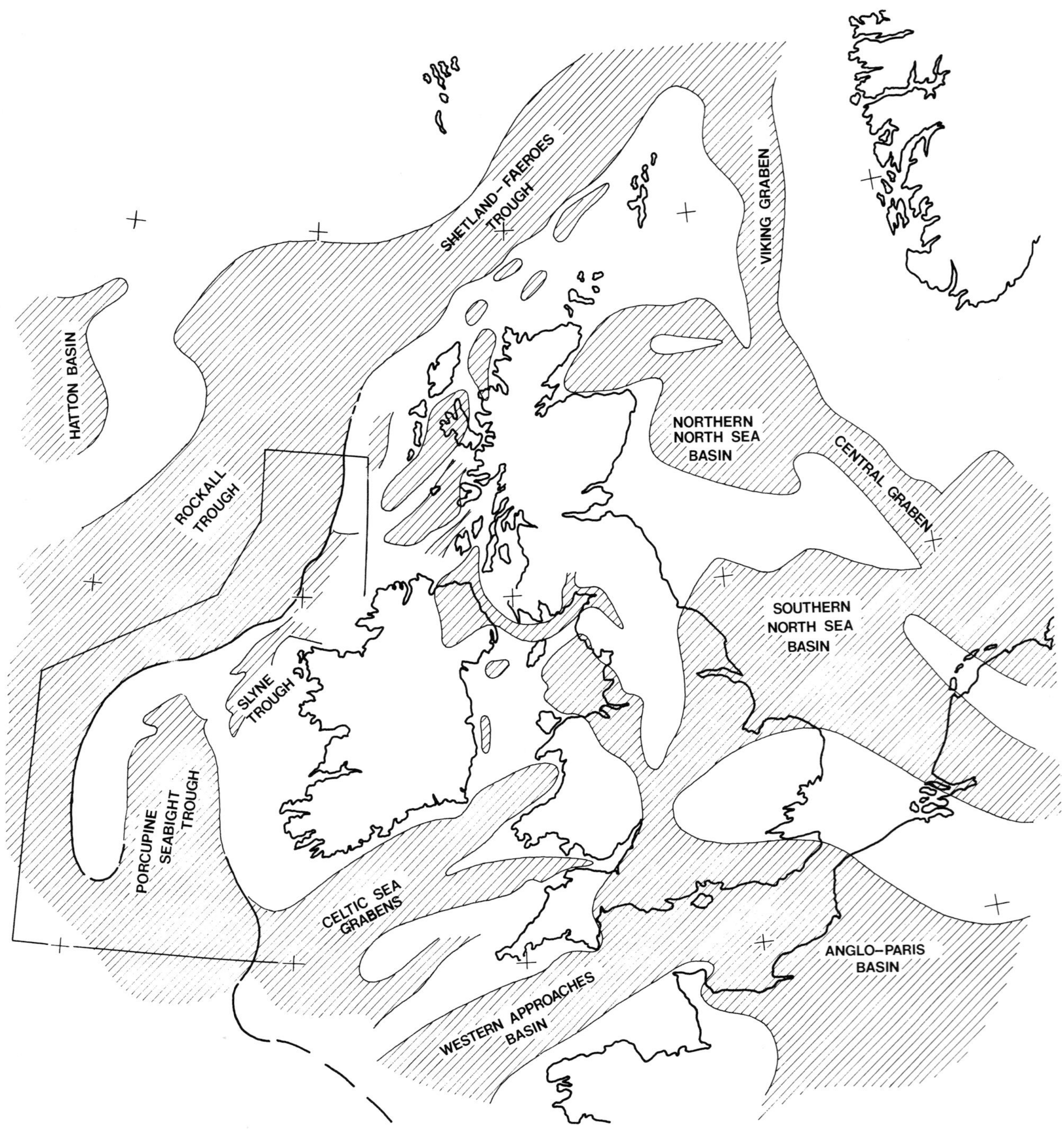

Fig. 2.1. Mesozoic and younger sedimentary basins (shaded) in and around the British Isles. The area described in this chapter is outlined.

the oceanic Rockall Trough (Bailey et al., 1975; Roberts, 1975a). They suggest a comparable mode of formation and have led to the postulate that the Seabight Trough came into being as a consequence of limited lithosphere plate separation involving the relative westerly translation of Porcupine Ridge (Stride et al., 1969; Bailey, 1975a; Riddihough and Max, 1976). Whatever its precise mode of origin, it seems reasonable that the formation of the Seabight Trough should be regarded as an adjunct to the Mesozoic opening of an oceanic rift between the British Isles and Rockall Plateau (Roberts, 1974, 1975a, b; and see discussion below).

Shallow structure and stratigraphy

The shallower geological structure and geological history of the area are read from marine seismic reflection profiles. Along the Western Irish Mainland Shelf (Fig. 2.2) these show an unconformity bevelling folded Upper Palaeozoic (?) sedimentary strata (Fig. 2.4). From the

Mainland, the unconformity surface inclines gently westward beneath an increasing thickness of prograded strata believed to be of Cenozoic age, but becomes obscured in the region of the "down-to-basin" faults along the margin of the Seabight Trough. Although the reflection profiles give no indication of its presence, it seems probable that the marginal basement ridge, mentioned above, lies buried beneath the sub-Cenozoic strata (Fig. 2.4, profiles a–c).

The bathymetry of the Seabight Trough (Fig. 2.2) and its inferred crustal structure suggest that it changes in character at about lat. 51°30′N – a view which is borne out by the reflection profiles. Thus, south of this latitude the thick succession (Fig. 2.4a) comprises more or less horizontal units, consistent with uniform subsidence across the width of the trough; but north of 51°30′N, the equivalent units form a catenary sequence, implying more active sedimentation and subsidence along the north–south bathymetric axis of the trough (Fig. 2.4b and c). This accords with a simple correlation between patterns of subsidence and the extent of the crustal thinning beneath the trough.

There is a growing consensus that the Seabight and Rockall troughs are products of late Jurassic and/or

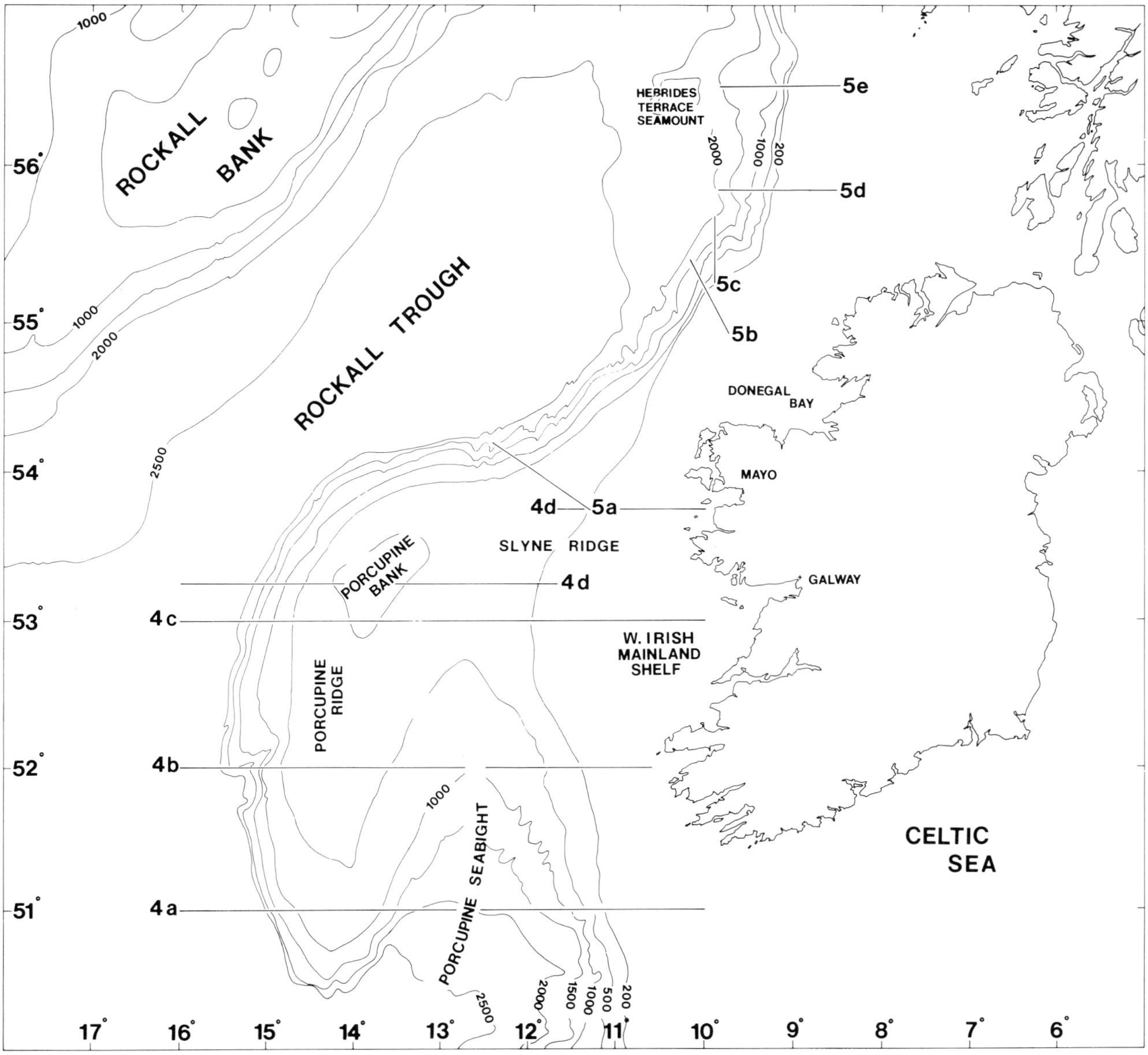

Fig. 2.2. Bathymetry of the margin (metres) with locations of profiles featured in Figs. 2.4 and 2.5.

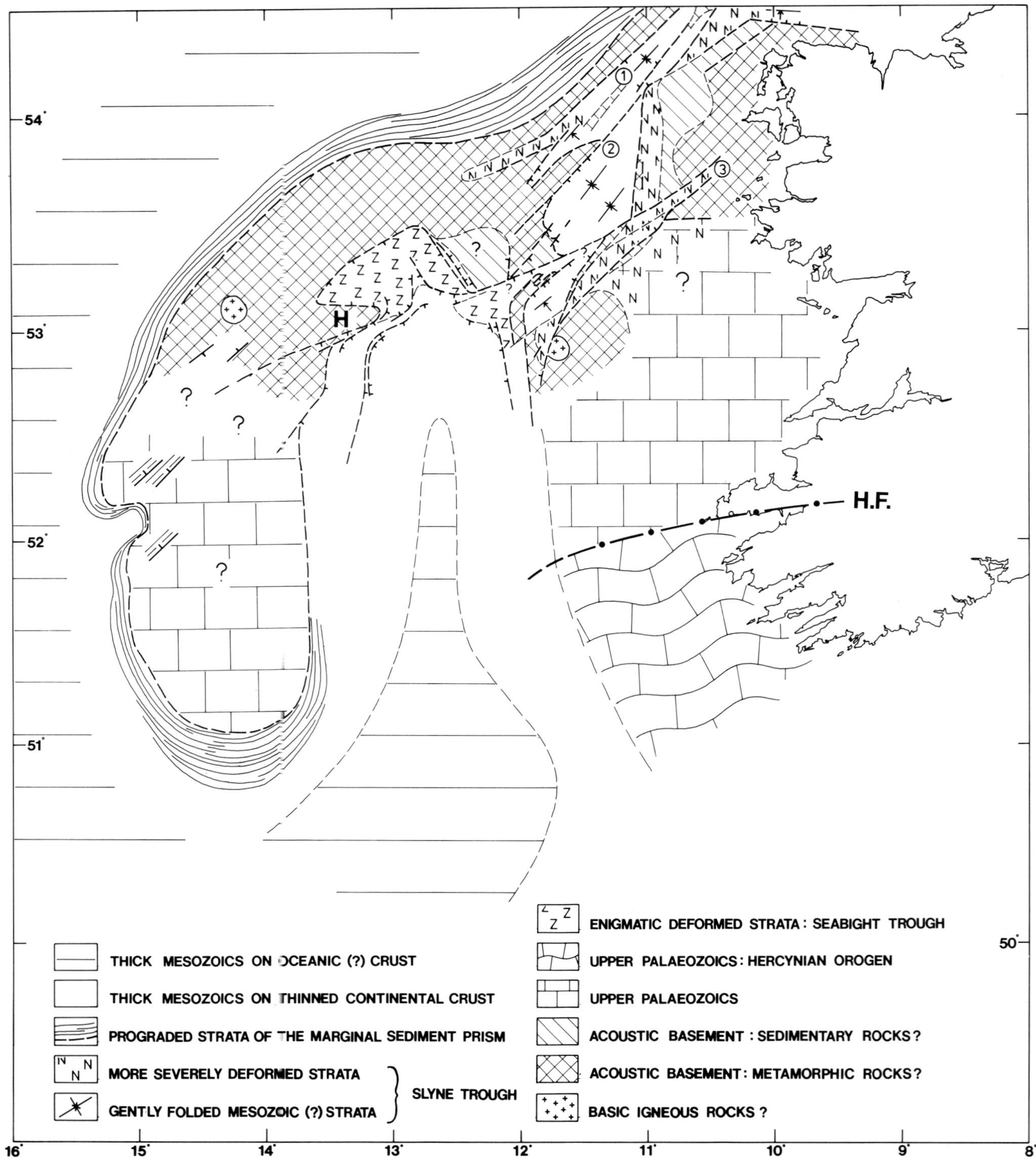

Fig. 2.3. Inferred geology to the west of Ireland as it might be seen if Cenozoic sediment cover were removed. Faults bounding the Slyne Trough's three grabens are identified by encircled numbers. *HF* = Hercynian Front, *H* = major horst.

early Cretaceous sea-floor spreading (Bailey et al., 1971; Laughton, 1972; Vogt and Avery, 1974; Roberts, 1974; Williams, 1975; Bailey, 1975a), though the antecedent crustal rifting may have begun in Triassic times. Thus the sequences revealed by reflection profiling for the most part record late Mesozoic to Recent sedimentation. Phases of sedimentation in Rockall Trough (Roberts, 1975a) are marked by prominent reflectors *Z, Y, X* and *R4*, the latter three representing possible regional unconformities of Senonian, Palaeocene and end-Eocene to early Oligocene date. Similarly, in the northern part of the Seabight

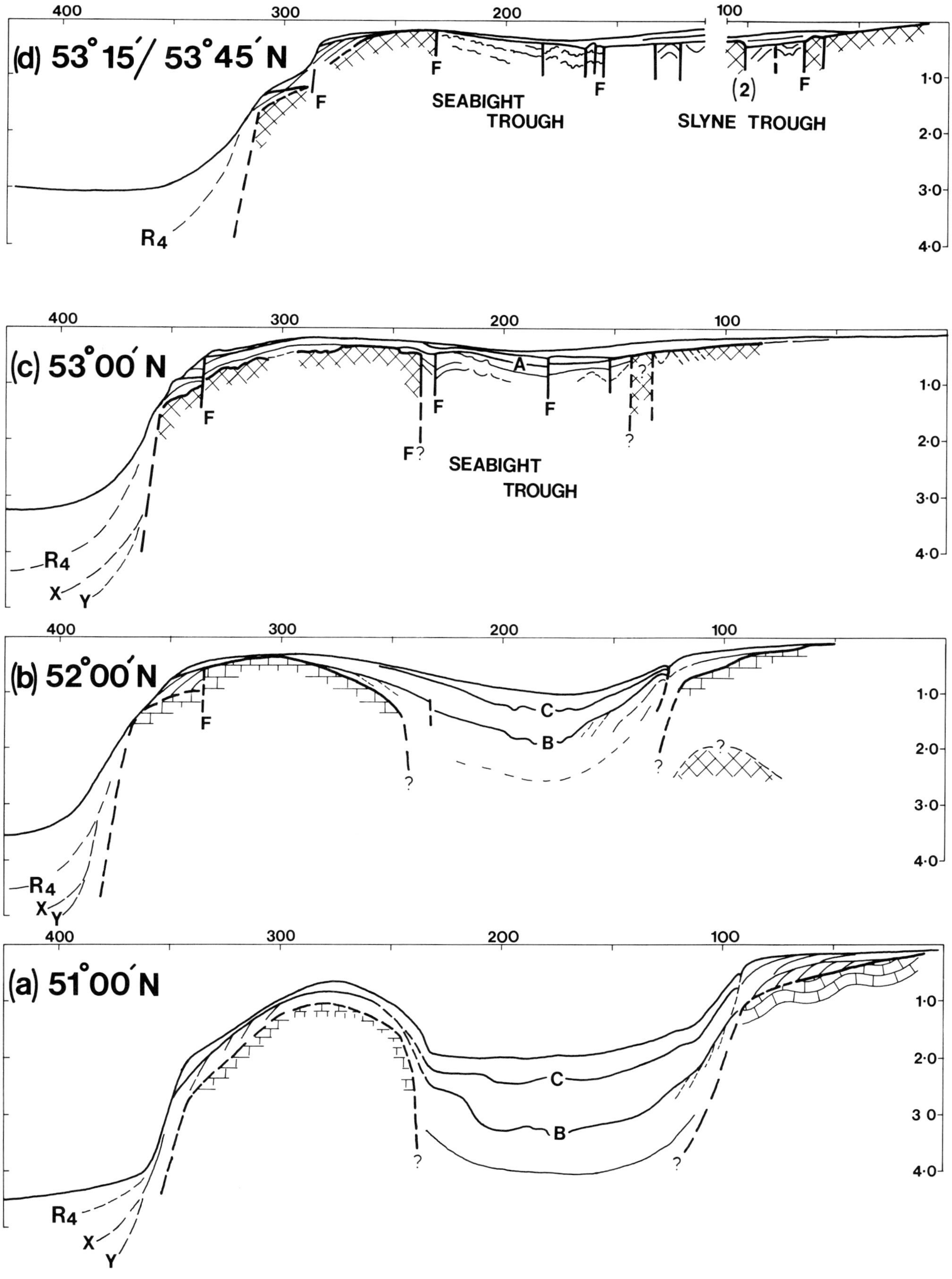

Fig. 2.4. Schematic profiles, located in Fig. 2.2. Scales in kilometres. Cross-hatching denotes basement. Upper Palaeozoics ornamented as in Fig. 2.3. *A, B* and *C* denote unconformities, as do *X, Y* and *R4* (Roberts 1975a). Profile 4(a) after Le Pichon et al. (1970).

Trough (Bailey, 1975a), it is possible to trace reflectors *A, B* and *C* (Fig. 2.4), corresponding to depositional non-sequences and marginal unconformities. Of these, *B* was assigned a tentative end-Cretaceous–Palaeocene date, and, if correlated with reflector *Y* of the Rockall Trough, could imply equivalence of reflectors *C* and *R4*. On the other hand, there is the possibility that reflector *B* correlates with *R4* of the Rockall Trough (Fig. 2.4) and this would imply a corresponding reappraisal of the age of reflectors *A* and *C*.

The published reflection profiles only allow correlation between the two troughs via the intervening Porcupine Ridge, but there are numerous faults along its flanks and the succession along its crest is much attenuated, with several obvious unconformities (Bailey, 1975a). The most obvious of these has been correlated with reflector *B* of the Seabight Trough (Fig. 2.4) and the main unconformity of the continental margin north of lat. 53° N, thus acquiring a tentative Palaeocene date (Bailey et al., 1974; Bailey, 1975a). However, viewed from Rockall Trough, it might appear that the unconformity developed across the crest of the ridge again best correlates with the early Oligocene (?) reflector *R4* (Roberts, 1975a). Actually, both interpretations may be valid in the sense that Porcupine Ridge has long been an elevated feature, starved of terrigenous detritus and sensitive to any marine regression. This combination of circumstances could well lead to cumulative non-sequences involving the superimposition of regional unconformities which are separated by thick sequences of strata in the adjacent troughs (Fig. 2.4). Accepting, for the time being, the Palaeocene age for the prominent unconformity developed across the Porcupine Ridge, the underlying prism of prograded strata, perched on its upper western flank (Fig. 2.4b and c), should be at least as old as late Cretaceous – a suggestion to some degree supported by the recovery of Senonian chalk from this scarp (Bailey and Haynes, 1974). However, on several profiles (Fig. 2.4c and d), these prograded strata have built westwards over still older marginal sediment wedges. These latter could conceivably record deposition taking place when Porcupine Ridge was emergent alongside the newly-formed Rockall Trough. If so, the overlying prograded strata might be seen in terms of carbonate sediment build-up subsequent to the (Cretaceous?) submergence of the ridge, the whole cycle being terminated by the Palaeocene regression. Certainly, it seems easiest to envisage the ridge's important contribution to the sedimentary fill of the Seabight Trough in terms of biogenic carbonate sediment production.

The continental margin, 53°–54°30′N – Slyne Ridge

Slyne Ridge (Fig. 2.2) is the most comprehensively surveyed and (therefore?) the most geologically complex section of the continental margin known. Like Porcupine Ridge, it is continental in crustal structure and has often been an emergent or shoal area in the past. Consequently, a mere "feather-edge" of Cenozoic sediment and a major sub-Cenozoic unconformity represent the several kilometres of Cretaceous–Recent sediment in the Rockall and Seabight troughs (Figs. 2.4d and

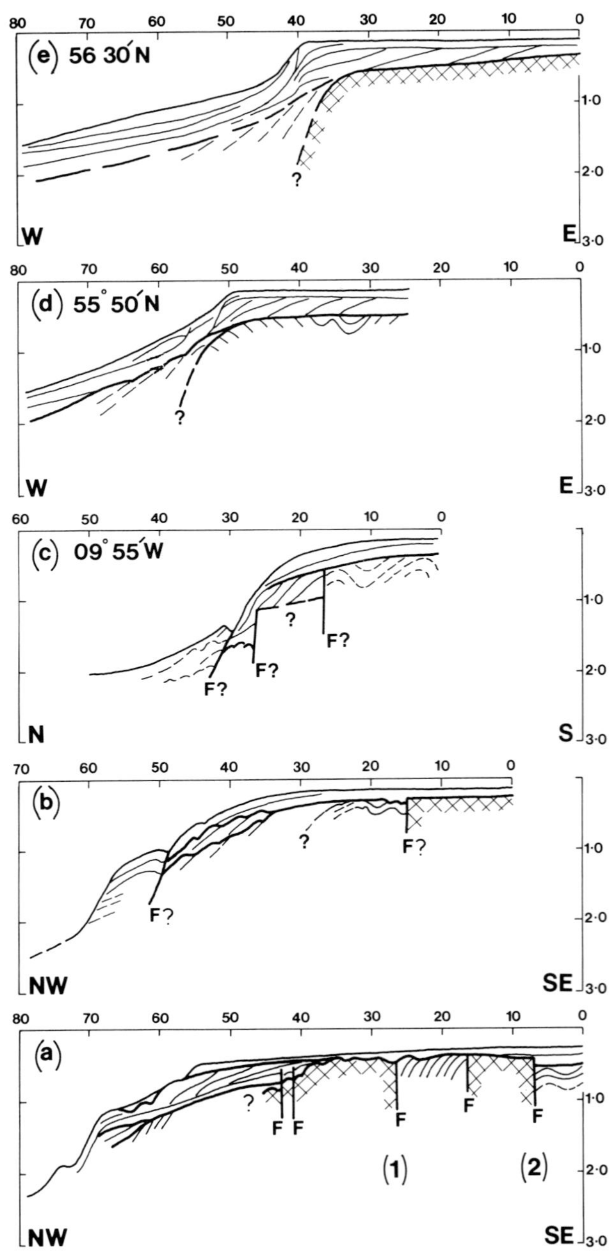

Fig. 2.5. Schematic profiles, located in Fig. 2.2. Scales in kilometres. Heavier lines denote unconformities, basement cross-hatched. *F* indicates possible faults; and numbered faults relate to Fig. 2.3.

2.5a). At its inshore end, folded Carboniferous (?) strata, such as lie beneath the Cenozoics to the south, and the variably metamorphosed rocks of the Irish Caledonian orogen, extend a few tens of kilometres offshore beneath the thin Cenozoic cover (Fig. 2.3). The metamorphics typically return no coherent seismic reflections – i.e., they form an "acoustic basement" – and have associated short-period magnetic anomalies. Judging from the geophysical data, similar kinds of rock are virtually at outcrop on the broad summit of Porcupine Bank and along the crest of the continental slope to the east (Fig. 2.3). Accordingly, Slyne Ridge is believed to represent the offshore part of the broad crustal swell wherein are exposed the Caledonian/Precambrian metamorphic basement rocks of the north-west British Isles. It is becoming increasingly clear that major Caledonoid–Hercynoid lines of weakness within this basement swell were reactivated during the regional crustal tension which characterised the early Mesozoic in north-western Europe. For example, grabens or asymmetric half-grabens developed along Caledonoid–Hercynoid faults in the Minch and the Sea of the Hebrides areas of Scotland's western shelf and in the Malin Sea (Binns et al., 1974; Figs. 2.6 and 2.7).

Similar features are encountered beneath the generally thin Cenozoic sediment cover of the Slyne Ridge. In

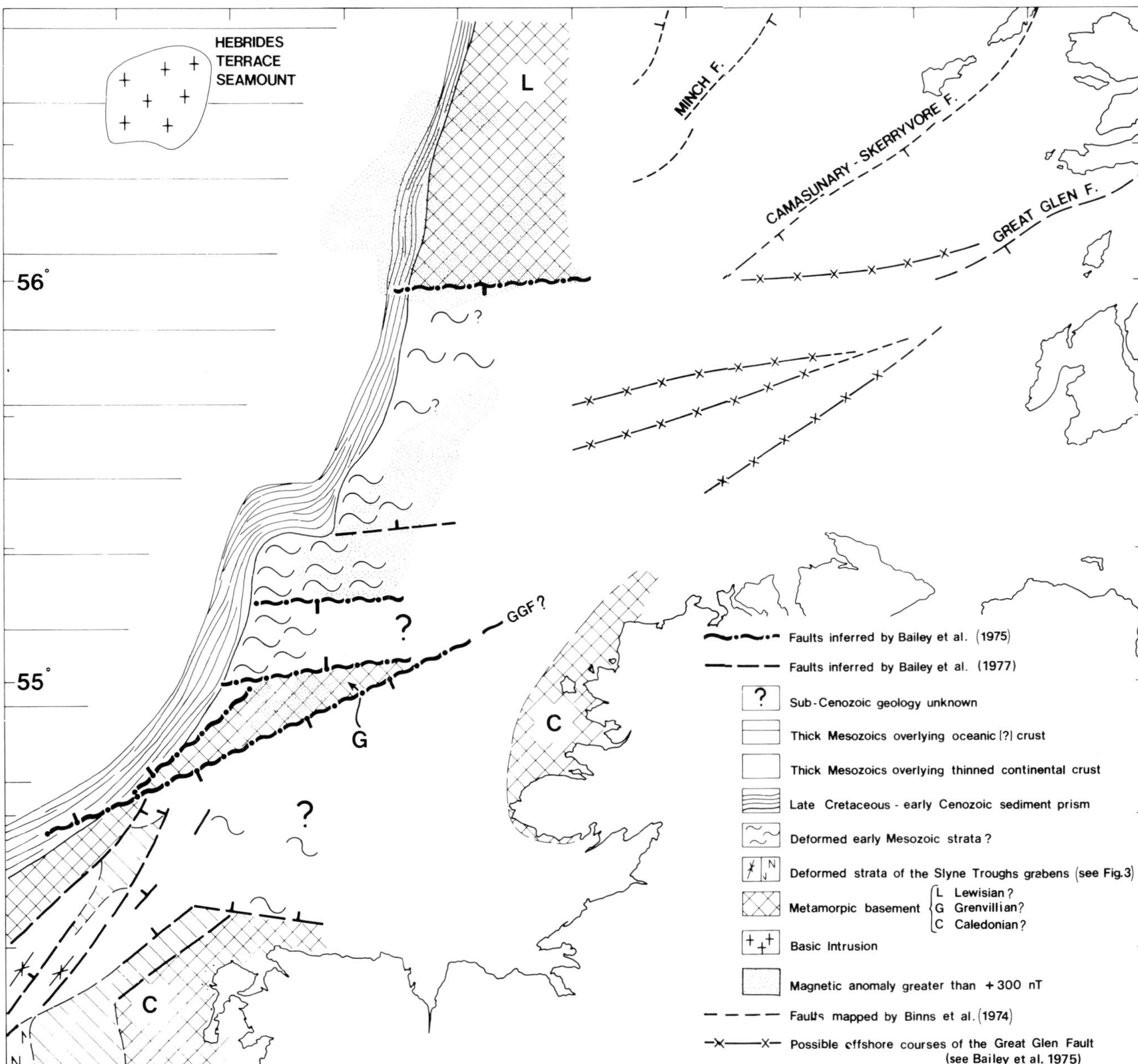

Fig. 2.6. Inferred geology to the north-west of Ireland as it might be seen if the Cenozoic sediments were removed. *GGF* = Great Glen Fault.

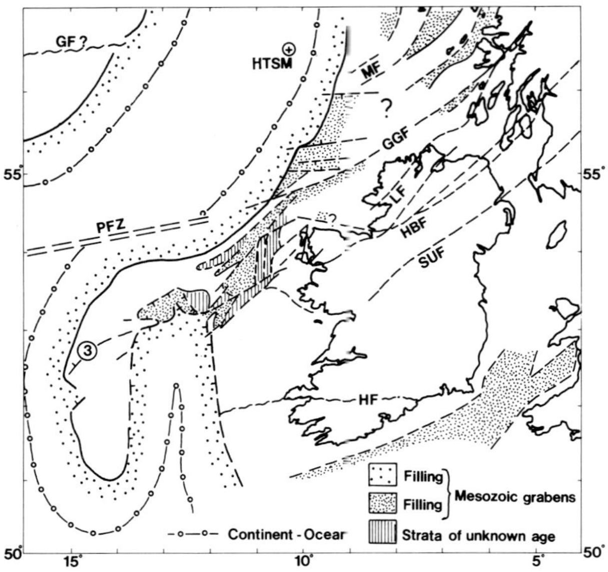

Fig. 2.7. Structural relationships. *MF* = Minch Fault, *HTSM* = Hebrides Terrace Seamount; *GF* = Grenville Front; *GGF* = Great Glen Fault; *PFZ* = Porcupine Fracture Zone; *LF* = Leannan Fault; *HBF* = Highland Boundary Fault; *SUF* = Southern Uplands Fault; *HF* = Hercynian Front. The (*3*) on Porcupine Ridge denotes the suggested continuations of fault lineament (*3*) of Fig. 2.3. The Charlie Gibbs Fracture Zone approaches the western margin of the map at about 52°25′N.

particular, new data reveal that the Slyne Trough (e.g. Bailey, 1975b) probably comprises three contiguous grabens or half-grabens disposed en échelon, from north to south, against Caledonoid faults (Figs. 2.3, 1–3; 2.4d and 2.7). The latter are broadly located by gravity gradients suggesting the juxtaposition of thick sedimentary sequences and higher density basement rocks (Buckley and Bailey, 1975), but are more accurately delineated by magnetic and seismic reflection profiles which indicate that somewhat deformed sedimentary strata are downthrown against the acoustically "massive" and magnetically variable basement rocks lying to the north-west (Riddihough, 1975; Riddihough and Max, 1976; Bailey et al., 1977). To the east, the mildly deformed strata of the grabens overlie, or are faulted against, more severely deformed strata, possibly the correlatives of the Upper Palaeozoics encountered off Galway Bay (Fig. 2.3). Profiles of the largest graben (between faults *2* and *3*, Fig. 2.3) suggest that subsidence accompanied the deposition of the Cenozoic sediment cover. Otherwise, the thin, undeformed Cenozoic strata overstep the grabens' boundary faults without detectable throw.

On the reflection profiles, the gently folded strata filling the grabens are registered as numerous parallel reflectors disposed in broad concentric folds with a probable north-east–south-west trend (Fig. 2.3). The strata are of uncertain age, but are likely to be early Mesozoic, in accordance with the North Atlantic's regional pattern of graben subsidence and filling (Binns et al., 1974; AMOCO and Imperial Oil Staffs, 1974; Kent, 1975; Ziegler, 1975; Ballard and Uchupi, 1975).

Gravity surveys suggest that the buried basement ridge following the eastern margin of the Seabight Trough extends northward at increasingly shallow depths to merge with the shallow basement of north Slyne Ridge. It thus separates the Slyne Trough's three grabens from the northern end of the Seabight Trough. There, the basement rocks of Slyne Ridge are again depressed some 2–3 km beneath deformed sub-Cenozoic strata (Buckley and Bailey, 1975; Figs. 2.3 and 2.4d). However, there appear to be no grounds, other than proximity and general similarity of structural setting, for correlating the two deformed successions, since that detected at the northern end of the Seabight Trough is characterised by unusual patterns of irregularly undulating, even disharmonic, widely spaced and strong reflectors. These are the "enigmatic deformed strata" of earlier papers (Bailey, 1975a, Fig. 6; 1975b).

The "enigmatic deformed strata" earn their name by virtue of their unusual seismic signature and equivocal geological relationships. They seem to occupy a continuation of the Seabight Trough (Fig. 2.3), but gravity and magnetic anomalies (Buckley and Bailey, 1975; Riddihough and Max, 1976) imply a deep-seated ENE–WSW structural discontinuity at about 53°N; the Seabight Trough's undeformed Mesozoic–Recent succession thins dramatically across this same line (Figs. 2.3 and 2.4d), permitting the uppermost Cenozoic units to overstep onto the "enigmatic deformed strata". The line in question can be correlated to the east with the boundary fault of the Slyne Trough's southernmost grabens (Fig. 2.3, *3*) and to the west with the trend of a major horst (Fig. 2.3, *H*). It is clearly one of the most significant lineaments in the structure of the Irish margin and can be traced onshore, where the relationships suggest important Caledonian antecedents (see below). The problem lies in assessing its role in the structuring of the Seabight Trough and the enclosing basement highs. Riddihough and Max (1976) suggested that its prime role was as a locus of the shear accompanying the lateral translation of Porcupine Ridge. It thus at once defines the northern margin of the resultant Seabight Trough and the southern margin of central Slyne Ridge. On the other hand, Bailey (1975a) saw the role of the line in terms of the reactivation of an old crustal weakness, occurring *after* the early rifting and sedimentation along the Seabight Trough. The reactivated line supposedly operated as a locus of differential vertical movement and thus isolated the "enigmatic deformed strata" from the profound subsidence and consequent burial affecting their presumed correlatives within the Seabight Trough to the south (but see below). While it is not yet possible to decide between these two alternatives, their common reliance on a major line of crustal weakness seems well founded. This line is now seen to have a west-southwesterly trend, rather than the westerly trend suggested by Young and Bailey (1974) and Bailey (1975a, b); but this only becomes important in the context of its landward correlatives (now more obvious) and its course across Porcupine Ridge (Fig. 2.3; Riddihough and Max, 1976).

The continental margin, 54° 30′–57° N

Deeper crustal structure

This section of the margin (Fig. 2.2) reverts to the classic morphology with a broad shallow shelf sea floor inclining gently westwards to the shelf break at the bathymetric limit of Rockall Trough. Detailed surveys in the Minches and the Sea of the Hebrides (Chapter 6; Binns et al., 1974) reveal that the inner shelf is traversed by concealed, sediment-filled grabens or half-grabens developed along old Caledonoid–Hercynoid lines during the Mesozoic, but locally reactivated in the aftermath of the Hebridean volcanicity (Smythe and Kenolty, 1975). This pattern of concealed grabens, often flanked by horsts of Precambrian metamorphic basement rocks, strikes southwestwards towards the shelf edge, and might be expected to have some geophysical expression there. In fact, the general pattern of magnetic and gravity anomalies aligns with the main structural feature, the margin of Rockall Trough, and, for the most part, it is only the local disruption of this pattern that suggests the influence of older lines of crustal weakness (Fig. 2.7).

Between 53° and 57° N broad, complex, positive magnetic and gravity anomalies develop along the continental margin (Fig. 2.6; Roberts, 1970, fig. 2). These are qualitatively interpreted in terms of the presence of buried ridges of ancient metamorphic basement rock, perhaps relics of the crustal arching along the nascent Rockall Trough (Bailey et al., 1975; Bailey, 1975a). The corollary of this simple interpretation is that closure of the anomalies as they are traced *along* the continental margin may indicate increasing depth of burial of the basement ridge, or, where magnetic/gravity gradients are steep, truncation by faults which bring the relatively dense and magnetic basement rocks into juxtaposition with younger sedimentary rocks. On this basis a crude pattern of deep-seated faults may be derived from the reconnaissance survey data (Fig. 2.6). Metamorphic basement appears less deeply buried north of lat. 56° N, is downthrown by a major east–west fault at this latitude, and then becomes steadily shallower to the south. Here an important east–west graben appears to isolate a still shallower, horst-like terminal section of the ridge, tenuously linked to the basement high bordering north Slyne Ridge (Fig. 2.6). Arguments have been advanced for regarding the important southern boundary fault of this horst-like feature as the main off-shore prolongation of the Great Glen Fault of Scotland (Fig. 2.6; Bailey et al., 1975). Others of the series of faults inferred to cross the margin between 55° and 57° N may well be branches of this important Caledonoid–Hercynoid line of crustal weakness, which evidently has a Mesozoic–Cenozoic history of movement (Holgate, 1969; Garson and Plant, 1972; Bailey et al., 1974; Binns et al., 1974; Smythe and Kenolty, 1975). However, the 56° N fault which shared in the Mesozoic reactivation may have come into being during the late Precambrian (Bailey, 1975b).

Shallow structure and stratigraphy

The later geological history of this section of the continental margin may be inferred from the seismic reflection profiles (Bailey et al., 1974). In the south, the succession recalls that of the western and northern slopes of Slyne Ridge (Fig. 2.5a and b). At the shelf break, prograded Cenozoic sediments, thickening northwards, rest unconformably on older prograded strata originally and tentatively assigned a late Cretaceous date. Bailey et al. (1974) suggested a Palaeocene age for this "main unconformity", equating it with regional regression and local volcanogenic uplift of this date. Roberts (1975a) also suggested its correlation with reflector *R4*, marking an early Oligocene hiatus in Rockall Trough. In fact, there are indications that the "main unconformity" landward of the shelf edge represents the superimposition of two breaks in the marginal succession (Fig. 2.5b, 35–50 km) and could thus record both Palaeocene (reflector *X* of the Rockall Trough) and early Oligocene erosion (also see Roberts, 1975a). The effect of this erosion was to strip any correlatives of the earlier marginal sediments from the outer shelf, and the main unconformity's relatively thin Cenozoic cover permits the shallow seismic reflection profiling system to register something of the nature of the underlying rocks. As on Slyne Ridge, the sub-crop features both gently folded strata and massive acoustic basement with a locally strong magnetic signature (Fig. 2.6). Both are seen on profiles north-west of Ireland (Fig. 2.5b), where the fault-bounded, acoustically massive rocks appear to represent the planated summit of the metamorphic basement horst inferred from the magnetic data (Fig. 2.6, *G*). North of this, however, thickening Cenozoics are underlain by gently folded strata (Fig. 2.5c and d), and the associated magnetic anomaly (Fig. 2.6) must relate to more deeply buried basement rocks. The 56°N fault apparently determines the northern limit of the marginal occurrences of the gently folded strata, none having been detected north of its inferred trace; a thick sedimentary section immediately south of the fault is implied by the purchase of exclusive hydrocarbon exploration rights to that locality (Anonymous, 1976).

The age of the gently folded strata north of lat. 54°30′N (~, Fig. 2.6) remains in doubt, but since they seemingly ante-date the marginal sediment prism of Rockall Trough, they are arguably pre-late Cretaceous, probably Triassic–Jurassic, in age (Bailey et al., 1974). The dating, in part evolved by reference to the known geology of the Sea of the Hebrides (Binns et al., 1974). places the gently folded strata, including those of west Donegal Bay, in the same age-bracket as the similarly structured fill of the Slyne Trough's three grabens. On a regional basis, this represents a plausible concordance of stratigraphy and structure, but precise correlations are beyond the power of the available data.

The margin in its oceanic context

Chronology

Current thought on the post-Mesozoic evolution of the oceans stems from the twin postulates that oceanic crust is produced continuously at active "mid"-ocean ridges, in the process known as sea-floor spreading, and that the crust so formed preserves the chronology of its growth in the form of an imprinted sequence of reversals in geomagnetic polarity, the so-called magnetic strip anomalies. In an early application of such ideas, Vine (1966) suggested that Rockall Trough was a product of Mesozoic Atlantic sea-floor spreading. Whilst the testing of this theory is far from complete, it does accord with the survey data acquired during the last ten years (Stride et al., 1969; Roberts, 1970; Bailey et al., 1971; Laughton, 1972; Scrutton, 1972; Vogt and Avery, 1974; Bailey, 1975a; Roberts, 1975a). The broad consensus would seem to be that regional tension evidenced by Triassic–Jurassic graben formation in and around the British Isles (Kent, 1975; Ziegler, 1975) culminated during the late Jurassic in more vigorous lithosphere plate separation, with the actively spreading Mid-Atlantic Ridge extending north-eastwards along the line of the Rockall Trough. Some 200 km width of oceanic crust were formed before the Trough was superceded as a locus of sea-floor spreading by a NW-trending branch of the ocean ridge system, working to open the Labrador Sea. This later Cretaceous pattern of lithosphere plate movement persisted until Eocene times, when a triple junction south of Greenland formed to accommodate the simultaneous opening in the Labrador Sea and along a new branch of the ocean ridge system lying between Greenland and Rockall Plateau. By early Oligocene times, the latter branch of the ridge system had become the sole locus of sea-floor spreading in the northernmost Atlantic, a situation which persists to this day.

This broad picture of successive revolutions in the pattern of North Atlantic lithosphere plate movement has suggested a time-frame for the interpretation of the sedimentary history revealed by reflection profiling. In effect, each major break in sedimentation is regarded as a possible manifestation of regional adjustment to a new pattern of lithosphere plate movement (Bailey et al., 1974; Bailey, 1975a; Roberts, 1975a; and above). However, there are other possible influences, such as global marine transgressions and regressions (Hays and Pitman, 1973; Bailey, 1975a), which may only indirectly relate to North Atlantic patterns of plate movement. In general, the equivocal nature of the seismo-stratigraphy underlines the need for local biostratigraphic standards, such as would be provided by the analysis of deepsea drilling cores from Rockall Trough.

Tectonics

Perhaps the most striking feature of the continental margin, when considered as a large-scale Mesozoic tectonic feature, is the manner in which it transects older lines of crustal weakness, even those Caledonoid–Hercynoid faults which show clear evidence of continuing activity in Mesozoic times (Fig. 2.7). Presumably the tectonics of the megascopic rifting which creates a new continental margin is such as to over-ride even the strongly-marked crustal anisotropies of ancient orogenic belts.

Roberts (1974) suggested that the existing tectonic fabric is influential in shaping the margin only insofar as it includes vertical faults affecting the whole lithosphere. Even so, the influence may be intermittent. For example, a powerful east–west fault is transected by the margin at lat. 56°N (Fig. 2.7). This has been correlated with a lineament on Rockall Plateau which may well represent the prolongation of the Grenville Front, a major zone of shearing between older and younger Precambrian orogenic belts in the eastern Canadian Sheild (Fig. 2.8; Roberts, 1974; Bailey et al., 1975; Bailey, 1975b). On the Canadian margin and south-west Rockall Plateau there is an associated off-set in the continent–ocean boundary, and this off-set seems to have propagated as a transform fault in the late Mesozoic oceanic crust of the Labrador Sea (Van der Linden, 1975). Yet neither the supposed Grenville line, nor, indeed the better known and convergent Minch Fault (Fig. 2.7; Binns et al., 1974) noticeably deflects the margins of the Rockall Trough. Only around Slyne Ridge does it appear that the form of the margin was influenced by the prevailing tectonic grain. The structures involved are plausibly identified with what could be regarded as the two most important lines of weakness in the north-western British Isles: the Great Glen Fault and the Highland Boundary Fault (Fig. 2.7).

The offshore continuation of the Great Glen line appears to follow the westerly trend of the continental margin north of Slyne Ridge. A congruent swing in the inferred continent–ocean boundary has been detected on the opposite flank of the Rockall Trough; the intervening oceanic basement features the buried, east–west, Porcupine Fracture Zone (Roberts, 1975a). Thus it appears that the Great Glen Fault determined an offset in the nascent ocean ridge propagating along Rockal Trough, and is ancestral to a transform fault in the Trough's Mesozoic oceanic crust (Fig. 2.7).

There are increasingly clear indications that the northern termination of the Seabight Trough expresses a major WSW-trending line of crustal weakness. Riddihough and Max (1976) suggested that it represents the Great Glen Fault's main westerly continuation, extending south-westwards via the Caledonoid boundary faulting of the Slyne Trough (Fig. 2.7). The view expressed here is that the structure in question continues as lineament (*3*) of the Slyne Ridge and makes its landfall in Mayo, where the Leannan branch of the Great Glen Fault system (Pitcher, 1969) and the Highland Boundary line of Ireland (Max and Riddihough, 1975) are convergent (Fig. 2.7). However, in the present context the tectonic affinities of the structure are secondary to its possible role in the early evolution of the Irish continental margin, and it will be assumed to represent the continuation of lineament (*3*).

Riddihough and Max (1976) adhere to the view that the Seabight Trough is a mini-ocean, formed by the relative westerly translation of the continental Porcupine Ridge (Stride et al., 1969; Bailey et al., 1971), but abandon the hypothesis of the pivotal movement of the Ridge in favour of one of rotation about a more distant pole, with lineament (*3*) functioning as a transform fault. This carries the implication that the lineament located a primal offset in the continental margin west of Porcupine Ridge – an offset which was somewhat fortuitously nullified by the Ridge's later westerly relative motion. Any such offset would be expected to propagate as a transform fault in oceanic crust; thus it may be significant that the major transform of the North Atlantic, the Charlie Gibbs Fracture Zone (Laughton, 1972), seemingly terminates in oceanic crust at about 52°15′N 16°00′W, precisely on the trend of lineament (*3*) (Fig. 2.7). Roberts (1975a, fig. 21) shows that the Charlie Gibbs Fracture Zone probably operated as a transform fault during the creation of the late Cretaceous oceanic crust of magnetic strip anomaly 32 (75 Ma), some 200 km west of Porcupine Ridge. However, there are older magnetic strip anomalies, with a similar NNW trend, closer to the Ridge; Roberts (1975a) shows that this older (pre-75 Ma) crust is very much wider in extent north of the Charlie Gibbs Fracture Zone than to the south. Hence, using the Riddihough and Max model, it could be argued that some of the "missing" pre-anomaly-32 oceanic crust south of the Fracture Zone may be represented by the thinned, quasi-oceanic crust of the northern Seabight Trough. The trough would, in effect, owe its formation to the fact that at some time in the late Cretaceous (pre-75 Ma) Porcupine Ridge lagged behind in the general north-easterly motion of the Eurasian lithosphere plate. Orphan Knoll, NE of Newfoundland (Grant, 1972), could represent the continental fragment formerly adjacent to Porcupine Ridge (Fig. 2.8), for the Charlie Gibbs Fracture Zone makes its North American landfall just north of the Knoll (Laughton, 1972) in association with a major offset in the continental margin (Grant, 1972). Flemish Cap, a second marginal "outlier" of North American continental crust lying off the Grand Banks, might then correspondingly key into the more oceanic southern section of the Seabight Trough (Fig. 2.8).

The suggested situation prior to the separation of the Orphan Knoll – Flemish Cap section of the North American margin from Porcupine Ridge and Slyne Ridge is shown in Fig. 2.8. The NNW-trending ocean ridge which brought about the separation was possibly that which accomplished the early Cretaceous opening of the Labrador Sea (Van der Linden, 1975; Williams, 1975). As already suggested, its links with the contemporary Mid-Atlantic Ridge may have embraced the primitive Charlie Gibbs Fracture Zone, propagating from

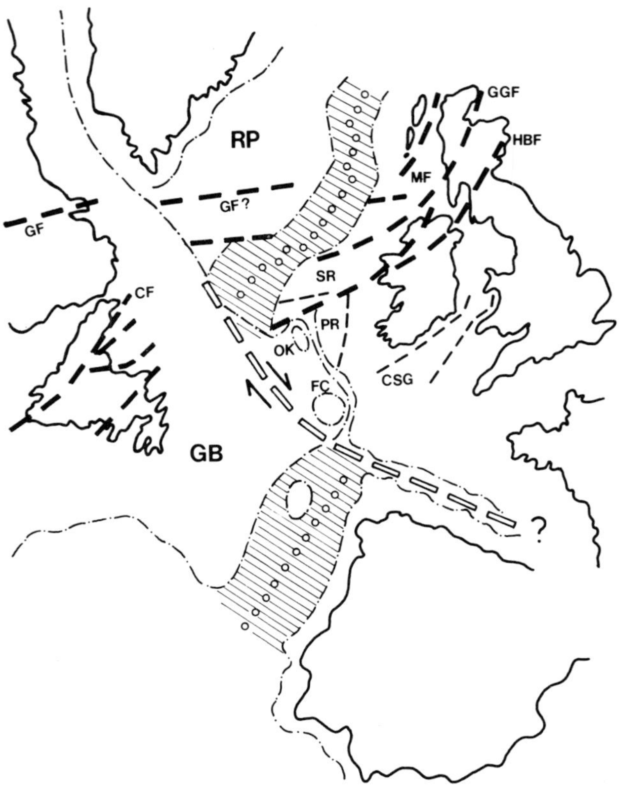

Fig. 2.8. A reconstruction of the late Jurassic–early Cretaceous North Atlantic (modified after Laughton, 1972). Hachured area = the then-existing oceanic crust, with the ocean ridge indicated by a chain of circles. Major faults and oceanic fracture zones shown by heavy dashed lines, minor or equivocal structures by light dashed lines. Open dashes show the trend of the inferred Biscay–Labrador Sea transform fault system.
Note the juxtaposition of Orphan Knoll (*OK*), Flemish Cap (*FC*), Slyne Ridge (*SR*) and Porcupine Ridge (*PR*). The separation of these features would generate an offset perhaps parental to the Charlie Gibbs Fracture Zone. (see text).
GGF = Great Glen Fault; *RP* = Rockall Plateau; *HBF* = Highland Boundary Fault; *MF* = Minch Fault; *GF* = Grenville Front; *CF* = Cabot Fault; *CSG* = Celtic Sea Graben; and *GB* = Grand Banks.

lineament (*3*) and its Appalachian counterpart off Newfoundland (Grant, 1972). Also, in view of the sharp change in trend of the ocean-ridge system south-west of Ireland, these links may demand a triple junction thereabouts, possibly with the Celtic Sea grabens representing the triple junction's "failed arm" (Burke and Dewey, 1973; see also Riddihough and Max, 1976).

Rockall Trough, north of the Porcupine Fracture Zone (Fig. 2.7), appears to show correlatable linear magnetic anomalies with a NNE trend (Roberts, 1974). These would seem to imply either a further early Cretaceous triple junction at the Trough's southern end, or a still earlier phase of sea-floor spreading. The latter explanation is featured in Fig. 2.8, but, as is shown in the figure, it demands that the (late Jurassic?) Mid-Atlantic Ridge propagated north-eastward along the Rockall Trough via a major north-westerly-trending transform fault system (Kay, 1969; Bailey et al., 1971; Laughton, 1972; Williams, 1975). Closure of the oceanic rift by rotation about this line (Fig. 2.8) would bring the major Caledonoid lines of crustal weakness into close conjunction with their supposed counterparts in the Appalachian orogen of Newfoundland, but the long-sought correlation of the Great Glen and Cabot faults (Wilson, 1962; Bailey, 1975b) remains difficult to achieve.

Finally, there are the broader aspects of marginal tectonics. For example, the generally accepted model for the creation of a new ocean (e.g. Lowell and Genik, 1972) involves thermal arching of the continental lithosphere. The resultant erosion and profound normal faulting combine to thin the continental crust along the crest of the arch. However, once plate separation is accomplished and oceanic lithosphere is being generated at an ocean ridge axial to the resultant rift, the thinned crust of the new continental margins is increasingly displaced from this locus of thermal activity. Displacement from the zone of arching, and cooling of the underlying lithosphere, are supposed to induce protracted slow subsidence, so that the marginal zone of thinned continental crust becomes buried beneath a sedimentary prism of "geosynclinal" dimensions.

Now, given this model, it is puzzling that the outermost shelf areas of Porcupine Ridge, west Slyne Ridge, the margin north-west of Ireland and, indeed, Rockall Bank all appear to feature Precambrian or Paleozoic basement rocks within a kilometre, or so, of sea level. To put these occurrences into context, it could be argued that the margins to which they belong were in being in the early Cretaceous, since which time the American outer shelf south of Grand Banks has subsided to accumulate as much as 5 km of sediment, mostly in shelf facies (Sheridan, 1974; Scott and Cole, 1975). This contrast in marginal subsidence and sedimentation is probably a function of the relative abruptness of the transition from the "normal" thickness, granitic, continental crust to the very much thinner, non-granitic, oceanic crust. If the transition is accomplished within a few tens of kilometres, the marginal zone of increasingly thin, subsiding continental crust becomes so narrow and steep that, in effect, it constitutes a continental slope, bypassed by the sediment input. On the other hand, if the transition zone is broad, the situations is far more appropriate to the slow seaward growth of a massive prism of prograded continental shelf sediments, such as characterises the U.S. Atlantic margin. Supposing that margins of the latter type constitute the norm, the explanation of an abnormally narrow continent–ocean transition, lacking an important sediment prism, must lie in the tectonics of the parental lithospheric rift. In the case of Biscay and South Labrador Sea, for example, an appeal might be made for the prior existence of a major zone of transform/transcurrent faulting (Fig. 2.8) which became the locus of virtually all the tensional strain associated with the lithosphere plate separation. A more widely applicable hypothesis, however, would be that the width of the continent–ocean transition zone is a function of the *rate* of strain during the initial lithosphere plate separation – high rates giving narrow zones, such as characterise the margins bordering Rockall Trough.

Acknowledgements

Much of the data used in this review derives from the "Western Approaches" cruise programme of the University College of North Wales Research Vessel "Prince Madog". The writer acknowledges his continuing debt to Captain T. Donovan, the Officers and Crew of the "Prince Madog" and the award of Natural Environment Research Council grant GR/3/75 which funded the first six years of the research.

Many scientists have participated in this work, but particular mention must be made of Denzil Taylor Smith, who began the work, Clive McCann and Richard Clarke, who supervised the early surveys, and John Buckley, Peter Simpkin, Peter Jackson and Jim Bennell, without whose help little of the survey work would have been accomplished.

Paddy Dewes, as always, coped admirably with the production of the figures.

References

Anonymous, 1976. Companies plan 10–12 exploration wells to test Irish prospects. Ocean Ind., 11: 113–114.

AMOCO and Imperial Oil Staffs, 1974. Regional geology of Grand Banks. Bull., Am. Assoc. Pet. Geol., 58: 1109–1123.

Bailey, R.J., 1975a. The geology of the Irish continental margin and some comparisons with offshore eastern Canada. In: C.J. Yorath, E.R. Parker, and D.J. Glas (Editors), Canada's Continental Margins and Offshore Petroleum Exploration. Can. Soc. Pet. Geol. Mem., 4: 313–340.

Bailey, R.J., 1975b. Sub-Cenozoic geology of the British continental margin (Lat. 50° to 57°N) and the reassembly of the North Atlantic late Paleozoic supercontinent. Geology, 3: 591–594.

Bailey, R.J. and Haynes, J.R., 1974. New dredge samples from the continental margin bordering Rockall Trough. Mar. Geol., 16: M57–M62.

Bailey, R.J., Buckley, J.S. and Clarke, R.H., 1971. A model for the early evolution of the Irish continental margin. Earth Planet. Sci. Lett., 13: 79–84.

Bailey, R.J., Grzywacz, J.M. and Buckley, J.S., 1974. Seismic reflection profiles of the continental margin bordering Rockall Trough. J. Geol. Soc. London, 130: 55–69.

Bailey, R.J., Buckley, J.S. and Kielmas, M.M., 1975. Geomagnetic reconnaissance on the continental margin of the British Isles between 54° and 57°N. J. Geol. Soc. London, 131: 275–282.

Bailey, R.J., Jackson, P.D. and Bennell J.D., 1977. Marine geology of Slyne Ridge. J. Geol. Soc. London, 133: 165–172.

Ballard, R.D. and Uchupi, E., 1975. Triassic rift structure in Gulf of Maine. Bull. Am. Assoc. Pet. Geol., 59: 1041–1072.

Binns, P.E., McQuillin, R. and Kenolty, N., 1974. The geology of the Sea of the Hebrides. Rep. Inst. Geol. Sci., 73/14: 43 pp.

Blundell, D.J., 1975. The geology of the Celtic Sea and Southwestern Approaches. In: C.J. Yorath, E.R. Parker and D.J. Glass (Editors), Canada's Continental Margins and Offshore Oil Exploration. Can. Soc. Pet. Geol. Mem., 4: 341–362.

Buckley, J.S. and Bailey, R.J., 1975. A free air gravity anomaly contour map of the Irish continental margin. Mar. Geophys. Res., 2: 185–194.

Burke, K. and Dewey, J.F., 1973. Plume generated triple junctions: key indicators in applying plate tectonics to old rocks. J. Geol., 81: 406–433.

Clarke, R.H., Bailey, R.J. and Taylor Smith, D., 1971. Seismic reflection profiles of the continental margin west of Ireland. In: F.M. Delany (Editor), The Geology of the East Atlantic Continental Margin. 2. Europe. Rep. Inst. Geol. Sci., No. 70/14: 67–76.

Garson, M.S. and Plant, J., 1972. Possible dextral movement on the Great Glen and Minch faults in Scotland. Nature (London), Phys. Sci., 240: 31–35.

Grant, A.C., 1972. The continental margin off Labrador and Eastern Newfoundland – morphology and geology. Can. J. Earth Sci., 9: 1394–1430.

Gray, F. and Stacey, A.P., 1970. Gravity and magnetic interpretation of Porcupine Bank and Porcupine Bight. Deep-Sea Res., 17: 467–475.

Hays, J.D. and Pitman, W.C. III, 1973. Lithosphere plate motion, sea level changes and climatic and ecological consequences. Nature, 246: 18–22.

Holgate, N., 1969. Palaeozoic and Tertiary transcurrent movements on the Great Glen Fault. Scott. J. Geol., 5: 97–139.

Kay, M., 1969. Continental drift in the North Atlantic. In: M. Kay (Editor), North Atlantic Geology and Continental Drift. Mem. Am. Assoc. Pet. Geol., 12: 965–973.

Kent, P.E., 1975. The tectonic development of Great Britain and the surrounding seas. In: A.W. Woodland (Editor). Petroleum and the Continental Shelf of North West Europe. 1. Geology. Applied Science Publishers, London, pp. 3–28.

Laughton, A.S., 1972. The southern Labrador Sea – a key to the Mesozoic and early Tertiary evolution of the North Atlantic. In: A.S. Laughton, W.A. Berggren et al. (Editors), Initial Reports of the Deep Sea Drilling Project. U.S. Government Printing Office, Washington, 12: 1155–1179.

Le Pichon, X.A., Cressard, A., Mascle, J., Pautot, G. and Sichler, B., 1970. Structures sous-marines des bassins sedimentaires de Porcupine et de Rockall. C. R. Acad. Sci. Paris, 270: 2903–2906.

Lowell, J.D. and Genik, G.J., 1972. Sea-floor spreading and structural evolution of southern Red Sea. Bull. Am. Assoc. Petrol. Geol., 56: 247–259.

Max, M.D. and Riddihough, R.P., 1975. The continuation of the Highland Boundary Fault in Ireland. Geology, 3: 206–210.

Pitcher, W.S., 1969. Northeast-trending faults of Scotland and Ireland and chronology of displacements. In: M. Kay (Editor) North Atlantic Geology and Continental Drift. Mem., Am. Assoc. Pet. Geol., 12: 724–733.

Riddihough, R.P., 1975. A magnetic map of the continental margin west of Ireland involving part of the Rockall Trough and the Faeroe Plateau. Dublin Inst. Adv. Stud. Geophys. Bull., 33 (map and explanation).

Riddihough, R.P. and Max, M.D., 1976. A geological framework for the continental margin west of Ireland. Geol. J., 11: 109–118.

Roberts, D.G., 1970. Recent geophysical studies on the Rockall Plateau and adjacent areas. Proc. Geol. Soc. London, 1662: 87–93.

Roberts, D.G., 1974. Structural development of the British Isles, the continental margin and Rockall Plateau. In: C.A. Burke and C.L. Drake (Editors), The Geology of Continental Margins. Springer, New York, N.Y., pp. 343–359.

Roberts, D.G., 1975a. Marine geology of the Rockall Plateau and Trough. Philos. Trans. R. Soc. London, A278: 447–509.

Roberts, D.G., 1975b. Tectonic evolution of the Rockall Plateau and Trough. In: A.W. Woodland (Editor), Petroleum and the Continental Shelf of North West Europe. 1. Geology. Applied Science Publishers, London, pp. 131–148.

Scott, K.R. and Cole, J.M., 1975. Geology of the U.S. continental margin from Maine to Florida – a résumé. In: C.J. Yorath, E.R. Parker and D.J. Glass (Editors) Canada's Continental Margins and Offshore Petroleum Exploration. Can. Soc. Pet. Geol. Mem., 4: 33–44.

Scrutton, R.A., 1972. The crustal structure of the Rockall Plateau microcontinent. Geophys. J. R. Astron. Soc., 27: 259–275.

Scrutton, R.A., Stacey, A.P. and Gray, F., 1971. Evidence for the mode of formation of Porcupine Seabight. Earth Planet. Sci. Lett., 11: 140–146.

Sheridan, R.E., 1974. Atlantic continental margin of North America. In: C.A. Burke and C.L. Drake (Editors), The Geology of Continental Margins. Springer, New York, pp. 391–407.

Smythe, D.K. and Kenolty, N., 1975. Tertiary sediments in the Sea of the Hebrides. J. Geol. Soc. London, 131: 227–233.

Stride, A.H., Curray, J.R., Moore, D.G. and Belderson, R.H., 1969. Marine geology of the Atlantic continental margin of Europe. Philos. Trans. R. Soc. London, A264: 31–75.

Van der Linden, W.J.M., 1975. Mesozoic and Cainozoic opening of the Labrador Sea, the North Atlantic and the Bay of Biscay. Nature (London), 253: 320–324.

Vine, F.J., 1966. Spreading of the ocean floor: new evidence. Science, 154: 1405–1415.

Vogt, P.R. and Avery, O.E., 1974. Detailed magnetic surveys in the northeast Atlantic and Labrador Sea. J. Geophys. Res., 79: 363–389.

Whitmarsh, R.B., Langford, J.J., Buckley, J.S., Bailey, R.J. and Blundell, D.J., 1974. The crustal structure beneath Porcupine Ridge as determined by explosion seismology. Earth Planet. Sci. Lett., 22: 197–204.

Williams, C.A., 1975. Seafloor spreading in the Bay of Biscay and its relationship to the North Atlantic. Earth Planet. Sci. Lett., 24: 440–456.

Wilson, J.T., 1962. Cabot Fault, an Appalachian equivalent of the San Andreas and Great Glen faults and some implications for continental displacement. Nature, 195: 135–138.

Young, D.G.G. and Bailey, R.J., 1974. An interpretation of some magnetic data off the west coast of Ireland. Geol. J., 9: 137–146.

Ziegler, P.A., 1975. North Sea basin in the tectonic framework of Northwestern Europe. In: A.W. Woodland, (Editor) Petroleum and the Continental Shelf of North West Europe. 1. Geology, Applied Science Publishers, London, pp. 131–148.

Chapter 3

Aspects of the Post-Permian History of the Aseismic Continental Shelf to the West of the British Isles

M.R. DOBSON

Introduction

Our understanding of the evolution of north-west Europe and particularly the British Isles since the Hercynian events has, until recently, been based on the "Alpine Storm" hypothesis. Whilst global tectonic theory, as applied to the region, clearly demanded a reappraisal of the accepted structural and sedimentary events, it was detailed studies, first of Holland (Heybroek, 1974) and later of the North Sea (Woodland, 1975), that finally exposed the inadequacy of the previously supposed to be all-embracing "Alpine" concept.

The British Isles, and specifically, its Western Shelf, up to and including the continental slope, possess one of the most complicated passive margins in the world. The interacting, conflicting and militating events which combined to account for the development of the region make any efforts to elucidate even a part of the history a daunting task. These events include, for example, episodic Atlantic evolution and lithospheric plate movement, continental volcanism, globally controlled transgressions and regressions and thermal tectonics; the latter provided, among other things, vertical movements and crustal thinning.

In order that some rationalisation of these events might be attempted, two widely separated shelf areas were selected by the Aberystwyth marine group for detailed study. One, the Malin Sea area, had a very limited post-Liassic sedimentary record, but a complex igneous and erosional history. The other, the North Celtic Sea Basin, principally comprising the modern South Irish Sea area, preserved a long, nearly continuous, history of incremental subsidence and sedimentation with almost no igneous activity.

From the geology of these two contrasting shelf regions tentative models are proposed to account for the observations made within an aseismic margin context. Unfortunately, our present knowledge of the age and sedimentary history of the margin is very imperfect as is our understanding of the crustal structure beneath much of the Continental Shelf. Necessarily, therefore, a few broad assumptions must be made; these are partly substantiated, in the south, by deep-seismic detail and some commercial geological information. One aspect of this shelf work which has received little attention involves the very significant contribution our knowledge of the sub-crop can make to coastal geomorphological studies. Therefore, the final part of the Chapter deals, if only briefly, with this porism.

General geology of the Malin Sea area

Apart from its employment on British weather maps, the term Malin Sea is rarely used to identify the sea area north of Ireland; however, the region may be considered to extend from the Irish Coast to lat. 56° 20′N and from the Continental Shelf break to Kintyre, comprising an area of more than 14,000 km². Figure 3.1 is a synopsis of current knowledge of the sub-Pleistocene subcrop and is based on gravity, magnetic and shallow seismic data. Direct evidence is limited to that from I.G.S. boreholes, close to the coast, and from sampling by Scuba divers and from rock dredging conducted by the Aberystwyth Marine Group. At the time of writing there is, with the exception of two samples from the Stanton Banks, an immense gap in the direct evidence from the area extending from the Colonsay Basin–Blackstones Bank westwards as far as the Rockall Trough. The petrographic descriptions of the samples together with the geophysical detail is contained in a Malin Sea report by Evans et al. (1978).

The margin sequence of block-faulted Mesozoic rocks is unconformably overlain by a prograding series of Palaeogene sediments. The "Tertiary", indicated on Fig. 3.1 to exist along the western edge, refers to this prograding series. The upper unconformity may now be

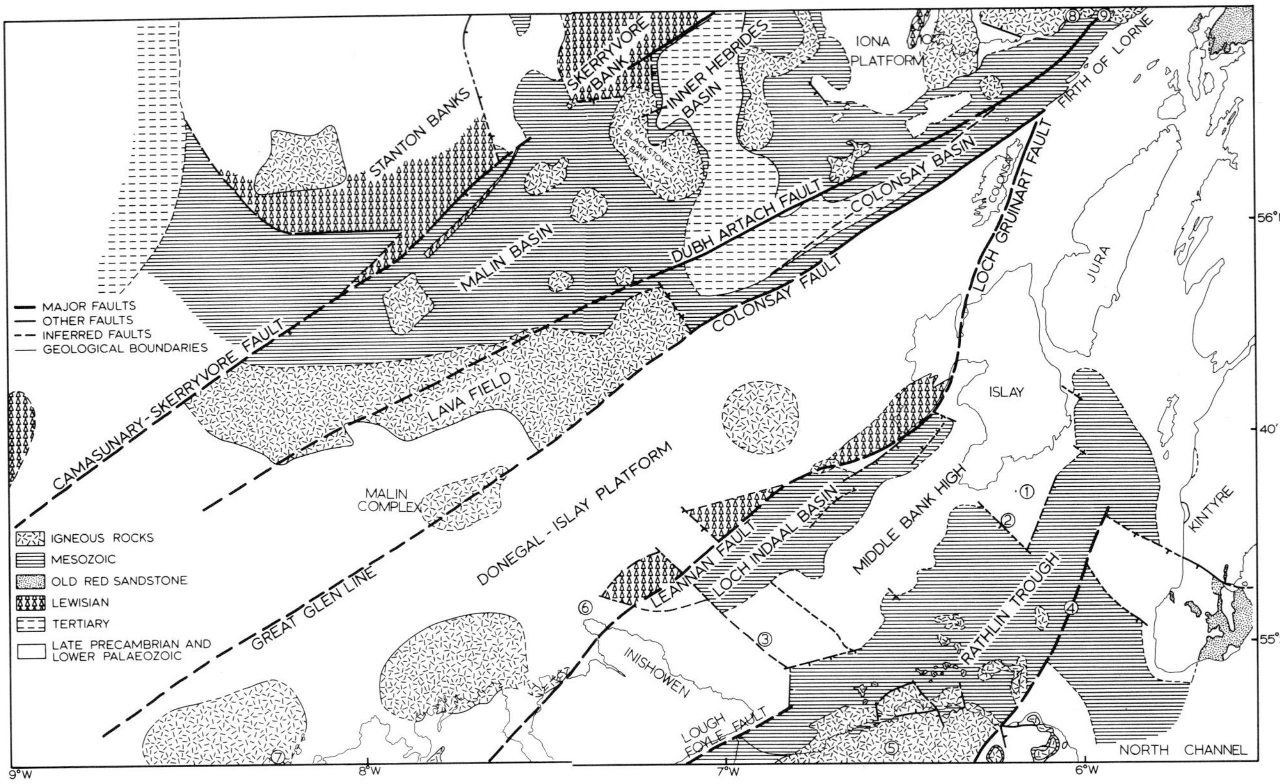

Fig. 3.1. Sub-Pleistocene outcrop map of the Malin Sea based on I.G.S. and U.C.W. Aberystwyth data. *1* = Otter Rock; *2* = Mull of Oa Fault; *3* = Inishowen Coastal Fault; *4* = Tow Valley Fault; *5* = Antrim Basin; *6* = Inistrahull Shoal; *7* = Bloody Foreland NW Ireland; *8* = Carsaig, Mull; *9* = Loch Buie, Mull; *10* = zone of the Moine Thrust.

equated with the *R4* interval of the Rockall Trough (Roberts, 1975) and, by extrapolation, to an horizon penetrated in Deep-Sea Drilling Project sites of the Hatton–Rockall Basin considered to be the Eocene–Oligocene boundary (Berggren, 1972). It is succeeded by an upper sedimentary sequence, draping the Margin, which includes sediments of Quarternary age that are widely developed in the Malin Sea (but have been excluded from the map; for the margin sedimentary sequence see also Bailey et al., 1974).

Thin Mesozoic rocks are also preserved between outcropping blocks on the Stanton–Skerryvore positive ridge which persists to the Margin (Bailey et al., 1975); otherwise, this ridge is almost entirely Lewisian. The south-eastern margin of this elongate block consists of a plexus of normal faults (collectively termed the Camasumary–Skerryvore Fault) which in a few places resolves to a single plane. Adjacent to the Stanton Banks, faulted inliers of basement parallel the major fracture line. Because Mesozoic rocks both overstep the line of the master fault and appear to be poorly preserved in the region to the south, the true nature of the structural low (here termed the Malin Basin) has for long remained in doubt (for example, in Naylor and Mounteney, 1975). In addition, a combination of intense block faulting and pronounced igneous activity (which includes granitic like bodies and bosses and sills, some, like those south of the Blackstones Bank, being very extensive) has further complicated the picture. Even the southern limits of the sedimentary subcrop, although recognisable in places, are difficult to establish. Riddihough (1968), on the basis of magnetic evidence, suggested that an extensive lava field occurred (see Fig. 3.1) on the southern basin – basement edge; this, like the Minch Basin to the north, probably terminates along a series of cross-faults. The lava field is mapped on Fig. 3.1 on the basis of magnetic evidence; it has not been possible, using shallow seismic data, to distinguish basement from the lava field, so the southern margin of this igneous body as shown on Fig. 3.1 has magnetic control only. In fact, deep seismic evidence indicates that the basin sediments extend further to the south. Thus, much of the strong magnetic effect recorded for this area may be due to shallow basement igneous material rather than that occurring near surface. The structurally complex nature of this margin means that detailed surveys will be required if these problems are to be resolved.

Towards the north-east the basinal structure becomes better developed. The distorting effect of the Blackstones Bank Centre on the gravity data makes it virtually impossible to estimate the likely thickness of the sediments of this Basin, but comparisons with adjacent basins suggests a figure between 3 and 4 km. North-east of the Blackstones Cretaceous centre (Durant et al., 1976), the structural continuation of the Malin Basin, termed here the Inner Hebrides Basin, narrows as a result of the presence of the Iona Platform but still may by traced to Skye (Binns et al., 1975, fig. 1). South of the Iona Platform, the low, here called the Colonsay Basin, extends into the Firth of Lorne at least as far as Loch Buie. It may be assumed for the sake of simplicity that the region between the Camasumary Fault and the line of the Great Glen Fracture remained negative throughout the Mesozoic, so that the sedimentary sequence on the west coast of Morven (consisting of Upper Cretaceous resting on Liassic) represents deposits of the eastern end of the basin complex. Later Tertiary faulting and igneous activity (the Mull Centres) uplifted the central area, effectively to divide the basin; the Dubh Artach Fault may have been associated with this event and may even owe its origin to it.

From the evidence of coastal exposures (particularly at Carsaig in Mull, Loch Don and Skye), I.G.S. boreholes and the samples collected by dredging and Scuba diving, a representative sequence of rocks can be recognised. At the base of the sequence is a variable thickness of Triassic rocks; these are overlain by Jurassic deposits which extend into the Kimmeridgian towards Skye (baked Kimmeridgian shales are also preserved on Mull). In the Colonsay Basin, the youngest Jurassic which has so far been dated is Bathonian; this is represented by shallow-water deltaic facies. Remnant Cretaceous rocks, largely post-Cenomanian are locally recorded above the Jurassic. The Tertiary is represented by bedded non-marine ?Oligocene deposits which dip towards the Camasumary fault, between the Blackstones Bank and the Iona Platform (Fig. 3.1) (Smythe and Kenolty, 1975). The basinal geology is further complicated by lava flows and sills. Finally the whole area is overlain by a substantial thickness of Quaternary deposits.

To the southeast of the Colonsay Fault lies the extensive and little known Donegal–Islay platform. This horst block may be traced from the Lewisian of northeast Colonsay to the Dalradian and granodiorites of Bloody Foreland north-west Ireland. The submarine extensions of these Irish plutons have been mapped previously (Evans and Whittington, 1976) as has the platform to the west and south of Colonsay. In the case of the latter feature, teams of divers have recovered rock of Torridonian (Colonsay facies) aspect; this recovery revises the original, largely geophysical, interpretation of Dobson and Evans (1974).

In addition to this evidence, sonar records allow the Lewisian of the Rhinns of Islay to be mapped offshore with some confidence (rocks of comparable aspect occur on the Inistrahull Shoal) but the overall extent of the Lewisian outcrop remains a mystery. The large circular igneous body (Fig. 3.1), to the north-west of the Loch Indaal Basin, has been delimited only on the basis of gravity data as no other information is currently available. The Malin Complex (Riddihough, 1968), further to the south-west and bisected by the Great Glen Line, forms a prominent magnetic feature and may be regarded tentatively as a Tertiary Igneous centre. The enigma that this platform presents to the geologist will persist for a long time to come, if only because basement mapping presently relies on recovery from Scuba diving and rather ineffective dredge sampling techniques.

In contrast to the previously described region, the Loch Indaal Basin is known in some detail, largely because of the recent drilling programme conducted by I.G.S. Samples 75/41, 75/42, 75/43, and 75/44 (see Evans et al., 1978, for details) indicated that the westward dipping strata over this area consist of Triassic, Rhaetic and Liassic. Gravity modelling, based on the control afforded by the boreholes and making an allowance for intra basinal faulting, suggests that this trough exceeds 3 km in overall thickness.

The Rathlin Trough, in reality the offshore section of the Antrim Basin, is delimited by the Lough Foyle Fault to the north-west and the Tow Valley Fault to the south-east. The Trough further extends longitudinally into the region of the North Channel. The overall configuration of the Trough is complicated, and finally terminated, by a succession of cross faults. The geological sequence (partially determined by the coastal boreholes and partially by the offshore samples) consists of Permo-Triassic resting on Carboniferous; this is followed by Liassic which is unconformably overlain by Upper Cretaceous. As in the Loch Indaal Basin, the majority of the 3–4 km thick fill is composed of Permo-Triassic rocks.

The Tertiary lavas of Antrim extend only a short distance offshore; their northern limits, as with Rathlin Island, are abrupt and spectacular. Within and beyond the lava field are intrusive bosses and sills, many of which have been sampled (Durant in Appendix to the Malin Sea Report). (It is now thought that the lava field covered all of the Antrim Basin–Rathlin Trough area, as far as the Islay Dalradian and the Lough Foyle Fault; it would have been thin in the north, because of the regional southerly slope induced by thermal doming (see later)). Separating these two sediment traps is a basement shoal, which is composed mainly of Dalradian and an associated epidioritic dyke swarm, referred to as the Middle Bank High (Fig. 3.1). This geological combination results in a rugged sea-bed topography; this, together with the area immediately to the southeast of the Oa Peninsula (Dobson and Evans, 1974) has been sampled in order to permit coastal extrapolation. (In the latter area, a pinnacle, termed the Otter Rock, was examined.) The geological high persists to the north as the islands of Islay and Jura. Here, in the shallow sea area to the east of Colonsay, pockets of Triassic are still preserved (see I.G.S. borehole 71/9); this suggests that what was originally conceived as a basin along the Loch Gruinart Fault extension (Dobson et al., 1975), is now thought to be only an impersistent veneer.

Structural history of the Malin Sea

The rocks of the Malin Sea demonstrate that the major phase of sediment accumulation, associated with substantial vertical movements, occurred during the Triassic and Lower Jurassic. The structural adjustments are particularly related to the Caledonoid trending master faults, which were responsible for a series of narrow horsts and asymmetric grabens. The latter have been sites of rapid continental-type sedimentation since early Triassic times. The preserved record notably displays the disparity in thickness between the Triassic of Western Mull and that to be found in, say, the Loch Indaal Basin or Rathlin Trough. This provides evidence for marked syn-depositional movement, along narrow north-east trending zones. Steel (1974) specifically refers to fault-controlled alluvial fan sedimentation along the margin of the Minch Basin. This mode of sedimentation process is locally displayed by the Stornoway Beds and repeatedly occurs throughout western Britain.

The well-documented major transgression which introduced marine Liassic rocks to the British Isles, is well represented in the Malin Sea. Although a regressive phase developed during the Middle Jurassic, localised warping has ensured preservation of these rocks in addition to much of the subsequent Upper Jurassic. Movements that occurred throughout the Triassic and Lower Jurassic are ascribed to the early Kimmerian phase, which has been widely recognised in the North Sea (Ziegler, 1975a, p. 143), the Irish Sea and Celtic Sea. This movement instigated the extensive marine phase of sedimentation, in combination with other global events, and permitted Liassic deposits to accumulate in thickness. Boreholes along the Northern Ireland coastline and the deep seismic offshore data both provide evidence that the total thickness of these deposits, of between 3 and 4 km is comparable with that of the troughs of the south.

Whilst there is a history of Mid-Jurassic movements, there is no evidence of any major unconformity between the Lower and Middle Jurassic because the main phase of tectonic adjustment off western Scotland occurred later (in the Kimmeridgian). There is strong, if not unequivocal, evidence of Upper Jurassic movements along the Camasunary Fracture (Fig. 3.1) in Skye; recently published surveys relating to the Moray Firth (Chester and Bacon, 1975), on the eastern side of Scotland, reveal a similar story. (In the Moray Firth, the Great Glen Fault was active throughout the Mesozoic as a normal fault; the final phase of activity occurred towards the end of the Lower Cretaceous.) Along the Helmsdale coast, the evidence is also of Kimmeridgian movements (Bailey and Weir, 1932). Supporting evidence for this phase of adjustment in the region is provided by the area to the south-east of Mull; here, normal faulting took place along this master structure. Deep seismic data (Binns et al., 1975) across a branch of the Great Glen Fault Zone to the south-west of Mull, known as the Dubh Artach Fault, provide further evidence of normal movements. The principal period of movement on the Colonsay Fault, a structure which is considered to be on the line of the Great Glen Fracture, occurred after the Middle Jurassic.

The especial significance of the *late* Kimmerian movements is that they adopted the existing fracture lines in direct contrast to the situation which occurred off West Wales. The movements marked the end of the period of persistent sedimentation; they initiated phases of erosion, tilting, of thin margin facies and extended periods of non-deposition. The shallow, open folds recorded for the Malin Sea are probably the response of the area to basement faulting. Lower Cretaceous rocks are not present in the Malin Sea troughs, implying that major vertical movements on the faults had ceased by this time: only in the area of the Sea of the Hebrides (see Site 71/10, Chester et al., 1972) is there some (poor) evidence for Lower Cretaceous preservation. The persistence of the regionally positive attitude in the structure has resulted in only thin Upper Cretaceous sedimentary deposits, with much of these belonging to the Senonian period. The major late Cretaceous event was the establishment of the Blackstones igneous centre (Durant et al., 1976), which is now dated at between 70 and 79 Ma. Its emplacement must have affected the palaeogeographic picture in the Upper Cretaceous, by restricting the transgressive events.

During the Tertiary era, pre-lava sediments persisted in Skye and provide evidence of the weakly positive, but low-lying nature, of the western Scottish Area. As George (1966) has indicated, evidence of post-Cretaceous pre-basalt movement is not impressive.

Significant Lower Tertiary sedimentation was restricted to the Continental Margin and the adjacent Rockall Trough where a prograding series (Bailey et al., 1974) may be recorded to the west of approximately 08°30′W (Uruski, 1978). The thick sequences of Tertiary sediments, recognised by Roberts (1975) in the Rockall Trough, provides evidence of the extended phase of erosion undergone by the mainland of Scotland and the adjacent shelf area during much of the Cretaceous and early Tertiary times. Clearly, much of what was once mainland Scotland now resides in the Rockall Trough.

The major igneous phase, during the Eocene, involved a significant thermal control over the region; this caused lateral crustal extension by doming and also allowed the dykes within the province to extend to the highest parts of the rock sequence. There is no evidence that any dykes exploited the north-west trending cross faults which might be indicative of the absence of such features: dyking also did not occur along the Caledonoid trends, although some "sill-like" bodies appear to terminate along the master-fault boundaries.

A period of pronounced fault adjustments in the Tertiary occurred after the lava phase and probably after the period of non-marine deposition in Lough Neagh and the Inner Hebrides Basin (Fig. 3.1) of supposed Upper Oligocene age. Two directions of movement can be recognised: the first occurred along the previously established Caledonoid master faults, as demonstrated by the displacement of lavas in the Tow Valley Fault, in Northern Ireland, and the Camasunary Fault; the second direction was along the north-west–south-east cross faults (e.g. the Mull of Oa Fault, the Inishowen coastal Fault and the fracture that terminates the Colonsay Basin in the west). A feature of these movements, which can be observed on the map (Fig. 3.1) is the preservation of the lava field to the south-west of the Malin Basin. However, such is the uncertainty surrounding the explanation of this extensive igneous subcrop that it might be of late Cretaceous age, and associated with the Blackstones event, rather than a Tertiary field.

The early Neogene faulting phase was succeeded by a long period of erosion which was a function of the positive aspect of much of the Shelf region and, in particular, the Scottish mainland. The geomorphological history of Scotland during the Neogene has been consummately analysed by George (1966) and it is sufficient to generalise by stating that only the Quaternary period provided a thick sequence of sediments and that most of these deposits are considered to be of glacio-marine origin.

Two additional features of the structural evolution of this part of the British Shelf are worthy of mention. The first is the Great Glen Fault Line, about which there has been a voluminous literature. On the map (Fig. 3.1) the position of the Line has been determined largely on the basis of analysis of the magnetic and gravity data for the region; the line extends north-east to the normal Colonsay Fault, which is a well documented basin margin fracture. The connection between this geophysically determined line in the south-west and the geologically appreciated structure in the Loch Buie area, is suggested by the dashed line passing through the Colonsay Basin (Fig. 3.1). The Dubh Artach Fault to the north possesses a marked normal component, but retains no geophysical signature beyond the basin: similarly, the Colonsay Fault appears to terminate at its junction with the Loch Gruinart Fault. The "Line" has been proposed by the writer in order to resolve this dilemma. One consequence of this proposal concerns the Malin Complex which is located along the line, i.e. the fact that this complex appears not to have suffered faulting, considered in relation to the proposed line of the Great Glen, might indicate that the complex is of Tertiary age. The second consideration concerns the Loch Gruinart Fault and the extension of the Leannan Fault. If the linearity of the Leannan Fault is maintained towards the north-east then the best course would be a line bisecting the island of Islay, along the Bonahaven Fracture of Durrance (1976), and extending along the north-west coast of Jura. The curving Loch Gruinart Fault might, in turn, be a displaced part of the Moine Thrust.

General geology of the South Irish Sea, including parts of the Celtic Sea

The South Irish Sea, a practical term particularly used by marine geologists, refers to the landlocked sea area between the latitudes of Anglesey (53°20′N) and the St. David's Peninsula (51°52′N); it incorporates the St. Georges Channel and the two Welsh coastal embayments of Caernarvon and Cardigan (see Fig. 3.3). It is an area about which extensive geological literature has been published. The northern half of the Celtic Sea is included in this discussion, partly because it is the structural extension of the South Irish Sea and partly because it provides very necessary additional geological support for the views expressed by the writer concerning the evolution of the Margin.

Viewed simplistically, the structure of the region may be imagined to consist of a deep narrow trough, orientated along the Caledonoid trend, bounded by equally narrow positive ridges of similar orientation (Fig. 3.2). This central trough, which extends west-south-west from the Cardigan Bay coastline and across the northern Celtic Sea to the outer Continental Margin, is in reality a complex of sub-basins, each with a unique geological history. The ridges are also complex and poorly understood because of their geological composition and unattractive commercial prospects.

Whilst the southerly ridge, which is a south-westerly extension of the St. David's Peninsula (Fig. 3.3), may be conveniently termed the Pembroke Arch and the central trough usefully referred to as the North Celtic Sea Basin, the northerly ridge is less easily named. Basically, it includes the Lleyn Peninsula, the Irish Sea Geanticline (Dobson et al., 1973), the Carnsore Point area and part of what is today the south-eastern margin of Eire (Fig. 3.3).

The geology of the northern ridge is best understood over its eastern section where it can be recognised as a fault-bounded positive block of Precambrian and Lower Palaeozoic rocks. The geology of part of this structure, the "Geanticline", has already been described elsewhere (Dobson et al., 1973, p. 18). The recognition of Monian metasediments in County Wexford allows, at least,

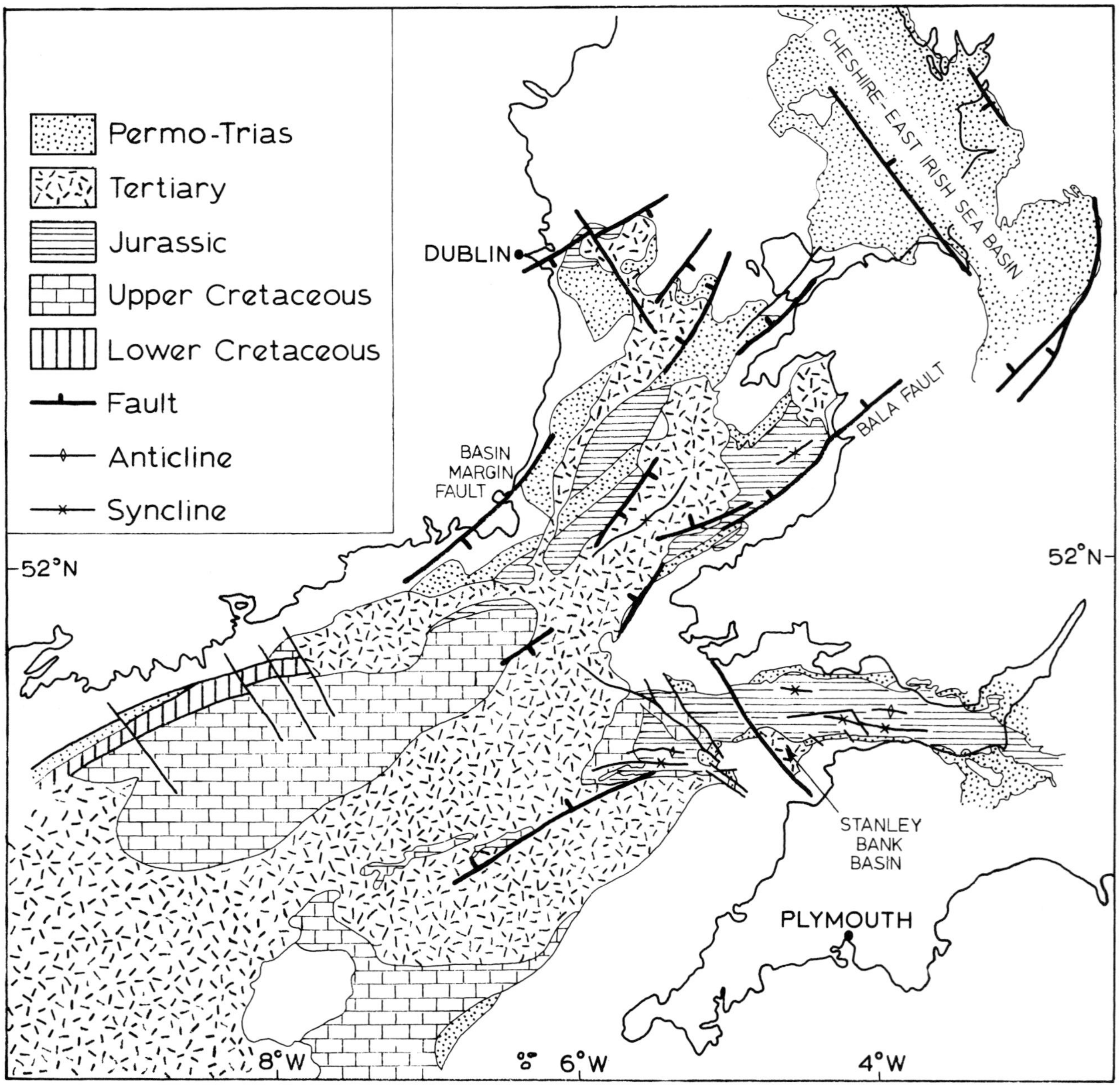

Fig. 3.2. Sub-Pleistocene outcrop map of the Irish Sea and parts of the Celtic Sea, based on U.C.W. Aberystwyth data and information from I.G.S. (Doré, 1976; Evans, M.S.; Ph.D. thesis, U.C. Swansea, 1973).

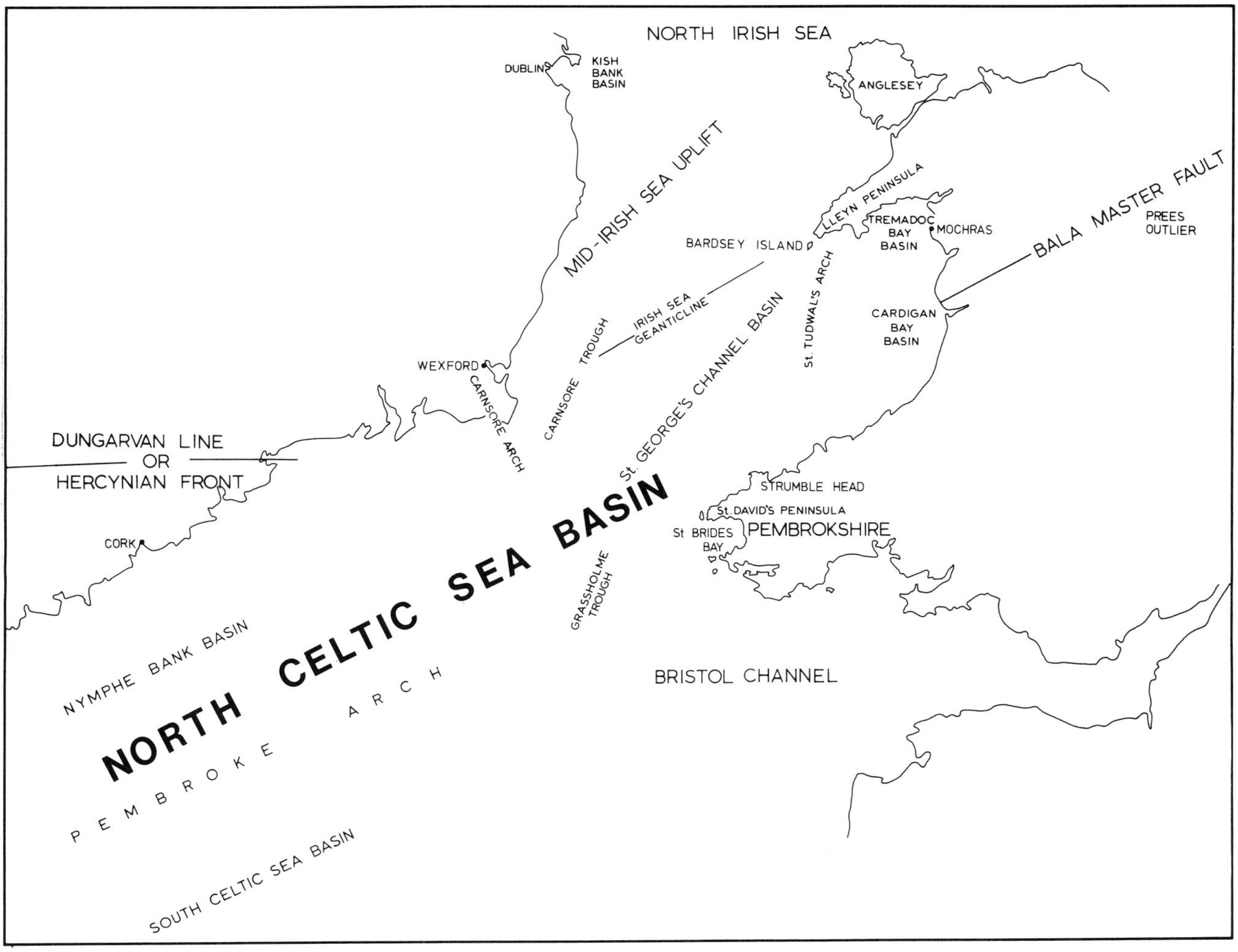

Fig. 3.3. Major structural regions in the South Irish Sea and North Celtic Sea.

tentative links to be made with the Lleyn Peninsula; a general description of the eastern part of the feature should, therefore, emphasise the Precambrian and Lower Palaeozoic nature of the core. Granitic bodies, like those at Carnsore and on Bardsey Island, provide additional support to this hypothesis; however, towards the south-west the ridge loses its clear linear identity. Although overlain by very thick Upper Palaeozoic rocks, deep seismic data for the continental slope end of the ridge (Whitbread, 1972) indicate that the feature is a major margining structure which is bounded by a series of large step faults.

Two southerly trending arches form part of this fault bounded block. One of these, the St. Tudwal's Arch, is a prominent and well-defined basement ridge which is located at the end of the Lleyn Peninsula (Fig. 3.3); it effectively isolates Cardigan Bay from the remainder of the main trough. The other, the Carnsore Arch, is less easily defined; however, its history of uplift, like that of the St. Tudwal's Arch, has meant that basement rocks penetrate southwards across the line of the main basin; this is again effective in basinal separation, but in this case between the St. George's Channel Basin and the Nymphe (Fig. 3.3). Apart from these arches, the block is also divided by two major, but narrow, graben structures which are known as the Bardsey Island Trough and the Carnsore Trough. These structures appear to contain both Mesozoic and Tertiary sediments and define the limits of the Geanticline portion of the ridge; this portion, unlike the rest of the structure, is veneered by Permo-Triassic–Lias, and in the east by thin Tertiary sediments. Palaeozoic basement rock outcrops along the ridge axis. Because this bounding structure is a complex association rather than a single feature, the structural controls are difficult to resolve. Clearly, the North Celtic Sea Basin margin fault, the fracture on the south-east side of the Mid-Irish Sea Uplift (see Dobson et al., 1973), and the complex of faults along the southern flank of the "Geanticline" all display normal movement with downthrow to the south-east. Evidence is accumulating that suggests that the two former faults are linked through Wexford and along the Irish coast. Although accommodation faulting along the southern flanks of the ridge is well documented, it is assumed that much of the associated basinal development was accomplished by block tilting towards the south-east.

The main trough in the area, or the North Celtic Sea Basin, is more than 350 km long but is rarely more than 35 km wide. It is contained wholly between the two basement ridges, except where small grabens (to the north), and a single trough across the Pembroke Arch (to the south – Fig. 3.3) distort the picture. The geology of the eastern portion of this low is known in some detail (Blundell et al., 1971; Dobson et al., 1973). The western end of the feature, although examined by commercial concerns, has not yet been fully reported in the literature. Recent cruises by U.C. Aberystwyth, have improved our knowledge of the shallow geology; however, although data on the deeper geology are still unavailable, enough can be gleaned from reports (see References), relating to both the Celtic Sea basins, to establish the general geological pattern. East of St. Tudwal's Arch lies the Cardigan Bay Basin and its sub-basin in Tremadoc Bay (Fig. 3.3). The geology of the bay is now well known, based on information from closely spaced seismic surveys, and both coastal and off-shore boreholes. The rock sequences in Tremadoc Bay were originally determined by reference to the stratigraphic test hole at Mochras (Woodland, 1971) which revealed the presence of a thick (1305 m) Liassic sequence overlying approximately 1000 m of Permo-Triassic rock. A profound unconformity, which reaches the Liassic at Mochras, allows Oligo-Miocene to overstep from the Middle Jurassic to the Palaeozoic. Both the Mesozoic and the Tertiary successions thicken towards the coastal fault, which is the present-day eastern margin of the North Celtic Sea Trough. Vertical movements on this fault plane, on a massive scale, have resulted in the juxtapositioning of Neogene and Lower Palaeozoic rocks. The Caıdigan Bay Basin, developed between the St. Tudwal's Arch and the Palaeozoic of the Cardigan coastline, contains a thick Mesozoic fill consisting principally of Triassic and Liassic rocks with a thinner sequence of Middle Jurassic. Minor local unconformities within the Jurassic succession bring the Middle Jurassic against the Lower Liassic rocks. The youngest Jurassic sequence from the Bay is of Bathonian age; no Upper Jurassic has been recovered. The approximate thickness of the Middle Jurassic is 1280 m, which gives a Triassic to Middle Jurassic fill of about 3500 m.

The south-east margin fault, part of which may be linked to the Bala master fault, is in reality a complex of at least three sub-parallel faults each of which has a normal throw component. Major salt features which are concentrated along this faulted edge as a salt wall (see I.G.S., Cardigan Bay Sheet), further complicate the geological picture; salt also occurs as deep pillow structures, although isolated high level diapirs have been mapped. Thin Tertiary sediments, extending eastwards from the St. George's Channel Basin, unconformably overlie the Mesozoic rocks; this notably occurs in the south, where the effects of the St. Tudwal's Arch have largely disappeared. It is in this southern section that the mergence of the two basins is best demonstrated by the westward thickening of the Tertiary rocks: this is a characteristic feature of the region.

The St. George's Channel Basin extends from the St. Tudwal's Arch to the Carnsore Arch, and from the Bardsey Island Trough in a south-westerly direction towards the Grassholm Trough (see Fig. 3.3), which crosses the Pembroke Arch. This basin is of particular interest as it contains a substantially greater thickness of sediment (> 6000 m) than adjacent structures. Along the western flank of the basin is a series of Mesozoic horst blocks, some of which have been affected by salt movements. Evidence of major salt tectonics is also present along the axis of the trough, which is offset to the north in comparison with the general Caledonoid trend of the North Celtic Sea Basin. The geological succession within the Basin can, at the present time, only be determined by extrapolation of data from the Cardigan Bay boreholes and close inspection of the deep seismic coverage of I.G.S. With the exception of two unconformities (one at the Jurassic–Cretaceous boundary and the other of Tertiary age) sedimentation processes appear, from the deep seismic data, to have been continuous. However, this evidence may be deceptive for whilst thick Triassic and Jurassic sequences undoubtedly do occur, representations of the Upper Mesozoic and Tertiary sequences are not as consistent. The isolation of this basin from the Nymphe section, as a result of the Kimmerian movements, was probably enhanced because of its northerly location with reference to the Hercynian Front, or Dungarvan Line (see I.G.S. Report 73/11). Because of the presence of thin Upper Chalk overstepping Lower Jurassic rocks at the longitude of Carnsore Point, it might be assumed that a major fracture has adopted the trend of the Hercynian Front. The effect on the St. George's Channel Basin of these two structural influences would be to restrict it to long periods of non-marine sedimentation.

It is understood that a very considerable thickness of Tertiary rocks was encountered, in (commercial) boreholes, lying on Jurassic; if this is correct, it implies that the Upper Cretaceous subcrop is either severely restricted or absent in the basin. A uniform layer, more than 500 m in thickness is recorded on the deep seismic records; in relation to its seismic signature, this had previously been referred to as chalk (see I.G.S. Report 73/11). However, it now seems that the basin sequence consists of Lower Tertiary rocks overlying Jurassic, with a high proportion of non-marine material. Difficulties in dating these non-marine sequences allow one to speculate that the bulk of the sequence above the Upper Jurassic, and recovered from the commercial borehole, is of Oligo-Miocene age. Two Neogene successions, referred to in the 1973 report of I.G.S. have now been reinterpreted, with the aid of information from marginal boreholes in the areas, as Pleistocene (Delanty and Whittington, 1977). On the western side of the basin, the prominent flanking blocks have seismic characteristics typical of those of Lower Mesozoic (particularly Liassic) sediments. Seabed outcrop of these rocks would support this contention, for the only mid-basin Mesozoic rocks that appear sufficiently indurated to withstand repeated ice attack and outcrop on the seabed are Liassic calcite mudstones (e.g., in the Kish Bank Basin – see Fig. 3.3, see Dobson and Whittington, in press).

The southwestern part of the North Celtic Sea Basin is that zone which lies beyond the Carnsore Arch and has been described as the Nymphe Basin. The terminology strictly relates to the Tertiary end but tacitly subsumes a thicker subcropping Mesozoic sequence. The Nymphe–North Celtic Sea Basin (Fig. 3.3) contains, in the east, a southward dipping Triassic–Lower Jurassic sequence, in fault contact with the Precambrian and Upper Palaeozoic of the Irish block. To the south, Upper Cretaceous chalk oversteps these rocks; in the west, the Triassic and Jurassic are progressively overstepped by both Lower Cretaceous sands and by the Upper Cretaceous although the picture is complicated by normal faulting. The Tertiary consists largely of Eocene with (probably) Oligocene non-marine sediments; both are preserved in a narrow fault bounded trough. This trough extends to the longitude of Cork, it is connected to the St. George's Channel sequence to the east-north-east. Further to the west, the Tertiary subcrop is largely restricted to the margin zone. No positive identification of younger Neogene deposits, which are thought to be present in the area, has yet been made.

Quaternary deposits are confined to the section to the east of Cork and are only found in any appreciable thickness overlying the Tertiary rocks. However, they are sufficient to prevent sampling, at least by inexpensive coring techniques, of the eastern Tertiary deposits. Basin – see Fig. 3.3, see Dobson and Wittington, in press). 3.2, is markedly affected by the NW–SE trending faults. These crossfaults are considered to have initiated, and by subsequent movement eventually to have preserved, thick Oligocene sedimentary deposits (this process can be compared with that in the Stanley Bank Basin (see Fletcher, 1975)). The extensive outcrop of chalk masks the underlying structural grain of the basement; it straddles the Pembroke Arch and extends on into the South Celtic Sea Basin.

With the exception of these restricted examples, the deeper geology of the Nymphe Bank Basin Zone (which is known in detail from commercial sources), has not been published; however, sufficient information may be abstracted from annual reports and published deep seismic data (see Bibliography) to make the following comments. The western end of the basin has a fill of sediment which consists of a 2500 m thick section of Permo-Triassic, Liassic and later Jurassic, thin deposits of Lower Cretaceous (100 m) and Chalk (130 m) and very thick Tertiary sequences (900 m) along the Margin. It would therefore appear that the sedimentary fill of the Nymphe Basin progressively thickens eastwards from a thickness of 2900 m to one of more than 6000 m in the St. George's Channel Basin; locally, extremely thick non-marine Tertiary deposits would complicate this general pattern. Further, it is clear from the evidence presented in most of the reports that a profound unconformity of between Purbeck and Wealden age is recorded throughout much of the Celtic Sea area such that on the flanks of the median Pembroke Arch erosion to the Lower Liassic and Triassic has occurred.

Above the unconformity, Lower Cretaceous deposits are known to vary between 400 m and 1000 m in thickness: the chalk, although very thin adjacent to the margin, thickens eastwards to nearly 800 m (at least in the South Celtic Sea Basin).

Structural history of the South Irish and Celtic Sea areas

According to Lorenz and Nicholls (1976), the structural setting in the South Irish sea area at the beginning of the Permo-Triassic (which has been loosely described as a series of grabens and horsts essentially orientated along a Caledonoid trend) was the crustal response to an Upper Mantle diapir, which formed above two subduction zones at the end of the Hercynian Orogeny. Whatever was the cause of the basic structure, data from the deep seismic sections clearly demonstrate that the Triassic and subsequent Liassic period were dominated by a series of sub-parallel Caledonoid trending basement ridges and lows. These controlled the sedimentation processes which, from the evidence of deep seismic records and microfaunal analysis, were maintaining a flat sea floor at shelf depths; this was in keeping with the rate of basement adjustments (a view shared by Sellwood and Jenkyns, 1975).

In the North Irish Sea, the Caledonoid influence both extends to and coalesces with the northerly trend of the Cheshire–East Irish Sea Basin. To the south, the eastern extent of the Caledonoid Basin trend is restricted by the (now positive) Welsh block. It is proposed by the writer that, at least by early Keuper times and certainly during the Liassic, the axis of the area of thick sedimentation between the Lleyn Peninsula Ridge and the Pembroke Peninsula Arch also extended into the Cheshire Basin. This view is supported by the geophysical data and that from coastal boreholes. The Triassic thickens towards the south-east, away from the Lleyn Carnsore Ridge, towards the offshore extensions of the Bala Fault

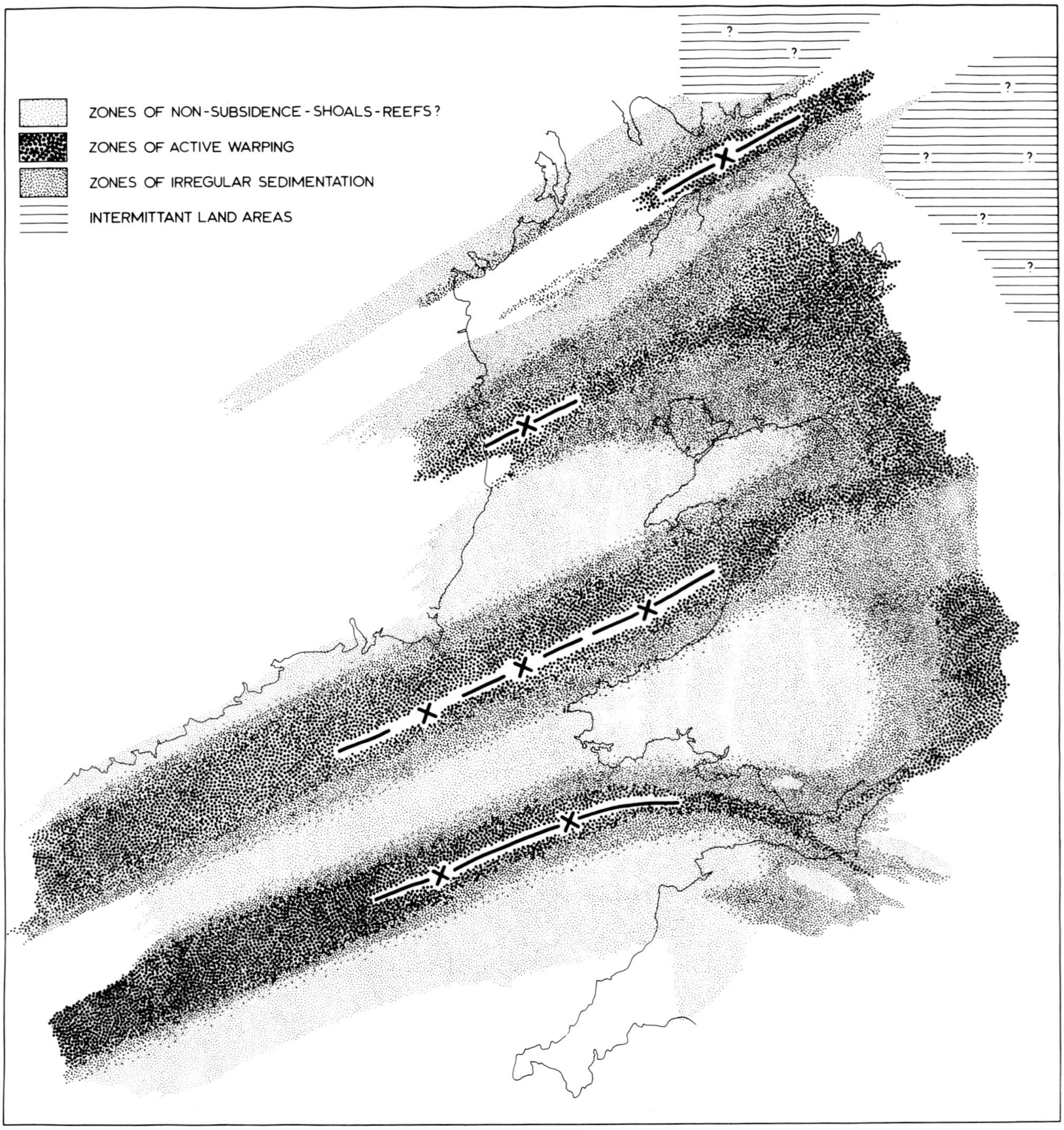

Zone. In fact, the axis of the maximum thickness of Triassic deposits lies along this margining fault system. It is probable, therefore, that in Triassic times the southern edge of the trough was what is now the Pembrokeshire coast. The Longmynd horsts may represent the eastern expression of this, now obliterated, trough margin. Figure 3.4 is a palaeogeographic reconstruction of this interpretation. The Permian has been deliberately excluded from the interpretation because of the transitional nature of this period which is supposedly one of only limited downwarping; for example, in the North Irish Sea, the Triassic is five times as thick as the Permian (Colter and Barr, 1975). Triassic thicknesses of between 1000 and 3000 m reveal the scale of the downwarping processes which were experienced by these troughs.

Within the context of the Margin Cycle Theory (Sleep, 1971), the Liassic represents the marine transgressive phase characterised by continued subsidence and thick axial sedimentation. Naylor and Mounteney (1975) suggest that the Welsh massif may have stretched to the Central Irish massif creating a partial marine barrier, thereby establishing two provinces. No evidence is available for this; on the contrary, all the investigations carried out by U.C. Aberystwyth, are indicative of almost total regional submergence, with the Palaeozoic ridges supporting carbonate-rich deposits. The sediment and faunal character of the Liassic sediments of the Kish Bank Basin, together with the overall structural configuration of the low, allows the adduction that a trough extended deep into what is now central Ireland. Only over the Lleyn Peninsula and Pembrokeshire would the cover have been very thin. The pronounced faulting along the trough flanks and later erosion (Lower Cretaceous) has surely been the cause of the restriction of the present Liassic sub crop, rather than peculiarities in the palaeogeography. The Trias–Lias palaeogeography, as presented here, has received further support from the detailed microfauna data (Dobson and Whittington, in press). This clearly indicates shelf-sea depths during deposition and a single faunal province throughout; pronounced shallowing is only seen to occur in the later Toarcian. Incremental basinal growth is indicated, with thicknesses in excess of 1000 m commonly detected in the North Celtic Sea Basin, together with regional subsidence of a very gradual character. Finally, the Prees outlier (Ager, 1974) which is known to include both Lower and Middle Liassic, contains no evidence that would support the idea of thinning towards a supposed Welsh land mass.

A major corollary of this palaeogeographic reconstruction concerns the Bala Fault. It is thought that the fault zone was inactive, at least as far as Strumble Head, during the Triassic and Liassic. The more westerly fracture which lies to the north-west of St. David's Head may, however, have been the margining trough fault. Further consideration is that the North Celtic Sea Trough was probably asymmetric with the depositional axis offset towards the south in the Cardigan Bay area; the same occurs in the Nymphe Bank sector.

With thicknesses in excess of 800 m common for the Middle and Upper Jurassic deposits in both the Celtic Sea troughs, it is clear that regional subsidence persisted throughout this period. There was, for example, an uninterrupted sequence of sedimentation in the Bristol Channel–South Celtic Sea Basin. However, major structural adjustments, noticeably in the North Celtic Sea Basin, were taking place: the St. Tudwals Arch rose during the Middle Jurassic period, causing marked tensing of the sequence; also, unconformities are recorded for the end of the Lias period, with minor erosion and warping occurring in the Cardigan Bay Basin. These "Kimmerian" movements were most pronounced along the western side of the South Irish Sea; they combined with the other fractures to produce a realignment of the St. George's Channel Basin. A regressive phase of sedimentation, closely linked with these movements, can be recognised in association with shallow water carbonates and deltaic sequences (see I.G.S. Report 76/6). Thus, by the latter part of the Upper Jurassic–Lower Cretaceous, a completely new structural regime had emerged in the South Irish Sea; this consisted of a narrow (30 km wide) graben orientated at N32°E and extending from the Bardsey Island Rift (which splits the Lleyn Carnsore Ridge) to the Grassholm Trough (on the Pembroke Ridge). Marginal faults, with adjacent uplift, also occurred. These late Kimmerian movements therefore resulted in the following: uplift of Southern Ireland and the Carnsore Arch; the Carnsore Trough; elongated Mesozoic blocks along the western edge of the St. George's Channel Basin; the final separation of the Cardigan Bay Basin from the rest of the North Celtic Sea Basin; uplift of the bay area and marked erosion; and, most significantly, movement on the Bala Fault which, with the Lleyn Peninsula uplift and minor rotation, represented the regional rise of the Welsh massif. It is thought that the attendant erosion mainly to the east of the Arch provided much of the material for the deepening Celtic Sea basins (see Fig. 3.5). The thickness of Lower Cretaceous sediments reflects the degree of downwarping that occurred during this non-marine and dominantly deltaic phase. The Cenomanian transgression, although initiating a period of widespread pelagic sedimentation, was constrained by the Kimmerian blocks;

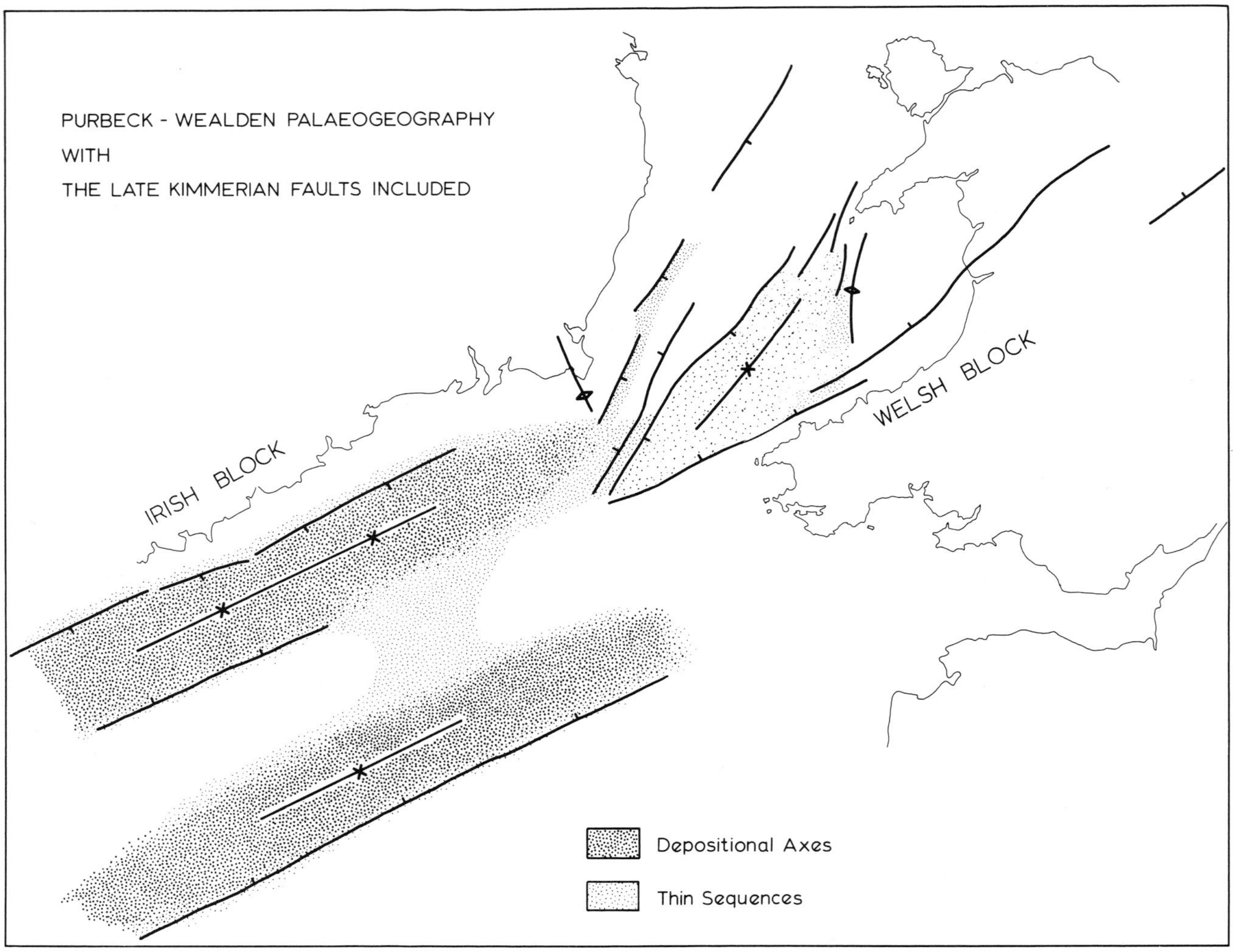

Fig. 3.5. Purbeck–Wealden palaeogeography.

the greatest thickness of sediments has been recorded in the east, notably the South Celtic Sea Basin. Although locally thick deposits of chalk are present in parts of the North Celtic Sea Basin they lie close to the Chalk Sea Margin; thus it is the succeeding Palaeogene era that emphasises the relative independence of the various sub-basins first developed in the Lower Cretaceous.

Lower Tertiary sedimentation produced thick, and at times rapid, deposition in all the sub-basins (with the exception of the Western Nymphe). Not only were thick Tertiary deposits laid down across the Pembroke Arch in the Grassholm Trough, but this was also a period of major faulting (with a pronounced wrench component (6 km) developing during the Oligocene). This reactivation of earlier Armorican–Hercynian fractures was sufficient to allow the very thick (600 m) non-marine sediments to accumulate in the "trapdoor type" of structural environments (Kent, 1975).

The structural sequence which emerges for this region may be summarised as follows:

(1) An Upper Carboniferous–Lower Permian horst–graben province.

(2) An early Kimmerian phase with Caledonoid trending basinal sagging.

(3) A late Kimmerian phase with main basin realignment and uplift.

(4) Mid-Tertiary faulting along a NW–SE trending axis, with both normal and wrench components.

Discussion

Probably the single most important tectonic event to affect western Britain since the late Palaeozoic was the establishment of the basin–horst (or rift-block) structural regime. This fundamental configuration, which developed mainly during the Permian and which still dominates much of the subcrop, presents a major interpretative puzzle. Can it be explained in terms of the result of incipient continental margin development or an aspect of post-orogenic adjustment? With the cessation of the compressive stresses during the late Carboniferous, tensional structures formed which were accompanied by uplift, igneous activity, tilting, erosion and differential subsidence. A series of new, distinctively narrow, mainly NE–SW trending horst–graben systems was developed; these linked locally with earlier formed N–S trending structures (like the Worcester–Cheshire graben (see Ziegler, W., 1975)) and resulted in a depositional axes complex. These new grabens were formed in response to early movements along the Arctic–North Atlantic rift zone (Ziegler, P., 1975). They form a continuous spectrum, which is consistent with the progressive breaking up and lateral drift of the north-west European plates and preserve a history of sedimentation which is unexceptional within this structural context (Schneider, 1972).

Loren and Nicholls (1976) prefer the geodynamic model of the Great Basin which was introduced by Scholz et al. (1971); this invokes the doming as the result of a rising diapir of mantle material. In order to include all the linear depressions of north-west Europe, this doming is considered to have been a regional event.

Whatever the precise nature of the initiating mechanism, the thickness of sediments which was subsequently preserved (firstly of continental type and later of marine character, with a marked final regressive phase) and related to a period of some 50 million years, is remarkably similar to the sequences described by Sleep and Snell (1976). These authors' model, partially a refinement of the interpretation by Vogt and Ostenso (1967), involves thermal tectonics in association with isostatic coupling which allows for marginal uplift and renewed subsidence.

Irrespective of the mechanism which sustained this structural regime from the Triassic to the end of the Jurassic, it was effective from northern Scotland to the English Channel. Towards the end of the Jurassic period, a substantial change occurred which was heralded by a widespread period of uplift and erosion which is now preserved as a profound unconformity. The cause of these structural adjustments is inextricably linked with the genesis of the Atlantic Ocean. The rifting sequence, well described by Williams (1975) provides the necessary explanation; this included both the Bay of Biscay and the Rockall Trough, and was no doubt preceded by doming and all the associated phenomena consistent with plate separation. Much of the subsequent sedimentation during the Lower Cretaceous period was non-marine; this only developed in thickness well away from the spreading ridge. In the west Scottish province, doming was altogether too close for sedimentation and, thus, marked erosion is the only legacy. Thermal contraction occurred, noticeably in the South Celtic Sea Basin, as distance from the ridge increased to the south. This phenomenon, in association with the world-wide Cenomanian transgression, ensured that a thick marine phase of sedimentation ensued. This phase of vertically extensive marine sedimentation appears to have occurred later in the North Celtic Sea Basin. With the major zone of active spreading remaining close to western Scotland, contraction was very limited and the land surface was subdued but not completely inundated. Thus, because of the complexity of the Atlantic opening around Greenland, the model proposed by Sleep (1971) and involving periods of approximately 50 million years, ceases to be applicable. Spreading along the Reykjanes Ridge, after precursory doming, again ensured that Scotland's western regions were uplifted; this caused further extensive erosion and sedimentation was restricted to the continental margin along the Rockall Trough.

Finally, in the late Palaeogene, wrench and normal faulting occurred along NW–SE and N–S trending lines; this appears to have affected most of western Britain and has allowed thick (essentially) non-marine sediments to be preserved. In the south, this depositionary sequence added to the already considerable Tertiary deposits; in the north, these occur as rather unique additions to an area otherwise dominated by erosion.

These attractive continental margin models must remain speculative; until such time as our knowledge of the deep crust to the west of the British Isles improves; nevertheless, enough is now known of the stratigraphy of the continental shelf to place some confidence in them.

Geological control of coastline configuration and the sea bed

Although both the Malin Sea and the South Irish Sea have been repeatedly glaciated with thick Quaternary deposits, now recorded in the overdeepened troughs, the present-day bathymetry reflects both the distribution of the troughs and ridges. This is particularly true of the Malin Sea, where the Precambrian ridges form the rough ground and the shoals; in contrast, the basins are generally smooth and deep. An example of this is provided by the bathymetry of the extensive Donegal–Islay Platform (Fig. 3.1); although this superficially gives the appearance of being uniform, it is, in fact, rugged with a veneer of glacial debris. Marked changes of bathymetry regularly relate to structural breaks; thus, the general disposition of basins and platforms has been known for a long time (see, for example, Stride et al., 1969). The pronounced shoals which occur in the basins are, almost without exception, due to igneous intrusive bodies or complexes like the Blackstones Bank.

The South Irish Sea, although structurally complex, does possess a geologically dominated bathymetry; for example, the shoal rock platforms are all part of ridge systems and Cardigan Bay is shallow, being largely free of Tertiary sediments. The north–south orientated depression, in the central part of the St. George's Channel, is only an exception to this generalisation where it transgresses the geological high of the Geanticline. This feature is related to a combination of very thick Quaternary sediments and the late glacial drainage pattern which is largely contained within these deposits (Dobson et al., 1971, and Whittington, 1977). The Celtic Sea proper, apart from the Palaeozoic platforms of its margin, is covered by Chalk and Tertiary deposits; the ridges are largely buried. Consequently, the bathymetry appears to exist independent of the geology (exceptions include the offshore extension of the Cornubian batholith).

A brief glance at the general geological map (Fig. 3.2) of the North Celtic Sea and associated South Irish Sea region, will serve to indicate the influence which geological structure has on the configuration of the present coastline. This influence is in part deceptive. The boundary faults of the major blocks, which record a long history of episodic adjustment, were most active during the late Kimmerian phase (which here approximates to the Upper Jurassic–Lower Cretaceous). Much of what is now Southern Ireland and Wales became persistently positive elements at that time. However, early coastal development did not immediately follow this adjustment, principally because the Lower Cretaceous sediments are non-marine in character. The Cenomanian transgression, which incidently may not have penetrated into the South Irish Sea area to any extent, probably initiated the shoreline development of the south of Ireland and around Cornubia. Irrefutable evidence is not presently available to substantiate the margins of this incursion. The Lower Tertiary was a period of rapid clastic sedimentation which occurred in narrow fault-bounded troughs; these were generally remote from the present coastline. Non-marine Oligocene deposits, preserved in Tremadoc Bay, St. George's Channel and the Stanley Bank Basin, attest to the probable extensive sub-aerial erosion and thick weathered mantle that subsequently covered the regions. Nevertheless, it is thought that the first significant elements of the present coast of configuration were being developed along the margins of southeast Ireland and around St. David's Head. The fact that the basic "Kimmerian" structure persisted throughout the Tertiary era indicates that the present coastline was etched out from this structural framework during the later Neogene (Upper Miocene to Pliocene) period by repeated transgressions and regressions. The major coastal platforms (e.g. the "200" foot platform of Pembrokeshire) were also probably cut during the Neogene. Certainly, if the Flimstone deposits are Upper Oligocene in age, some support for this contention could be presented. Unfortunately, dating of these structures is further hampered by the uncertainty surrounding the history of the vertical movements which these blocks endured.

Apart from the complexity of land-bound platforms, there are also the offshore rock-platforms which effectively isolate the present coastline from the basins. These platforms were undoubtedly cut by sub-aerial erosion followed by marine trimming in a similar manner to the higher ones; this mechanism is considered to be more appropriate than that involving cliff retreat under repeated storm attack. The widest platforms naturally occur in those areas directly exposed to the Atlantic where a combination of trimming and wave attack (notably in St. Brides Bay) ensured rapid cliffline withdrawal. The details of these later periods of etching are preserved in the platform to cliff profile. Recent work by Wood (1974, 1976) has done much to clarify the sequence of events in this region. Minor Pleistocene and Holocene adjustments complete the sequence of events up to the present time.

The Malin Sea coastline has a similar history to that of the Celtic Sea. Throughout much of the Upper Mesozoic and Lower Tertiary period, this region was situated above even the most profound transgressions. Non-marine sediments of Oligocene age (Smythe and Kenolty, 1975), located in the central part of the region, clearly support this contention. The mapping of the *R4* reflector in the adjacent Rockall Trough (Roberts, 1975) lends further support to this contention, as does the extent and nature of the marine Tertiary deposits along the margin (Dobson et al., 1976). Therefore, the general form of the present-day coastline is largely a result of marine activity during the late Miocene, Pliocene and Quaternary. With such a subdued landscape to inundate, the sea was not so constrained by basic structural control, as compared to the area in the south; even so, basement ridges were able to exercise some restraint.

The extensive Donegal Platform, lying to the north of Ireland and in excess of 50 km wide, may be taken as an example of an area which was not simply the result of marine action. As occurred with areas in the south, it was created by a combination of sub-aerial weathering and marine trimming. In contrast, the Antrim coast, for long protected by the Lewisian ridge of the Rhinns of Islay, reflects the resistance provided by the basaltic flows which was maintained during the later Neogene and Quaternary. In this region, the platform is very narrow. When full details of the coastal evolution of western Britain become known to scientists, it will be the great age of its conformation that will stir the imagination rather than its complexity.

Acknowledgements

These investigations have been made jointly with Mr. R. Whittington, Senior Research Associate at Aberystwyth, to whom I am especially grateful. I wish to express my appreciation to N.E.R.C. for providing shiptime and equipment and to several Oil Companies, including Ranger Oil and Texaco, who provided much needed support.

References (cited and additional)

General

Ager, D.V., 1974. The Jurassic period in Wales. In: T.R. Owen (Editor), The Upper Palaeozoic and Post-Palaeozoic rocks of Wales. University of Wales Press, Cardiff, pp. 323–339.

Agger, H.E. and Carpenter, E.W., 1965. A crustal study in the vicinity of the Eskadalemuir Seismological Array Station. Geophys. J. R. Astron. Soc., 9: 69.

Audley Charles, M.G., 1970. Stratigraphical correlation of the Triassic rocks of the British Isles. Q. J. Geol. Soc. London, 126: 19–47.

Bailey, E.B. and Weir, J., 1932. Submarine faulting in Kimmeridgian times: East Sutherland. Trans. R. Soc. Edinburgh, 47: 431–467.

Berggren, W.A., 1972. Cenozoic biostratigraphy and palaeobiogeography of the N. Atlantic. Initial Reports of the Deep Sea Drilling Project. U.S. Government Printing Office, Washington, D.C., 12: 965–76.

Bott, M.H.P., 1971. Evolution of young continental margins and the formation of shelf basins. Tectonophysics, 1: 319–327.

Brooks, M., 1973. Some aspects of the Palaeogene evolution of Western Britain in the context of an underlying mantle hot spot. J. Geol., 81(1): 81–88.

Chester, J.A. and Bacon, M., 1975. A deep seismic survey in the Moray Firth. Inst. Geol. Sci., London, Rep., 75/11.

Craig, G.Y. (Editor), 1965. The Geology of Scotland. Oliver and Boyd, Edinburgh, 556 pp.

Dobson, M.R., Garrett, P., Haynes, J.R., Jenkins, D.G. and Medani, A.H., 1976. Upper Cretaceous and Cenozoic carbonates from the margins of Rockall Trough, N. Atlantic. J. Geol. Soc. London, 132: 6Γ1–621.

Donovan, D.T. (Editor), 1968. Geology of Shelf Seas. Oliver and Boyd, Edinburgh, 160 pp.

Eden, R.A., Wright, J.E. and Bullerwell, W., 1970. The solid geology of the East Atlantic continental margin adjacent to the British Isles. In: F.M. Delany (Editor), The Geology of the East Atlantic Continental Margin. 2. Europe. Inst. Geol. Sci., London, Rep., 70/14: 111–128.

Gardiner, P.R.R., Riddihough, R.P. and Robinson, K.W., 1974. The low of the sea, seabed resources and Ireland. Technol. Ireland, 1–5.

George, T.N., 1958. Lower Carboniferous Palaeogeography of the British Isles. Proc. Yorks. Geol. Soc., 31(3), No. 10: 227–318.

George, T.N., 1966. Geomorphic evolution in Hebridean Scotland. Scott. J. Geol., 2: 1–34.

George, T.N., 1974a. Prologue to a geomorphology of Britain. Inst. Br. Geogr., Spec. Publ., 7: 113–125.

George, T.N., 1974b. The Cenozoic evolution of Wales. In: T.R. Owen (Editor), The Upper Palaeozoic and Post-Palaeozoic Rocks of Wales. University of Wales Press, Cardiff, pp. 341–371.

Hallam, A. and Sellwood, B.W., 1976. Middle Mesozoic sedimentation in relation to tectonics in the British area. J. Geol., 84: 301–321.

Heybroek, P., 1974. Explanation to tectonic maps of the Netherlands. Geol. Mijnbouw, 53: 43–50.

Jones, O.T., 1951. The drainage system of Wales and the adjacent regions. Q. J. Geol. Soc. London, 107: 201.

Kent, P.E., 1949. A structure contour map of the surface of the buried Pre-Permian rocks of England and Wales. Proc. Geol. Assoc., 60: 87–104.

Kent, P.E., 1975. The tectonic development of Great Britain and the surrounding seas. In: A.W. Woodland (Editor), Petroleum and the Continental Shelf of North West Europe. 1. Geology. Applied Science Publishers, London, pp. 3–28.

Lorenz, V. and Nicholls, I.A., 1976. The Permo-Carboniferous basin and range province of Europe. An application of plate tectonics. In: H. Falke (Editor), Continental Permian in Central, West and South Europe. Reidel, Dordrecht, pp. 313–343.

Moseley, F., 1973. A tectonic history of north-west England. J. Geol. Soc. London, 128: 561–598.

Naylor, D. and Mounteney, S.N., 1975. Geology of the North-West European Continental Shelf, 1. Graham Trotman Dudley, London, 162 pp.

Pitman, W.C., and Talwani, M., 1972. Sea floor spreading in the North Atlantic. Geol. Soc. Am. Bull., 83: 619–645.

Rhys, G.H. (Compiler), 1974. A proposed standard lithostratigraphic nomenclature for the southern North Sea and an outline structural nomenclature for the whole of the (UK) North Sea. Inst. Geol. Sci., London, Pap., 74/8: 14 pp.

Roberts, D.G., 1974. Structural development of the British Isles, the Continental margin and the Rockall Plateau. In: C.A. Burke and C.L. Drake, (Editors), The Geology of Continental Margins. Springer, New York.

Roberts, D.G., 1975. Marine geology of the Rockall Plateau and Trough. Philos. Trans. R. Soc. London, Ser. A., 278: 447–509.

Schneider, E.D., 1972. Sedimentary evolution of rifted continental margins. Geol. Soc. Am. Mem., 132: 109–118.

Scholz, C.H., Barazangi, M. and Star, M.L., 1971. Late Cenozoic evolution of the Great Basin Western United States, as an ensialic interarc basin. Geol. Soc. Am. Bull., 82: 2979–2990.

Sellwood, B.W. and Jenkyns, H.C., 1975. Basins and swells and the evolution of an epeiric sea (Pliensbachian–Bajocian of Great Britain). J. Geol. Soc. London, 131: 373.

Sleep, N.H., 1971. Thermal effects of the formation of Atlantic continental margins by continental break-up. Geophys. J. Roy. Astron. Soc., 24: 325–350.

Sleep, N.H. and Snell, N.S., 1976. Thermal contraction and flexure of mid-continent and Atlantic marginal basins. Geophys. J. Roy. Astron. Soc., 45: 125–154.

Stride, A.H., 1963. Current-swept sea floors near the Southern half of Great Britain. Q. J. Geol. Soc. London, 119: 175–199.

Stride, A.H., Curray, S.R., Moore, D.G. and Belderson, R.H., 1969. Marine geology of the Atlantic continental margin of Europe. Philos. Trans. R. Soc. London, Ser.A., 264: 31–75.

Vogt, P.R. and Ostenso, N.A., 1967. Steady state crustal spreading. Nature, 215: 810–817.

Wills, L.T., 1971. Palaeogeological Map of the Palaeozoic floor. Geol. Soc. London publication.

Whitbread, D.R., 1972. Hydrocarbon potential of Western Britain and Ireland. Financial Times, Petroleum Times and Investors Chronicle North Sea Conference (preprint).

Whitbread, D.R., 1975. Geology and petroleum possibilities west of the United Kingdom. In: A.W. Woodland (Editor), Petroleum and the Continental Shelf of North West Europe. 1. Applied Science Publishers, London, pp. 45–57.

Williams, C.A., 1975. Sea-floor spreading in the Bay of Biscay and its relationship to the North Atlantic. Earth Planet Sci. Lett., 24: 440–456.

Ziegler, P.A., 1975. North Sea basin history in the tectonic framework of North-Western Europe. In: A.W. Woodland (Editor). Petroleum and the Continental Shelf of North-West Europe. 1. Applied Science Publishers, London, pp. 138–148.

Ziegler, W.H., 1975. Outline of the geological history of the North Sea. In: A.W. Woodland (Editor), Petroleum and the Continental Shelf of North-West Europe. 1. Applied Science Publishers, London, pp. 165–187.

The Malin Sea

Bailey, R.J., Grzywacz, J.M. and Buckley, J.S., 1974. Seismic reflection profiles of the continental margin bordering Rockall Trough. J. Geol. Soc. London, 130: 55–69.

Bailey, R.J., Buckley, J.S. and Kielmas, M.M., 1975. Geomagne tic reconnaissance on the continental margin of the British Isles between 54° and 57°N. J. Geol. Soc. London, 131: 275–282.

Belderson, R.H., Kenyon, N.H. and Stride, A.H., 1970. Holocene sediments on the continental shelf west of the British Isles. In: F.M. Delany (Editor), Geology of the East Atlantic Continental Margin. Inst. Geol. Sci., London, Rep., 70/14: 161–169.

Binns, P.E., Harland, R. and Hughes, M.J., 1974a. Glacial and post-glacial sedimentation in the Sea of Hebrides. Nature (Phys. Sci.), 248: 751.

Binns, P.E., McQuillin, R. and Kenolty, N., 1974b. The geology of the sea of Hebrides. Inst. Geol. Sci., London, Rep., 73/14: 43 pp.

Binns, P.E., McQuillin, R., Fannin, N.G.T., Kenolty, N. and Ardus, D.A., 1975. Structure and stratigraphy of sedimentary basins in the Sea of the Hebrides and the Minches. In: A.W. Woodland (Editor), Petroleum and the Continental Shelf of North-West Europe. 1. Applied Science Publishers, London, pp. 93–102.

Chester, J.A., Deegan, C.E., Ardus, D.A., Binns, P.E. and Fannin, N.G.T., 1972. I.G.S. marine drilling with M.V. Whitethorn in Scottish Waters 1970–71. Inst. Geol. Sci., London, Rep., 72/10.

Dobson, M.R. and Evans, D., 1974. The geological structure of the Malin Sea. Q. J. Geol. Soc. London, 130: 475–478.

Dobson, M.R., Evans, D. and Whittington, R., 1975. The offshore extension of the Loch Gruinart Fault, Islay. Scott. J. Geol., 11(1): 23–35.

Durant, G.P., Dobson, M.R., Kokelaar, B.P., Macintyre, R.M. and Rea, W.J., 1976. Preliminary report of the nature and age of the Blacksones Bank Igneous Centre Western Scotland. J. Geol. Soc. London, 132: 319–326.

Durrance, E.M., 1976. A gravity survey of Islay, Scotland. Geol. Mag., 113(3), 251–261.

Eden, R.A., Ardus, D.A., Binns, P.E., McQuillin, R., & Wilson, J.B., 1971. Geological investigations with a manned submersible off the west coast of Scotland, 1969-1970. Inst. Geol. Sci., London, Rep., 71/16.

Eden, R.A., Deegan, C.E., Rhys, G.H., Wright, T.E. and Dobson, M.R., 1973. Geological investigations with a manned submersible in the Irish Sea and off Western Scotland, 1971. Inst. Geol. Sci., London, Rep., 73/1.

Evans, D., 1973. A shallow seismic survey in Lough Swilly and Trawbreaga Bay, Co. Donegal. Proc. R. Ir. Acad., 73: 207–216.

Evans, D. and Whittington, R., 1976. The submarine extensions of the Thorr and Fanad Plutons, County Donegal. Proc. R. Ir. Acad., 76: 111–120.

Evans, D., Kenolty, N., Dobson, M.R. and Whittington, R., 1978. The geology of the Malin Sea. Rep. Inst. Geol. Sci., 77 (in press).

Hallam, A., 1972. Relation of Palaeogene ridge and basin structures and vulcanicity in the Hebrides and Irish Sea regions of the British Isles to the opening of the North Atlantic. Earth Planet. Sci. Lett., 16: 171–177.

McQuillin, R. and Binns, P.E., 1973. Geological structure in the Sea of Hebrides. Nature (Phys. Sci.): 241: 2–4.

McQuillin, R., Bacon, M. and Binns, P.E., 1975. The Blackstones Tertiary igneous complex. Scott. J. Geol., 11: 179–192.

Pendlebury, D.C. and Dobson, M.R., 1976. Sediment and macro faunal distribution in the eastern Malin Sea, as determined by side-scan sonar and sampling. Scott. J. Geol., 11(4): 315–332.

Pitcher, W.S., 1969. Northeast-trending faults of Scotland and Ireland, and chronology of displacements. North Atlantic-Geology and Continental Drift. Am. Assoc. Pet. Geol., Mem., 12.

Riddihough, R.P., 1964. Magnetic survey off the north coast of Ireland. Nature, 203 (4946): 747–748.

Riddihough, R.P., 1968. Magnetic surveys off the North coast of Ireland. Proc. R. Ir. Acad., 66(B): 27–41.

Riddihough, R.O. and Young, D.G.G., 1971. Gravity and magnetic surveys off Inishowen and adjoining sea areas off the coast of Ireland. Proc. Geol. Soc. Lond., 1664: 215–220.

Smythe, D.K. and Kenolty, N., 1975. Tertiary sediments in the Sea of the Hebrides. Q.J. Geol. Soc. London, 131: 227–234.

Steel, R.J., 1974. New Red Sandstone flood plain and piedmont sedimentation in the Hebridean province Scotland. J. Sed Petrol., 44: 336–357.

Uruski, C., 1978. The Geology of the Stanton Banks Area. Proc. Geol. Soc. London, 134: 478.

The South Irish and Celtic Seas

Al-Shaikh, Z.D., 1969. Geophysical results from Caernarvon and Tremadoc Bays. Nature, 224(5222): 897–899.

Al-Shaikh, Z.D., 1970. The geological structure of part of the central Irish Sea. Geophys. J. R. Astron. Soc., 20: 233–237.

Bacon, M. and McQuillin, R., 1972. Refraction seismic surveys in the North Irish Sea. J. Geol. Soc. London, 128: 613–621.

Baker, J.W., 1969. Correlation problems of metamorphosed Pre-Cambrian rocks in Wales and S.E. Ireland. Geol. Mag., 106: 249–256.

Baker, J.W., 1970. Petrology of the metamorphosed Pre-Cambrian rocks of South-easternmost Co. Wexford. Proc. R. Ir. Acad., 69: 2–20.

Baker, J.W., 1971. Intra-Lower Palaeozoic faults in the Southern Irish Sea Area. Geol. Mag., 108(6): 501–509.

Blundell, D.J., 1975. The geology of the Celtic Sea and south-west approaches. In: C.J. Yorath, E.R. Parker and D.J. Glass

(Editors). Canada's Continental Margins and Offshore Petroleum Exploration. Can. Soc. Petr. Geol., Mem., 4: 341–362.

Blundell, D.J. and Parks, R., 1969. A study of the crustal structure beneath the Irish Sea. Geophys. J. R. Astron. Soc., 17: 45–62.

Blundell, D.J., Davey, F.J. and Graves, L.J., 1968. Sedimentary basin in the South Irish Sea. Nature, 219: 55–56.

Blundell, D.J., Griffiths, D.H. and King, R.F., 1969. Geophysical investigations of buried river valleys around Cardigan Bay. Geol. J., 6(2): 161–180.

Blundell, D.J., Davey, F.J. and Graves, L.J., 1971. Geophysical surveys over the south Irish Sea and Nymphe Bank. J. Geol. Soc. London, 127: 339–375.

Bott, M.H.P., 1964. Gravity measurements in the north-eastern part of the Irish Sea. Q. J. Geol. Soc. London, 120: 369–394.

Bott, M.H.P., 1965. The deep structure of the Northern Irish Sea – a problem of crustal dynamics. In: W.F. Whittard and R. Bradshaw (Editors), Submarine Geology and Geophysics. Colston Pap., 17: 179–204.

Bott, M.H.P., 1968. The geological structure of the Irish Sea Basin. In: D.T. Donovan (Editor), Geology of Shelf Seas. Oliver and Boyd, Edinburgh, pp. 93–115.

Bott, M.H.P., and Young, D.G.G., 1971. Gravity measurements in the North Irish Sea. J. Geol. Soc. London, 126: 413–434. 434.

Bowen, D.Q., 1973. The Pleistocene succession of the Irish Sea. Proc. Geol. Assoc., 84(3): 249–272.

Brooks, M. and James, D.G., 1975. The geological results of seismic refraction surveys in the Bristol Channel 1970–1973. Q. J. Geol. Soc. London, 131: 163–182.

Browne, B.C. and Cooper, R.I.B., 1950. The British submarine gravity surveys of 1938 and 1946. Philos. Trans. R. Soc. London, Ser.A., 141: 243–310.

Bullerwell, W. and McQuillin, R., 1968. Preliminary report on a seismic reflection survey in the southern Irish Sea. Inst. Geol. Sci., London, Rep., 69/2.

Colter, V.S. and Barr, K.W., 1975. Recent developments in the geology of the Irish Sea and Cheshire Basins. In: A.W. Woodland (Editor), Petroleum and the Continental Shelf of North West Europe, 1. Applied Science Publishers, London, pp. 61–73.

Curry, D., Gray, F., Hamilton, D. and Smith, A.J., 1967. Upper chalk from the sea-bed, south of Cork, Eire. Proc. Geol. Soc. London, No. 1640: 134–136.

Davey, F.J., 1970. Bouguer anomaly map of the North Celtic Sea and entrance to the Bristol Channel. Geophys. J. R. Astron. Soc., 22: 277–282.

Day, G.A. and Williams, C.A., 1970. Gravity compilation in the N.E. Atlantic and interpretation of gravity in the Celtic Sea. Earth Planet. Sci. Lett., 8: 205–213.

Delanty, L.J. and Whittington, R.J., 1977. A re-assessment of the "Neogene" deposits of the South Irish sea and Nymphe Bank. Mar. Geol., 24: M23–M30.

Dobson, M.R., 1969. The oblique asdic and its use in an investigation of a marine high-energy environment. Sedimentology, 13: 105–122.

Dobson, M.R., Evans, W.E. and James, K.H., 1971. The sediment on the floor of the Southern Irish Sea. Mar. Geol., 11: 27–69.

Dobson, M.R., Evans, W.E. and Whittington, R., 1973. The geology of the south Irish Sea. Inst. Geol. Sci., London, Rep., 73/11: 35 pp.

Dobson, M.R. and Whittington, R., in press. The geology of the Kish Bank Basin. J. Geol. Soc. London.

Doré, A.G., 1976. Preliminary geological interpretation of the Bristol Channel approaches. J. Geol. Soc. London, 132: 453–459.

Fletcher, B.N., 1975. A new Tertiary basin east of Lundy Island. J. Geol. Soc. London, 131: 223–225.

Garrard, R.A., 1977. The sediments of the South Irish Sea and Nymphe Bank area of the Celtic Sea. In: C. Kidson and M.J. Tooley (Editors), The Quaternary History of the Irish Sea. Geol. J., Spec. Issue, 7: 69–92.

Garrard, R.A. and Dobson, M.R., 1974. The nature and maximum extent of glacial sediments off the west coast of Wales. Mar. Geol., 16: 31–44.

Griffiths, D.H., King, R.F. and Wilson, C.D.V., 1961. Geophysical investigations in Tremadoc Bay, North Wales. Q. J. Geol. Soc. London, 117: 117–171.

Hamilton, D. and Blundell, D.J., 1971. Submarine geology of the approaches to the Bristol Channel. Proc. Geol. Soc. London, 1664: 297–300.

Kenyon, N.H., 1970. The origin of some transverse sand patches in the Celtic Sea. Geol. Mag., 107: 389–394.

Lloyd, A.J., Savage, R.J.G., Stride, A.H. and Donovan, D.T., 1973. The geology of the Bristol Channel floor. Philos. Trans. R. Soc. London, Ser.A. 274: 595–626.

McQuillin, R., Wright, J.E., Owens, B. and Lister, T.R., 1969. Recent geological investigations in the Irish Sea. Nature, 222: 365–366.

Moore, J.R., 1968. Recent sedimentation in Northern Cardigan Bay, Wales. Bull. Br. Mus. (Nat. Hist.), Mineral., 2(2).

Owen, T.R., 1971. The structural evolution of the Bristol Channel. Proc. Geol. Soc. London, 1664: 289–294.

Pegrum, R.M. and Mounteney, S.N., 1976. The Oil and Gas Prospects of Offshore Wales. Exploration Consultants Ltd., London, 15 pp.

Penn, I.E. and Evans, C.D.R., 1976. The Middle Jurassic (mainly Bathonian) of Cardigan Bay and its palaeogeographical significance. Inst. Geol. Sci., London, Rep., 76/6.

Thomas, T.N., 1973. This changing world: the geological basis of oil and natural gas prospects in the Celtic Sea. Geography, 58: 47–51.

Whittington, R.J., 1977. A late-glacial drainage pattern in the Kish Bank area and post-glacial sediments in the Central Irish Sea. In: C. Kidson and M.J. Tooley (Editors). The Quaternary History of the Irish Sea. Geol. J., Spec. Issue, 7: 55–68.

Wood, A., 1974. Submerged platform of marine abrasion around the coasts of south-western Britain. Nature, 252 (5483): 563.

Wood, A., 1976. Successive regressions and transgressions in the Neogene. Mar. Geol., 22: M23–M29.

Wood, A. and Woodland, A.W., 1968. Borehole at Mochras, West of Llanbedr, Merionethshire. Nature, 219 (5161): 1352–1354.

Woodland, A.W. (Editor), 1971. The Llanbedr (Mochras Farm) borehole. Inst. Geol. Sci., London, Rep., 71/18: 115 pp.

Woodland, A.W. (Editor), 1975. Petroleum and the Continental Shelf of North West Europe. 1. Geology. Applied Science Publishers, London, 501 pp.

Wright, J.E., Hull, J.H., McQuillin, R. and Arnold, S.E., 1971. Irish Sea investigations 1969–70. Inst. Geol. Sci., London, Rep., 71/19: 43 pp.

Chapter 4

The Geology and Structure of the Celtic Sea

D.J. BLUNDELL

Introduction

The geology of the land areas peripheral to the Celtic Sea has been extensively studied for many years and is consequently much better known than the geology of the sea area itself. Here, only a relatively small proportion of the sea bed has been sampled at all systematically and our understanding of the geology has to be largely inferred from the interpretation of various geophysical surveys. This chapter attempts to put together the various fragments of geophysical information, together with such direct geological information as is available, to provide an overall view of the geological history of the Celtic Sea.

As in other parts of the continental shelf around Britain, the Celtic Sea proves to have had a markedly different geological history from the adjacent land areas. Thus it is that we should examine the marine geology of the Celtic Sea in order to appreciate more fully the geological history of the adjacent land areas.

Direct geological observations

Land areas

The surface geology of the land areas fringing the Celtic Sea is almost exclusively concerned with Precambrian and Palaeozoic rocks. In the south-west of Ireland, Devonian and Carboniferous sandstones, siltstones and slates occur in alternating narrow outcrops extended E–W, having been folded in this alignment. In the south-east of Ireland, near Carnsore, a granite of 430 Ma age intrudes Precambrian gneisses of the Rosslare Complex, to the north of which lie Cambrian and Ordovician slates. From Dungarvon to Dingle Bay lies the "Variscan Front" marked by faults with downthrow to the north or by upthrusting from the south (Figs. 1.3, 2.7). South of this line the dominant structures are the E–W folds of Upper Carboniferous (Armorican = "Variscan") age, whereas to the north, the structures trend mainly NE–SW along Caledonian lines.

Further east, the "Variscan Front" can be traced in South Wales, where in southern Pembrokeshire, Precambrian and Lower Palaeozoic rocks are thrust over Carboniferous. South of the line, folds are dominantly of E–W (Armorican) trend and affect Devonian and Carboniferous rocks much as in south-west Ireland whereas, to the north, strongly disturbed Precambrian and Lower Palaeozoic rocks are structurally aligned along NE–SW Caledonian trends.

Along the north coast of Devon, structures strike E–W and involve Devonian sandstones and slates. These, together with the overlying Culm measures of Carboniferous age, form the main outcrops of the Cornish Peninsula. In the Devonian succession there is a general predominance of sandstones, although lithologies vary considerably, whereas the Carboniferous is generally finer grained, in just the same way as in south-west Ireland. But in Cornwall these sediments are intruded by granite batholiths of 290 Ma age, trending NE–SW and extending as far as the Scilly Isles. In the south, the Start–Lizard complexes are made up of a schistose basement that is much deformed and intruded by serpentinites, gabbros and granites and are thrust from the south over Devonian rocks. Further south, across the Channel in Brittany, Variscan trends dominate and here affect mainly Precambrian rocks that have been strongly deformed and metamorphosed and which are intruded by granite.

Sea areas

Systematic sampling of the geology of the sea floor has been carried out by dredging and gravity coring over

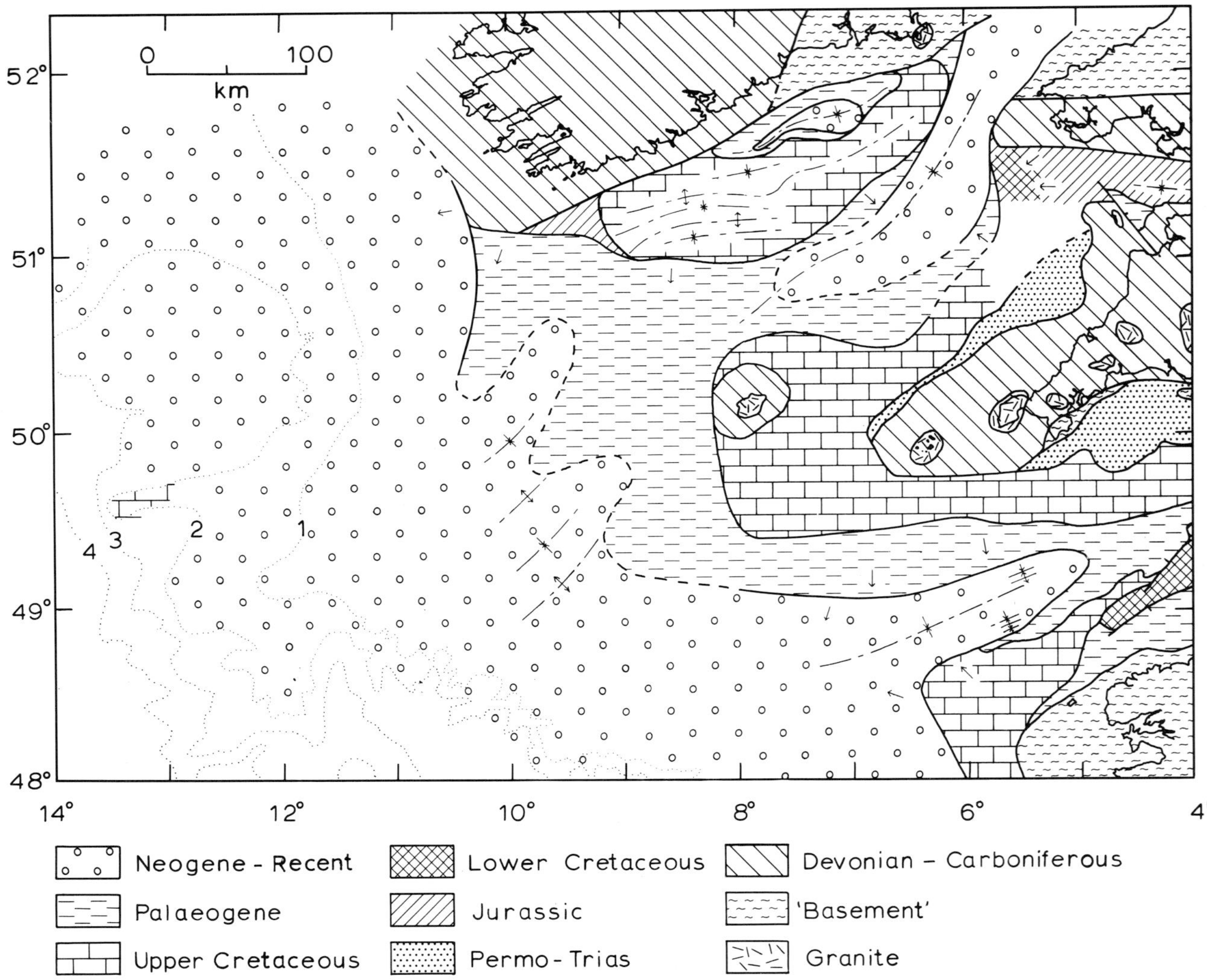

Fig. 4.1. Geological map of the Celtic Sea area. Compiled from Smith et al. (1965), Hersey and Whittard (1966), Stride et al. (1969), Curry et al. (1970), Blundell et al. (1971), Hamilton and Blundell (1971), Handley (1971), Lloyd et al. (1973) and Fletcher (1975). The dotted lines are isobaths at 1 km intervals.

a wide area of the western English Channel and the Bristol Channel, but only a relatively few scattered observations have been made elsewhere across the Celtic Sea. A group of geologists centred on the University of Bristol, under the original leadership of Professor Whittard, have over a period of twenty years mapped these areas in detail and built up a most impressive understanding of their stratigraphic and structural history. In more recent years, direct sampling has been augmented by side-scan sonar to trace sea-bed outcrops in detail and by continuous seismic profiling.

Curry, Hamilton and Smith (1970, 1971) have given a comprehensive account of the geology of the western English Channel and the relevant portion of their map, in slightly modified form, has been incorporated into Fig. 4.1. Further detail has subsequently been provided by these same authors in collaboration with Andreieff, Bouysse, Fletcher and Monciardini (1975). A major sediment trough fills the central portion of the western English Channel, of synclinal form but fault bounded on its southern flank. The structure trends ENE–WSW and both deepens and widens towards the west. Within the trough, the oldest formations are Permo-Triassic red sandstones and marls, very similar in lithology to those found elsewhere in Britain, which rest on an ancient eroded land surface of Palaeozoic basement rocks. These are overlain by a succession, up to 1000 m thick, of Jurassic sediments in which clays predominate but which otherwise resembles the succession to be found further east in southern England. These are succeeded by Lower Cretaceous deposits of terrestrial aspect, similar to those of the Weald. The Upper Cretaceous is strongly transgressive across these older strata and consists predominantly of chalk. Chalk from Cenomanian through to Danian (Palaeocene) has been found but the succession is condensed towards the east. A total thickness of Cretaceous and Palaeocene of 800 m has been recognised. Eocene clays with numerous limestone bands rest unconformably on Cretaceous strata but unlike the Chalk, are not found overlapping on to the basement. Eocene strata total up to 500 m thickness and are in turn overlain unconformably by up to 120 m of fine sandy, calcareous sediments of Miocene age. These thicken towards the west and show evidence of transgression towards the east.

Smith et al. (1965) reported the discovery, through dredging, of the Haig Fras granite situated some 100 km west-north-west of the Scilly Isles, of roughly circular outcrop on the sea bed and with 15 km diameter. The granite intrudes slate of Devonian or Carboniferous age and is a part of the Cornubian Variscan suite, with radiometric dates showing a 275 Ma age. The only other direct sign of more recent igneous activity in the vicinity is the Wolf Rock, just to the south of Lands End, which is composed of phonolite which has been dated by K-Ar methods at 111 ± 6 and 115 ± 7 Ma (whole rock) and 113 ± 7 Ma (sanidine). The whole rock analyses of the nearby Epsom shoal, composed of phonolitic lava, have yielded dates of 127 ± 7 Ma and 132 ± 8 Ma. As there is the possibility of leakage of Ar and because the rocks from the two localities are closely associated and thus contemporaneous, their likeliest age seems to be the oldest given, which is Lower Cretaceous.

Curry et al. regard the western English Channel trough as beginning as an inter-montane Permo-Trias basin in response to isostatic readjustment following the emplacement of the Cornubian granites as part of the Variscan orogeny. It continued to develop through the Jurassic and Lower Cretaceous with contemporaneous folding and faulting providing general subsidence north of a hinge line passing just north of Aldernay and Ushant. Pre-Upper Cretaceous faulting and folding is locally intense within the trough, in contrast with the more gentle folding that has affected Upper Cretaceous and Eocene strata. The Cenomanian transgression and deposition of Chalk facies mark a major change in the character of formation of the trough, from which time the structural history is dominated by subsidence to the west, with repeated marine transgression and regression, but there has at the same time been continued subsidence and tectonic activity within the trough with post-Eocene faulting evident along the Aldernay–Ushant line and some modern seismic activity recorded from the vicinity.

A similar story emerges from the account of the geology of the Bristol Channel given by Lloyd et al. (1973) and a simplified version of their map is also incorporated in Fig. 4.1. A trough of Mesozoic sediments is developed in an E–W trending syncline along the centre of the Bristol Channel. The southern limb of the syncline is the steeper of the two, with dips of 10–20°. There is considerable evidence of strike faulting, possibly on the northern limb as well, with one fault in particular, regarded as a thrust, controlling the southern boundary of the trough. The structures are parallel with those on land to north and south, developed in the Variscan orogeny. An anticlinorium along the northern part of the Bristol Channel separates the Bristol Channel syncline from the South Wales coal basin. Permo-Trias red sandstones and marls form the basal layers within the basin and are about 250 m in thickness. These are overlain by a Jurassic succession involving up to 700 m of Lias and 800 m of Mid to Upper Jurassic ranging just up to the Portlandian. The Jurassic succession differs from elsewhere in southern England by a conspicuous lack of carbonate rocks, although thin cementstone bands are found within the predominant shale, clay and marl sequence. Towards the upper part of the succession the Oxford Sands occur as interbedded sandstones and lignitic silty clays. The maximum subsidence occurs around 4°10′W, but the trough continues to 4°40′W where it terminates against a NW-trending basement ridge from Hartland, through Lundy, towards Pembroke. The north-eastern margin of this ridge is regarded by Brooks and Thompson (1973) as a major fault connected with the Sticklepath–Lustleigh line that crosses Devon which is reckoned to be a structure of Tertiary age. Brooks and Thompson consider there to be a close relationship between the intrusion of the Lundy granite (age 55 m.y.) and this tectonism.

The NW–SE basement ridge across the Bristol Channel is in alignment with a similar one recognised by Curry et al. (1970) between Start Point and Ushant and may well be a structural continuation, linked via the Sticklepath Fault.

Apart from these areas of detailed geological sampling, the only other direct indications of the geology come from a few scattered observations. These include samples of Eocene and Miocene age collected by Curry et al. (1962) from the continental slope west of the English Channel and Chalk samples collected by Bradley (1940) from Goban Spur. Cole and Crook (1910) collected Tertiary and Cretaceous limestone from the shelf off south-west Ireland and Curry et al. (1967) reported on four core stations along 8°W south of Cork in which they found Eocene gravels in the northernmost and Upper Chalk in the other three. More recently some information has been released following Oil Company drilling programmes south of Ireland. In May 1974 it was disclosed that reserves of natural gas amounting to 1 million cu. ft. had been outlined by Marathon Petroleum Company some 50 km south of Cobh at lat. 51°20′N, long. 8°0′W. This, the Kinsale gas field, is located in deltaic sandstones of Lower Cretaceous age at depths between 800 m and 900 m which must underlie a considerable thickness of Chalk (See Fig. 4.1). In November, Esso Petroleum Company revealed they had discovered oil from Jurassic sandstones at about 1800 m depth lat. 51°10′N, long. 8°20′W in a structure on trend with the Kinsale gas field and only 25 km distant from it. Gas was also found there in shallow sand at about the same levels as in the Kinsale field. Meanwhile Marathon have found indications of natural gas in what may be a third field south of Ireland further to the east at lat. 51°30′N, long. 7°21′W but bad weather called a halt to further exploration in 1974. Beyond these direct observations, the geology of the Celtic Sea is reliant upon the indirect evidence of various geophysical methods.

Gravity observations

Brown and Cooper (1950) made the first gravity observations within the Celtic Sea in a submarine and these remained the only information until, with the advent of the ship-borne gravimeter, it became possible to make detailed gravity surveys at sea reasonably rapidly. Day and Williams (1970) were the first to prepare a gravity map of the Celtic Sea, from which they were able to recognise the main structural features of the area. At the same time, Davey (1970) produced a more detailed gravity map of the northern Celtic Sea area including the approaches to the Bristol Channel. The gravity survey was extended by Handley (1971) from south of Ireland westwards across the southern part of the Porcupine Seabight. To the east, a gravity survey of the Bristol Channel has been completed by Brooks and Thompson (1973). More recently, the Hydrographic Department, M.O.D., have compiled a detailed gravity map of the Celtic Sea from these and other data. It is with their kind permission that I have used their maps as the main

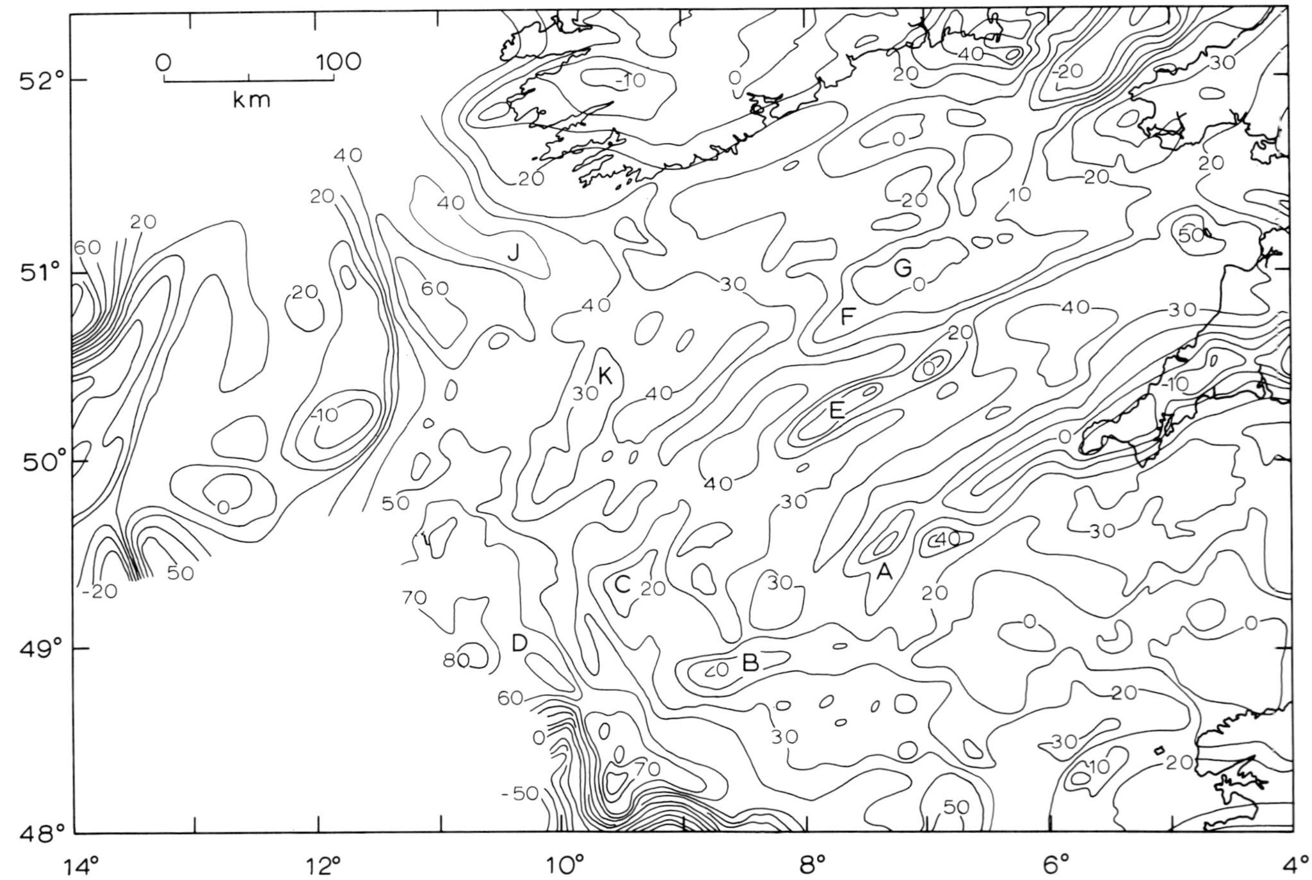

Fig. 4.2. Gravity map of the Celtic Sea area: Bouguer anomalies on land and free-air anomalies at sea, contoured at 10 mgal intervals. Compiled from data from Day and Williams (1970), Davey (1970), Handley (1971), Brooks and Thompson (1973) and Hydrographic Department, M.O.D. (unpubl. charts). (For explanation of symbols, see text.)

basis of the gravity map of the Celtic Sea area shown in Fig. 4.2. This shows Bouguer anomalies across the land areas and free air anomalies at sea. It is clear from this map, as Day and Williams appreciated from theirs, that the main structural trend of the Celtic Sea is ENE–WSW, more or less perpendicular to the continental margin. But a careful look at the gravity anomalies shows this main trend to be modified in various ways that are important to the geological development of the area. In addition to the gravity variations caused by density contrasts within the upper part of the crust there is a gravity anomaly associated with the continental margin ascribed as due to the change in crustal thickness from continental to oceanic regime. A zone of low gravity values is associated with the Cornubian granites and extends south-westwards as a continuous feature to the Scilly Isles, from which Bott et al. (1958) inferred the continuity of the batholith underlying the peninsula. The gravity anomaly continues to the south-west to a further low at 7°20′W, marked as *A* by Day and Williams (1970, fig. 5) and in Fig. 4.2, which appears to relate to an extension of the granite, and on towards a second low, marked as *B* at 8°40′W. This line of gravity lows is accompanied by a SW-trending ridge of high gravity on its north-western flank and another, less well defined, to the south-east. It appears to culminate at the continental margin at a particularly high feature at lat. 48°10′N, long. 9°30′W. The site of the Haig Fras granite is marked by another gravity low, *E*, which is also extended along a NE–SW trend. A secondary low to the north-east of Haig Fras is seen to be structurally associated with it. This group of NE–SW trending features occupies a zone which forms a single structural unit that separates two areas, the western English Channel and the north-eastern Celtic Sea, which are both characterised by gravity values close to zero and by broad anomalies of low relief with ENE-WSW trend. With similar gravity anomalies the inference is that these two areas are geologically similar and are each the site of a major sedimentary basin in which the gravity lows result from a combination of an increased thickness of low density sediments and an increased density contrast with the surrounding basement. Day and Williams (1970) considered the background value of gravity over basement areas to be about + 250 g.u. (+25 mgal) rising to +600 g.u. close to the margin whereas in the North Celtic Sea area Davey believed the background to be about +400 g.u. The problem in interpretation comes not only in assigning a background, regional field value but in deciding whether gravity lows relate to sedimentary basins or to granite intrusions and additional evidence from the seismic surveys is needed to distinguish between the alternatives. Thus gravity lows *C* and *D* were thought by Day and Williams to be due to buried granites, because they lay in alignment with the SW-trending anomalies known to relate to the Cornubian and Haig Fras granites, and another to the north-west of Haig Fras, marked *F*, was similarly explained, whereas *G* they regarded as due to a sedimentary trough. In Fig. 4.2, it is seen that *F* and *G* really relate to the same feature but in Day and Williams' map it was less clear. Further to the north, Davey regarded the gravity low extending across Nymphe Bank south of Ireland as due to a sediment trough that possibly connected with the deep gravity low trending NE–SW through to South Irish Sea that was known (Blundell et al., 1968) to be associated with a sedimentary basin. But he related the gravity low at Carnsore to the granite there and the surrounding high to the dense, basic nature of the basement it intruded. A gravity low beneath St. Bride's Bay was regarded as a similar feature by Blundell et al. (1971) and the gravity anomaly associated with Lundy shows a similar aspect with a surrounding high due presumably to dense basement or possibly to basic volcanic rocks associated with the Tertiary igneous activity that produced the granite.

In interpreting his gravity survey south-west of Ireland, Handley (1971) calculated the gravity effect of crustal thinning in direct proportion to the increase of water depth, maintaining isostatic equilibrium to obtain an indication of the background variation of gravity. He found superimposed on this a gravity low south-west of Fastnet which can be seen in Fig. 4.2 as a SE–NW-trending feature, marked *J*, but which is a more equidimensional feature when the background variation is removed. This he interpreted in terms of a sedimentary trough, in the light of the seismic survey evidence that he had obtained at the same time. The gravity map, Fig. 4.2, thus outlines two major sedimentary basins, the Western Approaches Basin and the Celtic Sea Basin, within each of which are subsidiary troughs and swells. In addition to providing a better control on the shapes of features already identified, it also gives a better view of the interactions of the differing structural trends. The north-west trending basement rise across the Bristol Channel through Lundy is evident and another appears to lie between the Scilly Isles and Fastnet. There is a further closed low (*K* on Fig. 4.2) over Great Sole Bank, but the interpretation of this and other features needs to be augmented with seismic evidence.

Seismic refraction observations

Early observations were made by Bullard and Gaskell (1941) using geophones on the sea bed connected by cable to a recording ship, with a second vessel firing explosive charges out to a maximum range of 18 km. Observations were recorded at nine stations along a line south-westwards from Cornwall as far as the shelf edge but the data from two of them were insufficient to be used on their own. Interpretation in terms of plane-layered models allowed them to produce a cross-section along the line of stations with an upper sedimentary layer having seismic velocities ranging between 1.7 and 2.2 km/s, within the top 200 m, up to 2.9 km/s where the layer thickens to 2500 m towards the shelf edge. Beneath this they recognised a lower layer having seismic velocities between 5.0 and 7.1 m/s which they considered to be crystalline basement.

This work was extended by Day et al. (1956) with a series of refraction lines in the Western Approaches to the English Channel, using sonobuoy recorders with hydrophone detectors. Unfortunately the majority of these lines were unreversed and several were aligned across the strike. Nonetheless, Day et al. were able to divide the seismic velocities that they obtained into four groups which could be related to specific layers. These they correlated with major geological units as shown in Table 4.1a. They found that the top surface of the lowest layer (*D*) is deepest at 3.6 km, along the axis of the western English Channel, trending ENE–WSW but becoming NE–SW further west in the Western Approaches. This trough is bounded to the north-west by a ridge extending south-westwards from the Cornish Peninsula. They were also able to show that the base of layer *B* forms a trough-like feature trending ENE–WSW through the western English Channel and maintains this trend across the Western Approaches to extend on to the south-westerly end of the ridge from Cornwall. Layer *C* is only present beneath this trough and is absent across the ridge, so that layer *B* is unconformable upon layers *C* and *D*.

Day et al. were not aware of the extent of Mesozoic and Tertiary sediments and did not have the benefit of Curry et al.'s map. In the light of this new information it seems better to correlate instead (Table 4.1b) the seismic layer *B* with Palaeogene (probably Eocene) and Chalk, unconformable on layer *C* which is correlated with Lower Cretaceous, Jurassic and Permo-Trias rocks. Layer *A* is retained as Recent and Upper Tertiary (probably Miocene) and layer *D* includes both Devonian–Carboniferous and older rocks as a basement layer. This modified correlation is consistent with the results from the Southern Irish Sea (Blundell et al., 1968). Day et al. showed that at their station Dy 7, located close to the centre of gravity low *B*, layer *D* is at only 1800 m depth and has a seismic velocity of 5.2 km/s. As it directly underlies layers *A* and *B* which cannot account for the gravity anomaly, it substantiates Day and Williams' association of the gravity low with the presence of a granite there which must lie within layer *D*.

TABLE 4.1

(a) Interpretation of seismic refraction surveys in the Western Approaches to the English Channel by Day et al. (1956)

Layer	A	B	C	D
Seismic velocity range (km/s)	1.7–2.5	2.7–3.3	3.6–4.8	5.2–7.0
Geological correlation	Recent to Mesozoic	Permo-Trias	Devonian to Carboniferous	metamorphic basement

(b) Reinterpretation of the same seismic results in the light of additional gravity, magnetic and geological information

Layer	A	B	C	D
Seismic velocity range (km/s)	1.7–2.5	2.7–3.3	3.6–4.8	5.2–7.0
Geological correlation	Neogene to Recent	Upper Cretaceous and Palaeogene	Permo-Trias to Lower Cretaceous	basement (inc. Dev.–Carb.)

(c) Interpretation of recent seismic refraction lines in the same area by Avedik (1975)

Layer	A	B	C	C′	D
Seismic velocity range (km/s)	1.9–2.4	2.8	3.2–3.7	4.2–4.3	4.6–6.0

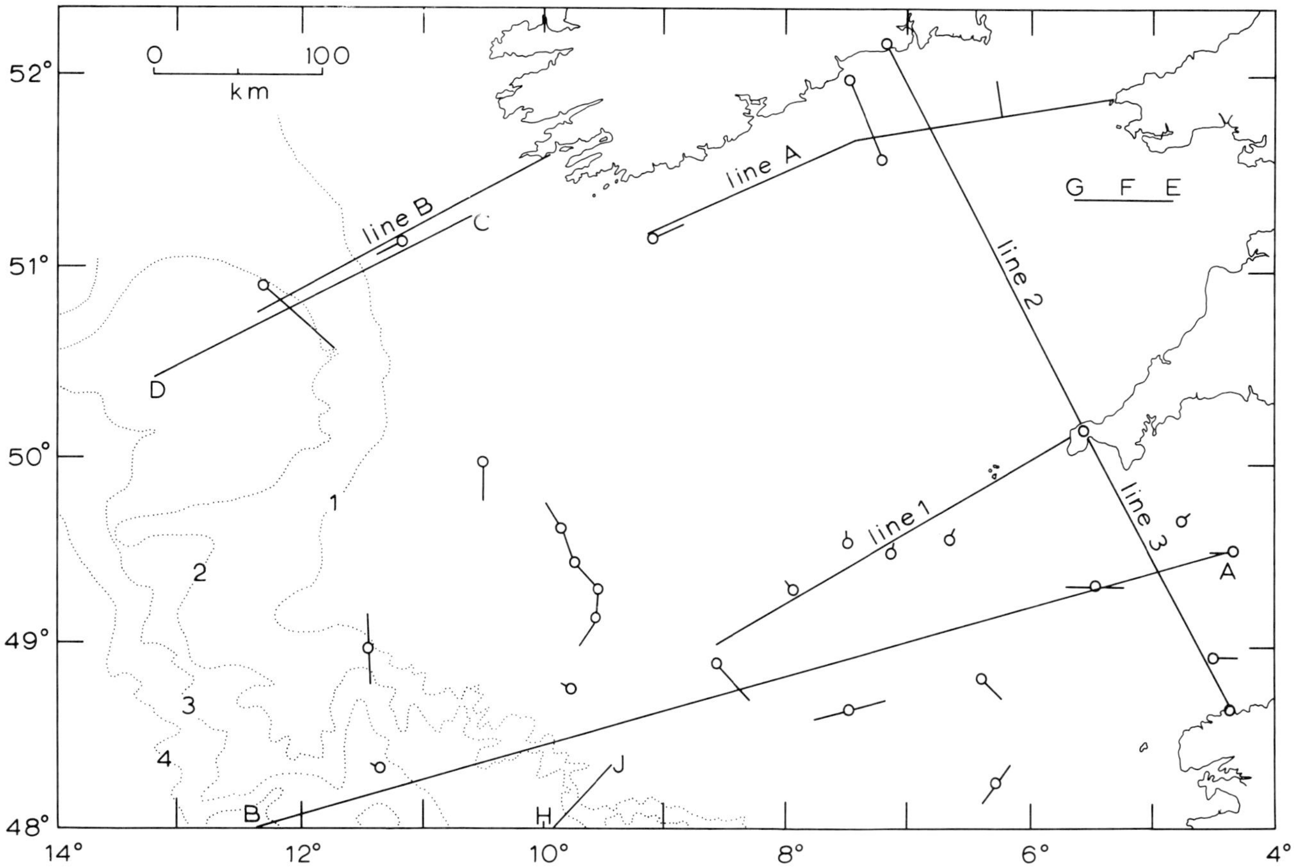

Fig. 4.3. Map of the Celtic Sea area showing the locations of seismic refraction lines and the lines of cross-section given in Figs. 4.4–4.7.

More recently, Avedik (1975) has described a seismic survey of the Western Approaches which involved 28 seismic refraction profiles averaging 30 km in length. In this he found that the Western Approaches Basin thins near to the shelf edge, but within it he recognised a layered structure as shown in Table 4.1c. The upper three layers display only shallow dips and together reach a maximum thickness of 3000 m. They are unconformable upon a layer *C* which is only found in "pockets" within layer *D*.

Results from a group of linked seismic refraction lines located just to the north of the ridge, at about 9°40′W, were presented by Bunce et al. (1964). They adopted the same classificiation of layers as Day et al. (1956) and interpreted their data to provide a N–S cross-section. The section cuts obliquely across the main structural trend so that, not surprisingly, it indicates a certain amount of topography on the boundaries. The ridge of layer *D* is shallow at the southern end but deepens to give an increased thickness of layers *A* and *B* coincident with gravity low *C* and provides for the presence of layer *C* beneath the northern half of the line.

Seismic refraction lines have been recorded south of Ireland (Blundell et al., 1971; Handley, 1971) that relate seismic velocities of 3.1–3.8 km/s to the Chalk outcrops. Refraction lines by Brooks and James (1975) in the Bristol Channel indicate velocities mainly in the range 2.8–3.2 km/s representing Mesozoic strata that are predominantly Jurassic, although velocities are as much as 4.2 km/s where the Jurassic is up to 1000 m thick in the centre of the Bristol Channel syncline. More recently, Brooks and Al-Saadi (1977) have observed velocities from refraction lines across the inner Bristol Channel of 4.0 km/s for Triassic, of 4.2–4.7 km/s for Coal Measures, 5.2 km/s for Carboniferous limestone and around 5.3 km/s for Devonian.

In Porcupine Seabight an unreversed refraction line by Scrutton et al. (1971) shows velocities of less than 3.1 km/s to 2 km depth and of 4.35 km/s to 9 km depth. Two long-range refraction experiments to determine crustal structure have been carried out in the area. One described by Holder and Bott (1971) involved three lines of shots south-west, south-east and north-west from Lands End (Fig. 4.3). Their line *1*, south-west from Lands End, indicates a relatively shallow sediment layer, consistent with the earlier refraction results, and a seismic velocity associated with the granite of 5.85 km/s. This velocity increases in the lower crust to 6.9 km/s. The Moho is horizontal at 27 km depth beneath the line with an upper mantle P_n velocity of 8.0 km/s. Their line *2*, north-west from Lands End to Waterford, indicates large time delays in the refracted waves due to a thick sedimentary layer between shots 43 and 26 but rather smaller ones at other points along it. This correlates with the region of lower gravity values (Fig. 4.2). Holder and Bott assumed a mean seismic velocity of 4 km/s for the sediments and 5.85 km/s for the underlying basement and showed these are consistent with the gravity anomaly in giving a sediment thickness of about 3000 m if a density contrast of 300 kg/m³ is used. Taking account of the variations of travel time due to the sediments, Holder and Bott found the delay times for refractions from the Moho to be constant along the line at 3.2 s, indicating a horizontal Moho, and estimated its depth to be between 27 and 31 km. Their interpretation of line *3* south-east from Lands End to Brest is similar to line *2*, since moderately large time delays are associated with shots across the English Channel which are attributed to the presence of sediments with relatively low seismic velocity. They are less successful in reconciling this with the gravity variation and suggest the higher gravity values than expected may be due to a 2 km rise in the Moho beneath the centre of the sediment trough. Otherwise they regard the Moho as horizontal at about 27 km depth as along line *1*. Their results are consistent with a long refraction profile across France south-eastwards from Brest from which Fuchs et al. (1972) concluded that although lateral variations occurred within the crust, the mean crustal thickness and velocity are nearly constant at 30 km and 6.2 km/s respectively. Bott and Watts (1971) incorporated this information together with the results of Day et al. (1956) and the gravity computations of Worzel (1968) to produce a self consistent crustal model cross-section along the centre line of the Western Approaches Basin. This I have further modified to include my geological correlation of seismic layers *A–D*, as shown in Fig. 4.4.

The other long range refraction experiment involved two lines of shots, line *A* parallel with the south coast of Ireland and about 20 km offshore and line *B*, south-southwest from Bantry Bay crossing into the Porcupine Seabight (Bamford and Blundell, 1971). Results from line *A* indicate time delays on refractions from lower layers due to sediments above which correlate with gravity variations along the line just as Holder and Bott found for their line *2*. The Moho is found to be at a constant depth of 28 km beneath the line. The mean crustal seismic velocity is 6.3 km/s and the upper mantle

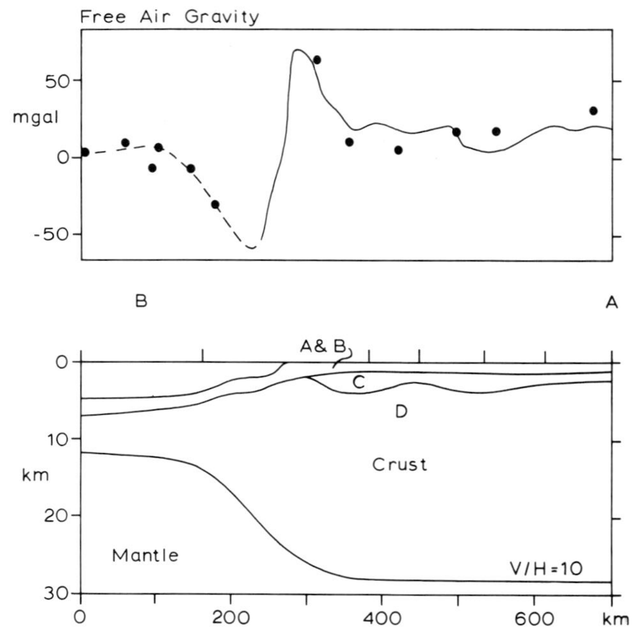

Fig. 4.4. Gravity profile and cross-section along line *AB* (Fig. 4.3) showing re-interpretation of line shown by Worzel (1968) using all available geophysical information. Layers are lettered as in Table 4.1.

P_n velocity is 8.0 km/s (Bamford, 1971, 1972).

Handley (1971) has considered the results of line *B* in conjunction with his other geophysical observations in the area. The line is effectively unreversed so that he had to assume the P_n velocity of line *A* is maintained further west. However, he found this assumption provides a model consistent with the gravity variations. He made use of the refraction line of Scrutton et al. (1971) which crosses line *B*, together with continuous seismic profiling data, to control the effects of the upper sedimentary layers and produced a cross-section to which has been added the seismic reflection information of Stride et al. (1969, profile 5), as shown in Fig. 4.5. Handley argued that the thinner crust beneath the Porcupine Seabight could be due either to substantial thinning of a continental crust or to the presence of oceanic crust and that neither magnetic nor gravity nor seismic data could distinguish between the two alternatives. It is clear, however, that Porcupine Seabight is not underlain by a foundered block of crust of normal continental thickness.

Bunce et al. (1964) attempted a long range refraction line along the full length of their linked set of shorter lines but in poor weather conditions they could obtain little information of value and these data have to be discounted. The conclusion from combining the results of the main crustal profiles is that crustal thickness is uniform at about 28 km beneath the continental shelf of the Celtic Sea and only close to the margin does the crust thin to an oceanic form.

Seismic reflection observations

Continuous seismic profiler surveys have been made using a 1000 J sparker source across the northern part of the Celtic Sea (Blundell et al., 1971; Hamilton and Blundell, 1971) and westwards to the Porcupine Seabight (Handley, 1971). These surveys involved a network of intersecting traverses to build up a three dimensional view of structures but were limited in penetration to about 300 m. I made use of the particular characteristics of the reflecting horizons to distinguish between differing lithologies (Blundell, 1970) and, in particular, I

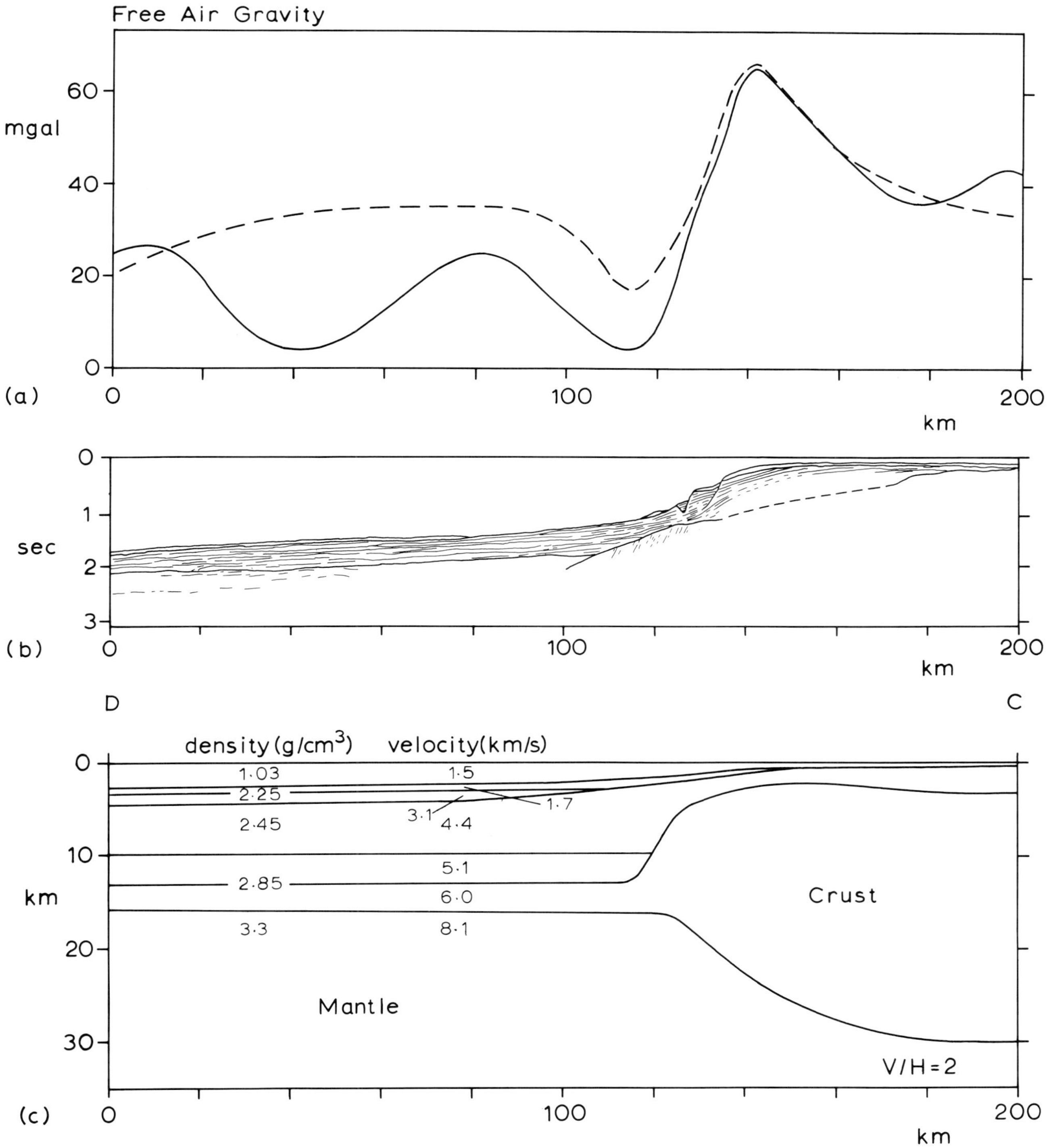

Fig. 4.5. (a) Gravity profile, (b) seismic reflection profile and (c) cross-section along line *CD* (Fig. 4.3) interpreted from combined gravity and seismic data. (Stride et al., 1969; Bamford and Blundell, 1971; Handley, 1971; Scrutton et al., 1971.)

was able to recognise the characteristic features of reflections from the Chalk, using traverses across the core stations of Curry et al. (1967) to make direct correlation. Handley and I were thus able to map the shallower sedimentary deposits across a wide area of the northern part of the Celtic Sea. We discovered an extensive outcrop of Chalk overlain unconformably by gently folded Tertiary deposits along the ENE–WSW trending axis of a syncline beneath Nymphe Bank and by similar deposits both to the south and to the west. We considered these to be a well bedded sequence, mainly clays and silts, and because they in turn are unconformably overlain by unfolded beds we regarded them as Palaeogene in age with the unfolded ones above as Neogene. As the northernmost core station of Curry et al. (1967) indicated an Eocene age to the deposits and as this would be consistent with the depositional history of the western English Channel it seemed likely that the Palaeogene deposits would be mainly of Eocene age. Delanty and Whittington (1977) have subsequently demonstrated that the so-called Neogene deposits (Fig. 4.1) are in fact of Pleistocene age and in-fill erosional features believed to have arisen from sub-glacial stream erosion at a time of reduced sea level.

I also found Tertiary–Recent sediments occupying the core of a shallow syncline trending south-west from St. George's Channel but, unconformably beneath its eastern limb, in the Bristol Channel Approaches, I found not Chalk but a thick sequence of bedded sediments dipping west at an average of 3° which I take to be Jurassic to Lower Cretaceous deposits (Fig. 4.6) similar to those of the Bristol Channel syncline. A traverse due west from Lundy shows these strata locally to be strongly faulted and folded. From the analysis of a detailed continuous seismic profiler survey across the Bristol Channel Approaches, Doré (1976) confirms the presence of a Jurassic to Lower Cretaceous succession in this area, folded along E–W axes, but he finds that further to the west it dips beneath Chalk. Doré also recognises faults to the west and to the south-west of Lundy that can be associated with the Lundy–Carnsore Ridge and the Lundy–Haig Fras line (Fig. 4.9). From seismic profiling and seabed cores, Fletcher (1975) has delineated a basin just to the east of Lundy filled with Oligocene silts, mudstones and lignitic clays and bounded by the north-westerly extension of the Lustleigh–Sticklepath Fault. The basin is similar to those at Petrockstow and at Bovey Tracey in Devon which are also associated with the Sticklepath Fault.

Handley (1971) found that Tertiary sediments increase in thickness westwards from the south coast of Ireland into the Porcupine Seabight. Dips are shallow

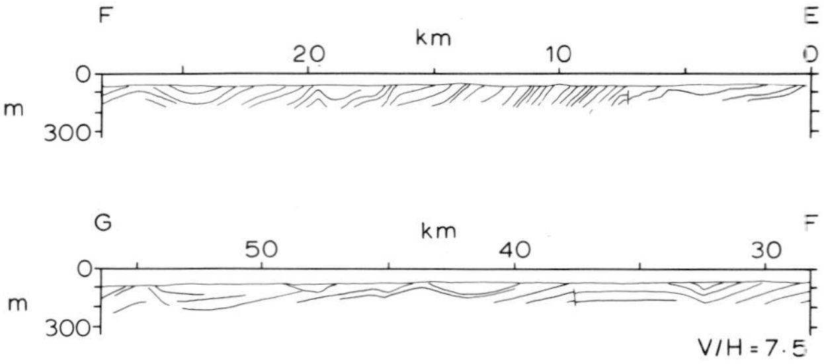

Fig. 4.6. Interpretation of continuous seismic profile (sparker) record along line *EFG* (Fig. 4.3) assuming an average seismic velocity of 2.5 km/s for the bedrock.

but increase towards the west and become parallel with the sea bed surface on the slope down to the seabight, the bedding there showing signs of disturbance and irregularity. He also noted a thin uppermost layer unconformably resting on the Tertiary deposits near to the continental margin, between sea depths of 150 m and 350 m, which showed no signs of bedding within it and which increased from zero to a maximum of 180 m thickness towards the west.

This uppermost layer was also recognised by Stride et al. (1969) in Arcer profiles across the continental margin. They presented three profiles across the margin in the Celtic Sea area. Their profile 5 is aligned through the Porcupine Seabight close to the seismic/gravity line produced by Handley and their data have been incorporated with his in Fig. 4.5. In addition to the unconformable base of the thin uppermost layer, they recognised a second lower unconformity, beneath which faint reflecting horizons were visible, indicating a westward dipping bedding in a formation they reckoned from the reflection characteristics to be Chalk near to the margin and, from an absence of reflections, to be basement nearer to shore. Immediately to the south, across Goban Spur, an abnormally thin cover of young sediment has also been reported by Dingle and Scrutton (1977). This has enabled them to investigate deeper structures with an airgun source seismic reflection survey. They have been able to map a set of horst and graben structures with block faulting both parallel with the continental margin and also ENE–WSW in alignment with structures across the shelf south of Ireland. They regard the basement as continental in composition and it includes granite, metamorphics and old sedimentary rocks. The graben are filled with relatively thick sedimentary sequences (probably in excess of 2000 m) that are presumably of Mesozoic age.

Stride et al.'s other two profiles, 6 and 7, cross the margin west of the Western Approaches Basin, close to where the basement ridge of Day et al. (1956) should be.

Their profiles indicate Tertiary deposits dipping parallel with the continental slope unconformably overlying Chalk with the suggestion that a basement ridge is present near the top of the continental slope, consistent with Avedik's seismic refraction evidence and Worzel's (1968) model based on gravity data. Their profiles also indicate slumping down the slope and faulting with downthrow on the continental side. They have no evidence of a deep sedimentary basin at the foot of the continental slope. But Grau et al. (1973) show a Flexotir profile across the margin close to Stride et al.'s (1969) profiles 6 and 7 with evidence of reflections from horizons that continue down to abyssal plain involving up to 3000 m of flat lying sediments reckoned as Lower Cretaceous, Upper Cretaceous and Tertiary. On the slope they recognise "important irregularities such as horsts and perhaps in some cases piercing structures which make up dams for sedimentary basins". In particular they see the existence of an ancient horst (interpreted as being of Early Cretaceous age) underneath the Trevelyan Escarpment. An interpretation of this section is given in Fig. 4.7. Using gravity and magnetic data as well, Bacon and Gray (1971) show the likelihood that this flanking zone beneath deep water is derived from continental crust. Leg 48 of I.P.O.D. included three drill sites in this region which have clearly established the presence near the continental margin of Upper Jurassic and Cretaceous limestones overlain by a succession of deep-sea carbonate sediments of Tertiary age (Montardert et al., 1977). Stride et al. (1969) also described in general terms their observations from several continuous seismic profile traverses they made, mainly across the southern part of the Celtic Sea. They used these to trace the continuity of the Miocene outcrop in the western English Channel westwards and north-westward towards the continental margin to link with the places where dredge samples of Miocene age were collected. They also recognised the presence of Tertiary "though not necessarily Miocene" strata north of Haig Fras up to 100 m thick unconformably overlying "folded strata which, when followed north, appear to extend as far as known Cretaceous outcrops" and thus link with Blundell et al.'s survey to the north.

Hersey and Whittard (1966) also carried out a series of continuous seismic profiles in the Western Approaches. Within the Western Approaches Basin they found important anticlines and synclines distributed throughout the main structural trough as well as minor folds involving Tertiary rocks. They observed that the top of the Chalk sequence is a pronounced unconformity and, in conjunction with their sampling programme, showed that the overlying Eocene (or possibly Oligocene) rocks are mainly coarse grained limestones. Traverses further north crossed the locality of the refraction lines of Bunce et al. (1964) and Day and Williams' (1970) gravity low. Hersey and Whittard reported the presence of a shallow syncline with ENE–WSW axis centred over the low. Combining the information from all three geophysical surveys provides an interpretation in which the gravity low is due not to a granite but to a

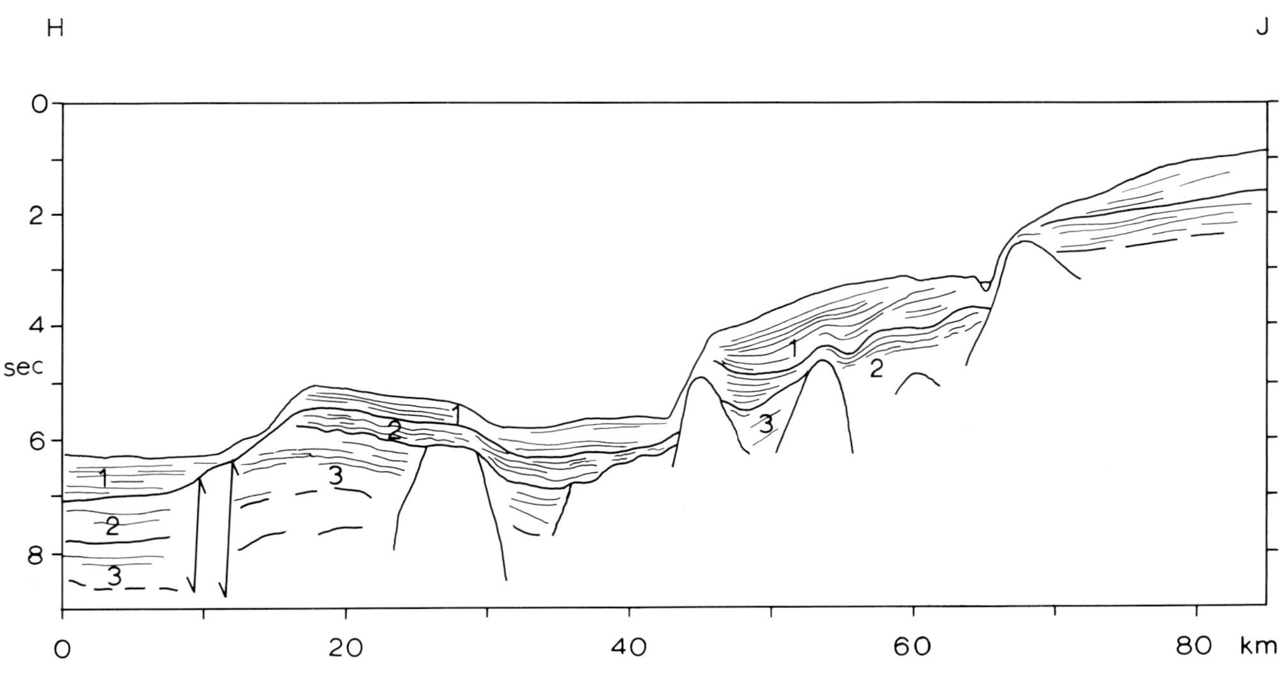

Fig. 4.7. Line drawing of seismic reflection profile along line *JH* (Fig. 4.3) given by Grau et al. (1973). Layer *1* is interpreted as Neogene, layer *2* as Palaeogene and Upper Cretaceous, layer *3* as Lower Cretaceous.

sedimentary trough. Using the modified correlation of seismic refraction layers given in Table 4.1b, the seismic and gravity information give consistent results with sediment thickness within the trough of up to 5000 m.

Structures within the Western Approaches Basin have also been worked out in detail by Bouysse et al. (1975) from interpretation of seismic reflection surveys across the western end of the English Channel. They confirm the predominant synclinal form of the basin and the important controlling influence of the faults bounding its southern flank.

Deep seismic reflection surveys have been made across the Celtic Sea in the search for hydrocarbons but the results from these remain confidential. Naylor (1972) gave some hint of the kind of structures found in presenting a record-section from an area probably related to the Fastnet gravity low. Flat lying, well bedded strata about 800 m thick overlie unconformably about 500 m of gently dipping, slightly faulted strata which in turn unconformably overlie a thick sequence of folded sediments within a fault controlled basin (or graben) structure which reaches at least 4000 m depth. Describing the area in general terms, Naylor and Mounteney (1975) see it as made up of fault-bounded graben structures of which the most important are the Nymphe Bank Trough (which they call the North Celtic Sea Basin) and the Celtic Sea Trough (which they call the South Celtic Sea Basin). From a seismic reflection section across the latter they interpret it as containing thick sequences of Mesozoic and Tertiary sediments in which salt structures similar to those of the southern North Sea may also be present. They also regard the basement rise between the two troughs as being continuous with the Pembrokeshire peninsula and consequently call the feature the Pembroke Ridge. Naylor also reported rumours of the deltaic lithology of Lower Cretaceous sediments in the northern part of the Celtic Sea.

Magnetic surveys

The aeromagnetic survey of Great Britain, flown in 1955 for the Institute of Geological Sciences (1965) covers the eastern part of the Celtic Sea but does not go sufficiently far west to provide much information

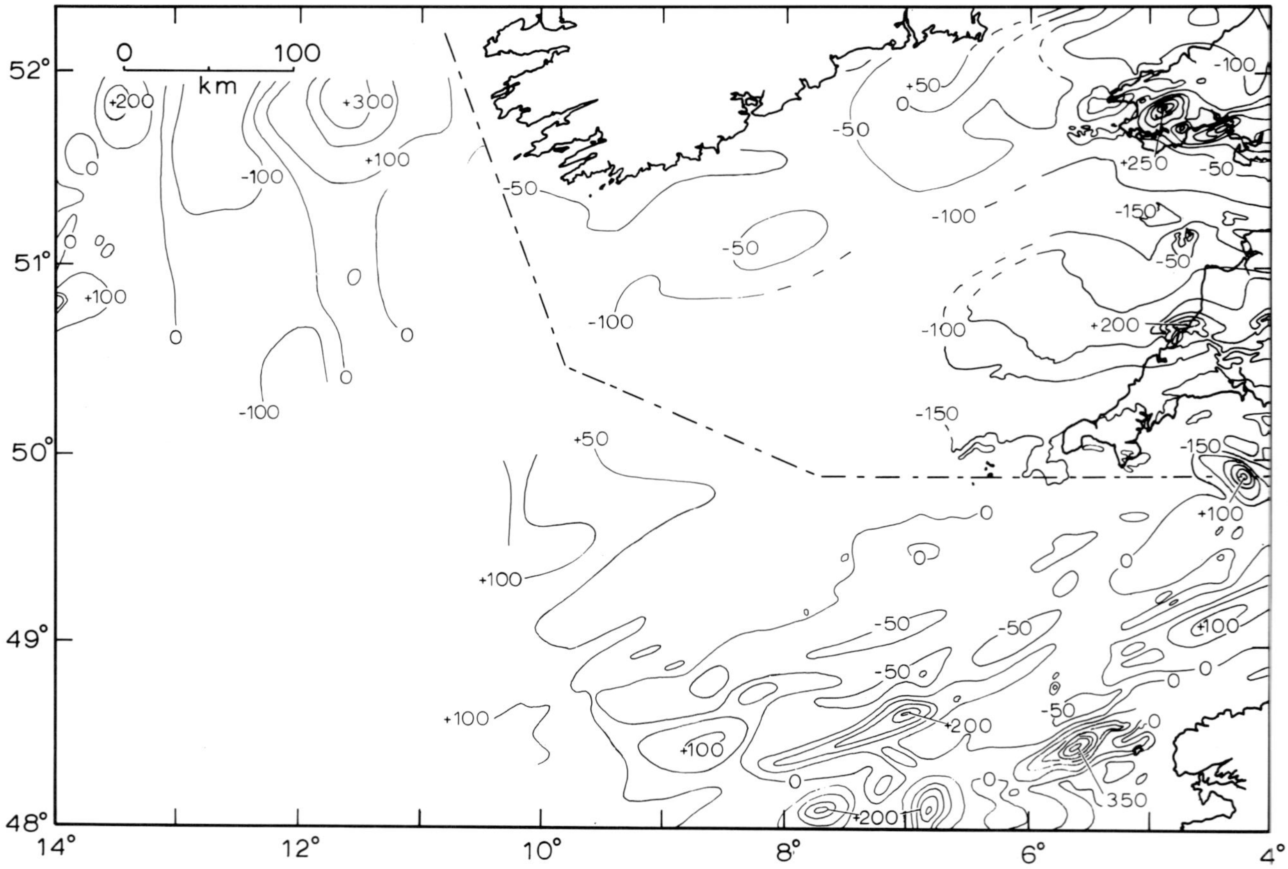

Fig. 4.8. Magnetic anomaly map of the Celtic Sea area compiled from surveys by Blundell and the Institute of Geological Sciences (1965), based on the I.G.S. reference field, and by Hill and Vine (1965) and Young and Bailey (1974) based on the IGRF values. The dashed line separates the two areas of differing reference field. Magnetic anomalies are shown in gammas (10^{-9} T).

about the sea area. Hill and Vine (1965) gave the first account of marine surveys in which they described the magnetic anomalies in the Western Approaches to the English Channel. They interpreted the anomalies, making use of the earlier seismic information from Day et al. (1956), as arising from the topography of the upper surface of a metamorphic basement underlying the sediments of the Western Approaches Basin. It was clear from their map that the WSW structural trend predominates and is maintained to the shelf edge. Beneath the main area of the Basin the magnetic anomalies are broad and of low relief with predominantly negative values but along the Aldernay–Ushant line there are several closed positive anomalies that stand out, one in particular being located at Lat. 48° 30′ W, which would seem to arise from something more than just basement surface topography.

In the course of the geophysical surveys south of Ireland, magnetic recordings were made. These have been reduced to the same regional field datum as the I.G.S. aeromagnetic survey so they could be directly compared in adjoining areas. The results of this work are given in Fig. 4.8 in compilation with the I.G.S. and Hill and Vine surveys together with a survey west of Ireland conducted by Young and Bailey (1974). Unfortunately, the survey data have been prepared relative to different reference fields which are difficult to reconcile without a great deal of effort. Young and Bailey have used the 1970 IGRF (International Geomagnetic Reference Field) and Hill and Vine used one that is compatible with it within the limits of accuracy of contouring at 50 gamma intervals. But the datum used by I.G.S. (and followed in my survey south of Ireland) differs from Hill and Vine's field by +150 gamma in the English Channel and from Young and Bailey's by about +50 gamma south-west of Ireland. Hence the map shown in Fig. 4.8 is divided into the two regions where the reference fields differ. Bearing this in mind, there is nonetheless a broad pattern of the magnetic variation that is consistent across the whole area. Across all those areas where a thick sedimentary cover is present (as deduced from the seismic and gravity data) magnetic anomalies are broad and of low relief, as is only to be expected, with predominantly negative values. The dominant structural trend is clearly seen as WSW except to the west of Ireland where a N–S trend is evident. There are a number of closed positive anomalies of several hundred gamma amplitudes but of relatively small extent such as the ones to the south of the English Channel already mentioned, and others over Cornwall, Lundy and Pembrokeshire. The Cornish anomalies have been ascribed to basic igneous rocks associated with the granites and the same association seems likely for Lundy. The direct source of the Pembrokeshire anomalies is uncertain but is probably basic igneous rocks that intrude the Precambrian basement there. Thus the positive anomalies along the Aldernay–Ushant line are probably also due to the presence of bodies of basic igneous rock but whether these are remnants of volcanic activity related to the tectonic activity associated with this structural line remains uncertain. Lefort (1977) has recently asserted that they represent an ophiolite suite marking the suture formed on impact of continental fragments in Variscan times. Certainly the character of magnetic anomalies south of the Aldernay–Ushant line, revealed from magnetic surveys by Segoufin (1975), differs from those of the Celtic Sea region, and this may well result from differences in basement structure between the two regions. Across the Celtic Sea there is no evidence from the magnetic surveys of any major volcanic activity associated with the continental margin or of any volcanic rocks interbedded with the sediments as discovered in the North Sea (Howitt, 1974). It is noticeable, for example, that there is no marked magnetic anomaly associated with the gravity high close to the shelf edge at lat. 48° 10′ N, long. 9° 30′ W. But of course there is still plenty of opportunity to discover such evidence as large tracts of the area, particularly across the shelf edge, remain to be surveyed.

Geology and structural history

Putting together all these fragments of geological and geophysical evidence leads to the overall view, given in Figs. 4.1 and 4.9, of two major sedimentary basins, each containing subsidiary fault controlled troughs, which developed along parallel lines commencing in Permo-Trias times. As Curry et al. (1970) have shown, the Western Approaches Basin is bounded to the south-east by the Aldernay–Ushant line and this can be traced south-westwards to the shelf edge. The north-western flank of the Basin is made up of the Cornubian granite ridge and its south-westerly extension, but the younger sediments override this in continuity with the Celtic Sea Basin to the north. This Basin has its main structural controls in the Carnsore–Fastnet line that marks its northern boundary and the Lundy–Haig Fras line that marks the north-western edge of the Cornubian Ridge. To the north-east it abuts against the Lundy–Carnsore Ridge and the related north-west extension of the Sticklepath Fault. Both basins contain smaller structurally controlled troughs (Fig. 4.9) separated by basement swells although the Western Approaches Basin is the

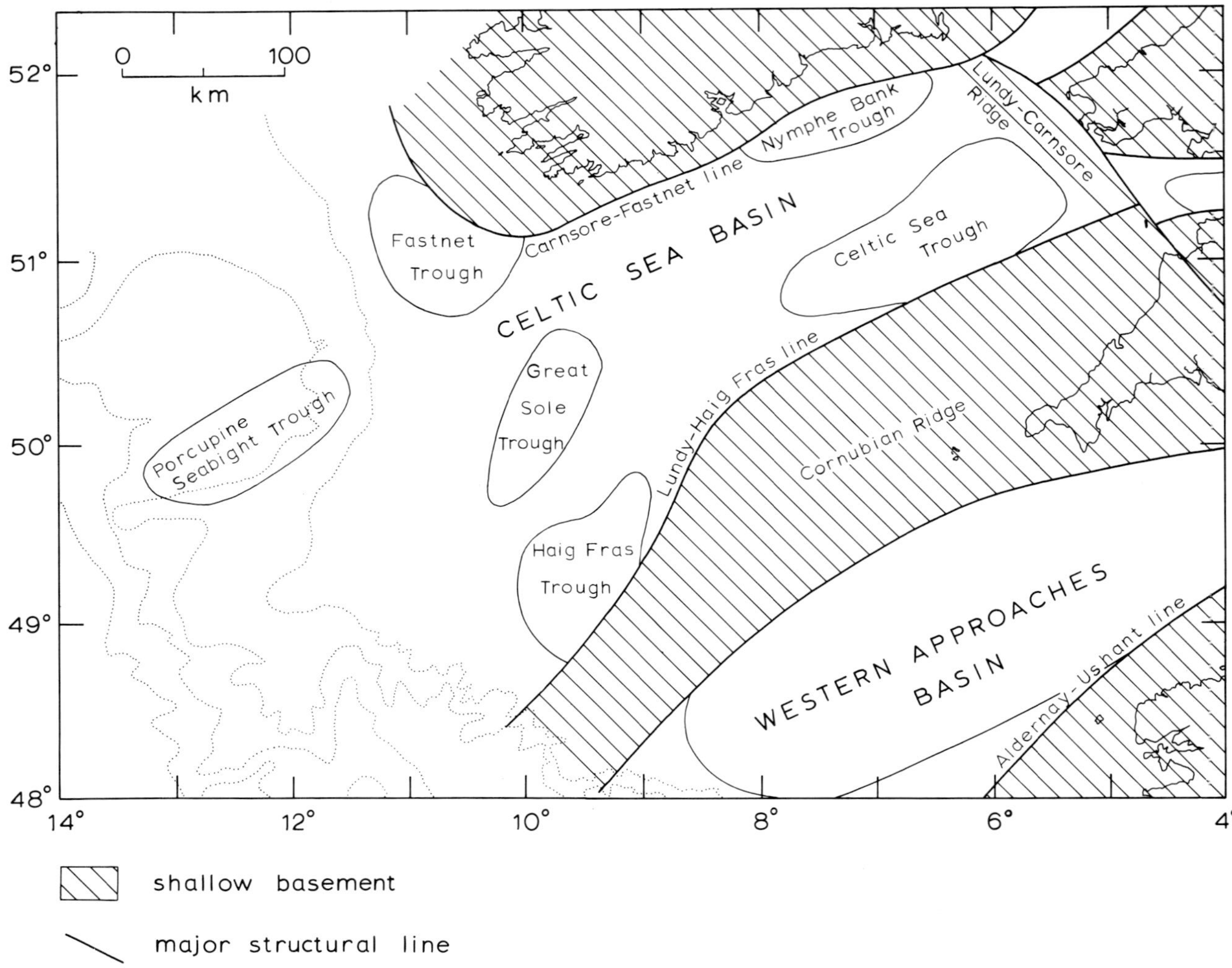

Fig. 4.9. Map of the Celtic Sea area outlining the troughs within the Celtic Sea Basin and Western Approaches Basin and indicating the major structural lines.

simpler and is of broadly synclinal form. Included in the Celtic Sea Basin are the Nymphe Bank Trough south of Ireland, the Celtic Sea trough which lies to the south-east, and which is separated from it by the Pembroke Ridge, and three troughs further to the west, the Fastnet Trough, the Great Sole Trough and the Haig Fras Trough. The two basins both appear to thin towards the shelf edge where various profiles indicate a shallow seismic basement (which could be Lower Palaeozoic or Precambrian) beneath the continental slope. Further basins have developed in the deep-water areas of the southern part of the Porcupine Seabight and at the base of the continental slope off the Western Approaches. The history of development of the Western Approaches Basin described by Curry et al. seems to be matched in the Celtic Sea Basin. From the sea areas there is virtually no evidence to add to the terrestrial evidence concerning the Pre-Variscan history of the area, but the granitic intrusion at a late stage of that orogeny is found to extend south-west from Cornwall to the shelf edge. All these older rocks are lumped into the term "basement" to describe the material whose eroded surface forms the floor of the two sediment basins. Within the basement, the only distinction that can be made is in terms of magnetic properties and it would seem from these that the Western Approaches Basin may be underlain in the south by metamorphic rocks similar to those of Brittany whereas to the north and beneath the Celtic Sea Basin a considerable thickness of relatively unmetamorphosed Lower Palaeozoic rocks, possibly similar to the Devonian and Culm of Devon and Cornwall, may extend to the shelf edge.

At the close of the Variscan (Armorican, Hercynian) orogeny it seems that a number of structural basins must have developed, as elsewhere around Britain, peripheral to blocks like the Cornubian Ridge elevated through granitic emplacement. Into these intermontane basins were deposited coarse clastic sediments, later with finer counterparts, that gave rise to successions with lithologies well known from the Permo-Trias of Devon, the Midlands and north-west England. Naylor's observations suggest that salt horizons, which have later produced "halo-kinetic" structures such as diapirs and the like, may be present within a succession that must total in excess of 1000 m beneath parts of the Celtic Sea Basin. In the Western Approaches Basin, the overlying Jurassic sequence is marine with clays predominating and Curry et al. suggest the succession is sufficiently close to that further east in southern England for them to have been in continuity. The only report of Jurassic rocks found in the Celtic Sea Basin is of the oil-bearing sands recorded at 1800 m depth in Esso's borehole. The succession observed on the c.s.p. (sparker) profile dipping into the Celtic Sea trough, however, infers a substantial thickness of Jurassic rocks with well bedded characteristics within the total sequence of 2000 m of presumed Jurassic and Lower Cretaceous strata. The presence of oil in Jurassic sands can also be used to guess at the likely presence of Kimmeridge Clay or its equivalent within the succession, by analogy with southern England and the North Sea where it is considered to be the source material for the oil. Such a guess must be treated rather cautiously, though, as the palaeogeographic situation here differs from the North Sea, but it would seem less likely for the oil to have formed here from Carboniferous deposits. With a total Jurassic thickness recorded in the western English Channel of up to 1000 m and in the Bristol Channel of about 1500 m it must be presumed that thicknesses of at least these amounts can be expected to be developed within the troughs of the Western Approaches and Celtic Sea basins. In the western English Channel and in the Bristol Channel the Jurassic is succeeded by Lower Cretaceous of terrestrial aspect with deltaic, fluvial and swamp deposits similar to those of the Weald. In the Marathon boreholes south of Cork, finds of natural gas are reported from deltaic sands of Lower Cretaceous at 800–900 m depth, so it must be presumed that the history of deposition in the Celtic Sea Basin is of similar character. Seismic profiling evidence from both basins suggests that fault movements occurred contemporaneously with Jurassic and Lower Cretaceous deposition in subsiding troughs and that near to faults the strata are considerably deformed. Both seismic and direct geological evidence are clear in showing the Mid-Cretaceous unconformity. Chalk is deposited across large tracts of the Western Approaches Basin and the Celtic Sea Basin where it forms large outcrops on the seabed, continuous from one to the other. In both basins the Chalk transgresses from the west to overlap on to successively older strata down to the Devonian. The Chalk is gently folded in broad synclines and anticlines and is occasionally faulted. It has particularly characteristic reflecting properties and so is easy to follow on sparker profiles. With Chalk facies continuing through to the Danian a total thickness of 620 m was estimated in the western English Channel by Curry et al. South of Ireland, Chalk thickness estimated from measuring the extent of limbs on folds from sparker profiles is about 600 m in the vicinity of the Marathon boreholes to the south of Nymphe Bank Trough. The seismic profiler evidence suggests the Chalk is not only continuous to the shelf edge but also extends down the continental slope and may be present in the flanking deep water areas. If so, it is involved in block faulting parallel with the shelf edge that gives rise to the horst and graben structures recognised by Grau et al. (1973) and by Dingle and Scrutton (1977). From the evidence of the various samples of Chalk obtained from the Celtic Sea area it looks to have been very uniform in character laterally and may well be of fairly uniform thickness over much of the area. It was, however, eroded in part prior to the deposition of Tertiary sediments which are found to overly it unconformably, with transgression from the west. These early Tertiary, probably Eocene, sediments are well stratified with alternating beds of marine sandstones, clays and limestones. They thicken to the west and show bedding that has shallow synclinal structures within the Nymphe Bank Trough, the Celtic Sea Trough (where they are continuous through St. George's Channel into the South Irish Sea) and the main syncline of the Western Approaches Basin.

In turn these are overlain unconformably by younger Tertiary (probably Miocene) sediments which likewise thicken and dip paralleling the sea bed surface of the slope, down to the deep water areas beyond.

Thus the structural character of the Celtic Sea area changed dramatically during the Lower Cretaceous from predominantly vertical movements involving relative subsidence or uplift of troughs or swells within the basin areas, first to provide a more general uplift giving rise to erosion of the higher areas followed by downwarping from the west and the eastward transgression of the Chalk sea. Thereafter, through the Tertiary, the pattern of subsidence of the continental margin with repeated regressions and transgressions in early Eocene and Oligocene times was maintained along with some further and gentle subsidence of the trough structures within the basins. Curry et al. ascribed this change as due to the formation of a new continental margin to the west of Britain during the Lower Cretaceous and the eventual ocean floor spreading to enlarge the North Atlantic, with the several regressions and transgressions being due in part to changes in the style and direction of spreading and also in part due to eustatic sea-level changes related to the spreading episodes. Of the general validity of this explanation there can now be little doubt but many details, particularly in relation to the early development of the margin and new ocean remain obscure. In the North Atlantic to the west of the Celtic Sea, Williams and McKenzie (1971) have recognised the oceanic magnetic anomaly pattern back to anomaly 32, dated at about 80 Ma, and they and Laughton (1971) and various others have worked out the spreading history of the Atlantic since that time. Changes in the spreading pattern at 60 Ma ago and at about 45 Ma ago could account for the two depositional breaks within the Tertiary. If the movements during the past 80 Ma are taken into account, Laughton showed that it is possible to reconstruct the positions of North America relative to Britain at that time. Doing this, it is found that the continental shelf off Newfoundland and Eastern Canada matches quite well against the western side of the Rockall Plateau but that an oceanic gap remains between Flemish Cap, on the Canadian side, and the Celtic Sea shelf edge of some 400 km (Blundell, 1975). Anomaly 32 is itself some 200 km from the Celtic Sea shelf edge and the intervening zone is marked by a lack of significant magnetic anomalies. This quiet zone has been recognised elsewhere in the North Atlantic peripheral to the edge of the continents and has been explained by some as due to a quiet period during the Cretaceous when no geomagnetic field reversals occurred so that sea-floor spreading could produce new oceanic crust clear of magnetic anomalies. Others, for example Drake et al. (1968), have suggested that the quiet zone is a fringe of collapsed and thinned continental crust and not the product of sea-floor spreading, so that the true continental margin should be placed at the edge of the quiet zone. This view is upheld, at least to some extent, by the evidence from Bacon and Gray (1971) of a fringe of continentally derived crust beneath deep water at the foot of the continental slope south-west of the Celtic Sea. Reconstructing the continental positions to give the best fit of continental margins defined by the 500 m isobath, Bullard et al. (1965) placed Flemish Cap in juxtaposition against the Celtic Sea shelf edge. A comparison of the Celtic Sea geology (Fig. 4.1) with that of Flemish Cap (Grant, 1972) indicates a clear mismatch. The Variscan Front of Southern Ireland should, on this reconstruction, have to pass north of Newfoundland, but there is no evidence of Newfoundland having been so affected and the continuation of the Front is better placed further south leading into the Alleghenian Front through the Chedabucto Fault in New Brunswick (Riding, 1974). The granite batholiths of the Cornubian Ridge lead south-westwards to the margin but find no equivalent on this reconstruction in Flemish Cap unless the granite there is substantially younger than the material from within its aureole that is dated at 590 Ma. It would thus seem necessary, in order to avoid this problem of mismatch, to allow for some mobility between the two units, North America and Britain, over the period of the Variscan orogeny. Riding has suggested as much in his model of the Hercynian fold-belt in which he distinguishes between the collision of North America and Africa on the one hand and the collison of north-west Europe with a purely oceanic plate (being subducted) on the other, with Iberia forming a microcontinent on the southerly plate. In such circumstances it might prove easier to find an extension of the Cornubian granite ridge in the north-west of Spain where granites of similar age have been emplaced in Galicia. This view is in keeping with the model recently put forward by Williams (1975) to explain the sea floor spreading in the Bay of Biscay, which starts with the northern margin of Spain and the Galicia Bank close to the Celtic Sea margin. With the first spreading movement coming between the western margin of Spain and Flemish Cap, lateral (transform) movement of the northern margin of Spain south-eastwards relative to the Celtic Sea margin thus exposes the latter to the early Atlantic Ocean. This model has two further advantages. First, it can provide a reason why the North Atlantic–Europe plate should split between Canada and Britain rather than between Britain and the rest of Europe along the line of weakness provided by the North Sea Graben system, if there were already in existence a line of weakness between North America and Europe remaining from the Variscan orogeny. Secondly, it shows how the Celtic Sea margin could have formed without going through the rift phase that is normally expected at the commencement of ocean development by spreading (Vogt, 1970), as seems to be the case since there are no signs of the volcanicity or regional uplift consistent with rifting there.

The only post-Variscan vulcanicity is the very localised episode of the Wolf Rock and the Tertiary (55 Ma) volcanicity associated with Lundy situated at the intersection of the Sticklepath Fault with the Lundy–Haig Fras line bounding the Celtic Sea Basin.

What still remains in doubt is how the horst and graben structures and the development of thick sediments at the foot of the continental slope south-west of the Celtic Sea could have formed or how far oceanwards they extend. Nor is it fully understood how the Goban

Spur came to be isolated from the rest of the continental shelf, although an underlying granite may be the cause, or how the Porcupine Seabight developed (Bailey, 1975). Perhaps these are the features of a continental margin formed as a result of transform faulting. No doubt it will not be long before the answers are found.

References

Andreieff, P., Bouysse, P., Curry, D., Fletcher, B.N., Hamilton, D., Monciardini, C. and Smith, A.J., 1975. The stratigraphy of the post-Palaeozoic sequences in part of the western Channel. Philos. Trans. R. Soc. London, Ser. A, 279: 79–98.

Avedik, F., 1975. Seismic refraction survey in the Western Approaches to the English Channel: preliminary results. Philos. Trans. R. Soc. London, Ser. A, 279: 29–40.

Bacon, M. and Gray, F., 1971. Evidence for crust in the deep ocean derived from continental crust. Nature, 229: 331–332.

Bailey, R.J., 1975. The geology of the Irish continental margin and some comparisons with offshore eastern canada. In: C.J. Yorath, E.R. Parker and D.J. Glass (Editors), Canada's Continental Margins and Offshore Petroleum Exploration. Can. Soc. Pet. Geol., Mem., 4: 313–340.

Bamford, S.A.D., 1971. An interpretation of first arrival data from the Continental Margin Refraction Experiment. Geophys. J. R. Astron. Soc., 24: 213–229.

Bamford, S.A.D., 1972. Evidence for a low-velocity zone in the crust beneath the western British Isles. Geophys. J. R. Astron. Soc., 30: 101–105.

Bamford, S.A.D. and Blundell, D.J., 1971. South-west Britain Continental Margin Experiment. Inst. Geol. Sci., Rep., 70/14: 143–156.

Blundell, D.J., 1970. Interpretation of profiler records from the southern Irish Sea and the Celtic Sea. Proc. Geol. Soc. London, 1663: 183–185.

Blundell, D.J., 1975. The geology of the Celtic Sea and South-western Approaches. Can. Soc. Pet. Geol., Mem., 4: 341–362.

Blundell, D.J., Davey, F.J. and Graves, L.J., 1968. Sedimentary basin in the South Irish Sea. Nature, 219: 55–56.

Blundell, D.J., Davey, F.J. and Graves, L.J., 1971. Geophysical surveys over the south Irish Sea and Nymphe Bank. J. Geol. Soc. London, 127: 339–375.

Bott, M.H.P. and Watts, A.B., 1971. Deep structure of the continental margin adjacent to the British Isles. Inst. Geol. Sci., London, Rep., 70/14: 89–110.

Bott, M.H.P., Day, A.A. and Masson Smith, D., 1958. The geological interpretation of gravity and magnetic surveys in Devon and Cornwall. Philos. Trans. R. Soc. London, Ser. A., 251: 161–191.

Bouysse, P., Horn, R., LeFort, J.P. and Le Lann, F., 1975. Tectonique et structures post-Palaeozoiques en Manche occidentale. Philos. Trans. R. Soc. London, Ser. A, 279: 41–54.

Bradley, W.H., 1940. Geology and biology of North Atlantic deep-sea cores between Newfoundland and Ireland. U.S. Geol. Surv., Prof. Pap., 196A: 13–15.

Brooks, M. and Al-Saadi, R.H., 1977. Seismic refraction studies of geological structure in the inner part of the Bristol Channel. J. Geol. Soc. London, 133: 433–445.

Brooks, M. and James, D.G., 1975. The geological results of seismic refraction surveys in the Bristol Channel, 1970–1973. J. Geol. Soc. London, 131: 163–182.

Brooks, M. and Thompson, M.S., 1973. The geological interpretation a gravity survey of the Bristol Channel. J. Geol. Soc. London, 129: 245–274.

Brown, B.C. and Cooper, R.I.B., 1950. The British submarine gravity surveys of 1938 and 1946. Philos. Trans. R. Soc. London., Ser. A., 242: 243–310.

Bullard, E.C. and Gaskell, T.F., 1941. Submarine seismic investigations. Proc. R. Soc. London, Ser. A., 177: 476–499.

Bullard, E.C., Everett, J.E. and Smith, A.G., 1965. The fit of the continents around the Atlantic. Philos. Trans. R. Soc. London, Ser. A., 258: 41–51.

Bunce, E.T., Crampin, S., Hersey, J.B. and Hill, M.N., 1964. Seismic refraction observations on the continental boundary west of Britain. J. Geophys. Res., 69: 3853–3863.

Cole, G.A.J. and Crook, T., 1910. On rock specimens dredged from the floor of the Atlantic off the coast of Ireland, and their bearing on submarine geology. Mem. Geol. Surv. Ireland, pp. 1–34.

Curry, D. and Hamilton, D., 1971. Geological evolution of the western English Channel and its relation to the nearby continental margin. Inst. Geol. Sci., Rep., 70/14: 129–142.

Curry, D., Martini, E., Smith, A.J. and Whittard, W.F., 1962. The geology of the western approaches of the English Channel. 1. Chalky rocks from the upper reaches of the continental slope. Philos. Trans. R. Soc. London, Ser. B., 245: 267–290.

Curry, D., Gray, F., Hamilton, D. and Smith, A.J., 1967. Upper chalk from the sea-bed, south of Cork, Eire. Proc. Geol. Soc. London, 1640: 134–136.

Curry, D., Hamilton, D. and Smith, A.J., 1970. Geological and shallow subsurface geophysical investigations in the western approaches to the English Channel. Inst. Geol. Sci., Rep., 70/3: 1–12.

Davey, F.J., 1970. Bouguer anomaly map of the north Celtic Sea and entrance to the Bristol Channel. Geophys. J. R. Astron. Soc., 22: 277–282.

Day, A.A., Hill, M.N., Laughton, A.S. and Swallow, J.C., 1956. Seismic prospecting in the western approaches of the English Channel. Q. J. Geol. Soc. London, 112: 15–44.

Day, G.A. and Williams, C.A., 1970. Gravity compilation in the N.E. Atlantic and interpretation of gravity in the Celtic Sea. Earth Planet. Sci. Lett., 8: 205–213.

Delanty, L.J. and Whittington, R.J., 1977. A re-assessment of the "Neogene" deposits of the South Irish Sea and Nymphe Bank. Marine Geol., 24: M23–M30.

Dingle, R.V. and Scrutton, R.A., 1977. Continental margin fault pattern mapped south-west of Ireland. Nature, 268: 720–722.

Doré, A., 1976. Preliminary geological interpretation of the Bristol Channel Approaches. J. Geol. Soc. London, 132: 453–459.

Drake, C.L., Ewing, J.I. and Stockard, H., 1968. The continental margin of the eastern United States. Can. J. Earth Sci., 5: 993–1010.

Fletcher, B.N., 1975. A new Tertiary basin east of Lundy Island. J. Geol. Soc. London, 131: 223–225.

Fuchs, K., Hirn, A., Muller, S., Prodehl, C., Sapin, M. and Steinmetz, L., 1972. A long-range seismic profile in France from the Bretagne to the Provence. Ann. Geophys., 28: 247–256.

Grant, A.C., 1972. The continental margin off Labrador and eastern Newfoundland – morphology and geology. Can. J. Earth Sci., 9: 1394–1430.

Grau, G., Montadert, L., Delteil, J.R. and Winrock, K., 1973. Structure of the European continental margin between Portugal and Ireland, from Seismic data. Tectonophysics, 20: 319–340.

Hamilton, D. and Blundell, D.J., 1971. Submarine geology of the approaches to the Bristol Channel. Proc. Geol. Soc. London, 1664: 297–300.

Handley, R.L., 1971. A Geophysical Study of the Porcupine Seabight. Ph. D. Thesis, University of Birmingham, 179 pp. (unpublished).

Hersey, J.B. and Whittard, W.F., 1966. The geology of the western approaches of the English Channel. V. The continental margin and shelf under the Celtic Sea. Geol. Surv. Can. pap., 66-15: 80–106.

Hill, M.N. and Vine, F.J., 1965. A preliminary magnetic survey of the western approaches of the English Channel. Q. J. Geol. Soc. London, 121: 463–476.

Holder, A.P. and Bott, M.H.P., 1971. Crustal structure in the vicinity of south-west England. Geophys. J. R. Astron. Soc., 23: 465–489.

Howitt, F., 1974. North Sea oil in a world context. Nature, 249: 700–703.

Institute of Geological Sciences, 1965. Aeromagnetic map of Great Britain, sheet 2. Publ. U.K. Ordnance Survey.

Laughton, A.S., 1971. South Labrador Sea and the evolution of the North Atlantic. Nature, 232: 612–617.

Lefort, J.P., 1977. Possible "Caledonian" subduction under the Dannonean domain, North Armorican area. Geology, 5: 523–526.

Lloyd, A.J., Savage, R.J.G., Stride, A.H. and Donovan, D.T., 1973. The geology of the Bristol Channel floor. Philos. Trans. R. Soc. London, Ser. A., 274: 595–626.

Montardert, L., Roberts, D.G., Auffret, G.A., Bock, W., Du Peuble, P.A., Hailwood, E.A., Harrison, W., Kagami, H., Lumsden, D.N., Muller, G., Schnitker, D., Thompson, R.W., Thompson, T.L. and Timofeev, P.P., 1977. Rifting and subsidence on passive continental margins in the northeast Atlantic. Nature, 268: 305–307.

Naylor, D., 1972. Hydrocarbon potential of offshore West Britain and Ireland. Financial Times, Second North Sea Conference, 13.

Naylor, D. and Mounteney, S.N., 1975. Geology of the North-West European Continental Shelf. 1. Geology. Graham, Trotman, Dudley, London, 162 pp.

Riding, R., 1974. Model of the Hercynian foldbelt. Earth Planet. Sci. Lett., 24: 125–135.

Scrutton, R.A., Stacey, A.P. and Gray, F., 1971. Evidence for the mode of formation of the Porcupine Seabight. Earth Planet. Sci. Lett., 11: 140–146.

Segoufin, J., 1975. Structure du plateau continental armoricain. Philos. Trans. R. Soc. London, Ser. A, 279: 109–122.

Smith, A.J., Stride, A.H. and Whittard, W.F., 1965. The geology of the western approaches of the English Channel. IV. A recently discovered Variscan granite west-north-west of the Scilly Isles. Colston Pap., 17: 287–301.

Stride, A.H., Curray, J.R., Moore, D.G. and Belderson, R.H., 1969. Marine geology of the Atlantic continental margin of Europe. Philos. Trans. R. Soc. London, Ser. A., 264: 31–75.

Vogt, P.R., 1970. Magnetized basement outcrops on the south-east Greenland continental shelf. Nature, 226: 743–744.

Williams, C.S., 1975. Sea floor spreading in the Bay of Biscay and its relationship to the North Atlantic. Earth Planet. Sci. Lett., 24: 440–456.

Williams, C.A. and McKenzie, D., 1971. The evolution of the north-east Atlantic. Nature, 232: 168–173.

Worzel, J.L., 1968. Survey of continental margins. In: Donovan, D.T. (Editor), Geology of Shelf Seas. Oliver and Boyd, Edinburgh, pp. 117–154.

Young D.G.G. and Bailey, R.J., 1974. An interpretation of some magnetic data off the west coast of Ireland. Geol. J., 9: 137–146.

Chapter 5

The Geology of the English Channel, South Celtic Sea and Continental Margin, South Western Approaches

D. HAMILTON

Introduction

The South Western Approaches (Fig. 5.1) have long been the maritime gateway to northern Europe. However, an understanding of the nature of the submarine geology of the region begins with the pioneering researches of Dangeard (1923, 1925) and the important syntheses of King (1948, 1954). Work prior to this date, mainly by British and French workers is summarised by Dangeard (1972). Early work suffered from the limited accuracy of the navigational methods in use at the time and also from the use of dredging, which was the main method of collecting samples. The more effective gravity corer, used in the first Channel Tunnel survey in 1866 (Donovan, 1967) became widely used after the mid-fifties. Early geophysical surveys, using seismic methods (Bullard and Gaskell, 1941; Hill and King, 1953; Hill and Laughton, 1954; Day et al., 1956; Bunce et al., 1964) and magnetics (Hill and Vine, 1965) gave a first insight into the deep structure of the area. With the development and refinement of acoustic techniques, such as echosounding, seismic reflection and refraction, and side scan sonar (Asdic), a considerable volume of data is now available that is precisely located through the use of modern navigational aids.

In the following chapter the nature of the sea floor in the English Channel, Celtic Sea and South Western Approaches will be discussed from a geological point of view. Firstly, this will include a brief description of the morphology of the area. Secondly, major aspects of the solid rock geology will be discussed with the main emphasis on sea floor and shallow-subsurface geology. Brief reference only will be made to deeper structure as this is dealt with in another chapter (see Blundell, Chapter 4, this volume).

In this short synthesis no attempt has been made to include a complete list of references.

Morphology and topography

The continental shelf

The sea floor to the south and west of Britain can be divided into the English Channel and the Celtic Sea respectively (Fig. 5.1). The outer western English Channel and the southern Celtic Sea are known generally as the South Western Approaches, in contrast to the North Western Approaches to Britain.

Except for the linear banks and the outcrop of the Haig Fras granite in the Celtic Sea (Fig. 5.1), the sea floor of both the English Channel and Celtic Sea is remarkably flat and is bounded by a sharp break in bathymetry in the west, at the continental margin. The area under consideration is thus an extensive continental shelf, which, westwards of the continental margin, passes into the long continental slope. The features of this long slope are discussed in a following section.

The continental shelf of the South Western Approaches is one of the widest in north-western Europe. From the continental margin, the shelf surface extends some 900 km eastwards along the English Channel to the Strait of Dover and some 550 km north-east across the Celtic Sea to the Bristol Channel and St. George's Channel. The regional gradient is extremely low. From the shelf break, at about 200 m depth, to Dover Strait, the gradient is about 1 : 5500, whilst the regional gradient across the Celtic Sea is close to 1 : 3500.

Sections transversely across the English Channel and Celtic Sea show similarly low gradients except close to the present coast or where localised depressions or deeps occur, such as on the southern side of the Channel (Fig. 5.1). Such deeps are virtually absent from the Celtic Sea (see section on Buried Channel systems). Close to the shore, the sea-bed profile may be cliffed or show stepwise changes in gradient. These are related to

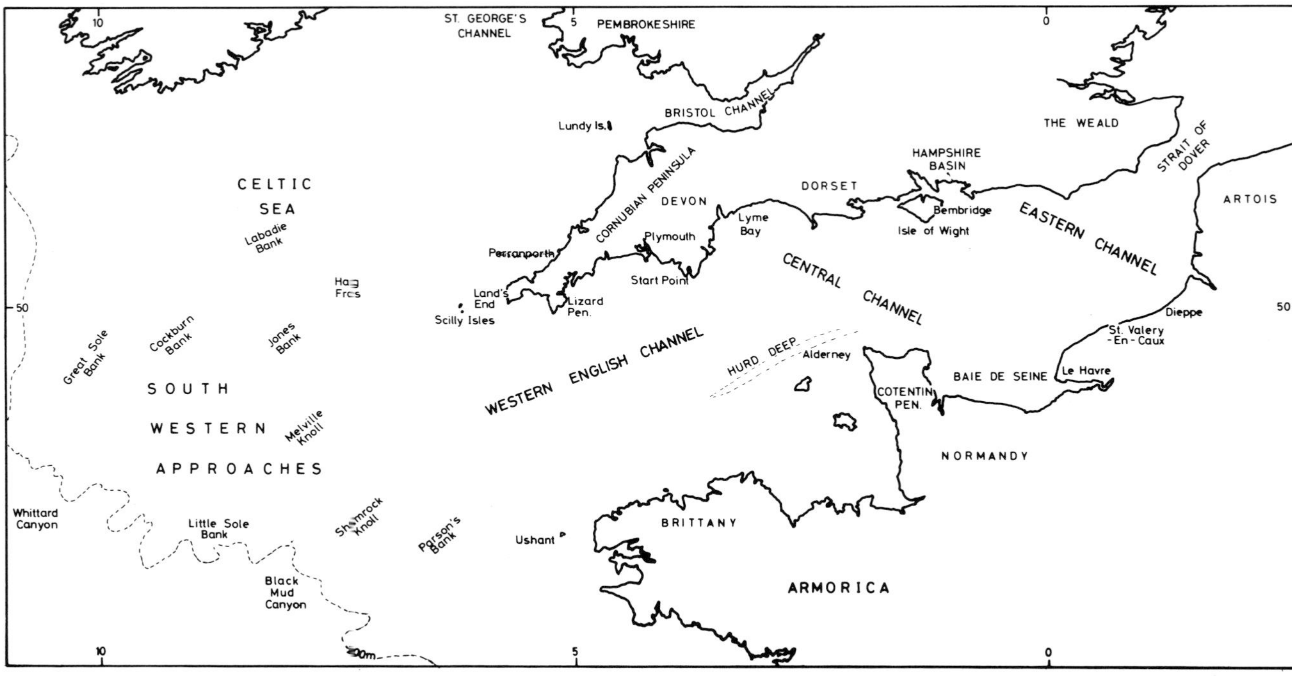

Fig. 5.1. Location map.

marine coastal erosion during periods of lower sea-level during the Late Tertiary and especially during the Pleistocene glaciations (Cooper, 1948; Wood, 1974, 1976; Donovan and Stride, 1975).

Though the general gradient of the sea floor is very low, there are quite considerable local contrasts in the nature of the surface of the sea floor in the area. In addition to the presence of banks and depressions, the English Channel is characterised by extensive areas of solid rock outcrop on the sea floor. Such rock floors occur mainly in the middle part of the English Channel where it is most constricted by the Cotentin Peninsula. This constriction, together with other influences (see section on Tides Vol. II, Chapters 9 and 13), increases the tidal currents sufficiently to maintain a virtually clean-swept rock floor (see also section on Holocene sediments).

In the Celtic Sea, by contrast, there are few areas of clean-swept solid rock. Exceptions occur where irregular topography is associated with granite masses, such as the Haig Fras granite, or off the Atlantic coast of the Cornubian Peninsula, where marine erosion of resistant Palaeozoic rocks has created wide marginal platforms. The erosion producing these platforms is due primarily to the full exposure of this coast to wave activity during severe Atlantic storms, not only at the present time, but during former periods of slightly lower sea-level (Wood, 1974, 1976). Elsewhere in the Celtic Sea, the solid rocks are buried beneath glacially derived sediments, which may range up to 100 m thick in places, as shown on continuous seismic profiles.

Solid rocks again form extensive outcrops on the floor of the Bristol Channel (Donovan et al., 1961; Lloyd et al., 1973) due to the very strong currents (see Admiralty Tidal Stream Atlas and Vol. II, Chapters 9 and 13).

Though echograms across the continental shelf in this area can be monotonously featureless, this belies the great diversity in surface sediment type and the variation in the solid rock geology of the sea floor. The present low regional gradient of the sea floor is not of simple origin. For instance, the current swept rock floor of the mid-English Channel merges westwards, without a major break, into a region of Pleistocene and Recent sediments. These sediments rest on the continuation of the solid rock surface which dips west-south-westward to the continental margin.

These superficial sediments form a wedge, becoming thicker, and of progressively younger age, towards the west.

The surface of the wedge of unconsolidated sediments is being readjusted in response to the present hydrodynamic regime of tides and waves (Stride, 1963; Belderson and Stride, 1966; Channon and Hamilton, 1976). Readjustment is taking place similarly over most of the Celtic Sea (McCave, 1971; Hamilton et al., 1974, 1975b). See Section below on Holocene Sediments, and Chapter 8, this volume.

The continental slope

The shelf break at the outer edge of the continental shelf is remarkably well defined in this area. The break occurs at almost 200 m depth, and below this the continental slope leads down to the continental rise and the adjacent abyssal plain, at depths of > 4,000 m. Hence, the continental slope involves a change in relief of about 3,800 m so it is a major topographic feature. As is generally the case, the continental slope in the South Western Approaches is steeper in the upper portion, the Upper Continental Slope, whilst the gentler lower slope is termed the Continental Rise. The features of the area are shown in Fig. 5.2.

The direction of the continental margin and slope changes significantly at Goban Spur. The continental margin runs almost due north from Goban Spur, to form the eastern boundary of the Porcupine Sea Bight (see Chapter 4). Here the gradient of the continental slope into the sea bight is very low.

South of Goban Spur, the continental margin swings to the south-east, across the western continuation of the English Channel, and continues into the Bay of Biscay. The gradient of the continental slope changes noticeably between these latter two areas. West of the English Channel the width of the slope embraces a horizontal distance of 55–75 km, with an average gradient of about 1 : 16, whilst the slope in the Bay of Biscay has a gradient of about 1 : 10.

In both these areas, the continental slope is incised by submarine canyons. Paradoxically, the largest and deepest submarine canyons do not occur on the steeper gradients in the Bay of Biscay but on the relatively more gentle slopes to the west of the English Channel. No major canyons occur on the very gentle slopes to the north of Goban Spur.

Small scale charts (Day, 1959; Brenot and Berthois, 1962; Hadley, 1964a; Laughton et al., 1975), and detailed surveys of Black Mud Canyon (Francis, 1962) and of Whittard Canyon by the Bristol Group, show that some major canyons have tributary systems in their upper reaches, like a river pattern. However, these tributary systems are virtually confined to the upper continental slope, for even the largest of the canyon

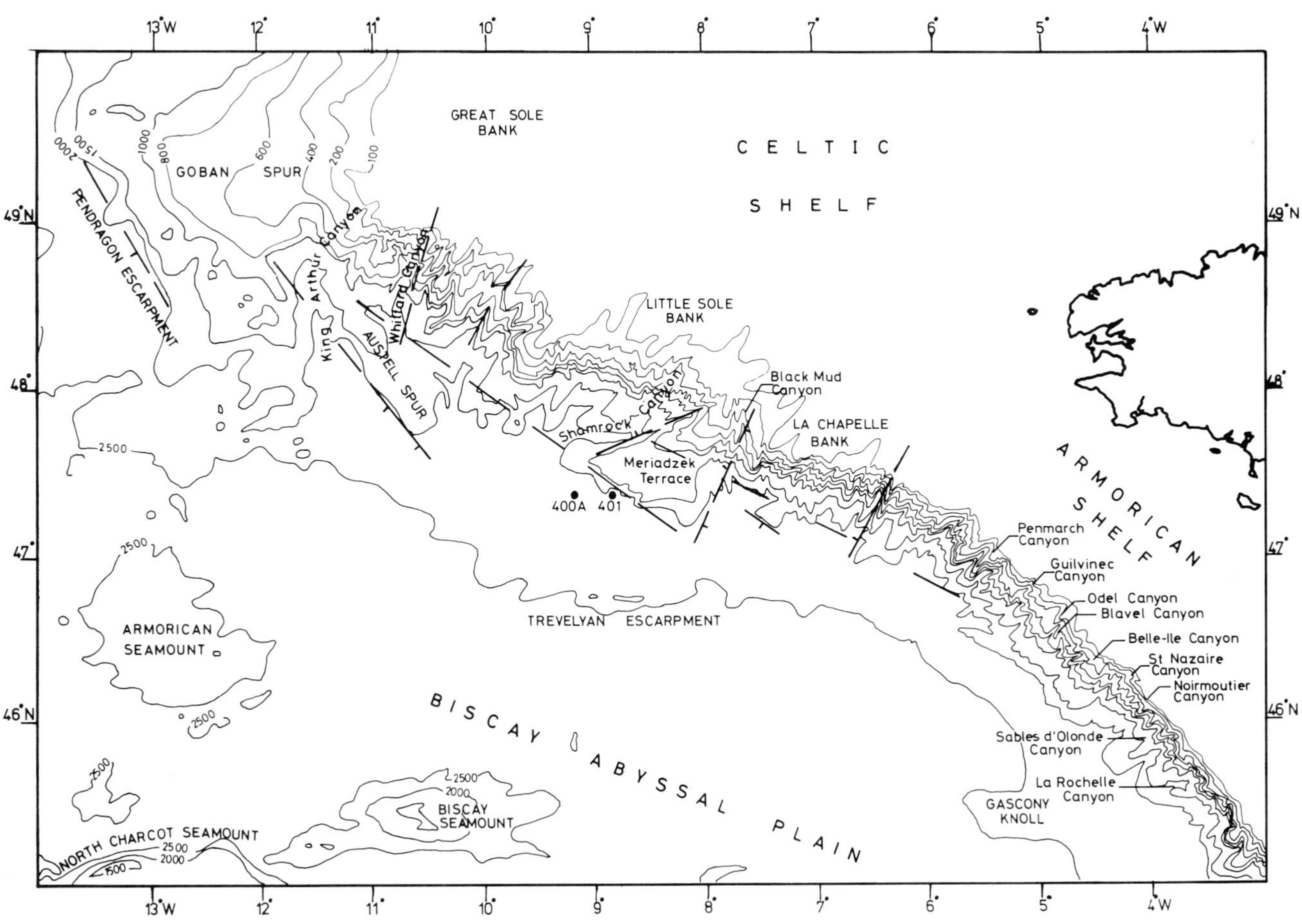

Fig. 5.2. Features of the continental margin of the South Western Approaches to Britain (modified after Laughton et al., 1975). Note the absence of submarine canyons north of Goban Spur, but the presence of large canyons south-west of the Celtic Shelf, and smaller canyons on the western edge of the Armorican Shelf. Heavy lines are probable faults. Depths in fathoms.

systems cuts back, at the present time, only a short distance (12–14 km) into the surface of the continental shelf.

These canyon heads form steep-sided amphitheatres, from which samples of Miocene and Eocene limestones have been dredged. These horizons are widely recognised in continuous seismic profiles across this continental margin. The shapes of the canyon heads are shown clearly in Asdic records (Belderson and Stride, 1969; Belderson and Kenyon, 1976). They liken the canyon heads to bad-land topography and suggest this is typical of an active canyon, i.e. one still acting as a transporting channel for sediment down the continental slope.

It is interesting to note that none of these large canyons in the South Western Approaches lies adjacent to a present land drainage system, as is frequently the case elsewhere in the world. The possible mode of origin of the canyons is discussed in the section on Buried Channel systems.

The downslope gradient of the canyons differs along their course, as shown in profiles of Black Mud and Whittard canyons (Fig. 5.3A). Changes in gradient are probably related to the occurrence of harder and softer rocks into which the canyons are eroded. No reversed gradients or barred basins have been detected in the upper or middle reaches of the canyons but gradients decrease markedly on the continental rise. Valley sides are remarkably steep, with slopes in solid rock reaching up to 15°–20° and canyon walls may rise 300–400 m above the canyon floor. The width of the canyon floors generally increases down canyon from the narrow floors of the V-shaped upper canyon to broad floors, up to 2 km wide, on the continental rise. A transverse profile in the Whittard Canyon at 1100 m depth (Fig. 5.3B) shows a distinct terrace about 50 m wide and an incised channel. Incised channels also occur in the broad canyon floor at the top of the continental rise.

Generally, the canyons follow a slightly sinuous course downslope. The Whittard Canyon, however, turns abruptly at 3500 m depth from a south-south-west course to the south-east. This change in direction is probably due to the influence of a NW–SE fault. The Meriadzek Terrace is an upfaulted block with a scarp on its south-west side (Day, 1959; Smith and Van Riessen, 1973). Day (1959) traces the Pendragon Scarp south-east from Goban Spur and this could well be on the same fault line as the scarp of the Meriadzek Terrace. This line is cut by two large embayments, which may also be fault controlled.

Solid-rock geology of the sea floor

Introduction

Before information on the sea-floor geology of the English Channel and Celtic Sea was available, it was not possible to understand easily the relationships of southern England to northern Europe, nor of western Britain to Ireland. With increasing knowledge of the geological history of these offshore areas, many of the gaps in the history of not only the offshore areas, but also the adjacent land areas are filled in, for the geological record preserved in offshore areas tends to be more complete than that on land. This is particularly so in the younger (Upper Mesozoic and Tertiary) parts of

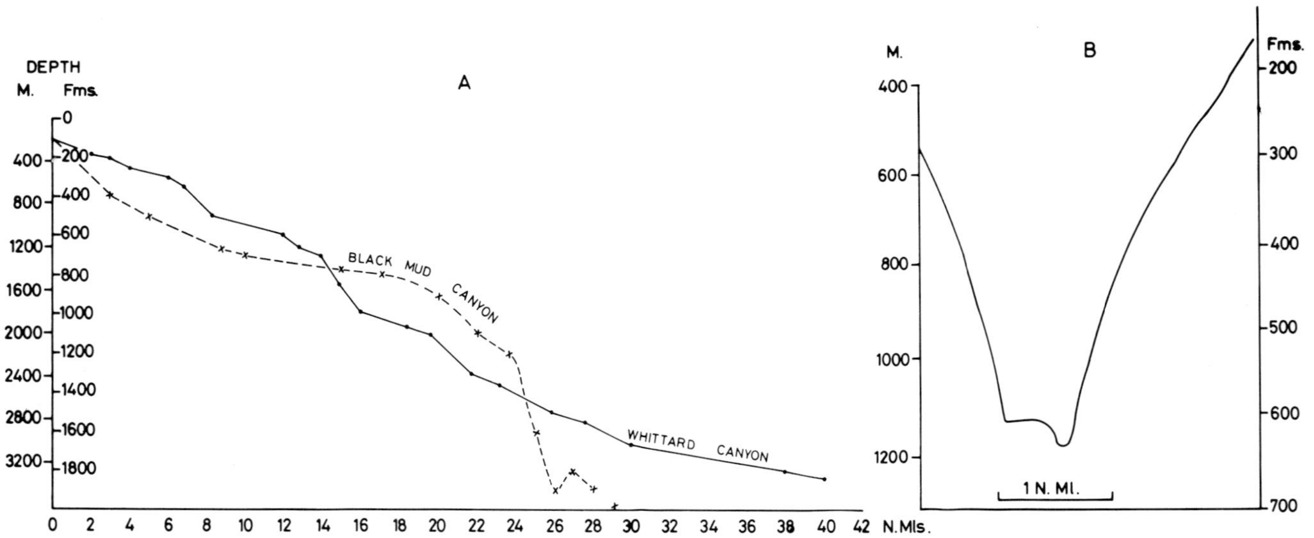

Fig. 5.3. A. Changes in gradient down the axes of the Whittard and Black Mud Canyons, South Western Approaches. B. Transverse profile across the Whittard Canyon at a depth of almost 1,200 m, showing the terrace and incised channel. The locations of Whittard and Black Mud Canyons are shown on Fig. 5.2.

the stratigraphic succession. The more complete record results from the dominance of subsidence in the offshore basins in contrast to the dominance of elevation in the land areas.

This contrast in subsidence and elevation is one of degree however, for both land and offshore areas have suffered complex deformational histories. Both areas have undergone periods of subsidence with the deposition of sediments, or elevation and erosion, producing widespread unconformities in the stratigraphic column. The pattern of deformation, leading to structural highs and basins, is the result of large stress–strain patterns imposed from outside the region. In the most general sense, such tectonic forces arise from the relationships between lithospheric plates or platelets, as well as the stress conditions prevailing within these segments.

The geological history of the English Channel and Celtic Sea can be traced with some certainty from Triassic times onwards. This is the history of intra-cratonic deformation (Ziegler, 1975b) of the European plate fragment during successive orogenic episodes induced by lithospheric readjustments. Intra-cratonic deformation applies particularly to the mid and eastern English Channel and towards the Bristol Channel in the Celtic Sea. The geological history of the western English Channel and the western Celtic Sea is influenced also, in post-Atlantic-rifting times, by subsidence–elevation episodes of the newly formed passive continental margin. Thus the western continental shelf has a basement structure, in part inherited from pre-rifting times and subsequently modified by rifting and post-rifting events. It is the effects of these later events (Upper Mesozoic and Tertiary) which are the main theme of the discussion which follows.

Results of investigations into the regional sea-floor geology are much more numerous for the English Channel than for the Celtic Sea. Contributions, amongst many others, on the shallow subsurface geology of the Channel are by Andreieff et al. (1970, 1972, 1975), Boillot (1964), Boillot et al. (1972) and a series of papers by the Bristol University Group, including Whittard (1962), Curry et al. (1962, 1965, 1970, 1971, 1972), Hersey and Whittard (1966), and Smith and Curry (1975). Regional syntheses, including this area, are made in Owen (1974), Roberts (1974), Blundell (1975) and Whitbread (1975), whilst comprehensive bibliographies on the geology of the English Channel (or La Manche) are given in Smith et al. (1972) and Hamilton et al. (1975a).

The English Channel

Introduction

From the geological map of the English Channel (Fig. 5.4) it is apparent that there are three areas, each having a distinctive structural grain. These areas have been called the eastern, central and western basins (Boillot and Musellec, 1975) or provinces (Smith and Curry, 1975). In each province, the sea-floor geology has a different relationship with its adjacent landward area. The general structure of the provinces is shown in Fig. 5.5.

The oldest rocks on the sea floor of the Channel occur on the flanks of the western province, offshore from Brittany and from the Cornubian Peninsula, but the major part of the whole Channel is floored by Jurassic, Cretaceous and Tertiary sediments. Samples of these rocks generally bear a very close lithological similarity to their stratigraphic equivalents on land, suggesting that during periods of widespread subsidence and deposition, there was little difference between present land and sea areas. Major differences in sedimentary history occur where relative uplift of blocks or massifs occurred. Two massifs, the Cornubian Peninsula and Armorica, both now exposed and in the western province, have a long history as positive areas and have acted as sources of sediment to the adjacent basins. Both these positive areas are characterised by metasediments and intrusion of granites, which are now exposed in the south-west of England, in Brittany and Normandy. Offshore, granites now form outcrops on the sea floor at Roscoff, between Start Point and Cotentin Peninsula, many small outcrops north-west of Lands End, the Scilly Isles, and Haig Fras in the Celtic Sea. Gravity and magnetic data also suggest the presence of buried granitic masses in the South Western Approaches (see Chapter 4). These granite masses are all associated with structural highs, which play an important role in the sedimentary and structural history of the offshore basins.

The geological history of the three English Channel provinces have much in common as well as having some notable differences. Some major points are discussed in the following sections.

Eastern English Channel

In the eastern province, a single shallow syncline extends across the Channel, joining the structures on the coasts adjacent to Brighton and Dieppe (Auffret et al., 1975; Curry and Smith, 1975). The syncline is aligned NW–SE and exposes a thin succession of Palaeogene sediments (380 m) and Cretaceous sediments. The latter

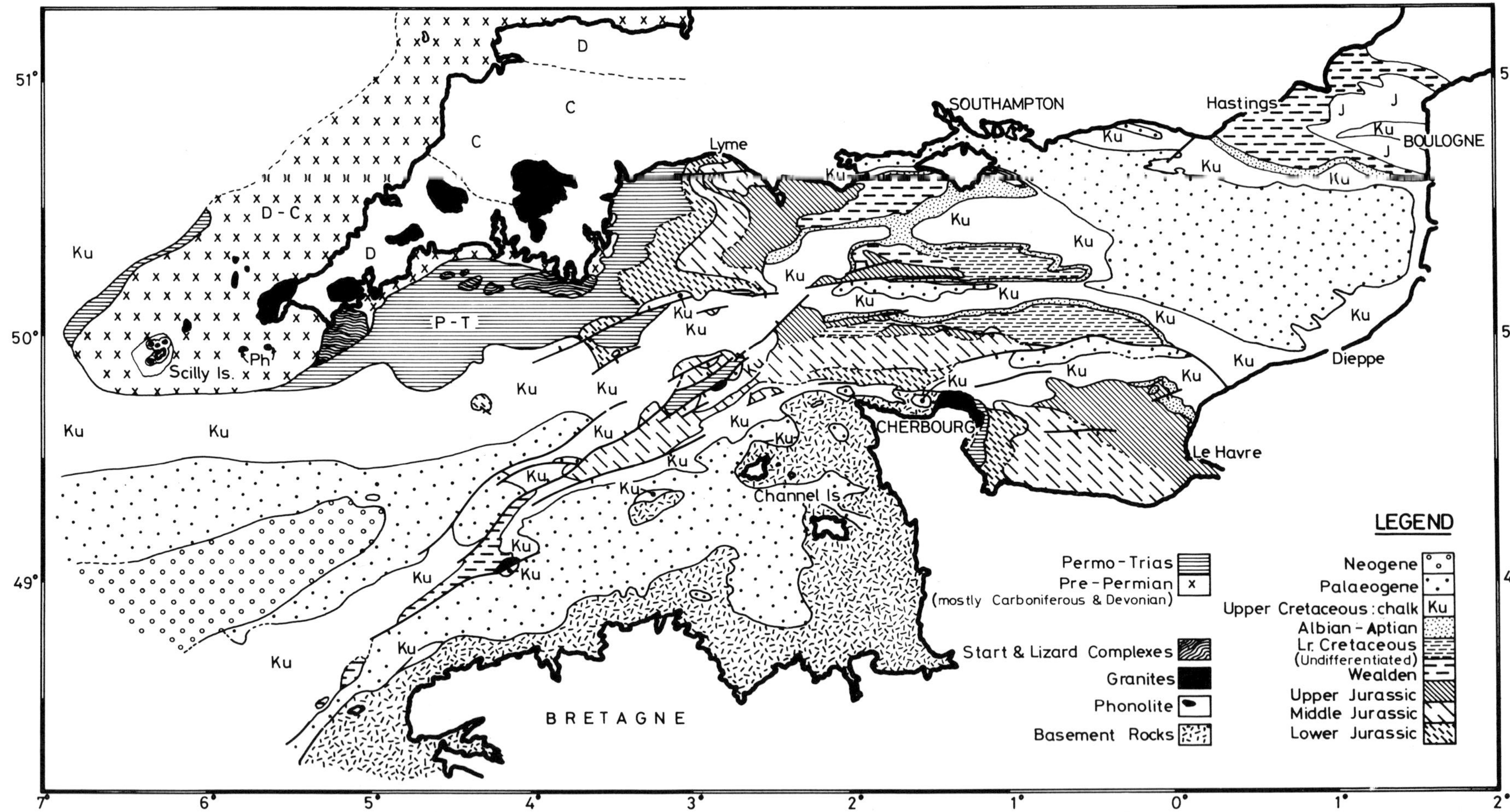

Fig. 5.4. Simplified geological map of the English Channel (after many authors), showing that the direction of the structural trend in the eastern Channel is NW–SE, in the central Channel it is E–W, and in the western Channel it is NE–SW.

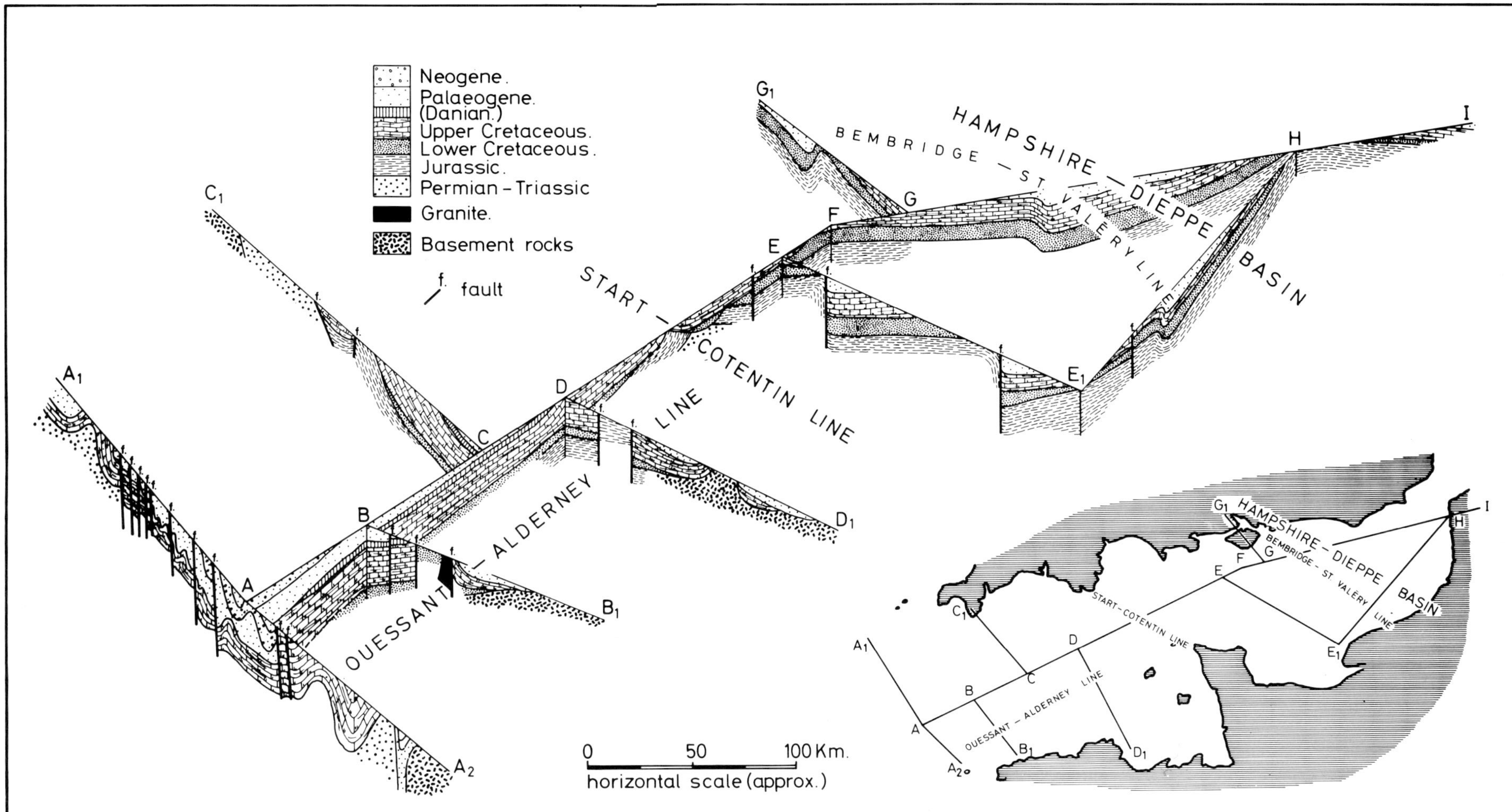

Fig. 5.5. Fence diagram showing the general geological structure of the English Channel (modified after Smith and Curry, 1975).

include Chalk, Gault and Upper Greensand, and the total thickness increases from 250 m in the south-east to 420 m in the north-west. Lithologically, they are similar to the sediments in the Paris Basin.

The syncline is asymmetric for it splays to the south-east due to the contrasting directions of the north-eastern and south-western boundaries. Toward the Strait of Dover, the syncline is bounded by the ESE–WNW Weald–Artois anticlinorium. The south-western boundary is the Bembridge–St. Valery line (Curry and Smith, 1975). This is a strong downturn formed by the offshore continuation of the Isle of Wight monocline, which first continues to the east and then curves gently to the south-east toward St. Valery-en-Caux. The downturn on this flexure is 1250 m at Culver Cliff and decreases south-eastwards. Near mid-Channel, the feature bifurcates to define a slightly uplifted block. The northern flexure, with downturn to the north, continues towards Dieppe, whilst the southern flexure, with downturn to the south, continues to St. Valery-en-Caux. Both these flexures join important NW–SE structural lines in France (Smith and Curry, 1975).

The syncline in the eastern English Channel is part of the Wessex and Anglo-Paris basins (Kent, 1975). Though there is little evidence about the Permo-Triassic history, virtually uninterrupted deposition took place in the central parts of the basin during Jurassic times (Strachan, 1913; Lees and Taitt, 1945). After minor uplift in the Portlandian, major uplift, producing folding, faulting and later erosion, occurred in the Aptian. The Albian Chalk succession was followed by regression. Marine transgression occurred in the Lower Eocene. Cuisian and Lutetian sediments (Auffret et al., 1975) are overlain unconformably by Bartonian beds closely similar to those of the type locality. Though the Mesozoic sediments are similar throughout the Hampshire–Paris Basin, the Tertiary sediments become confined to the onshore Hampshire Basin and also the eastern English Channel Basin.

The Eastern Channel syncline has a NW–SE orientation but to the north it merges with the E–W Hampshire Basin. The latter is part of the Hampshire–Weald Basin, lying on the southern margin of the London–Brabant Massif. The offshore structural grain parallels that of the Lower Palaeozoic basement of the London-Brabant platform in the Strait of Dover area, as shown by gravity surveys (Shephard-Thorn et al., 1972).

The confluence of the Variscan Front with the platform can be traced in the Pas de Calais area (Wallace, 1968). Shephard-Thorn et al. (1972) suggest that the front may cross the Channel and swing westward beneath the northern structures of the Late Tertiary Weald anticlinorium. Continued movement on the NW–SE Lower Palaeozoic basement lineaments in pre- and post-Jurassic times has been demonstrated by Shephard-Thorn et al. (1972) and in Jurassic strata in the Strait of Dover (Boillot and Musellec, 1975).

Laramide and Alpine movements, resulting from the opening of the North Atlantic from Mid-Cretaceous times and the relative closure of the Mediterranean Tethys during the Tertiary, gave rise to compression, wrench faulting and inversion of basins on the north flank of the Hercynides (Ziegler, 1975b). Old tectonic lines, especially those parallel to the NW–SE Tornquist line, were reactivated during these readjustments. They are probably the structural controls leading to the formation of the eastern English Channel basins.

Central English Channel

Structural pattern. The characteristic feature of the central province is the presence, to the north of Cotentin Peninsula, of three E–W trending monoclines, involving Jurassic, Cretaceous and Eocene sediments on the sea floor.

The E–W trend of these structures parallels those of south Dorset and indeed those farther westward, in Devon and Cornwall, as will be discussed later. The E–W trend is maintained southwards across the Channel until the northern end of the Cotentin Peninsula is reached. This peninsula is on the northern edge of the Armorican Massif, as shown by the magnetic pattern. South of this boundary, the N–S outcrop pattern of Triassic and Jurassic sediments in the Baie de Seine is controlled by the tectonics to the east of the Armorican Massif rather than the later movements in the Channel. N–S outcrops of Triassic and Jurassic sediments also occur in the west of Lyme Bay and may have a similar relationship to the Cornubian Massif.

Seismic and gravity evidence (Avedik, 1975; Bacon, 1975) shows the presence of a deep basin in the central Channel, the Central Channel Basin of Kent (1975). The western boundary of this basin is the NW–SE trending basement high, the Anglo-Norman ridge (Kent, 1975) forming the important Start–Cotentin line (Smith and Curry, 1975). Larsonneur (1972) placed this boundary slightly to the east, on the Portland–Caen line, to pass through the easterly dipping Triassic–Jurassic successions in Lyme Bay and Baie de Seine, whilst Bouysse et al. (1975) suggest the presence of a structural high between Start and Jersey. The Anglo-Norman ridge is characterised by low gravity (Bacon, 1975) due to the shallow depth to basement and the presence of granite intrusions. The eastern boundary of the Central Channel Basin is probably the less well defined Bembridge–St.

Valery line, also with a NW–SE trend.

Though the present surface structure of the Central Channel Basin is clearly E–W, it seems possible that the basement structure could be influenced strongly by NW–SE elements. The Triassic–Jurassic sediments of the Baie de Seine and western Lyme Bay are pre-Atlantic-rifting sediments. They could well have formed in the western part of a NW–SE oriented basin, forming the western extension of the Anglo-Paris Basin. At this time, both the Cornubian and Armorican massifs were probably positive areas. The NW–SE alignment continued into Cretaceous times (Donovan, 1972; Larsonneur, 1972; Destombes et al., 1974).

The present E–W structures in the central province involve Mesozoic and also lower Eocene (Cuisian) sediments, so that post lower Eocene deformation could imprint this pattern. However, younger tectonic movements generally reactivate older lines of weakness, for Hercynian deformation had imparted an E–W structural grain to the basement rocks. The presence of E–W lineaments in the gravity pattern (Bacon, 1975) suggests that relief in the basement is strong and such basement structure could give rise to the monoclines in the younger sediments by draping over the deeper structures. Major E–W monoclines have downturns to the south (Smith and Curry, 1975) and the eastern end of each flexure either abuts or swings into a NW–SE structural feature. Such arcuate structural lines are to be regarded as the intersection of persistent tectonic directions and do not, per se, indicate extensive thrusting movements.

Stratigraphic succession. The sequence of sediments in the province is known in some detail. On land, lower Palaeozoic rocks form outcrops in the Start Point area and in the Cotentin Peninsula. These are the exposed eastern margins of the Cornubian and Armorican massifs respectively. Triassic and Jurassic sediments dip away eastward from both these massifs into western Lyme Bay and Baie de Seine. The Triassic sediments include breccias and conglomerates, sandstones, gypsiferous and dolomitic marls, these being the continental deposits of the progressive denudation of the adjacent uplands during possible Bunter and Keuper times. The thickness of the upper Triassic sediments decreases towards the south (Donovan, 1972).

Jurassic sediments, of similar lithologies to coeval strata on land, form extensive areas of outcrop on the sea floor in this central province. Outcrops in Lyme Bay (Donovan and Stride, 1961) and to the north and north-east of Cotentin Peninsula (Larsonneur et al., 1975) are separated by a narrow strip of Upper Cretaceous, which continues the NE–SW direction of the western province structures. The continuation southwestwards of marine Liassic sediments offshore from Start Point, and the presence of Middle Jurassic sediments north of the Channel Islands, indicates that the Start–Cotentin ridge was submerged at these times. The persistence of this structural high produced a thinning over it of the sediments on the western side of the Anglo-Paris Basin, in a similar way to the thinning of sediments in the western province (Curry et al., 1970). Oscillations in the submergence and emergence of this ridge appear to be an important control of connections, migration routes and sedimentation patterns between the Anglo-Paris Basin and the west.

The stratigraphic succession in the western part of the Anglo-Paris Basin can be summarised as follows:

(1) Sedimentation during the Jurassic until the Bathonian. Deposition of clay alternated with sandy and shallow-water carbonates, possibly as a result of eustatic changes.

(2) Regression in the Portlandian gave rise to the lacustrine Purbeckean facies, which now crops out between the Isle of Wight and Cherbourg (Larsonneur and Rioult, 1969).

(3) Emergence during the Lower Cretaceous and the deposition of Wealden and other sediments, which now occur extensively offshore as a belt to the east of Jurassic sediments. An important feature of the pre-Albian earth movements was an easterly tilt, which probably continued into Tertiary times (Donovan, 1972), together with folding along E–W axes.

(4) Marine conditions, in which the Lower Greensand was deposited, transgressed the Central Province in the Aptian.

(5) During the Upper Cretaceous, the sea again transgressed the area. The successive overstep to the west along the Dorset–Devon coast of England during the Albian is widely known, but the transgression in the Cotentin–Britanny areas occurred during Cenomanian–Maastrichtian times. In spite of the widespread and generally uniform nature of these Chalk deposits, local variation is caused by the growth of tectonic swells (Drummond, 1970; Larsonneur et al., 1975).

(6) Withdrawal of the sea from the Anglo-Paris Basin by the end of the Campanian (Larsonneur et al., 1975).

(7) The Lower Eocene transgression, so extensive in the Lutetian in the western province, is represented by a transgression during the Ypresian over a restricted area in the central province. Eocene strata now form E–W outcrops following the re-activation of E–W structural lines during or after Oligocene times. Overturning to the north is evident in some of these structures.

Thus the sedimentary and structural history of the central province is controlled probably by an initial

NW–SE orientation of the western edge of the basin, together with an E–W regional structural direction of Hercynian age. This latter has been re-activated repeatedly until Tertiary times, so that the present structures are the surface manifestations of differential movement of basement blocks buried beneath Mesozoic and Tertiary sediments.

Western English Channel

Introduction. The western portion of the Channel comprises the area west of a line joining the Cotentin Peninsula and the Start–Bolt peninsula on the southern coast of England. Westwards of this line, the sea-floor geology is dominated by a major NE–SW syncline, the western English Channel Basin (Fig. 5.6).

The western part of the English Channel has a structural unity which is distinctive from the central and eastern parts of the English Channel. Also, the geology of the land to the north and south of the western English Channel gives little clue as to that of the intervening sea-floor.

On the northern side, the Cornubian Peninsula, which trends in a NE–SW direction, has a succession of E–W striking Devonian and Carboniferous strata, intruded by Variscan (Hercynian, Armorican) granites. Ultra-basic and metamorphic complexes occur at the Lizard and gneisses and schists occur in the Eddystone–Start Point area (Phillips, 1964). On land, the earlier Palaeozoic succession of the Cornubian Peninsula is overlain unconformably to the east by Permo-Triassic and younger sediments, north of Lyme Bay.

The Armorican Massif, forming the southern border of the Channel, is comprised of Precambrian and Palaeozoic sediments intruded by granitic rocks, the whole being strongly metamorphosed and tectonised. The structural trend is predominantly E–W to ENE–WSW, with prominent NW–SE faulting in the west of the Armorican Massif.

Sea-floor geology. Folded Devonian and Carboniferous rocks crop out for a short distance offshore from the south coast of Cornwall, around Land's End and westwards beyond the Scilly Isles (Curry et al., 1970). On the south side of the Channel, Devonian palynomorphs have been recognised (Le Fort and Deunff, 1970) in offshore Armorican outcrops.

The post-Hercynian development of the English Channel is discussed by Curry et al. (1971), Donovan (1972), Larsonneur (1972) and Le Fort (1975), among others.

Westwards from Start Point, Permo-Triassic strata occur to the south of the Devonian–Carboniferous offshore outcrops, apparently thickening rapidly south from Plymouth. No rocks of Permo-Triassic age are recognised on the sea floor immediately to the north of Brittany, though King (1954) suggests they occur at depth beneath the major syncline.

Jurassic strata are of restricted occurrence in the western English Channel, when compared with their extensive sea-floor outcrop in the central part of the Channel. To the west of the Start-Cotentin line, Middle Jurassic sediments are exposed in and adjacent to the Hurd Deep. South of Start Point and south of Plymouth small areas of fault-bounded Lower Jurassic sediments occur and these are of similar lithology to onshore outcrops. No sea-floor outcrops of Jurassic are known in the English Channel westward of 4°30′W, the location of the inlier south of Plymouth, but Jurassic strata are to be expected to occur westwards towards the shelf margin, beneath Cretaceous sediments.

In the central part of the western English Channel a broad syncline is comprised of Cretaceous and Tertiary strata. Lower Cretaceous strata are exposed on an upfaulted block to the south-west of the Hurd Deep. These Lower Cretaceous sands have a terrestrial aspect, with a microflora comparable with the Wealden of England, the Low Countries and Germany. Though there is no clear evidence of the stratigraphic relationship between the Lower and Upper Cretaceous beds, they are probably unconformable. Richly glauconitic marine sand, of doubtful Albian age, occurs in the cores of anticlines in Chalk, north of Ushant (Andreieff et al., 1972).

The Upper Cretaceous is strongly transgressive and rests with marked unconformity on earlier beds (Fig. 5.7A). Chalk is overlain by Danian and Tertiary strata on the north and south of the western English Channel synclinal axis. Thus, the submarine outcrop of Chalk extends far westwards of the last onshore outcrops at Beer, in Devonshire, and swings north around the Scilly Isles. The lithology of the Chalk succession over this area retains a close similarity with outcrops in southern England.

The thickness of the Chalk varies between 600 m and 700 m. There is strong evidence of a transgression from west to east. In the west, the Chalk samples are of Santonian age or older. In the east, Chalks older than Santonian are rare, suggesting a condensed or incomplete sequence as in Devon. The Chalk seas also transgressed on to and across the basement off north Brittany (Boillot and Le Calvez, 1961; Curry et al., 1970). Since Santonian rocks predominate in the west of the area, there must have been extensive removal by erosion of the younger Chalk zones or else this was an area of non-deposition.

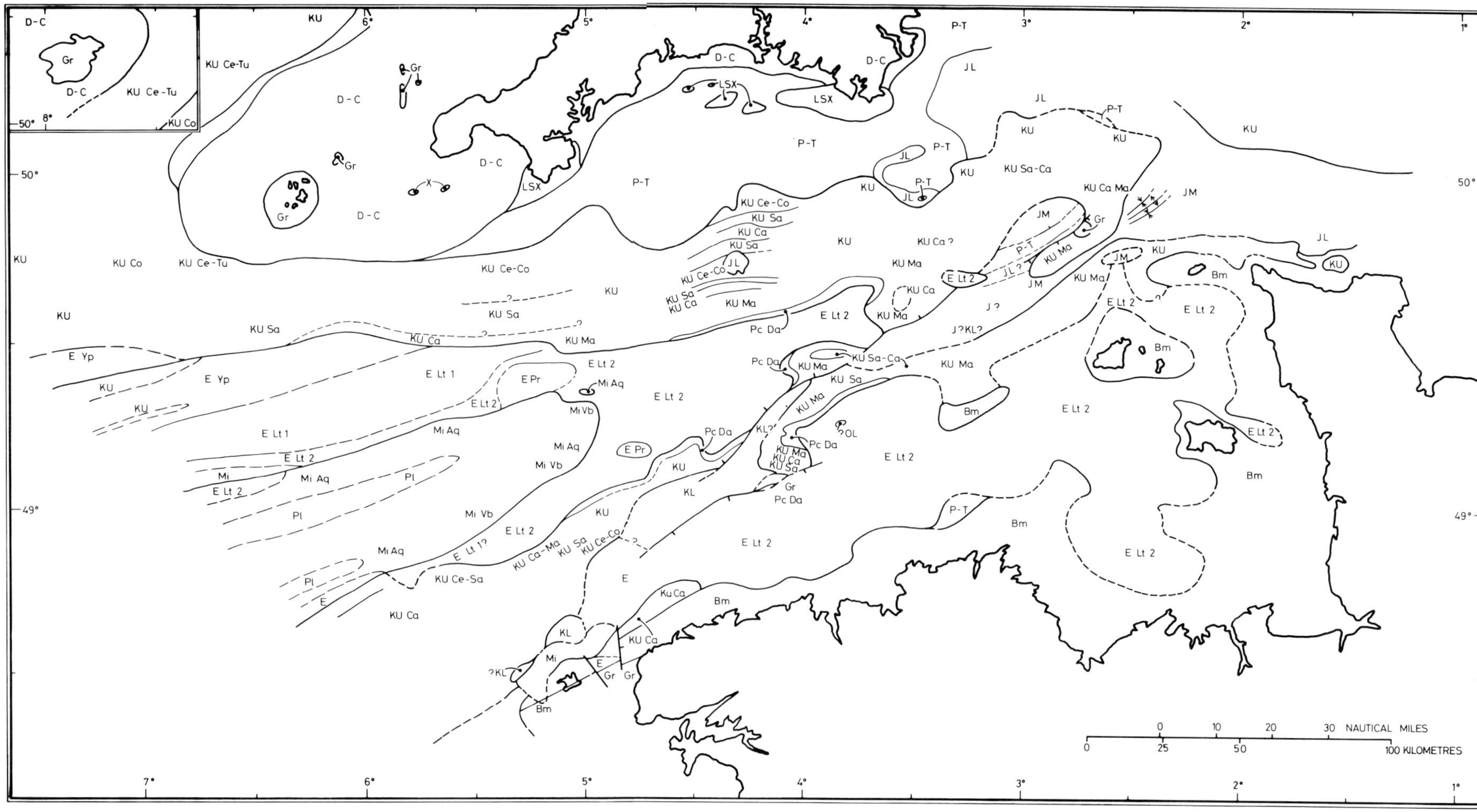

Fig. 5.6. Map of the pre-Pleistocene geology of the western English Channel (modified after Curry et al., 1970). *Pl* = Plio-Pleistocene, (unconformity); *Mi* = Miocene; *Vb* = Vindobonian; *Aq* = Aquitanian, (unconformity); *Ol* = Oligocene; *E* = Eocene; *Pr* = Priobonian; *Lt2* and *Lt1* = Upper and Lower Lutetian; *Yp* = Ypresian, (unconformity); *PcDa* = Palaeocene Danian;*KU* = Upper Cretaceous; *Ma* = Maastrichtian; *Ca* = Campanian; *Sa* = Santonian; *Co* = Coniacian; *Tu* = Turonian; *Ce* = Cenomanian, (unconformity); *KL* = Lower Cretaceous; *JM* = Middle Jurassic; *JL* = Lower Jurassic; *P-T* = Permo-Triassic, (unconformity); *D-C* = Devonian, Carboniferous; *X* = phonolite; *Gr* = granite; *LSX* = Lizard–Start complex; *Bm* = basement.

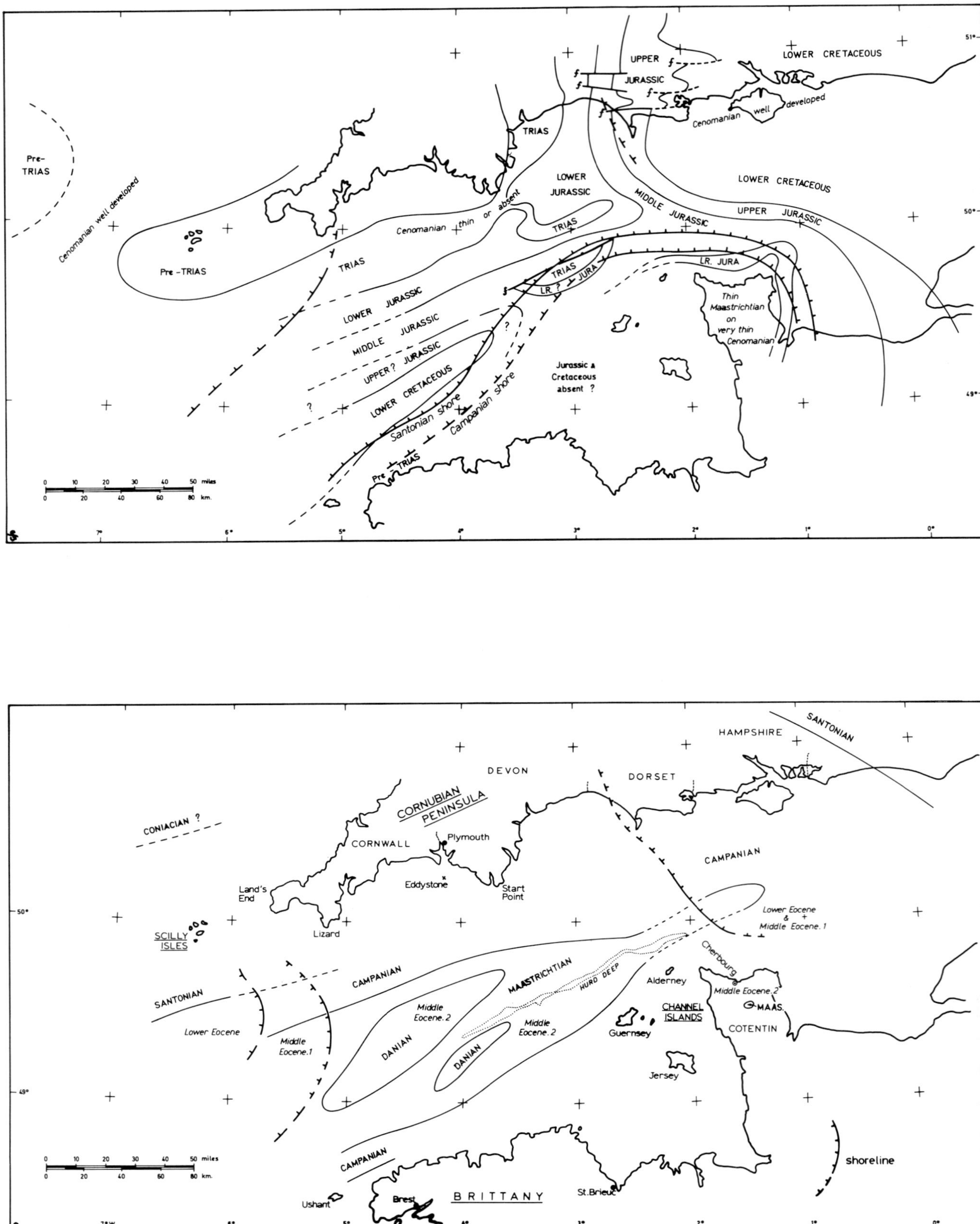

Fig. 5.7. A. The sub-Chalk unconformity in the western English Channel (after Curry et al., 1971). The surface over which the Chalk seas transgressed has considerable stratigraphic and structural diversity, due to Albian–Aptian deformation and erosion.
B. The sub-Eocene unconformity in the western English Channel (after Curry et al., 1971). The intensity of pre-Lutetian (Eocene) deformation is less than that of pre-Chalk movements, shown in Fig. 7A.

Though there is no clear evidence, as yet, of the relationship between the Lower and Upper Cretaceous, the succeeding Danian and Thanetian follow conformably on the Upper Maastrichtian. The Danian beds, which reach a thickness of 120 m, are mostly biomicrites.

A regional unconformity occurs at the base of the overlying Eocene, indicating extensive folding, uplift and erosion prior to transgression by Eocene seas (Fig. 5.7B). Like the Chalk transgression, the Eocene transgression was also from the west to the east. Ypresian and Lutetian I strata, occurring to the south and south-west of the Scilly Isles are locally unconformable (Andreieff et al., 1975) whilst extensive areas of Lutetian II occur to the east, extending as a thin transgressive sheet on to the northern margin of the Armorican platform. This apppears to be a major transgressive phase.

The thickness of Eocene strata is greatest in the west, attaining about 500 m. Here the lithology is almost pure biosparite and biomicrite whilst the carbonate content generally decreases towards the east. The Eocene sediments of the westernEnglish Channel suggest a carbonate producing embayment, in contrast to the coeval clays and sands of the Anglo-Paris Basin.

Wright and Murray (1972) have interpreted the Eocene foraminiferid faunas of the western Channel as the deposits of a shallow, nearshore hyposaline shelf sea, through hypersaline shelf to hypersaline lagoon. From detailed comparisons of the foraminiferid faunas of the English Channel, Hampshire and Paris basins, Murray and Wright (1974) related the differences to a lack of terriginous input in the Western Approaches area.

Oligocene sediments are known from a small area, centred on 49°10′N 3°50′W, between the Hurd Deep and the Lower Cretaceous outcrop. These Oligocene sediments include a freshwater limestone (Curry et al., 1970). To the southwest, at 4°50′W, marine sediments with a Lower Oligocene microfauna have been recognised (Andreieff and Le Fort, 1972).

Miocene sediments outcrop extensively to the west of 5°W in the core of the major syncline. They lie unconformably on Eocene and Upper Cretaceous strata. Again, Miocene sediments are more calcareous and are older to the west, suggesting transgression from the west. Thicknesses range up to 120 m.

Plio-Pleistocene siltstones are folded in the major syncline (Curry et al., 1965). A recent seismic profile by the writer reveals the presence of two areas of infolded sediments, now considered to be of Pliocene age, as shown in section A_1–A of Fig. 5.5. These Pliocene sediments are characterised by widespread low angle cross-bedding.

Major unconformities in the offshore sequence occur at the following horizons:

(a) Between metamorphic basement and Lower to Middle Palaeozoic strata, as distinguished by seismic evidence (Day et al., 1956; Avedik, 1975).

(b) Between Devonian–Carboniferous and Permo-Triassic strata, due to Variscan (Hercynian or Armorican) tectonic events.

(c) Between Jurassic or Lower Cretaceous, and Upper Cretaceous strata, due to pre-Cenomanian (?Cimmerian) movements.

(d) Between Danian and Eocene sediments, due to Laramide tectonic events.

(e) Between Eocene and Miocene sediments, due to Alpine tectonic movements.

The most important of these unconformities is that beneath the Upper Cretaceous. This erosion surface cuts across the sedimentary fill of the basins and also their bounding horsts or ridges. The widespread Upper Cretaceous (Cenomanian) transgression deposited sediments of surprisingly uniform lithology and thickness, covering the structural and sedimentary diversity of the basins.

Geophysical evidence indicates that four basins occur in the metamorphic basement, along the present site of the western English Channel syncline (Avedik, 1975). The basin floors generally dip westward and the basins are separated by structural highs, directed to the north-west and south-east. In consequence, there is considerable variation in the total thickness of sediment along the axis of the regional syncline. This variation in thickness takes place largely in the Permian to pre-Chalk succession. The basin to the south of Plymouth, bounded by the Start-Cotentin ridge and the structural high between the Lizard and Ile de Batz (Le Fort, 1975) might possibly include a thick succession of pre-Westphalian beds as Matthews (1977) has demonstrated the presence of separate basinal development in the Lower Carboniferous. The sedimentary history of each small basin could well be quite parochial, being controlled by the local tectonic history. Some basins may well have become closed, especially in their early history.

The formation of the regional basinal system resulted from a tensional stress pattern which post-dates the compressive (Upper Carboniferous) orogenic phase of the Variscan orogeny. The Lower Permian tensional phase (Saalian) may be related to the initial phases of the developing rift mechanisms of the North Atlantic (Russell, 1976), giving rise to intra-cratonic depressions in which low-density (?)Permian evaporitic sediments were probably deposited (Avedik, 1975). Ziegler, 1975a), suggests that the main graben development took place in Triassic times, reactivating older lines of tectonic weakness as the North Atlantic Seaway (Hallam, 1971) developed. Continental sedimentation, probably with Keuper evaporities, continued during the Triassic.

Though the sediments deposited in these depressions are now deeply buried in the western part of the shelf, some indications of their possible nature can be gained from coeval rocks, where present, on land. In pre-Atlantic rifting times, the Grand Banks off eastern Canada and the present north-west areas of the Iberian Peninsula, lay adjacent to the South Western Approaches. In the Grand Banks, shallow water Jurassic sediments overlie thick (?)Jurassic salt (Ayrton et al., 1973), and this could extend into the South Western Approaches (cf. Pautot et al., 1970). The nature of the Jurassic is enigmatic however as, on land, there is a general decrease from east to west in sand content and grain size. Avedik (1975) in discussion, suggests it is unlikely that thick Mesozoic carbonate sections are present in the South Western Approaches. Though Lower Cretaceous sediments outcrop only in the eastern part of the regional syncline, Whitbread (1975) notes the recurrence of deltaic sandstones in the Celtic Sea and coeval sediments in Portugal, northern Spain and the Scotian Shelf of eastern Canada. Allen (1972) ascribes a western source for detrital tourmaline in the Wealden facies.

The line of rifting between the African and North American plates lay directly on the line of the South Western Approaches, so this was a potential line of rifting (Curry et al., 1971). This line was terminated by the opening of the Bay of Biscay (Williams and McKenzie, 1971).

The subsequent history of the newly formed shelf is influenced by subsidence (transgression) during periods of rapid ocean spreading rates (Upper Cretaceous, Lower Eocene and Miocene) and elevation during slow rates of spreading (Bott, 1971). These latter give rise to the unconformities noted above. The late Tertiary–Quaternary history of the shelf is discussed in a later section.

Structure of the western English Channel. The southern margin of this major syncline is the zone of faults, the Alderney–Ushant fault zone (Curry et al., 1971), which forms the north-western edge of the Armorican platform. This fault zone probably continues south-westwards to the continental margin, where magnetic highs suggest the presence of basic igneous rocks (Gerard, 1975).

Movements along this fault zone have, in the main, been vertical, giving rise to elongated, local horst and graben structures. Though the amount of vertical displacement shown by the latest movements is generally small, persistent subsidence to the north of the fault zone has produced the major basin of the western English Channel whilst to the south, the Armorican Massif has remained generally a positive feature, receiving only thin transgressive deposits. The south-westerly strike of the fault zone is parallel to other major faults within the basin as well as to structural trends in north-west Brittany.

The northern side of the western English Channel basin is also bounded by persistent faults, associated with the graben structure on the south side of the Cornubian structural high. Strong south-westerly oriented folds and faults appear to be absent from the Cornubian Peninsula, except in the southwesterly tip, south of the Perranporth–Pentewan line (Dearman, 1971). The orientation of the peninsula itself is NE–SW, possibly controlled by the granite bodies, which are probably intruded along a Caledonoid direction. This is parallel to the regional offshore trend. However, the dominant structural trend in the Palaeozoic strata of the Cornubian Peninsula is E–W. This E–W trend is similar to, and may be an extension of, the structural grain in the central province of the English Channel.

In addition to the NE–SW and E–W structural trends already noted, a NW–SE trend, with active dextral movements in Tertiary times, is well developed in both the Cornubian Peninsula and north-west Brittany. Andreieff et al. (1975) have mapped offshore faults with the latter trend in the region of Ushant, whilst Bouysse et al. (1975) recognise north-westerly trending faults in the western part of the Channel.

The northern side of the western English Channel Basin is bounded by the continuation of the Cornubian Peninsula to the south-west. Metamorphic basement was considered by Day et al. (1956) to crop out south-west of the Scilly Isles but shallow gravity coring and "Sparker" profiling has not revealed this. However, metamorphic rocks may well approach near the surface. A seismic refraction survey along the line of low gravity (Day and Williams, 1970; Bott and Scott, 1964) gave a preferred model in which the upper crust was thought to be an extension of the Cornubian granite batholith (Holder and Bott, 1971).

The structural relationships of the metamorphic complexes of the Lizard, Eddystone and Start Point, to the land lying to the north, have long been a subject of controversy (Hendriks, 1959; Flett, 1946). Bott and Scott (1964) supported the interpretation that the metamorphic exposures are part of an extensive northerly directed thrust sheet, whilst Lambert (1965) demonstrated that many features in the Lizard Boundary Fault zone could be explained in terms of sedimentary features and normal faulting. Isotopic age-determinations, using K-Ar methods, indicate ages for Lizard gneiss, amphibolite and granulite to range between 375 ± 70 to 492 ± 26 Ma, whilst gneiss from Eddystone yielded a date of 375 ± 17 Ma. These dates indicate Variscan and pre-Variscan metamorphism of sediments that could be Precambrian in age. In a review, refuting the existence of a major Lizard–Dodman–Start thrust, Sadler (1975) suggests that the Lizard complex may include anticlinal folds, the cores of which accommodate slices of basement, and the present outcrop distribution relates to later normal faulting. Thus the structures affecting the basement could be due to pre-Variscan as well as Variscan movements. Sadler also presents stratigraphic and structural evidence to suggest that the Lizard metamorphic rocks are of pre-Ordovician age.

It has already been noted that the E–W structure of the central English Channel continues westward into the Cornubian Peninsula, so that this trend in the Start Point–Eddystone area may well be part of the Variscan E–W overprinting of older Caledonoid or even Cadomian trends. This may be the case in the Perranporth–Pentewan line to the west. To the south of this line, folds and faults with a NE–SW orientation, including the boundary of the Lizard complex, are important features of the structural pattern. The southern tip of the peninsula, together with the whole of the western English Channel, lay to the south of the zone of strong E–W Variscan overprinting. The NE–SW lines of weakness, dating possibly from Cadomian times in this area, have continued to be reactivated up to the present time, for an earthquake occurred in 1926 on the fault south-west from the Hurd Deep (Mourant, 1931).

The sea floor geology in the immediate vicinity of the Lizard complex has yielded little evidence in relation to its tectonic setting. This is due to the onlap of Permo-Triassic sediments to the south-east of the Lizard and also to the lack of structural detail in continuous seismic records made over the Devonian–Carboniferous rocks to the south-west. From present mapping, the major NE–SW faults, probably associated with the northern margin of the offshore basement grabens, pass to the south of the Lizard.

Two unusual offshore outcrops of phonolite occur in the vicinity of Lands End. Epsom Shoal yields an isotopic age of 127 ± 7 Ma and Wolf Rock is dated as 131 Ma (Harrison et al., 1977). These intracratonic alkaline basalts are probably related to periods of tension in the Lower Cretaceous, which may have been associated with the rotation of the Iberian peninsula from France, which took place about 130 Ma (Williams, 1975).

Structure of the continental margin

The well defined margin, from the Bay of Biscay to Goban Spur, suggests that the line of separation of continental segments by rifting was influenced very little by the existing NE–SW structures in the South Western Approaches. With the physical separation of the continental segments, the structural and sedimentary evolution of the new margin was controlled by further movements on both NW–SE and NE–SW directions of movement.

In the most simplistic view, the newly formed edge of the continental segment could have become progressively buried by a prograding apron of sediment deposited in post-rifting times. However, progradation did not take place to any great extent. Curry et al. (1962) reported on the occurrence of outcrops of Eocene rocks on the continental slope, whilst Hersey and Whittard (1966) saw little evidence in Thumper records for progradation of the margin. The latter considered the present margin to be controlled largely by NW–SE faulting, which has continued into uppermost Tertiary times. Only the youngest Pliocene–Quaternary sediments show a pronounced thickening towards the continental margin.

The nature of the NW–SE rift is not readily apparent. It is parallel to well developed structural trends in the western and eastern English Channel provinces. In the Cornubian Peninsula and Brittany these are faults with dextral transcurrent movement. Le Pichon and Sibuet (1971) and Ziegler (1975b) show the fracture at the continental margin to be of similar character and also parallel to the more distant Tornquist line. In the South Western Approaches, the Pendragon and Trevelyan escarpments (Fig. 5.2) are parts of the probably complex fracture zone. The intersection of this zone by the NE–SW fault systems (Fig. 5.2) has given rise to a series of fault bounded tilted blocks. The largest of these forms the Meriadzek Terrace, which is tilted to the north-east (Smith and Van Riessen 1973).

Though the interpretation that the margin could be a transcurrent fault system has been mentioned above, the evidence available at the present time provides support only for the fact that this so-called passive or aseismic margin has undergone subsidence in the order of some 4,000 m.

The results of the Deep-Sea Drilling Project, Leg 48 (see Initial Core Descriptions), have provided much valuable data on the conditions in the area just prior to rifting and after rifting. Hole 400A (see Fig. 5.2) was drilled just south of the Meriadzek Terrace, in a water depth of 4,399 m. After passing through Tertiary chalky rocks, a late Eocene hiatus was encountered at a core depth of 516 m. Then, Upper Palaeocene chalky marl,

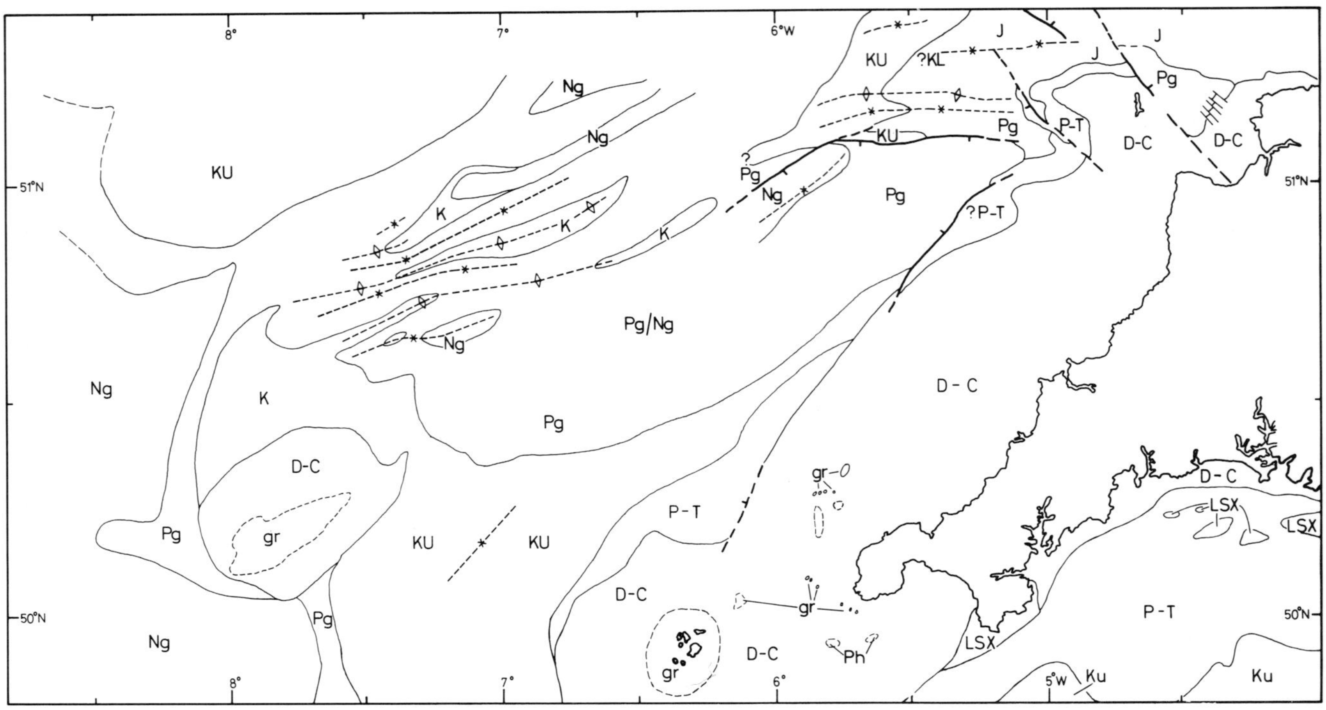

Fig. 5.8. Map (provisional) of the pre-Pleistocene geology of the south Celtic Sea (based on data collected by the author, Doré, 1976; Fletcher, 1975 and Lloyd et al., 1973). Devonian–Carboniferous (D–C) strata, with E–W structural trend, are truncated by the NE–SW South Celtic Sea Basin, with en echelon folding. Tertiary sediments overlie complexly faulted older strata.
Ng = Neogene, (unconformity); *Pg* = Palaeogene, (unconformity); *K* = Cretaceous undifferentiated; *KU* = Upper Cretaceous, (unconformity); *KL* = Lower Cretaceous; *J* = Jurassic; *P-T* = Permo-Triassic, (unconformity); *D-C* = Devonian, Carboniferous; *gr* = granite; *Ph* = phonolite; *LSX* = Lizard–Start Complex.

124 m thick, rests on only 20 m of Upper Cretaceous (Maastrichtian–Campanian) sediments and these rest on rhythmically interbedded carbonaceous mudstones and marly chalks of Early Cretaceous (Albian–Aptian) age. The organic matter in the Albian–Aptian sediments is probably derived from terrestrial plants and was deposited apparently in water depths of about 200 m. At Site 401, on a horst beneath the southern edge of Meriadzek Terrace, the borehole passed through a similar sequence and terminated in Upper Jurassic (Kimmeridgian–Portlandian) reefal limestones and shallow-water carbonates.

Thus, the presence of Upper Jurassic and Early Cretaceous shallow-water sediments indicates that the area had high relief (i.e. was a shallow-water shelf) at the end of the rifting phase, which took place between 113 Ma (Barremian) and 79 Ma ago (Williams, 1975). Subsidence took place after the Albian, for the succeeding Campanian–Maastrichtian chalks were deposited at about 1500 m depth, whilst the Eocene sediments are bathyal. The disconformity between the Albian and Campanian, representing about 30 Ma, corresponds with the widespread unconformity about this time on the adjacent shelf areas. The delay between rifting and the onset of subsidence is probably related to thermal contraction on cooling of the lithosphere (Sleep, 1969).

Summary of the structural history of the South Western Approaches

The Variscan movements, which gave rise to the horst and graben basement topography, were a major control of sedimentation probably during the Permo-Trias, the Jurassic and Lower Cretaceous. Vigorous faulting, uplift and erosion during the Aptian then occurred (cf. Larsonneur et al., 1975). The diversity of rock types and structural pattern exposed on this erosion surface was masked, however, by the widespread transgression of the Upper Cretaceous Chalk seas, followed by Lower and Middle Eocene transgressions (Curry et al., 1971). Successively younger Cenozoic sediments tend to become confined nearer to the continental margin.

Deformation of the Chalk and younger sediments may be intense in the major fault zones but away from strong faults the beds are either flat lying or are thrown into open folds. Those in the Chalk are often of shorter wavelength than the more open folds affecting the Eocene. The youngest folding has been ascribed to a compressional phase (Grau et al., 1973) arising from the Alpine orogenic movements. However, strong steeply dipping faults may displace Upper Cretaceous and younger sediments, so vertical displacement along reactivated basement lineations has played a significant role.

Hence, continued subsidence of the basins may well have been an important control of the younger phases of folding.

It is probable that the periods of most rapid rifting gave rise to uplift and then subsidence, resulting in the widespread Chalk and Eocene transgressions.

Bott (1971) has discussed the deep crustal mechanisms which control the behaviour of young aseismic continental margins, such as that of the South Western Approaches. He concludes that crustal creep from beneath the continent to the oceanic area will induce movements mainly of subsidence of the continental margin. These movements are parallel to the shelf margin. Though the results of these movements are evident in the western English Channel Basin, the major tectonic activity has been a continuing subsidence of a crustal segment along ancient NE–SW lines, lying almost at right angles to the continental margin, through the continued reactivation of older lines of weakness.

The South Celtic Sea

Introduction

The Celtic Sea extends southwards from the south coast of Ireland and embraces two distinctive sedimentary basins. The North Celtic Sea Basin, described by Dobson in Chapter 3 of this volume, is separated from the South Celtic Sea Basin by the submerged Pembrokeshire ridge. To the south, the South Celtic Sea Basin is bounded by the submerged Cornubian ridge.

Thus, for this discussion, the boundaries of the South Celtic Sea Basin are chosen rather arbitrarily at about 51½°N in the north, to include the approaches to the Bristol Channel, and to the south, about 50°N, in the vicinity of Lands End and the Scilly Isles. To the west of these islands, the sea-floor outcrop pattern of the South Celtic Sea swings round to become confluent with that of the western English Channel.

After a brief description of the shallow regional structure, the following topics will be discussed: the eastern margin of the basin, Haig Fras, Lundy Island, the relationship between the South Celtic Sea Basin and the Bristol Channel, and the evolution of the South Celtic Sea Basin.

Shallow regional structure

Due to the widespread veneer of Pleistocene sediments in the Celtic Sea, sea-floor outcrops of solid rock are of restricted occurrence when compared with the English Channel. Hence, the geological map presented here (Fig. 5.8), is based largely on the interpretation of Sparker records, together with a few gravity core samples (Day, 1959; Hamilton and Blundell, 1971; Hamilton, unpublished report). Wherever possible, interpretation of seismic records has been guided by profiles extending northwards from areas of known geological character in the western English Channel, where interpretation is based on successful coring operations (Curry et al., 1970, 1971). However, there are considerable differences in the stratigraphic succession, especially in the Cretaceous, between the English Channel and Celtic Sea areas, so the map must be considered as provisional only.

The major structure is a syncline (Stride et al., 1969; Hamilton and Blundell, 1971), the South Celtic Sea Basin, oriented WSW–ENE in the area adjacent to Haig Fras and plunging gently southwestwards to the Fastnet Basin (see Blundell, Chapter 4, this volume). From Haig Fras, the syncline is directed more to the north-east until it meets the east–west structures of the Bristol Channel (Dore, 1976).

The structure of the syncline is not simple, for two well developed en echelon folds lie parallel to the axis of the regional syncline to the north of Haig Fras (Fig. 5.9). These anticlines expose Cretaceous (possibly Lower Cretaceous) sediments in their cores, whilst Palaeogene and Neogene sediments are folded in the intervening synclines. This indicates considerable late Tertiary tectonic activity, mainly folding, in the area. In the fence diagram (Fig. 5.9) Palaeogene is equated with the Eocene, whilst the Neogene is divided into Miocene, characterised by well bedded reflections on continuous seismic records, and Pliocene, with consistent low angle cross-bedding. This latter is often difficult to differentiate from the Miocene where there is any deformation. For this reason they are not separated on the provisional geological map. The Pliocene beds appear to underlie the pre-Devensian Pleistocene beds west of the Bristol Channel (Delantey and Whittington, 1977).

Eastern margin of the Basin

As already discussed in the section on the western English Channel, the structural grain of the Cornubian Peninsula, to the north of the Perranporth–Pentewan line, is strongly E–W. The Devonian–Carboniferous rocks are structurally very complex (see the closely argued review by Matthews, 1977) and these rocks continue westwards offshore to form an extensive platform some 45 km wide. Seismic profiles over this platform fail to indicate any structural patterns, due to the very tight folding, but it is apparent that the E–W trend is terminated abruptly at the eastern margin of the South Celtic Sea Basin.

Magnetic and gravity data suggest a very rapid

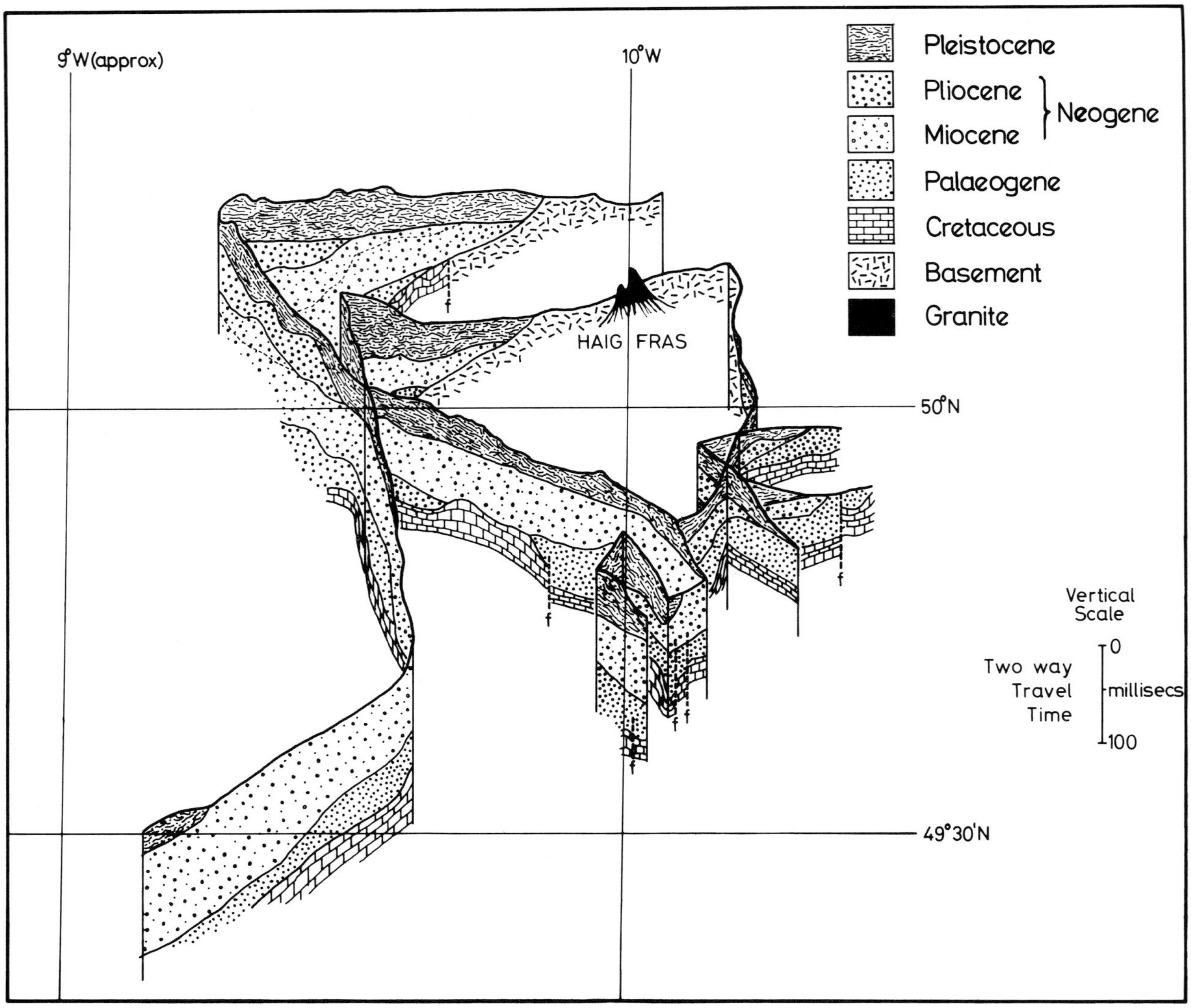

Fig. 5.9. Fence diagram showing the shallow subsurface structure of the south Celtic Sea (based on continuous seismic profiles by the author). Note the folding of Tertiary strata and the structure adjacent to Haig Fras. Only major faults are shown.

thickening of the post-Carboniferous sedimentary pile (Blundell, 1975, and Chapter 4, this volume) implying a major faulted margin along the Lundy–Haig Fras line (Blundell, 1975) analogous in some ways to the Alderney–Ushant line (Curry et al., 1971). The position of this deep seated margin lies to the north-west of the present line of contact on the sea bed between the Devonian–Carboniferous rocks and the younger sediments. North-east of the Scilly Isles, core samples indicate the presence of Permo-Triassic conglomerates and marls which could be in sedimentary contact with the Devonian–Carboniferous rocks, so representing a simple transgressive relationship. However, "slates" have been recovered to the west of this contact, so that faulting, probably in a NE–SW direction, affects these New Red Sandstones. To the north along this margin, Chalk is transgressive over the slates whilst, still farther north, small infaulted areas of possibly Tertiary sediments occur. These are probably of similar origin to those of the Stanley Bank Basin, east of Lundy Island (Fletcher, 1975).

To the south, in the Lands End–Scilly Isles area, Chalk directly overlies the Devonian–Carboniferous slates, which here form a wide platform, some 55 km across, through which rise the granites of the Scilly Isles and Seven Stones. In addition, several smaller apophyses, probably of granite, occur to the west and north-west of Lands End.

Haig Fras

The sea-floor outcrop at 50° 20′ N 7° 50′ W, of a granite mass, Haig Fras (Smith et al., 1966) with associated permeation gneisses, forms a north-easterly elongated area of noticeably rough topography. The map of the area has been revised in the light of additional Sparker traverses.

The granite is intruded into dark grey slates which have yielded palynomorphs indicating a Devonian age. Though no detailed structure in these rocks has been discerned, they are a westerly outcrop of slates lithologically similar to those of the Cornubian Peninsula. The south-eastern, north-western and south-western boundaries of the Haig Fras slates appear to be fault bounded, confirming that this is an uplifted block or horst (Fig. 5.9). Chalk is in fault contact along the north-west and south-east margins, whilst Chalk overlaps a fault scarp onto the slate along the south-west margin. Chalk, overlying possibly Lower Cretaceous (or ?Jurassic) sediments, occurs along the north-eastern margin of the block. Eocene and Miocene transgress the faulted Chalk contacts.

The commonest rock type in the granite mass is a leucocratic tourmaline granite (Sabine, in Smith et al., 1966), for which the date of emplacement is 277 ± 10 Ma. Exley (1966) has pointed out that the closest similarity is with the youngest leucogranites of the south-west.

The Haig Fras granite is offset to the north-west from the NE–SW line of the Cornubian granites. Bott and Scott (1964) suggest that this is due possibly to dextral displacement on a NW–SE transcurrent fault, passing to the west of the Scilly Isles. Such a fault would be consistent with the suite of NW–SE transcurrent faults active in Tertiary times in the Cornubian Peninsula (Dearman, 1971).

Seismic profiling, together with gravity and magnetic evidence suggests, however, that Haig Fras is the exposed part of a separate, uplifted, NE–SW trending basement ridge, lying parallel to the Cornubian Massif. The presence of a small closed, NE–SW trending sedimentary basin, between the Haig Fras block and the Cornubian Peninsula, is suggested by the broad syncline in the Chalk and the gravity anomaly pattern. The presence of the low density granite may well have contributed to the uplift of the Haig Fras block (Curry et al., 1971).

Lundy Island

On a line north-east from Haig Fras lies the granite mass of Lundy Island, at the entrance to the Bristol Channel. Though the two granite outcrops of Haig Fras and Lundy lie on a NE–SW line, the Lundy–Haig Fras line of Blundell (1975), the Lundy granite mass is of a different age and origin from Haig Fras and also the other Cornubian granites.

Lundy Island is comprised of four granite sheets, intruding slates similar to the Morte Slates (Dollar, 1941). Though petrographically the granites were compared with the Variscan granites of Cornwall, isotopic dating of the first granite with large phenocrysts of white orthoclase gave ages of 50 ± 3 Ma and 55 ± 3 Ma (Miller and Fitch, 1962). The granites and slates are intruded by swarms of basic, intermediate and acid dykes. Blundell (1957) showed that those basic olivine dolerite dykes which had stable remnant magnetization, had reversed polarity, and this was consistent with an Eocene age for the igneous activity.

Lundy Island occurs on the interactions of several important structural trends. As already noted, the eastern margin of the north-easterly directed South Celtic Sea Basin passes through Lundy, whilst the Lundy horst is bounded by the west-north-west extension of the Mere Fault and the NW–SE striking Sticklepath and Cambeak faults (Owen, 1971). These faults have been active in Tertiary times (Dearman, 1963; Shearman, 1967).

Marine magnetic surveys adjacent to Lundy Island show that locally the dyke swarm extends to the south-west and north-west from the island, with a major concentration of dykes to the south-west. Only a few dykes occur to the north-east and south-east of the island and these appear to terminate at the Sticklepath Fault. This fault forms a conspicuous boundary in the regional negative anomaly pattern adjacent to Lundy (Geological Survey Aeromagnetic Map, 1965). The pattern is also influenced by the presence of a large dyke (Cornwell, 1971), whilst the granite gives rise to a small positive anomaly at the south-eastern tip of the island.

A gravity survey (Brooks and Thompson, 1973) and a seismic refraction survey (Brooks and James, 1975), reveal the presence of a basic intrusion, centred to the north-west of Lundy Island, whilst the granite is a small high-level pluton within the slates (Miller and Fitch, 1962) and spreads at shallow depth less than a mile eastwards of its surface outcrop (Brooks and James, 1975).

Thus, the Lundy complex is to be considered as the most southern occurrence of activity in the Palaeogene igneous province in Britain. Bott and Scott (1964) suggested the granite was the base of a cone sheet and Brooks (1973) proposed that the igneous activity was due to a NW–SE elongated hot spot underlying western Britain and was associated with the opening of the north-eastern Atlantic.

Relationship between the South Celtic Sea Basin and the Bristol Channel

The structure of the Bristol Channel (Donovan et al., 1961; Owen, 1971; Brooks and Thompson, 1973; Evans, 1973; Lloyd et al., 1973; Brooks and James, 1975) continues as a WNW–ESE downfolded and downfaulted syncline westwards of Lundy Island. Kimmeridgian and Portlandian sediments occur in the axis of the syncline within the Bristol Channel where some 1500 m of Mesozoics are involved. To the west, continuous seismic profiles indicate a westerly dipping succession of probably Middle Jurassic and Lower Cretaceous sediments, approaching 2000 m in thickness (Hamilton and Blundell, 1971), before passing unconformably beneath Chalk. Further to the west, the Permian to Jurassic succession may reach 3 km in thickness (Doré, 1976). Dore has mapped east–west folds, affecting Jurassic to Upper Cretaceous sediments, westward of Lundy Island to about 6°W long. An east–west fault, separating Chalk from Palaeogene sediments occurs to the south of these folds and is the southernmost east–west feature mapped on the sea floor. This fault is close to a noticeable change in gradient of the Bouguer anomalies on the south margin of a deep Mesozoic basin (Davey, 1970), suggesting this is related to a deep seated structure. This might even modify considerably the submerged Pembroke ridge to the west. Near the western end of this fault, a diapir of Keuper salt rises through Chalk to penetrate the base of the Palaeogene sediments (Doré, 1976). This is part of a widespread occurrence of salt in the South Celtic Sea Basin, which gives rise to presumed halokinetic structures (Thomas, 1973; Naylor and Mounteney, 1975).

Evolution of the South Celtic Sea Basin

As previously discussed in relation to the western English Channel Basin, the E–W structural trend is probably related to the compressional phase of the Variscan orogeny, whilst the NE–SW structures, which truncate the E–W structures are lines of weakness reactivated probably in Permian and Triassic times.

The resulting basins were depocentres for sediments at least from Upper Triassic to Middle Cretaceous times, as a result of active subsidence, probably in the Mochras style. This tectonic activity affected both the South Celtic Sea and the Bristol Channel areas, but there is insufficient stratigraphic information available to determine if there is any real difference in the sedimentary history related to these contrasting structural trends. In pre-rifting times, the influence of the palaeogeography to the west on the sedimentary history of the Celtic Sea area, could well have been modified by the development of basins parallel and close to the present continental shelf margin. However, the reported widespread occurrence of thick Lower Cretacous beds of predominantly Wealden facies, suggests a high rate of sediment input to an actively subsiding area.

As in the English Channel, a widespread unconformity occurs beneath the Chalk, suggesting a Middle Cretaceous (Aptian–Albian) orogenic phase. Chalk, usually about 200–300 m thick, is transgressive over faulted margins of basement highs, such as Haig Fras and the buried Pembrokeshire ridge. Thus the Chalk sea extended far to the west and north of the Scilly Isles (Curry et al., 1967), with the youngest Chalks occurring far from the shelf margin. Chalk, however, is not present over the folded Jurassics in the Bristol Channel.

Eocene (Palaeogene) sediments are unconformable on Chalk and a pre-Miocene unconformity is also present. From the strong seismic reflections, both Eocene and Miocene strata appear to be strongly calcareous. They attain thicknesses of 200 m and 150 m, respectively.

Tertiary sediments are deformed into shallow, open synclines and anticlines, with little faulting. Faulting is intense in pre-Chalk sequences near the margins of basement highs and it is moderately common in the Chalk but may die out within the Chalk. Thus the orogenic phase of the Middle Cretaceous marks the end of fault controlled basinal subsidence and the onset of epeirogenic episodes (Kent, 1975).

Pliocene–Quaternary history

Pliocene events

As some of the morphological features of the present sea-floor may be attributed to early Pliocene sea-level changes (Wood, 1974, 1976; Donovan and Stride, 1975) this time may be regarded as the start of the latest phase in the sedimentary history of this shelf area.

In both the western English Channel and the Celtic Sea, Miocene sediments are thrown into open folds, along zones related to basement blocks. An erosion surface truncates some of these Miocene folds and this surface has also suffered slight local warping in late Miocene–early Pliocene times, for Pliocene sediments are folded in the complex axis of the western English Channel syncline (Fig. 5.5). Pliocene sediments here are confined to this axial area, attaining a thickness of 30–40 m and they continue westwards as a slightly thickening wedge.

Preliminary interpretations of sub-bottom profiles in the Celtic Sea suggest a similar occurrence of probably Pliocene sediments, which are gently folded in the synclines. These sediments have the same acoustic signature as those in the western Channel and both show westerly to south-westerly low angle (about 1 degree) cross-bedding. A core sample from the folded area in the western Channel at 48°08′N 5°40′W, yielded a grey calcareous siltstone with *Globigerina* cf. *pachyderma* and abundant *G. bulloides bulloides,* indicating a Pliocene to Pleistocene age and deposition in deep water (Curry et al., 1965). Thus marine conditions over much of this shelf area in the Pliocene are indicated (Bignot, 1974). However, oceanic planktonic foraminiferids may be transported some distance over the shelf (Murray, 1976), so the presence of cross-bedding would suggest, at least, a high rate of sediment influx during Pliocene to both the mid-Celtic Sea and western English Channel areas.

Though the synclinal areas have slightly deformed Pliocene (dips of 4°–5°), probably induced by continued subsidence of the basins, the basin margins on the Cornubian and Armorican massifs were eroded into a series of marine platforms. The oldest of these is suggested to be of Early Pliocene age. Cooper (1948) and Donovan and Stride (1975) describe three cliff levels on the south coast of England at depths of 38–49 m, 49–58 m 58–69 m. Wood (1976) describes a 58 m surface north of Lundy Island and a sea-cliff off the west Cornish coast between 36 m and 52 m depth, whilst Le Fort (1970) describes the flight of submerged erosion surfaces off the coast of Tregor, Brittany. Wood (1976) noted that the buried channels leading into Start Bay (Kelland, 1975) terminate at the base of the upper cliff and Dyer (1975) shows that the buried channels of the "Solent River" are incised to a base level of –46 m O.D. From a review of the occurrence of palaeovalleys around the English Channel, Kellaway et al. (1975) conclude that the rivers were probably graded to a base level of – 55 m O.D. Wood (1974) suggested that the oldest of the marine erosion surfaces could be related to the regression of uppermost Miocene–early Pliocene age, corresponding to the Messinian desiccation of the Mediterranean.

Seawards from the submerged cliffs and platforms, the sea floor of the English Channel becomes a very low-gradient surface, cut, for the most part, across lithified sediments of the central and western areas of the Channel. The history of this surface probably starts with the early Pliocene erosion, but the surface has been modified subsequently by the interplay of warping (which has folded the Pliocene sediments), together with marine and subaerial erosion during the Quaternary regressions and transgressions. Alternatively, this extensive flat surface could have been produced by erosion by grounded Quaternary ice (Kellaway et al., 1975), flowing around the Scilly Isles and eastward up the Channel, as first elaborated by Sandford (1929). During periods of maximum ice advance, in Saalian II times, ice covered the whole of the continental shelf south to the Bay of Biscay and eastwards into the English Channel to Dover Strait. For a discussion of Quaternary features the reader is referred to Kellaway et al. (1975).

Buried channels

In the English Channel a number of deeps occur in the sea floor (Fig. 5.10). The largest is the Hurd Deep, about 200 km long, 5 km across and it is 50–80 m deeper than the adjacent sea floor. Its shape is rather sinuous and it lies parallel to the structural grain of the Jurassic and Chalk rocks into which it is incised. At its north-eastern end, three linear depressions are tributary to the deep. Other deeps in the English Channel are of very much smaller size.

Initially, the interest in these features was to explain the origin of these depressions in the sea floor. However, investigations since the mid-1960's, using sub-bottom acoustic profiling (e.g., Hamilton and Smith, 1972), have revealed that these depressions are the partly infilled remnants of much more deeply eroded features. In addition, sub-bottom profiling has revealed the presence of other features, now completely infilled by younger sediments. Hence the present interest is to elucidate the origin of these much more extensive buried systems.

As a general rule, only partial filling has occurred in the middle and southern regions of the Channel, where relatively strong tidal currents occur. Elsewhere, in the eastern and western Channel and in the Celtic Sea, the Quaternary sediment supply and depositional conditions have resulted generally in the complete infilling of the depressions.

The deeps may occur as (1) small, closed basins, deeps or pits, and (2) linear deeps, with tributary systems.

(1) The small basins may be of several different types:

(a) One mode of origin appears to be due to erosion beneath or marginal to an ice sheet. Fosse Dangeard, in the eastern Channel, is elongated parallel to the structural grain and was eroded in soft Gault Clay (Destombes et al., 1975) by streams marginal to the Saalian ice sheet. A sub-glacial tunnel valley origin is invoked for other NNE–SSW deeps near by, these being eroded during the later Saalian glaciation. St. Catherine's Deep, south of the Isle of Wight, is also attributed to glacial scour (Kellaway et al., 1975).

(b) Some small closed deeps may occur off prominent headlands, as Cap de la Hague and Ushant. Typically these deeps are crescentic around the promontory suggesting they were eroded by fast currents, probably

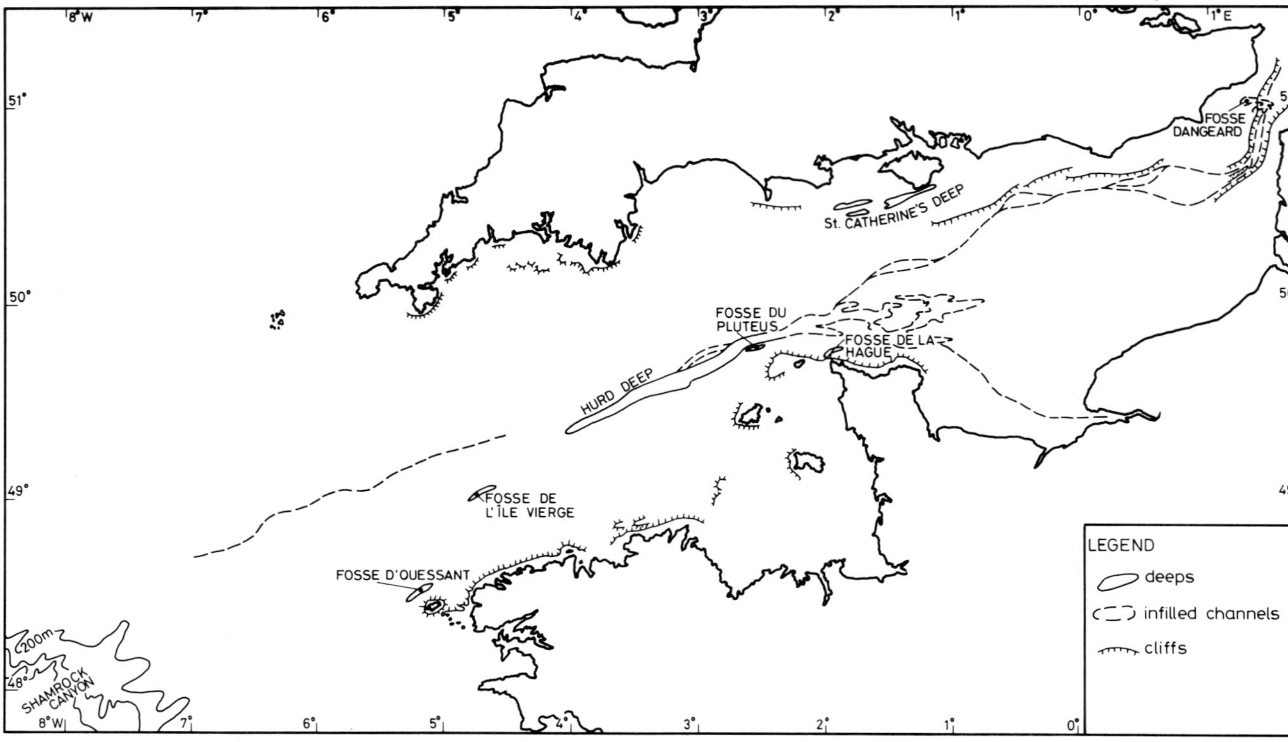

Fig. 5.10. The deeps and infilled channels of the English Channel. The major channel system, including the Hurd Deep (a partially infilled channel) forms an integrated system of drainage across the continental shelf. Other smaller deeps are structural, karstic or sub-glacial erosional (see text). (After Andreieff et al., 1972; Hamilton and Smith, 1972; Kellaway et al., 1975).

during phases of lower sea-level. Both the Fosse de la Hague (Hamilton and Smith, 1972) and Fosse d'Ouessant (Andreieff et al., 1972) have been partially filled by younger sediments. Structural and lithological controls have also played a part in shaping their morphology, as is the case with the Fosse de l'Ile Vierge. The absence of any adjacent palaeovalleys shows that these deeps are not part of a larger system.

(c) Isolated pits or holes, as Trou Santic and Trou Marcel, occur in limestone areas, and may be associated with irregular topography (Boillot, 1964), indicating a karstic origin for these features.

(2) The largest linear deep, with a tributary system is the Hurd Deep:

Palaeovalleys from the Dover Strait and River Seine, now infilled, join the main partially buried channel, just north of the Cotentin Peninsula. Hamilton and Smith (1972) demonstrated that this was a large integrated fluvial system, receiving drainage from southern England, Dover Strait area and north-western Europe. They suggested that this was eroded during the maximum withdrawal of the sea during the Quaternary and that tidal scour led to the deepening of the main channel. Subsequently Dingwall (1975) has suggested the drainage system could have become established during the Miocene (possible early Pliocene?) and was subsequently modified during the Quaternary. Kellaway et al. (1975) suggest that the whole system is due to sub-glacial erosion. This probably did lead to the local over-deepening of some palaeovalleys but it cannot give rise to such an extensive tributary system, related to major land drainage lines.

One puzzling feature about the Hurd Deep is that the ancestral valley terminates south-westwards at about 4°W. This may be related to a former shoreline position representing the stage of maximum river flow and erosion during interglacial transgression (Hamilton and Smith, 1972), or perhaps warping of the shelf, parallel to the shelf margin, may have caused erosion to eliminate the palaeovalley for some distance. The latter seems more likely, as a palaeovalley apparently a continuation but not connecting with the Hurd Deep, has been mapped almost to the shelf edge (Andreieff et al., 1972).

Closed basins or pits occur in the northern Celtic Sea (Delanty and Whittington, 1977) and in the Bristol Channel (Banner et al., 1971; Al-Saadi and Brooks, 1973). Sub-bottom profiling in the Celtic Sea reveals the presence of several infilled channels which appear to be tributary from the St. George's and Bristol Channel areas to a single large channel, directed to the south-west, near the continental margin.

There is apparently an integrated drainage system crossing south-west over the continental shelf of the Celtic Sea and joining into one large channel before reaching the continental margin. This Celtic Sea palaeovalley system is almost completely infilled with sediment, unlike its counterpart in the English Channel, since tidal current and sedimentary conditions are conducive to deposition. The erosional history of these two palaeovalley systems, however, is probably closely similar.

The evidence of such large drainage systems reaching the continental margin has considerable significance when considering the sedimentary history of the submarine canyons on the adjacent continental slopes. The English Channel palaeovalley leads into the Shamrock Canyon whilst the Celtic Sea palaeovalley leads into the Whittard Canyon. Both these are very well developed canyon systems, suggesting they have acted as important routes for sediment transport. Should the establishment of the shelf palaeovalley systems date from late Miocene–early Pliocene, then active canyon formation would probably occur at the same time. However, the canyons have suffered later modification by the considerable changes in sedimentary conditions during the Quaternary, for sub-bottom profiles near the continental margin show that some canyon heads have been infilled.

Holocene sediments

Following the Flandrian transgression the Pleistocene sediments, which are predominantly detrital in composition, have been reworked by tidal currents and oscillatory currents induced by storm waves. The influence of currents induced by internal waves, density currents, etc. is considered to be minor in relation to sediment movement.

In contrast to the detrital nature of the Pleistocene sediments, a high proportion of the present sea-floor sediments is calcium carbonate derived from comminuted shells, tests and other organic structures. The proportion of calcium carbonate may range higher than 90% so that, if preserved, these sediments could give rise to bioclastic limestones. This carbonate sediment is, however, being generated in a cool water continental shelf environment. The high proportion of calcium carbonate is the result, not of an exceptionally high rate of organic productivity, but rather, from the lack of dilution of the carbonate fraction by a terriginous input from the adjacent land surfaces. The low rate of terriginous input is consequent on the raising of base-level for the land drainage systems by the last marine transgression, so reducing stream transport capacity. Hence much of this continental shelf is starved of detrital sediment (Fig. 13.8). However, where easily eroded solid rock is exposed on the sea floor in areas of strong currents, active erosion takes place. Thus areas of Triassic Keuper sediments often yield sands (Clarke, 1970; Channon and Hamilton, 1976) whilst the areas of Chalk outcrops in the western English Channel are well known for the abundance of flints on the sea floor.

There are three sources for the sediments on the present sea floor:

(1) From the erosion of soft rocks exposed on the sea floor. This is a restricted source and is only locally important, as in the western English Channel.

(2) From organisms living on the shelf (see also Chapter 8 of this volume).

The most obvious contributors of calcium carbonate are the abundant pelecypods, echinoderms and bryozoa, but in addition, the dense populations of brittle stars and ophiuroids in the central Channel and western parts of the shelf, provide a ready source of organic debris. These fragments are of diverse shapes and are often perforate. The hydrodynamic behaviour of such irregularly shaped particles is complex. For sediment transport studies, use has been made of the particle settling velocities (Channon, 1971; Channon and Hamilton, 1976).

(3) The most extensive sources of sea-floor sediments are Pleistocene gravels, sands and clays, particularly in the Celtic Sea and the western English Channel (see also Chapter 8).

In these areas, reworking during and after the last transgression has winnowed out much of the sand and finer fractions, producing a basal pebble bed (Belderson and Stride, 1966). Where the bottom tidal currents are competent to move sand (Hamilton et al., 1974), sand sheets and sand waves move over this pavement of immobile pebbles and shells (Stride, 1963). In the central and western Celtic Sea and the western English Channel the sand waves are asymmetric, with the steeper face usually towards the continental margin. This is taken to indicate the direction of sediment transport (Stride, 1963; Kenyon and Stride, 1970; Belderson et al., 1971). Measurements of tidal currents (Figs. 13.6 and 13.7 and, at bottom, by the writer) confirm that the tidal ellipse over much of this shelf is asymmetric with a residual in the ebb direction of flow. This will gradually move sand toward the shelf edge (Fig. 5.11), where it may travel down submarine canyons as a grain flow.

Although the diurnal tidal currents are the main mechanism of transport of sand on the shelf, storm waves induce oscillatory currents (Hadley, 1964b; Ewing, 1973) which can be of sufficiently high speed to entrain particles at mid-shelf depths (McCave, 1971; Hamilton et al., 1975b; Channon and Hamilton, 1976).

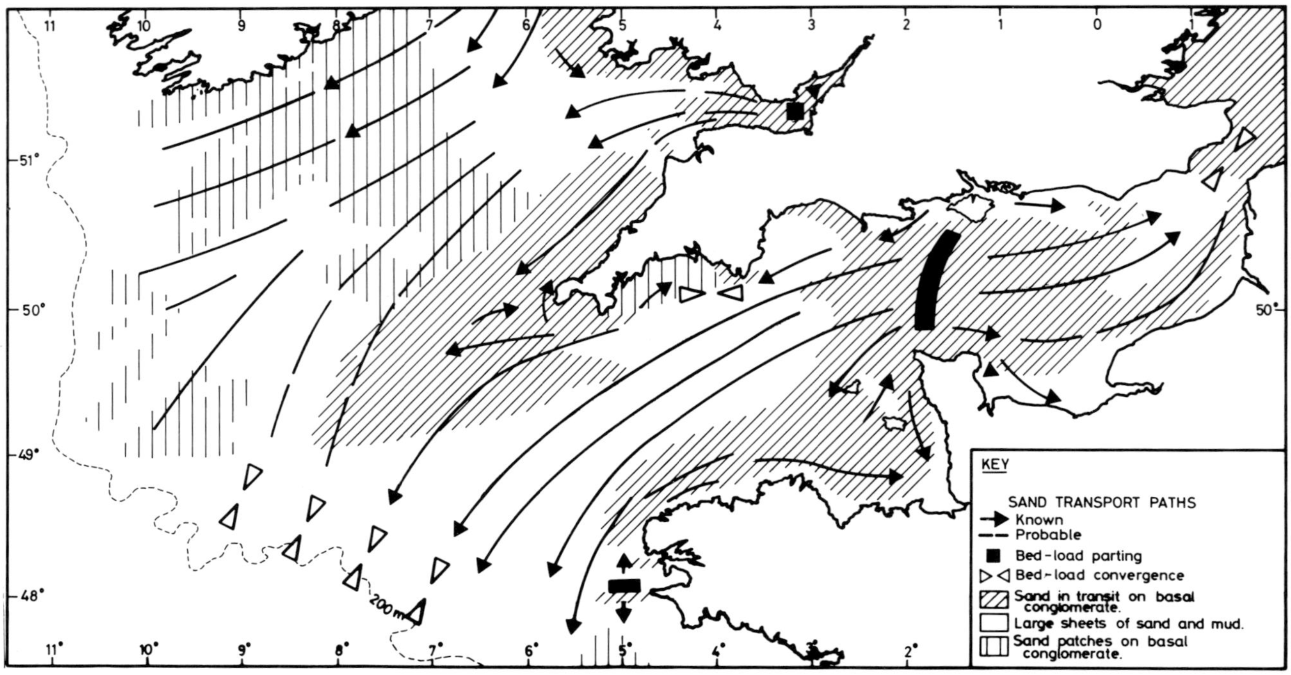

Fig. 5.11. The transport paths of sand over the continental shelf, south-west of Britain. Mobile sand forms patches, ridges and waves that move over a basal conglomerate (after Stride, 1963; Belderson et al., 1971; Kenyon and Stride, 1970).

Thus severe storms will probably destroy or modify the small and large ripples and dunes over much of the shelf but these will be reconstructed again by tidal flow, especially during spring tides.

The large banks in the outer part of the shelf (Fig. 5.1), however, are not destroyed by storms. In some cases the banks rise about 40 m above the adjacent sea floor. Underwater photography shows the surface of the banks to be composed of strongly current-rippled fine sand. Acoustic profiling reveals that the internal structure has low angle cross-bedding and also that the banks rest upon Pliocene or even Pleistocene sediments. The banks are related to the last transgression of the sea (Bouysse et al., 1976) and those banks situated on the shelf away from the continental margin, are aligned parallel to the long axis of the tidal elipse. They do not exhibit the sinuosity of the banks in the North Sea (see Chapter 7, this volume), but are linear throughout their length.

Acknowledgements

A summary such as this inevitably suffers from omissions and over-simplification, for which the writer is solely responsible. Thanks are due to Drs. B. Fletcher, S.C. Matthews and H.M. Pantin for reading and improving the manuscript; Mrs. J. Rowland for typing assistance; Mrs. J. Bees for the drafting; the Natural Environment Research Council, British Petroleum, Shell and Esso for support of the writer's research in the South Western Approaches; and finally, all those who have given assistance on our research cruises.

References

Allen, P., 1972. Wealden detrital tourmaline: implications for northwestern Europe. J. Geol. Soc. London, 128: 273–294.

Al-Saadi, R.H. and Brooks, M., 1973. A geophysical study of Pleistocene buried valleys in the lower Swansea Valley, Vale of Neath and Swansea Bay. Proc. Geol. Assoc., 84: 135–153.

Andreieff, P. and Le Fort, J.P., 1972. Contribution à l'étude stratigraphique des terrains secondaires et tertiaires affleurant en Manche occidentale. Mém. Bur Rech. Géol. Min., 79: 49–56.

Andreieff, P., Bouysse, P., Horn, R. and Monciardine, C., 1970. Contribution a l'étude géologique des approaches occidentales de la Manche. Mém. Bur. Rech. Géol. Min., 79: 31–48.

Andreieff, P., Bouyesse, P., Curry, D., Fletcher, B.N., Hamilton, D., Monciardini, C. and Smith, A.J., 1975. The stratigraphy of post-Palaeozoic sequences in part of the Western Channel. Philos. Trans. R. Soc. London, Ser. A., 179: 79–97.

Auffret, J.P., Bignot, G. and Blondeau, A., 1975. Géologie du bassin tertiaire de la Manche orientale au large du Pays de Caux. Philos. Trans. R. Soc. London, Ser. A., 279: 169–176.

Avedik, F., 1975. The seismic structure of the Western Approaches and the South Armorican continental shelf and its geological interpretation. In: A.W. Woodland (Editor), Petroleum and Continental Shelf of North West Europe. I. Geology. Applied Science Publishers, London, pp. 29–43.

Ayrton, W.G., Binnie, D.E., Swift, J.N., Wellman, H.R., Harrison, R.A., Stenulak, J.G., Waylett, D.C. and Wilkinson, R.A.F., 1973. Grand Banks regional geology. East Coast Offshore Symp. – Baffin Bay to the Bahamas, A.A.P.G. Atlantic City, p. 5 (abstract).

Bacon, M., 1975. A gravity survey of the western English Channel between Lyme Bay and St. Brieuc Bay. Philos. Trans. R. Soc. London, Ser. A., 279: 69–78.

Banner, F.T., Brooks, M. and Williams, E., 1971. The geology of the approaches to Barry, Glamorgan. Proc. Geol. Assoc., 82: 231–248.

Belderson, R.H. and Kenyon, N.H., 1976. Long range sonar views of submarine canyons. Mar. Geol., 22: M69–M74.

Belderson, R.H. and Stride, A.H., 1966. Tidal current fashioning of a basal bed. Mar. Geol., 4: 237–257.

Belderson, R.H. and Stride, A.H., 1969. The shape of submarine canyon heads revealed by Asdic. Deep-Sea Res., 16: 103–104.

Belderson, R.H., Kenyon, N.H. and Stride, A.H., 1971. Holocene sediments on the continental shelf west of the British Isles. In: S.M. Delany (Editor), The Geology of the East Atlantic Continental Margin, Pt. 2: Europe. S.C.O.R. Symp., Cambridge 1970. Inst. Geol. Sci., Rep., 70/14: 157–170.

Bignot, G., 1974. Esquisse stratigraphique et paléographique du tertiaire de Haute-Normandie. Bull. Soc. Géol. Normandie, 61: 23–47.

Blundell, D.J., 1957. A palaeomagnetic investigation of the Lundy dyke swarm. Geol. Mag., 94: 187–193.

Blundell, D.J., 1975. The geology of the Celtic Sea and Southwestern Approaches. In C.J. Yorath, E.R. Parker and D.J. Glass (Editors), Offshore Petroleum Exploration. Can. Soc. Pet. Geol. Mem., 4: 341–362.

Boillot, G., 1964. Géologie de la Manche occidentale. Ann. Inst. Océanogr. (Paris), 42: 1–220.

Boillot, G. and Le Calvez, Y., 1961. Etude de l'Eocène au large de Roscoff (Finistère) et au Sud de la Manche occidentale. Rev. Géogr. Phys. Géol. Dyn., 4(2): 15–30.

Boillot, G. and Musellec, P., 1975. Origine de la Manche d'après une carte géologique au 1 : 1000000e. Philos. Trans. R. Soc. London, Ser. A., 279: 21–27.

Boillot, G., Horn, R. and Le Fort, J.P., 1972. Evolution structurale de la Manche occidentale au secondaire et au tertiaire. Mém. Bur. Rech. Géol. Min., 79: 79–86.

Bott, M.H.P., 1971. Evolution of young continental margins and formation of shelf basins. Tectonophysics, 11: 319–327.

Bott, M.H.P. and Scott, P., 1964. Recent geophysical studies in south-west England. In: K.F.G. Hosking and G.J. Shrimpton (Editors), Present Views on Some Aspects of the Geology of Cornwall and Devon. R. Geol. Soc. Cornwall, pp. 27–44.

Bouysse, P., Horn, R., Le Fort, J.P. and Le Lann, F., 1975. Tectonique et structures post-paléozoiques en Manche occidentale. Philos. Trans. R. Soc. London, Ser. A, 279: 41–54.

Bouysse, P., Horn, R., Lapierre, F. and Le Lann, F., 1976. Etude des grands bancs de sable du sud-est de la mer celtique. Mar. Geol., 20: 251–275.

Brenot, R. and Berthois, L., 1962. Bathymetrie du Secteur Allantique du Banc Porcupine (ouest de L'Irelande) au Cap Finistère (Espagne). Rev. Trav. Off. Pêches Marit., 26: 219–246.

Brooks, M., 1973. Some aspects of the Paleogene evolution of western Britain in the context of an underlying mantle hot spot. J. Geol., 81: 81–88.

Brooks, M. and James, D.G., 1975. The geological results of seismic refraction surveys in the Bristol Channel, 1970–1973. J. Geol. Soc. London, 131: 163–182.

Brooks, M. and Thompson, M.S., 1973. The geological interpretation of a gravity survey of the Bristol Channel. J. Geol. Soc. London, 129: 245–274.

Bullard, E.C. and Gaskell, T.F., 1941. Submarine seismic investigations. Proc. R. Soc. London, Ser. A., 177: 476–499.

Bunce, E.T., Crampton, S., Hersey, J.B. and Hill, M.N., 1964. Seismic refraction observations on the continental boundary west of Great Britain. J. Geophys. Res., 69(18): 3853–3863.

Channon, R.D., 1971. The Bristol Fall Column for coarse sediment grading. J. Sediment Petrol., 41: 867–870.

Channon, R.D. and Hamilton. D., 1976. Wave and tidal current sorting of shelf sediments southwest of England. Sedimentology, 23: 17–42.

Clarke, R.H., 1970. Quaternary sediments off shoutheast Devon. Q. J. Geol. Soc. London, 125: 277–318.

Cooper, L.H.N., 1948. A submerged ancient cliff near Plymouth. Nature, 161: 280.

Cornwell, J.D., 1971. Geophysics of the Bristol Channel area. Proc. Geol. Soc. London, No. 1664: 286–289.

Curry, D. and Smith, A.J., 1975. New discoveries concerning the geology of the central and eastern parts of the English Channel. Philos. Trans. R. Soc. London, Ser. A., 279, 155–169.

Curry, D., Martini, E., Smith, A.J. and Whittard, W.F., 1962. The geology of the Western Approaches of the English Channel. I. Chalky rocks from the upper reaches of the continental slope. Philos. Trans. R. Soc. London, Ser. B., 245: 267–290.

Curry, D., Murray, J.W. and Whittard, W.F., 1965. The geology of the Western Approaches to the English Channel. III. The Globigerina silts and associated rocks. Colston Pap., 17: 239–264.

Curry, D., Gray, F., Hamilton, D. and Smith, A.J., 1967. Upper Chalk from the sea-bed, south of Cork, Eire. Proc. Geol. Soc. London, No. 1640: 134–136.

Curry, D., Hamilton, D. and Smith, A.J., 1970. Geological and shallow subsurface geophysical investigations in the Western Approaches to the English Channel. Inst. Geol. Sci., London, Rep., 70/3: 1–12.

Curry, D., Hamilton, D. and Smith, A.J., 1971. Geological evolution of the western English Channel and its relation to the nearby continental margin. Inst. Geol. Sci., London, Rep., 70/14: 129–142.

Curry, D., Hamilton, D., Smith, A.J., Murray, J.W., Channon, R.D., Williams, D.N. and Wright, C.A., 1972. Some research projects in submarine geology of the Bristol University group. Mém. Bur. Rech. Géol. Min., 79: 221–228.

Dangeard, L., 1923. Quelques observations sur la nature géologique et l'origine des surfaces pierreuses de la Manche. Bull. Soc. Géol. Mineral. Bretagne, 4: 44–46.

Dangeard, L., 1925. Carte géologique du fond du Manche. Rev. Gen. Sci., 30 avril, pp. 3–5.

Dangeard, L., 1972. Historique des premières recherches de géologie sous-marine dans la Manche. Mém. Bur. Rech. Géol. Min., 79: 13–15.

Davey, F.J., 1970. Bouguer anomaly map of the North Celtic Sea and entrance to the Bristol Channel. Geophys. J. R. Astron. Soc., 22: 277–282.

Day, A.A., 1959. The continental margin between Brittany and Ireland. Deep-Sea Res., 5: 249–265.

Day, A.A., Hill, M.N., Laughton, A.S. and Swallow, J.C., 1956. Seismic prospecting in the Western Approaches of the English Channel. Q. J. Geol. Soc. London, 112: 15–44.

Day, G.A. and Williams, C.A., 1970. Gravity compilation in the N.E. Atlantic and the interpretation of gravity in the Celtic Sea. Earth Planet. Sci. Lett. 8(3): 205–213.

Dearman, W.R., 1963. Wrench-faulting in Cornwall and South Devon. Proc. Geol. Assoc., 74: 265–287.

Dearman, W.R., 1971. A general view of the structure of Cornubia. Proc. Ussher Soc., 2(4): 220–236.

Delantey, L.J. and Whittington, R.J., 1977. A re-assessment of the "Neogene" deposits of the South Irish Sea and Nymphe Bank. Mar. Geol., 24: M23–M30.

Destombes, J.P., Juignet, P. and Rioult, M., 1974. Ammonites de l'Aptien-Albien du Bec de Caux, Normandie (N.W. France). Bull. Soc. Géol. Normandie, 61: 41–106.

Destombes, J.-P., Shephard-Thorn, E.R. and Redding, J.H., 1975. A buried valley system in the Dover Strait. Philos. Trans. R. Soc. London, Ser. A., 279: 243–256.

Dingwall, R.G., 1975. Sub-bottom infilled channels in an area of the eastern English Channel. Philos. Trans. R. Soc. London, Ser. A., 279: 233–241.

Dollar, A.T.J., 1941. The Lundy complex: its petrology and tectonics. Q. J. Geol. Soc. London, 97: 39–77.

Donovan, D.T., 1967. Henry Marc Brunel: The first submarine geological survey and the invention of the gravity corer. Mar. Geol., 5: 5–14.

Donovan, D.T., 1972. Geology of the central English Channel. Mem. Bur. Rech. Géol. Min., 79: 215–220.

Donovan, E.T. and Stride, A.H., 1961. An acoustic survey of the sea floor of Dorset and its geological interpretation. Philos. Trans. R. Soc. London, Ser. B., 244: 299–333.

Donovan, D.T. and Stride, A.H., 1975. Three drowned coast lines of probable Late Tertiary age around Devon and Cornwall. Mar. Geol., 19: M35–M40.

Donovan, D.T., Savage, R.J.G., Stride, A.H. and Stubbs, A.R., 1961. Geology of the floor of the Bristol Channel. Nature (London), 189: 51–52.

Doré, A.G., 1976. Preliminary geological interpretation of the Bristol Channel approaches. J. Geol. Soc. London, 132: 453–459.

Drummond, P.V.O., 1970. The Mid-Dorset Swell. Evidence of Albian–Cenomanian movements in Wessex. Proc. Geol. Assoc., 81: 679–714.

Dyer, K.R., 1975. The buried channels of the 'Solent River', southern England. Proc. Geol. Assoc., 86: 239–245.

Evans, D.J., 1973. The Stratigraphy of the Central Part of the Bristol Channel. Ph.D. Thesis, University of Wales, (manuscript).

Ewing, J.A., 1973. Wave induced bottom currents on the outer shelf. Mar. Geol., 15: M31–M35.

Exley, C.S., 1966. The granite rocks of Haig Fras. Nature, April 23, 210 (5034): 365–367.

Fletcher, B.N., 1975. A new Tertiary basin east of Lundy Island. J. Geol. Soc. London, 131: 223–225.

Flett, J.S., 1946. Geology of the Lizard and Meneage. (2nd ed.) Mem. Geol. Surv. Engl. Wales.

Francis, T.J.G., 1962. Black Mud Canyon. Deep-Sea Res., 9: 457–464.

Gerard, A., 1975. La tectonique du socle sous la Manche occidentale d'après les données du magnetisme aeroporte. Philos. Trans. R. Soc. London, Ser. A., 279: 55–68.

Grau, G., Montadert, L., Delteil, R. and Winnock, E., 1973. Structure of the European continental margin between Portugal and Ireland, from seismic data. Tectonophysics, 20: 319–339.

Hadley, M.L., 1964a. Wave induced bottom currents in the Celtic Sea. Mar. Geol., 2: 164–167.

Hadley, M.L., 1964b. The continental margin southwest of the English Channel. Deep-Sea Res., 11: 767–779.

Hallam, A., 1971. Mesozoic geology and opening of the North Atlantic. J. Geol., 79: 129–157.

Hamilton, D. and Blundell, D.J., 1971. Submarine geology of the approaches to the Bristol Channel. Proc. Geol. Soc. London, No. 1664: 297–300.

Hamilton, D. and Smith, A.J., 1972. The origin and sedimentary history of the Hurd Deep, English Channel, with additional notes on other deeps in the western English Channel. Mem. Bur. Rech. Geol. Min., 79: 59–78.

Hamilton, D., Channon, R.D., Griffiths, E.C. and Sommerville, J.H., 1974. Sediment and tidal energy in the South Western Approaches to Britain and the Severn Estuary. Proc. Int. Symp. on Interrelationships of Estuarine and Continental Sedimentation, Bordeaux, 9–14 July, 1973, pp. 259–266.

Hamilton, D., Hommeril, P., Larsonneur, C. and Smith A.J., 1975a. Geological bibliography for the English Channel (Part 2). Philos. Trans. R. Soc. London, Ser. A., 279: 289–295.

Hamilton, D., Sommerville, J.H. and Griffiths, E.C., 1975b. Sediment movement in the South Western Approaches to Britain, with a contrast in tidal energy of three sands. Rep. Congr. Int. Sedimentol., 9me, Nice, 1975, 6: 69–72.

Harrison, R.K., Snelling, N.J., Merriman, R.J., Morgan, G.E. and Goode, A.J.J., 1977. The Wolf Rock, Cornwall: new chemical, isotopic age and palaeomagnetic data. Geol. Mag., 114: 249–328.

Hendriks, E.M.L., 1959. A summary of present views on the structure of Cornwall and Devon. Geol. Mag., 96: 253–257.

Hersey, J.B. and Whittard, W.F., 1966. The geology of the Western Approaches to the English Channel. V. The Continental margin and shelf under the south Celtic Sea. Geol. Surv. Can, Pap., 66-15: 80–106.

Hill, M.N. and King, W.B.R., 1953. Seismic prospecting in the English Channel and its geological interpretation. Q. J. Geol. Soc. London, 109: 1–19.

Hill, M.N. and Laughton, A.S., 1954. Seismic observations in the eastern Atlantic, 1952. Proc. R. Soc. London, Ser. A., 222: 348–356.

Hill, M.N. and Vine, F.J., 1965. A preliminary magnetic survey of the Western Approaches to the English Channel. Q. J. Geol. Soc., London, 121: 463–475.

Holder, A.P. and Bott, M.H.P., 1971. Crustal structure in the vicinity of south-west England. Geophys. J. R. Astron. Soc., 23: 465–489.

Kelland, N.C., 1975. Submarine geology of Start Bay determined by continuous seismic profiling and core sampling. J. Geol. Soc. London, 131: 7–17.

Kellaway, G.A., Redding, J.H., Shepard-Thorn, E.R. and Destombes, J.-P., 1975. The Quaternary history of the English Channel. Philos. Trans. R. Soc. London, Ser. A., 279: 189–218.

Kent, P.E., 1975. The tectonic development of Great Britain and the surrounding seas. In: A.W. Woodland, Petroleum and the Continental Shelf of North-West Europe. I. Geology. Applied Science Publishers, London, pp. 3–28.

Kenyon, N.H. and Stride, A.H., 1970. The tide-swept continental shelf sediments between the Shetland Isles and France. Sedimentology, 14: 159–173.

King, W.B.R., 1948. The geology of the eastern part of the English Channel. Q. J. Geol. Soc. London, 104: 327–337.

King, W.B.R., 1954. The geological history of the English Channel. Q. J. Geol. Soc. London, 110: 77–101.

Lambert, J.L.M., 1965. A reinterpretation of the breccias in the Meneage crush zone of the Lizard boundary, southwest England. Q. J. Geol. Soc. London, 121: 339–358.

Larsonneur, C., 1972. Données sur l'évolution paleogéographique posthercynienne de la Manche. Mém. Bur. Rech. Géol. Min., 79: 203–214.

Larsonneur, C. and Rioult, M., 1969. Le Trias et le Lias des côtes de la Manche, au Nord et a l'Est du Cotentin. C.R. Acad. Sci. Paris, Ser. D., 268: 2019–2022.

Larsonneur, C., Horn, R. and Auffret, J.P., 1975. Géologie de la partie méridionale de la Manche centrale. Philos. Trans. R. Soc. London, Ser. A., 279: 145–153.

Laughton, A.S., Roberts, D.G. and Graves, R., 1975. Bathymetry of the North-east Atlantic Ridge to south-west Europe. Deep-Sea Res., 22: 791–810.

Lees, G.M. and Taitt, A.H., 1945. The geological results of the search for oilfields in Great Britain. Q.J. Geol. Soc. London, 101: 255–317.

Le Fort, J.P., 1970. Etude géologique de la Manche au Nord du Tregor. III. Géologie du substrat rocheux et morphologie. Bull. Soc. Géol. Min. Bretagne, (c) 11(2): 89–103.
Le Fort, J.P., 1973. La 'zonale' Biscaye–Labrador: mise en évidence de cisaillements dextres antérieurs à l'ouverture de l'Atlantique Nord. Mar. Géol., 14: M33–M38.
Le Fort, J.P., 1975. Le controle du socle dans l'évolutin de la sedimentation en Manche occidentale après le Paléozoique. Philos. Trans. R. Soc. Ser. A., 279: 137–143.
Le Fort, J.P. and Deunff, J., 1970. Découverte de Paléozoique à microplancton au Sud de la Manche occidentale. C. R. Acad. Sci. Paris, Ser. D., 270: 271–274.
Le Pichon, X. and Sibuet, J-C., 1971. Western extension of boundary between European and Iberian plates during the Pyrenean Orogeny. Earth Planet. Sci. Lett., 12: 83–88.
Lloyd, A.J., Savage, R.J.G., Stride, A.H. and Donovan, D.T., 1973. The geology of the Bristol Channel floor. Philos. Trans. R. Soc. London, Ser. A., 274: 595–626.
Matthews, S.C., 1977. The Variscan foldbelt in southwest England. Neues Jahrb. Geol. Palaeont., Abh., 154: 94–127.
McCave, I.N., 1971. Wave effectiveness at the sea bed and its relationship to bedforms and the deposition of mud. J. Sediment. Petrol., 41: 89–96.
Miller, J.A. and Fitch, F.J., 1962. Age of the Lundy granites. Nature (London), 195: 553–555.
Mourant, A.E., 1931. A study of the seismograms of the English Channel earthquakes. Mon. Not. R. Astron. Soc., Geophys. Suppl., 2: 374–383.
Murray, J.W., 1976. A method of determining proximity of marginal seas to an ocean. Mar. Geol., 22: 103–119.
Murray, J.W. and Wright, C.A., 1974. Palaeogene Foraminiferida and palaeoecology, Hampshire and Paris Basins and the English Channel. Spec. Pap. Palaeontol., 14: 1–129.
National Science Foundation, 1977. Initial Core Descriptions; Deep Sea Drilling Project, Leg 48, North Atlantic, 217 pp.
Naylor, D. and Mounteney, S.N., 1975. Geology of the North-West European Continental Shelf, 1. Graham, Trotman, Dudley Publishers, London, 162 pp.
Owen, T.R., 1971. The structural evolution of the Bristol Channel. Proc. Geol. Soc. London, No. 1664: 289–293.
Owen, T.R., 1974. The geology of the Western Approaches. In: A.E.M. Nairn and F.G. Stehli (Editors), The Ocean Basins and Margins. 2. The North Atlantic. Plenum, New York, N.Y., pp. 233–272.
Pautot, G., Auzende, J.M. and Le Pichon, X., 1970. Continuous deep-sea salt layer along the North Atlantic margins related to early phase of rifting. Nature, 227 (5256): 351–354.
Phillips, F.C., 1964. Metamorphic rocks of the sea floor between Start Point and Dodman Point, S.W. England. J. Mar. Biol. Assoc. U.K., 44: 655–663.
Roberts, D.G., 1974. Structural development of the British Isles, the continental margin, and the Rockall Plateau. In: C.A. Burk and C.L. Drake, (Editors), The Geology of Continental Margins. Springer, Berlin, pp. 343–361.
Russell, M.J., 1976. A possible Lower Permian age for the onset of ocean floor spreading in the northern North Atlantic. Scott. J. Geol., 12: 315–323.
Sadler, P.M., 1975. A re-appraisal of the "Lizard-Dodman-Start Thrust" concept. Proc. Ussher Soc., 3: 71–81.
Sanderson, D. and Dearman, W.R., 1973. Structural zones of the Variscan fold belt in S.W. England, their location and development. J. Geol. Soc. London, 129: 527–536.
Sandford, K.S., 1929. Erratic rocks and the age of the southern limit of glaciation in the Oxford district. Q.J. Geol. Soc. London, 85: 359–384.
Segougfin, J., 1975. Structure du plateau continental armoricain. Philos. Trans. R. Soc. London, Ser. A., 279: 109–121.
Shearman, D.J., 1967. On Tertiary fault movements in North Devonshire. Proc. Geol. Assoc., 78: 555–566.
Shephard-Thorn, E.R., Lake, R.D. and Atitullah, E.A., 1972. Basement control of structures in the Mesozoic rocks in the Strait of Dover region, and its reflexion in certain features of the present land and submarine topography. (With an Appendix – Faulting in the Kent Coalfield, by P.L. Rumsby.) Philos. Trans. R. Soc. London, Ser. A., 272: 99–113.
Sleep, N.H., 1969. Sensitivity of heat flow and gravity to the mechanism of seafloor spreading. J. Geophys. Res., 74: 542–549.
Smith, A.J. and Curry, D., 1975. Structure and geological evolution of the English Channel. Philos. Trans. R. Soc. London, Ser. A., 279: 3–20.
Smith, A.J., Stride, A.H. and Whittard, W.F., 1966. The Geology of the Western Approaches of the English Channel. IV. A recently discovered Variscan granite wes-north-west of the Scilly Isles. (With an Appendix – The petrology of specimens from Haig Fras, by P.A. Sabine.) Colston Pap., 17: 287–301.
Smith, A.J., Hamilton, D., Williams, D.N. and Hommeril, P., 1972. Bibliographie géologique de la manche. Mem. Bur. Rech. Géol. Min., 79: 303–323.
Smith, S.G. and Van Riessen, E.D., 1973. The Meriadzek Terrace: a margin plateau. Mar. Geophys. Res., 2: 83–94.
Strachan, A., 1913. The form and structure of the Palaeozoic platform upon which the secondary rocks of England rest. Q. J. Geol. Soc. London, 69: 70–91.
Stride, A.H., 1963. Current swept sea floors near the southern half of Great Britain. Q.J. Geol. Soc. London, 119: 175–200.
Stride, A.H., Curray, J.R., Moore, D.G. and Belderson, R.H., 1969. Marine geology of the Atlantic continental margin of Europe. Philos. Trans. R. Soc. London, 264: 31–75.
Thomas, T.M., 1973. The geological basis of oil and natural gas prospects in the Celtic Sea. Geography, 58: 47–51.
Wallace, P., 1968. The sub-Mesozoic palaeogeology and palaeogeography of northwestern France and the Straits of Dover. Palaeogeogr., Palaeoclimatol., Palaeoecol., 4: 241–255.
Whitbread, D.R., 1975. Geology and petroleum possibilities west of the United Kingdom. In: A.W. Woodland (Editor), Petroleum and the Continental Shelf of North West Europe. I. Geology. Applied Science Publishers, London, pp. 45–61.
Whittard, W.F., 1962. Geology of the Western Approaches of the English Channel: a progress report. Proc. R. Soc. London, Ser. A., 265: 395–405.
Williams, C.A., 1975. The opening of the Bay of Biscay. In: C.J. Yorath, E.R. Parker and D.J. Glass (Editors), Canada's Continental Margins and Offshore Petroleum Exploration. Can. Soc. Petrol. Geol., Mem., 4: 363–374.
Williams, C.A. and McKenzie, D., 1971. The evolution of the North-East Atlantic. Nature, 232: 168–173.
Wood, A., 1974. Submerged platform of marine abrasion around the coasts of south-western Britain. Nature, 252: 263.
Wood, A., 1976. Successive regressions and transgressions in the Neogene. Mar. Geol., 22: M23–M29.
Wright, C.A. and Murray, J.W., 1972. Comparisons of modern and palaeogene foraminiferid distributions and their environmental implications, Mém. Bur. Rech. Géol. Min., 79: 87–96.
Ziegler, P.A., 1975a. North Sea basin history in the tectonic framework of North-Western Europe. In: A.W. Woodland (Editor), Petroleum and the Continental Shelf of North-West Europe. Applied Science Publishers, London, pp. 131–148.
Ziegler, W.H., 1975b. Outline of the geological history of the North Sea. In: A.W. Woodland (Editor), Petroleum and the Continental Shelf of North-West Europe. Applied Science Publishers, London, pp. 165–190.

Chapter 6

Patterns of Glacial and Post-Glacial Sedimentation in the Minches, North-West Scotland

PHILIP BISHOP and E. JOHN W. JONES

Introduction

The Minches (Figs. 1.2 and 6.1) are two large interconnected channels, with adjoining lochs and sounds, between the mainland of Scotland and the Hebridean Islands. Being close to the mountains that nourished Britain's largest and most powerful Pleistocene glaciers, this part of the Scottish continental shelf has suffered extensive erosion and is now covered in many places with glacial deposits. Following the retreat of the ice sheets, the post-glacial rise in sea level has submerged much of the Minches floor and has subjected it to a contrasting environment. In this contribution we present a description of the pattern of glacial sedimentation in the Minches, based on sediment samples and seismic profiles, and we discuss the nature of the present surface sediments in relation to the modern oceanographic regime.

Sources of sediment samples and seismic profiles

The sediment samples and seismic profiles used in this study were mainly collected by the Continental Shelf and Marine Geophysics Units of the Institute of Geological Sciences, Edinburgh, between August 1969 and July 1973. Additional information was obtained in December 1973 by a party from University College London, aboard R.R.S. John Murray. The distribution of sample stations, sidescan sonar and continuous seismic profile lines is shown in Figs. 6.2, 6.3 and 6.4. The seismic profiles were generally recorded using a sparker sound source which gives a high resolution of the Quaternary sediment cover.

Although Decca Hi-Fix was employed by the I.G.S. for some of the geophysical work, Decca Main Chain was principally used for navigation. In the best conditions, the latter gives fixes to an accuracy of better than 0.25 n.m. (460 m) in this region. A combination of night and land effects occasionally became a problem in more enclosed bodies of water, as for instance, in the Inner Sound, where it was sometimes necessary to use radar rather than Decca Main Chain. Cross-over errors in the Little Minch and North Minch, estimated from bathymetry and sediment thickness on continuous seismic profiles, were generally insignificant except in the Inner Sound where the positions of some lines had to be shifted over distances up to about ½ km.

Geological structure and stratigraphy of the Minches area

The geology of North-West Scotland has been summarised by Phemister (1960) and Richey (1961) and its relationship to geomorphology discussed by Sissons (1967, 1976). Naylor and Mounteney (1975) have described the geology of the western Scottish continental shelf. Figure 6.5 shows a geological map of the Minches based on the work of the Institute of Geological Sciences (Binns et al., 1975).

The Minches form part of a chain of Mesozoic sedimentary basins on the western side of Britain, extending from the Celtic Sea to the Shetland Shelf. The basins appear to be tensional features associated with the opening of the Atlantic (Bott, 1974). Parts of two major sediment accumulations occur in the Little Minch–Sea of the Hebrides region and in the North Minch. These basins are separated by a basement ridge that runs from Rubh'Re to Kebock Head, through the Shiant East Bank (Fig. 6.5) and are infilled with Mesozoic and younger sediments, reaching about 3 km in thickness. The Minches Fault forms their western boundary, where a thick Mesozoic succession abuts against the upstanding Precambrian basement of the Outer Hebrides. The sedimentary rocks thin eastwards towards the mainland. The

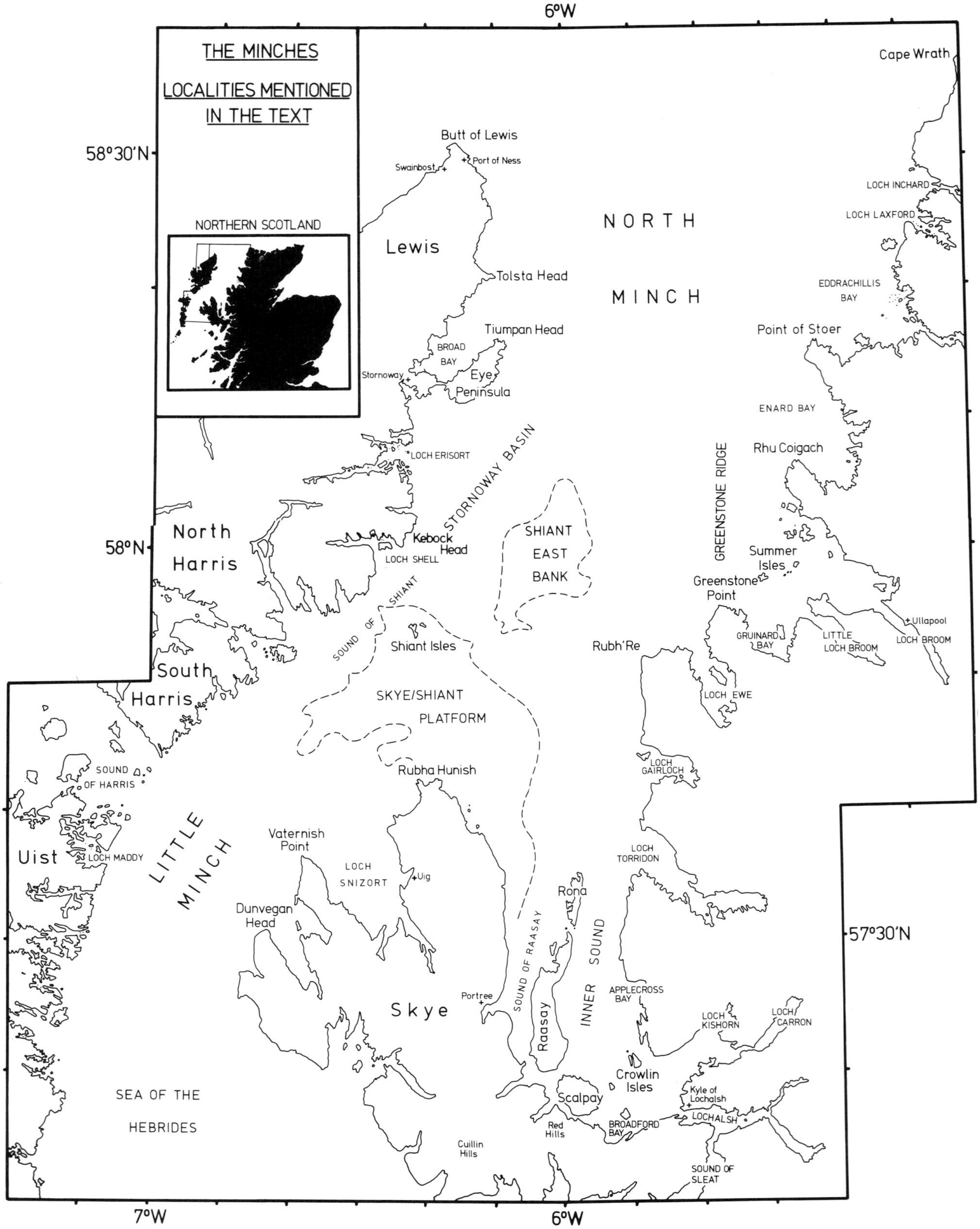

Fig. 6.1. Locations mentioned in the text. The geographic co-ordinates shown are those for all succeeding maps, except where otherwise indicated. Ten minutes of latitude (10 nautical miles) = 18.53 km.

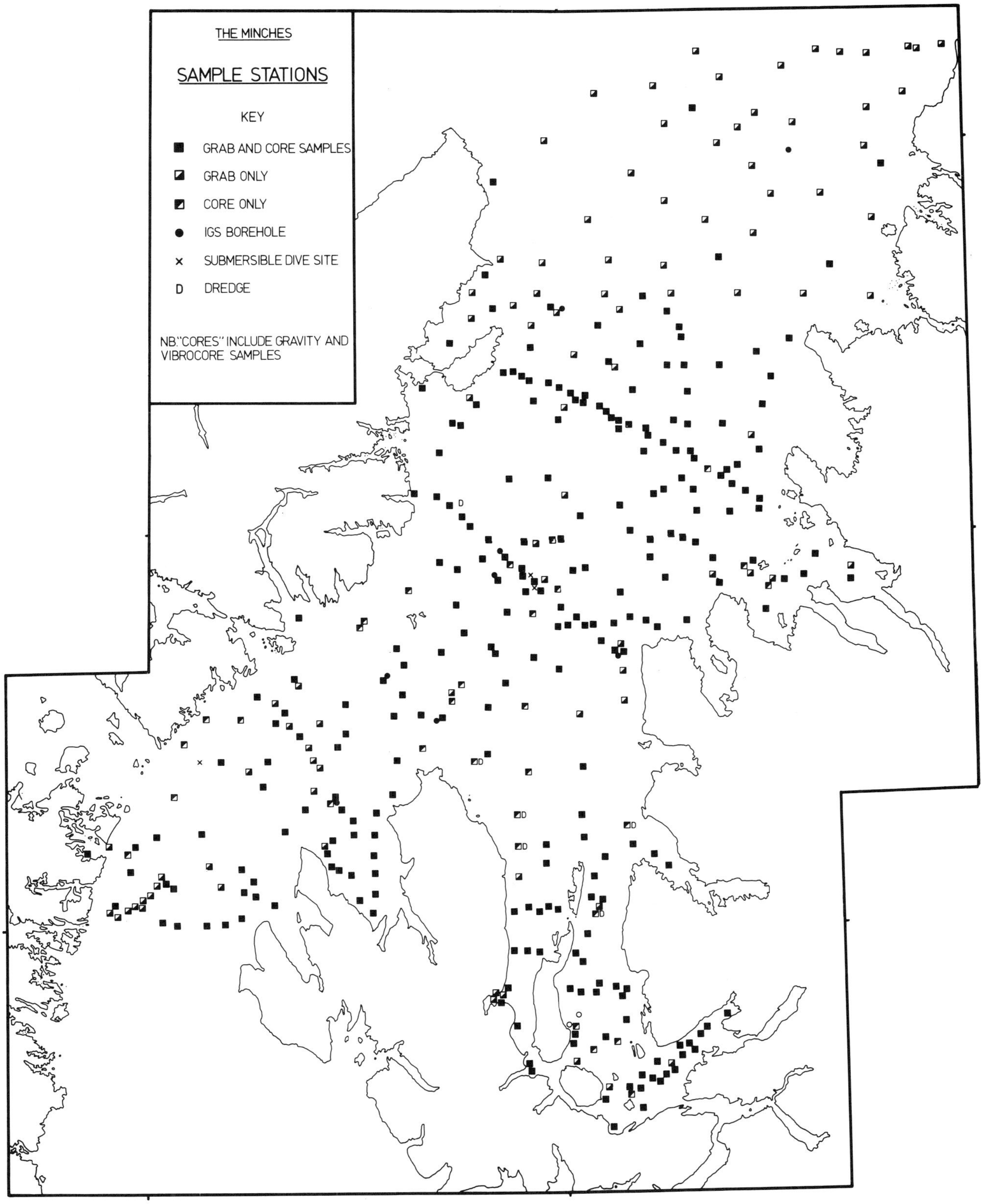
THE MINCHES
SAMPLE STATIONS
KEY
GRAB AND CORE SAMPLES
GRAB ONLY
CORE ONLY
IGS BOREHOLE
SUBMERSIBLE DIVE SITE
D DREDGE
NB."CORES" INCLUDE GRAVITY AND VIBROCORE SAMPLES

Fig. 6.2.

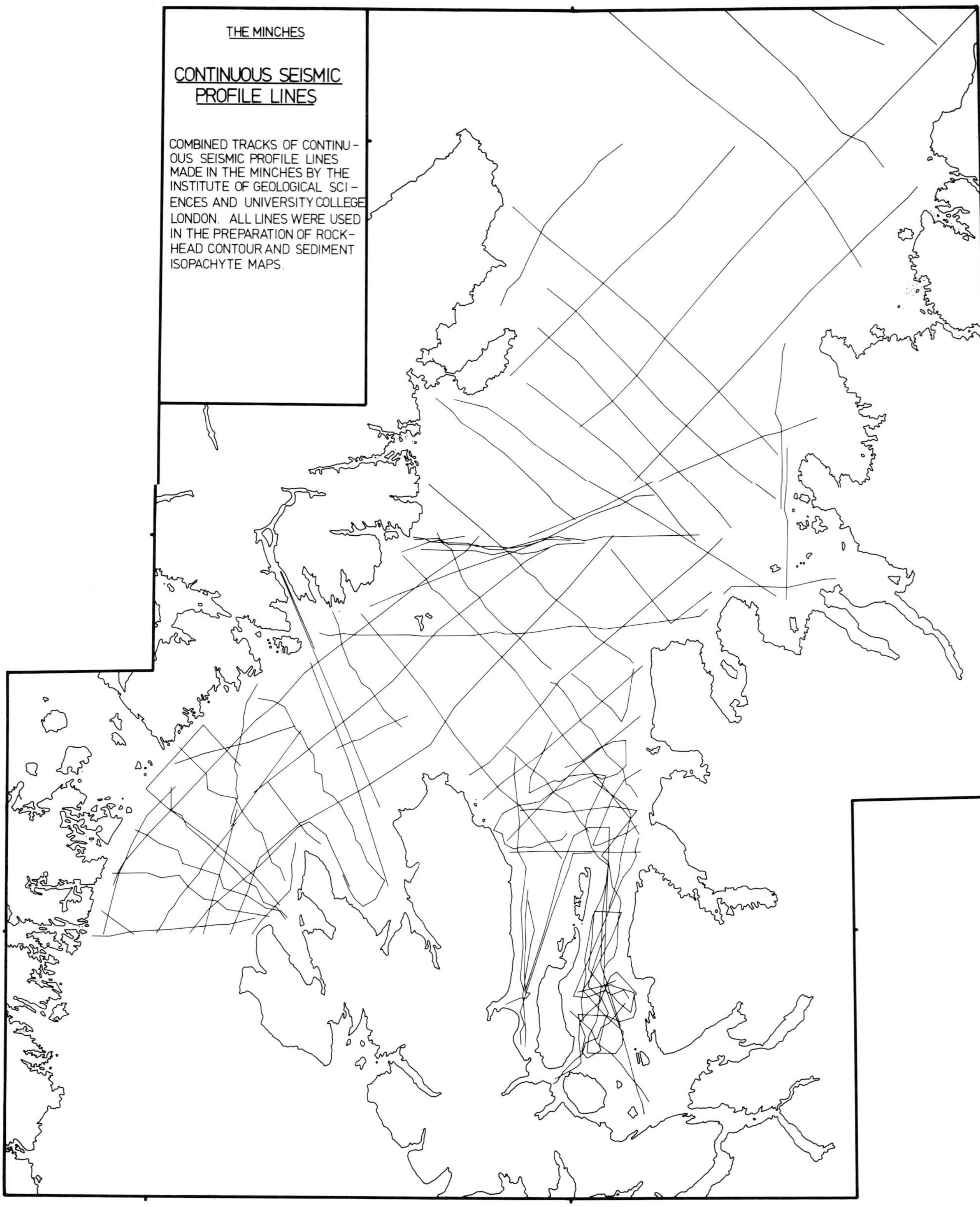
THE MINCHES
CONTINUOUS SEISMIC PROFILE LINES
COMBINED TRACKS OF CONTINUOUS SEISMIC PROFILE LINES MADE IN THE MINCHES BY THE INSTITUTE OF GEOLOGICAL SCIENCES AND UNIVERSITY COLLEGE LONDON. ALL LINES WERE USED IN THE PREPARATION OF ROCK-HEAD CONTOUR AND SEDIMENT ISOPACHYTE MAPS.

Fig. 6.3.

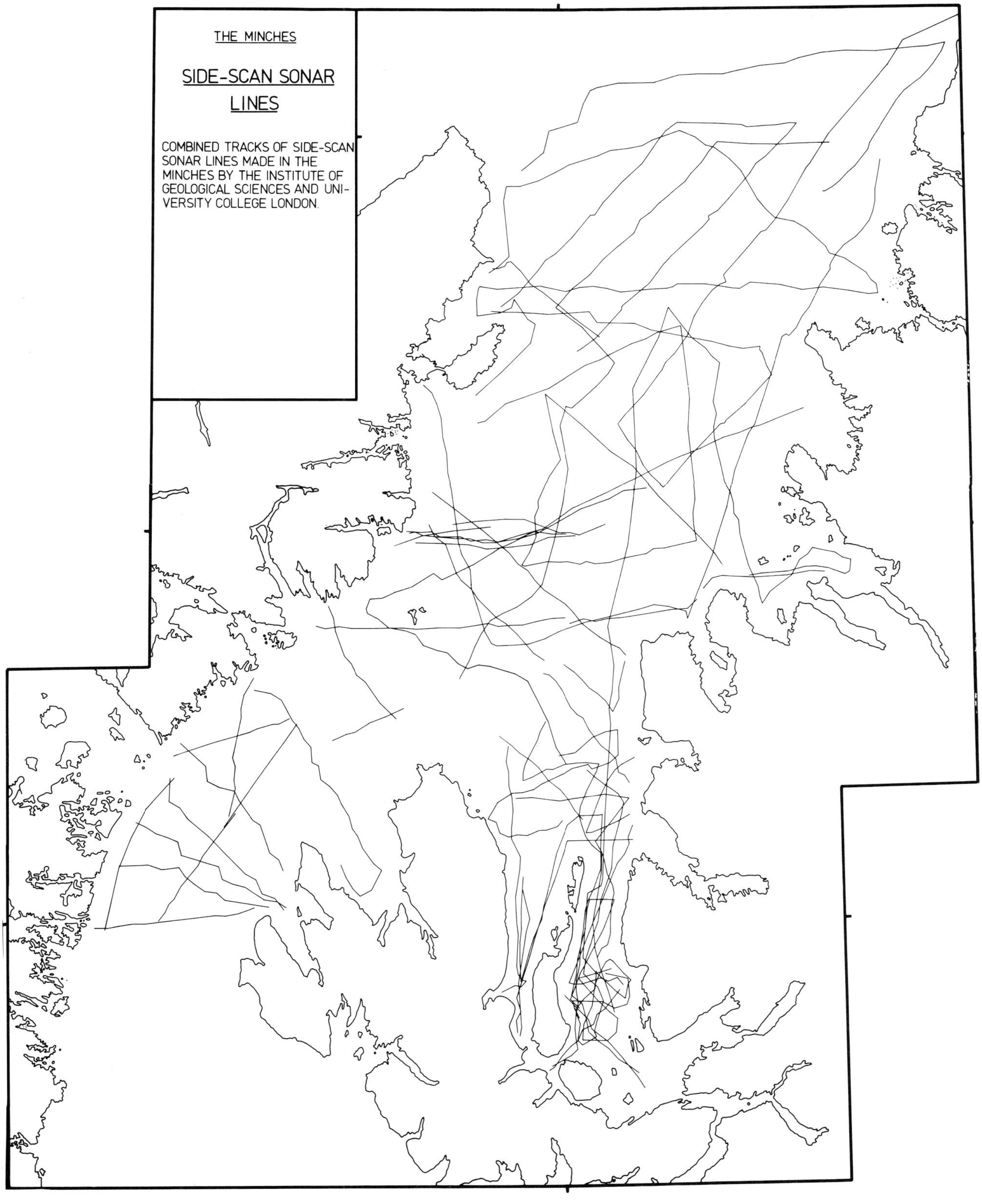
THE MINCHES
SIDE-SCAN SONAR LINES
COMBINED TRACKS OF SIDE-SCAN SONAR LINES MADE IN THE MINCHES BY THE INSTITUTE OF GEOLOGICAL SCIENCES AND UNIVERSITY COLLEGE LONDON.

Fig. 6.4.

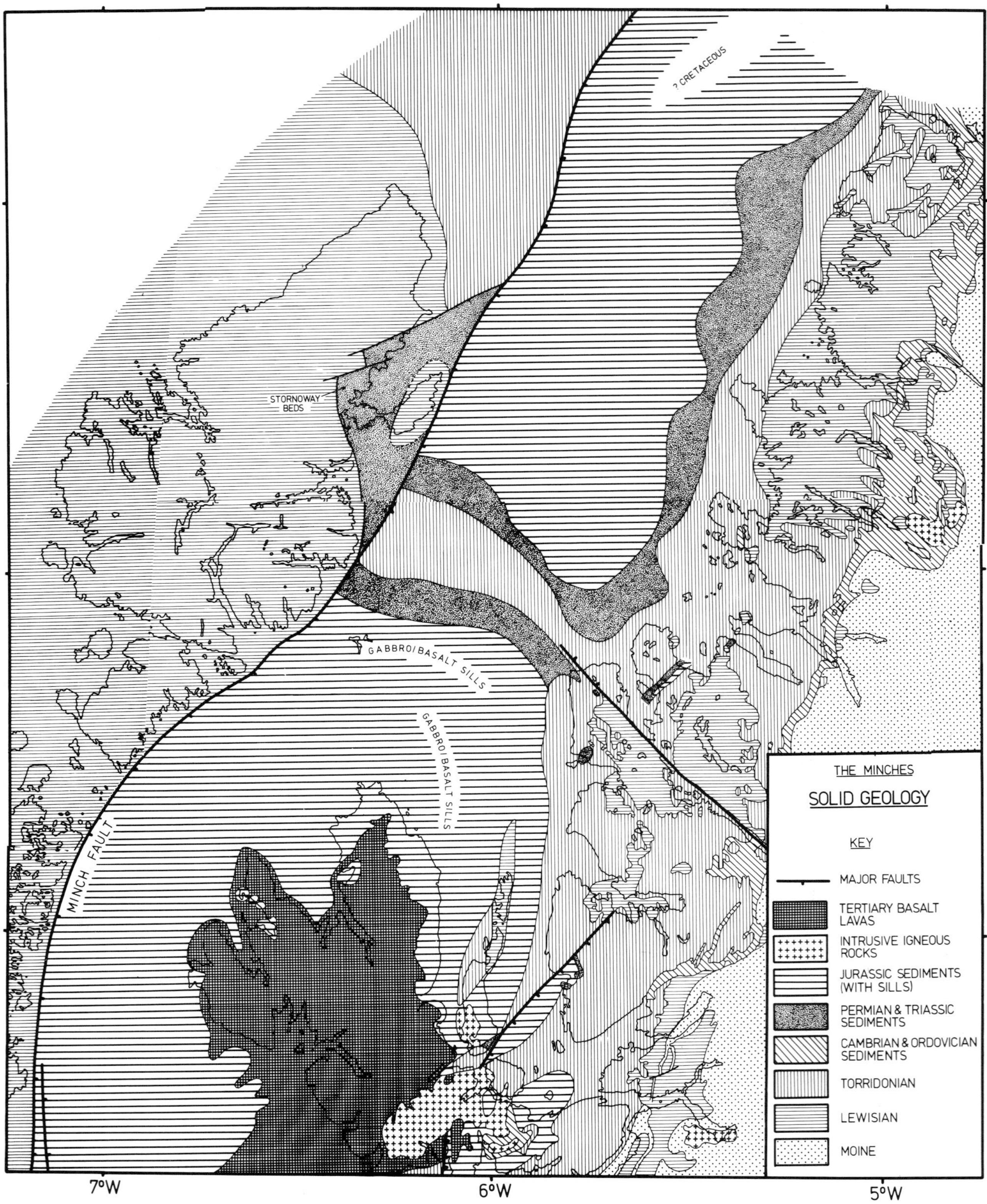
? CRETACEOUS
STORNOWAY BEDS
GABBRO/BASALT SILLS
GABBRO/BASALT SILLS
MINCH FAULT
58°30'N
58°N
57°30'N
7°W
6°W
5°W
THE MINCHES
SOLID GEOLOGY
KEY
MAJOR FAULTS
TERTIARY BASALT LAVAS
INTRUSIVE IGNEOUS ROCKS
JURASSIC SEDIMENTS (WITH SILLS)
PERMIAN & TRIASSIC SEDIMENTS
CAMBRIAN & ORDOVICIAN SEDIMENTS
TORRIDONIAN
LEWISIAN
MOINE

Fig. 6.5.

Inner Sound represents part of a Mesozoic trough, similarly fault-bounded to the west.

The oldest rocks in the region are those of the much folded and metamorphosed Lewisian, which generally outcrops on lower ground near the mainland coast, and extensively on the island of Lewis itself. Where exposed, the Lewisian has been subjected to severe glacial erosion and forms an irregular terrain of hollows, occupied by lakes and bogs, and ice-smoothed and plucked rocky hills. The geomorphological features show a marked lineation from south-east to north-west, along the trend of the principal faults, shatter belts, dykes and igneous sheets within the formation. The greater part of the Lewisian consists of orthogneiss, but paragneisses also occur. There are also intrusions of granite, pegmatite, basic and ultra-basic rocks, of pre-Torridonian age.

The Lewisian has a highly irregular, eroded surface, upon which rest the rocks of the Torridonian formation. Although also Precambrian in age, the Torridonian is practically unaltered and consists of feldspathic sandsstones and grits, overlying the Lewisian metamorphics in massive beds with only a slight dip. Conglomerates and carbonaceous shales occur in places, and the beds are thought to have accumulated in shallow water close to a desert, explaining the freshness of feldspars and iron coating on quartz grains. This staining gives the formation its characteristic reddish brown colour. Common vertical faults and joints promote the development of steep mountain blocks, sitting conspicuously upon the lower and flatter Lewisian surface.

Both the Torridonian and Lewisian formations are cut by a smooth inclined unconformity, above which are found Lower Palaeozoic rocks. The Cambrian succession commences with tough conglomerates and quartzites, which form protective cappings on many Torridonian mountains. These then pass up into shales and mudstones, and then into limestones, in which karst scenery is developed around Durness in the north. The Cambrian–Ordovician boundary lies within the Durness Limestone.

This is the succession that is to be found on the inner islands and along a belt extending inland between 15 and 35 km from the mainland coast. East of this belt, rocks deformed by Caledonian movements form a steep mountain front rising over the Moine Thrust Zone. Here, the Moine Series, consisting of siliceous flaggy granulites with pelite and conglomerate bands, has been thrust over a zone of intensely disturbed Cambrian, Torridonian and Lewisian rocks, from an original position some distance to the east. In general, the area of study lies to the west of the Moine Thrust Zone, but the innermost parts of Lochs Carron, Kishorn, Broom and Cairnbawn penetrate into the Moinian.

The sedimentary basins of the Minches were initiated in Permian times and the oldest rocks found in them are Permo-Triassic red beds, coarse continental sediments deposited in great thickness. They are now only exposed around the rims of the two basins, and are, for the most part, below sea level, although isolated fragments have long been known along the neighbouring coasts, for instance, at Applecross, Rubh'Re and Gairloch, where locally protected by harder Precambrian rocks. The significance of these small exposures has only been fully realised now the geology of the whole of the Minches area is better understood. The Stornoway Beds, over 3000 m of red conglomerates and sandstones, can now be more confidently dated as Permo-Triassic in age, after having been assigned by some previous authorities to the Torridonian or even the Devonian. These are the only Mesozoic sediments preserved west of the Minches Fault.

About 900 m of Jurassic strata have been protected on Skye by a capping of Tertiary basalt lavas. These are now seen to be representative of the Jurassic sequence from the central part of the Little Minch–Sea of the Hebrides Basin. A lower marine series, of Lower Lias age, consisting of limestones, calcareous sandstones and sandy and dark shales resulted from the increasingly marine environment that replaced Permo-Triassic continental conditions. The Middle Lias is characterised by massive white sandstones, and the Upper Lias by dark micaceous shales with pyrites and jet. Sandstones of Inferior Oolite age are succeeded by a thick estuarine series of shales, thin limestones and sandstones, which comprises the remainder of the Middle Jurassic succession. The Upper Jurassic marks a return to marine sedimentation, with Callovian sands, Oxfordian shales and micaceous sandstones and a similar Corallian series. Little Kimmeridgian sediment is present, for a major break in the sedimentary sequence occurs between late Jurassic and early Cretaceous. During the Upper Cretaceous a thin sandstone and chalk cover was laid down unconformably upon an eroded surface of Jurassic, Permo-Triassic and older rocks. Subsequent erosion has removed Cretaceous strata from the land around the Minches, but their presence is suspected in the centre of the North Minch Basin west of Cape Wrath. As well as on Skye, fragments of the Jurassic sequence can be seen on Raasay, the Shiant Isles, and at various sites on the mainland.

Although no sediments of Tertiary age occur in the Minches area, igneous rocks formed at that time are much in evidence. The most conspicuous are the plateau basalts of Skye. Some 600 m of basalt lavas are exposed, along with similar thicknesses of intermediate and acid flows. Associated with the lavas are abundant and varied intrusive igneous rocks, in the form of sills, dykes and plutonic complexes. Massive gabbros make up the Cuillin Hills, while the nearby Red Hills consist of both acid and basic rocks in complex intrusive relationships. To the north, the Jurassic strata are penetrated by large basic sills. The intrusions can be examined on Skye, Raasay and the Shiant Isles. There are no indications at outcrop that Tertiary igneous activity extended north of the Rubh'Re–Kebock Head basement ridge. Many of the scattered islets between Skye and the Shiant Isles are composed of Tertiary dolerites.

Bathymetric controls and glacial history

The bathymetry of the Minches floor is shown in Fig. 6.6. The map was prepared at University College with 1 : 100,000 scale charts made available by the I.G.S.; the area north of latitude 58°30′N was completed by the authors directly from Admiralty Hydrographic Office plotting sheets, which are also the basis of the I.G.S. maps.

After primary geological controls, the most important influence on the development of the bathymetry of the Minches has been a pre-glacial drainage system of great antiquity. Fluvial features on the Minches floor were first mapped by Ting (1937) and then in greater detail by Godard (1965). Figure 6.7 is based on Godard's work although the finer detail has been modified to align with the more accurate bathymetric map now available. A mechanism to account for the submergence of the drainage system since pre-glacial times has been described by Sissons (1967, 1976), who supports the theory that the rivers of northern Scotland were initiated upon an arched cover of Mesozoic (Cretaceous) rocks that has subsequently been removed from the land areas but preserved in the Minches, where the central part of the arch collapsed under tension at the time of the Tertiary igneous activity. Sissons argues that the plateau basalts of Skye were extruded onto a land surface now much lower in relation to the Highlands than in the Tertiary, and evidence can be produced from beach studies of a continued regional subsidence throughout Tertiary into Quaternary times, superimposed upon shorter glacially controlled fluctuations of land and sea-level. Subsidence along the Minches coasts could account for the pronounced knick points, discovered by Ting, where the drowned stream courses meet their landward extensions. The rivers of the Northern Highlands today are the short and fast flowing upper courses of the drowned mature sections.

Both Ting and Godard realised that the Isle of Skye formed part of a fundamental watershed in the pre-glacial drainage pattern, which led away from the main Scottish watershed on the mainland and can be traced as far as the continental shelf west of the Outer Hebrides. This watershed lies some distance south of the Rubh'Re–Kebock Head basement ridge, which does not appear to have been an important factor in the development of the pre-glacial drainage, and is still expressed as a topographic high in the Skye–Shiant Platform, a complex area of shoals and reefs extending north of Skye towards the Shiant Isles. The finer details of the bathymetry upon this platform and the rest of the area around Skye, from the Shiant East Bank to the south of the Little Minch, can be related to the presence of gabbroic sills within the Jurassic sediments that floor this region. The resistance of the sills to erosion has protected the softer sediments and accounts for the elevation of the numerous banks, the sills themselves often outcropping as narrow and broken arcuate ridges. Some rise above sea level to form rocks and islets between Skye and the Shiant Isles. Figures 6.8 and 6.9 offer a comparison of the bathymetry with descriptions of geological samples obtained by various methods from sill outcrops and flatter, sediment-floored areas between. Not all of the geological samples are confirmed as examples of bedrock, but it is evident that the micro-gabbro–dolerite sills of northern Skye are continued offshore and control bathymetry. Shallow Tertiary igneous activity does not appear to have continued north of the Rubh'Re–Kebock Head ridge, and the North Minch floor has no sill-protected banks like those in the south. Although a sill forms the southern part of the Shiant East Bank, its northern half has a much older geological foundation, for it stands astride and exposes the basement ridge itself. Erosion of softer Mesozoic sediments on either side of the ridge during glaciation has exhumed this ancient feature of the Minches geology.

The Quaternary glaciation has considerably modified the fluvial features that once dominated the Minches floor. Very little is known about the sequence of Quaternary events in Scotland prior to the decay of the last ice sheets. From studies of the drift deposits of southern Britain it is clear that inter-glacial periods, with temperatures at least as high as at present, must have occurred, but Scottish glaciers have been powerfully erosive and almost all evidence of inter-glacial events has been subsequently removed. Most depositional features so far described in Scotland relate only to the later history of the most recent glaciation. Pollen analysis has yielded only two sites in the whole of Scotland, neither of them in the north-west, that have demonstrable inter-glacial

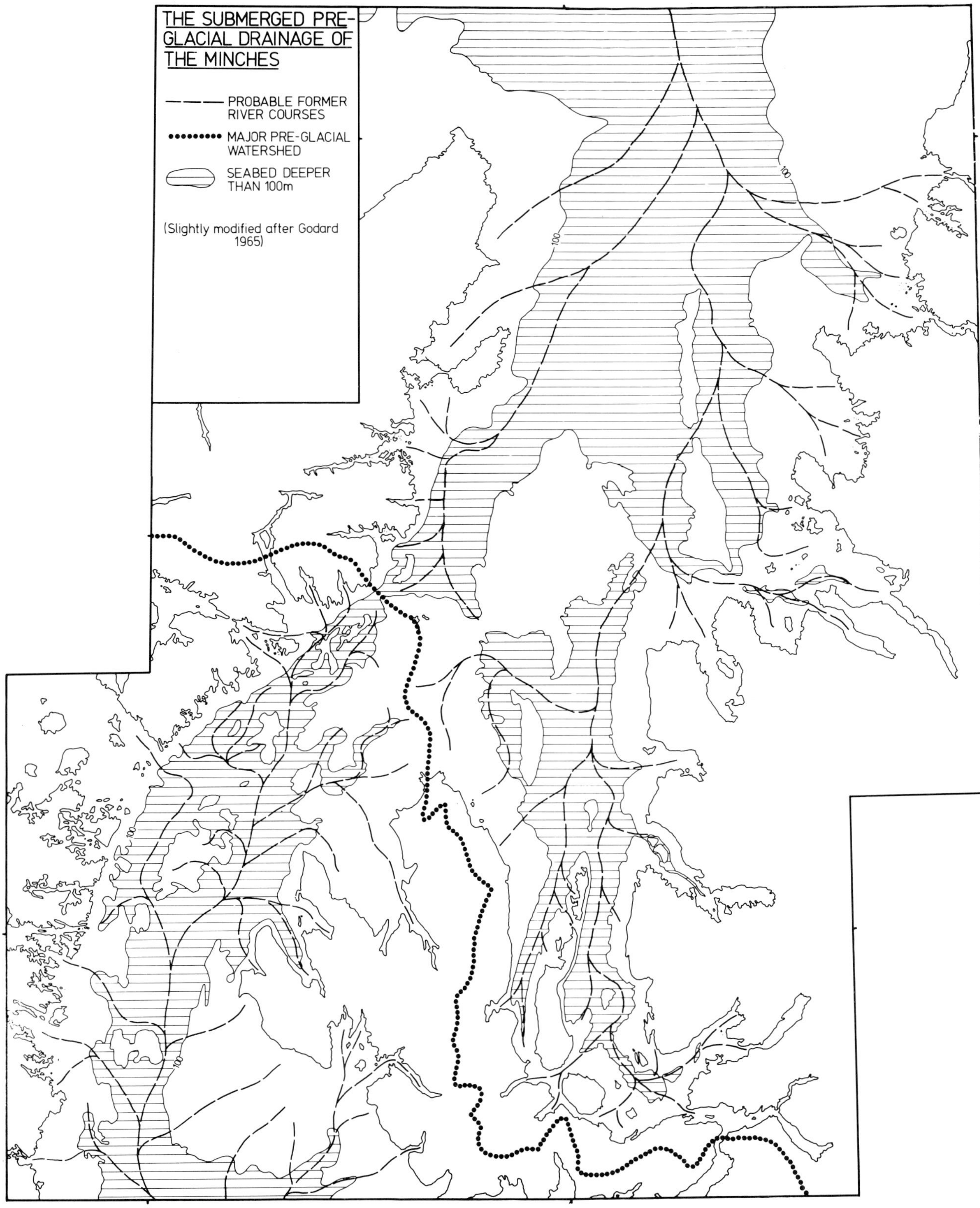
THE SUBMERGED PRE-GLACIAL DRAINAGE OF THE MINCHES
PROBABLE FORMER RIVER COURSES
MAJOR PRE-GLACIAL WATERSHED
SEABED DEEPER THAN 100m
(Slightly modified after Godard 1965)
100

Fig. 6.7.

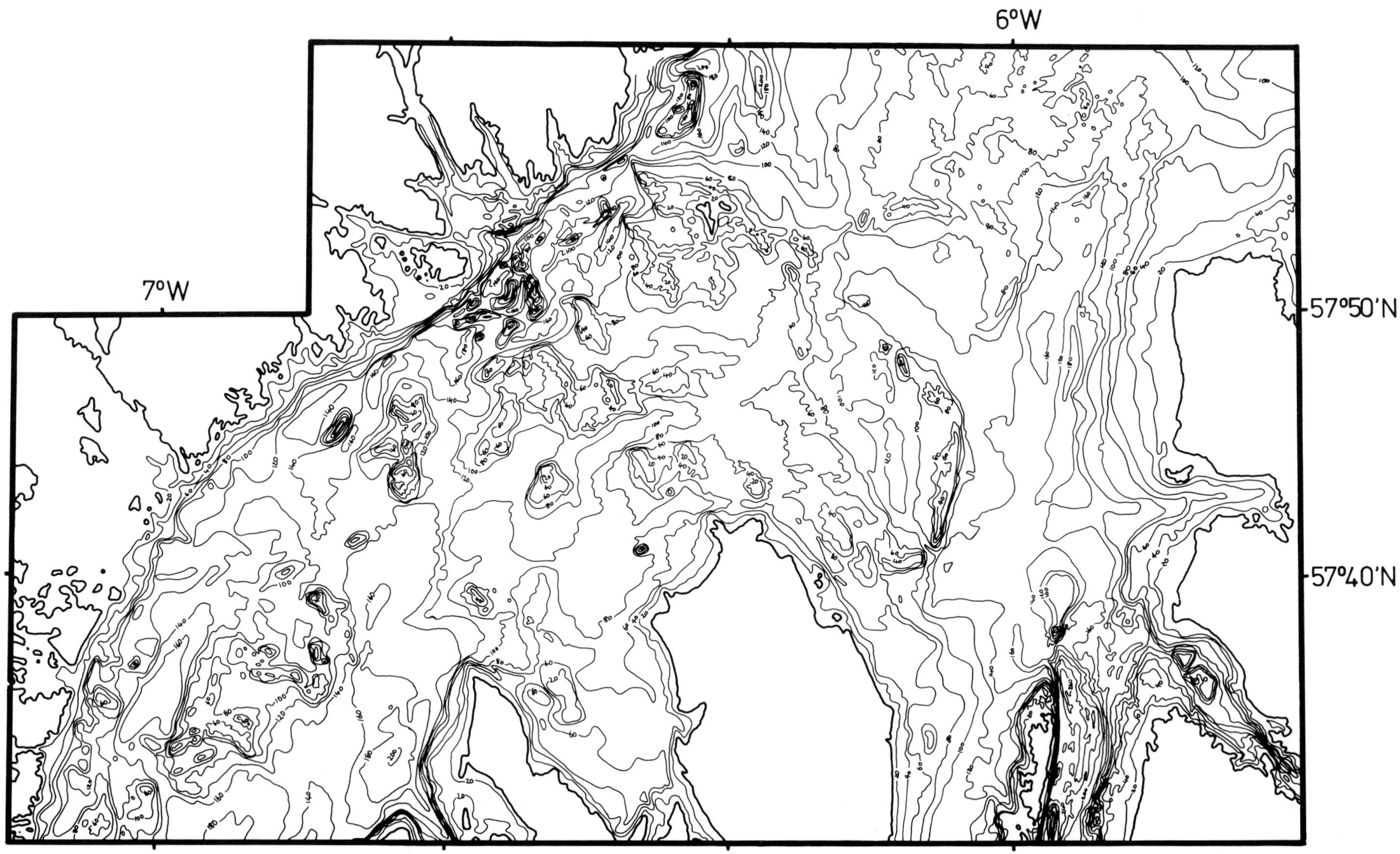

THE CENTRAL MINCHES — BATHYMETRY CONTOURS AT 20 METRE INTERVALS

Fig. 6.8.

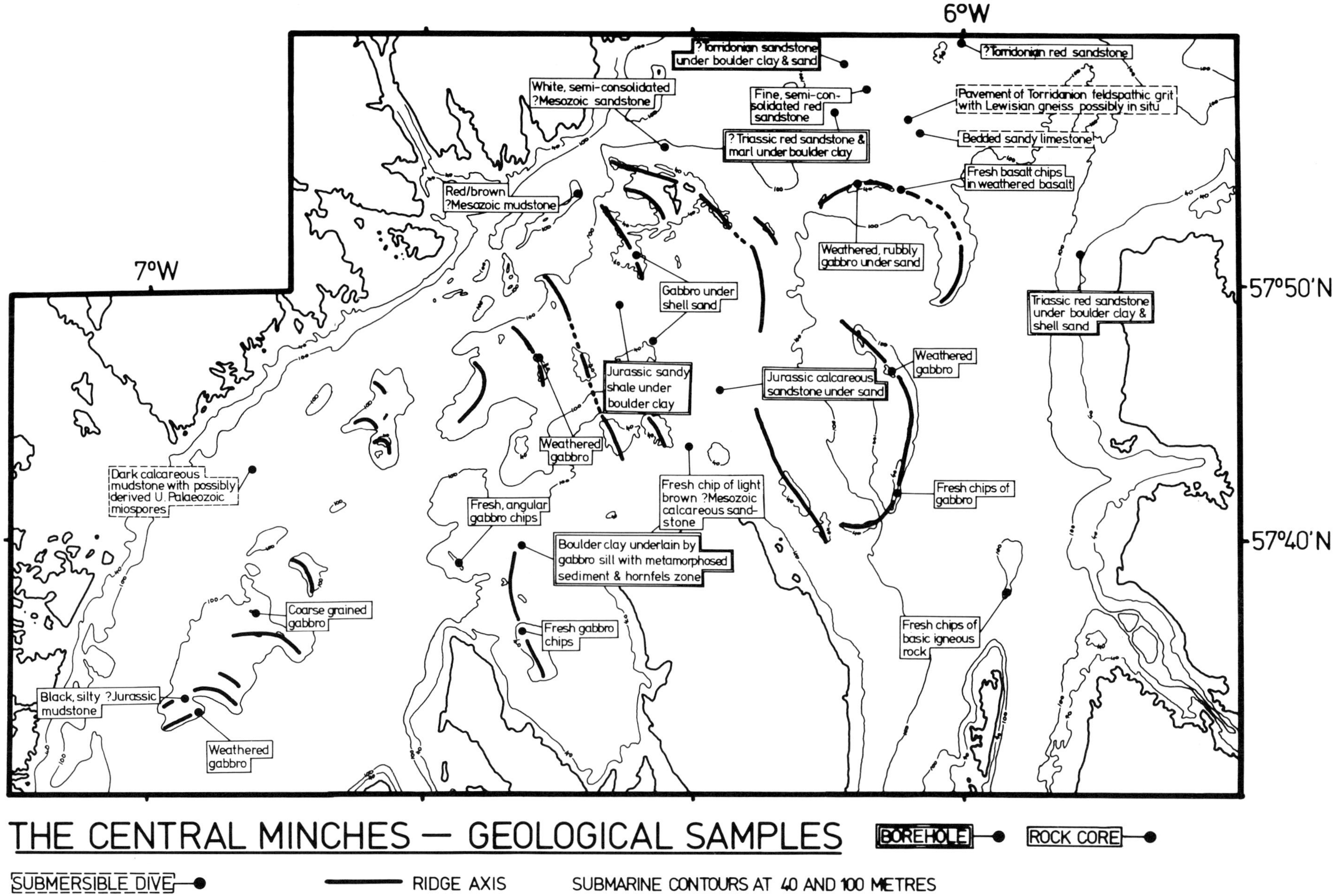
6°W
7°W
57°50'N
57°40'N
?Torridonian sandstone under boulder clay & sand
?Torridonian red sandstone
White, semi-consolidated ?Mesozoic sandstone
Fine, semi-consolidated red sandstone
Pavement of Torridonian feldspathic grit with Lewisian gneiss possibly in situ
Bedded sandy limestone
? Triassic red sandstone & marl under boulder clay
Fresh basalt chips in weathered basalt
Red/brown ?Mesozoic mudstone
Weathered, rubbly gabbro under sand
Gabbro under shell sand
Triassic red sandstone under boulder clay & shell sand
Weathered gabbro
Jurassic sandy shale under boulder clay
Jurassic calcareous sandstone under sand
Weathered gabbro
Dark calcareous mudstone with possibly derived U. Palaeozoic miospores
Fresh chip of light brown ?Mesozoic calcareous sandstone
Fresh chips of gabbro
Fresh, angular gabbro chips
Boulder clay underlain by gabbro sill with metamorphosed sediment & hornfels zone
Coarse grained gabbro
Fresh gabbro chips
Fresh chips of basic igneous rock
Black, silty ?Jurassic mudstone
Weathered gabbro
THE CENTRAL MINCHES — GEOLOGICAL SAMPLES
BOREHOLE
ROCK CORE
SUBMERSIBLE DIVE
RIDGE AXIS
SUBMARINE CONTOURS AT 40 AND 100 METRES

Fig. 6.9.

deposits (Sissons, 1976). However, a supposedly inter-glacial deposit occurs in the north of Lewis (Geikie, 1878, 1894; Baden-Powell, 1938) where a mixed fauna of warm and cold water species is found in a horizon of marine sediment between two boulder clays. The significance of this deposit will be discussed later. Most of the literature concerning Scottish glacial sediments involves a postulated sequence of glacial re-advance stages, in which many moraine deposits in northern Britain have been correlated as belonging to one of three periods when the boundaries of the last ice sheets ceased to retreat towards the Highlands and temporarily advanced again. As recent evidence has accumulated, the validity of two of the three re-advances has been called into question, and in 1976 Sissons (who himself helped to postulate this late glacial sequence) abandoned the idea of the two earliest re-advance stages. There now appears to be firm evidence for only the most recent, Loch Lomond re-advance, the limits of which lie almost entirely outside the area covered by this study, apart from minor ice accumulations at the upper ends of Lochs Broom and Torridon, and in the Cuillin Hills of Skye. Total deglaciation took place around 12,500–13,000 years B.P., while the Loch Lomond re-advance occurred about 10,300–10,800 years B.P. The limits of the Loch Lomond re-advance and the other major glacial features of north-west Scotland are shown on Fig. 6.10.

The main area of ice accumulation in northern Scotland was the Western Highlands, from Loch Lomond to Loch Broom (Sissons, 1967). Smaller centres included the mountains of Skye and Harris, and the narrow mountain belt of northern Sutherland. At maximum glaciation, an unbroken ice sheet extended beyond the Outer Hebrides to cover the Flannan Isles, which are 30 km west of Lewis, but not as far as St. Kilda. To the east, the Scottish ice sheet was confluent with the North Sea ice from Scandinavia. At the centre of the ice cap in the Western Highlands, ice domes rose up above the tallest mountain summits. Glacial striae and erratics have been reported on the peaks at altitudes of well over 1000 m (Sissons, 1976), although some sources suggest individual mountain tops may have stood out as nunataks (Phemister, 1960). Over North Harris, the surface of the ice cap attained an altitude of 600 m and only the tallest peaks were exposed. From the main Scottish ice-shed, ice streamed in all directions. The course of the flow is known from erratics and striae on the mountain sides. Maps of glacial stream lines (Fig. 6.11) have been constructed by various authors, using the high level striae found near the mountain peaks, to give an indication of the direction of flow at maximum glaciation. These maps show a general north-westerly movement across the Minches, with a northerly deflection caused by the obstruction of Skye and the pressure of the island's own ice source in the Cuillin Hills. In many places, particularly between Loch Ewe and Cape Wrath on the mainland, this north-westerly flow is congruous with the grain of the topography (and Lewisian geological structure), but at other localities the high level ice flowed obliquely to the direction of glacial troughs and lochs. Low level striae found in the valley floors, indicate directions of ice flow in accordance with the trends of the topography. Generally, the low level striae have been attributed to a later, valley-glacier stage of glaciation, when the great ice sheets that flowed more or less oblivious of the underlying topography degenerated into a number of individual glaciers occupying corries, troughs and lochs. However, the contemporary formation of angularly discordant high and low level striae is possible if a two-layer flow of ice is postulated, where the upper levels do not necessarily move in the same direction as the ground-ice. (A later valley-glacier stage would, of course, still be capable of forming more low level striae.)

As early as 1878, James Geikie demonstrated a two-layer flow of ice in the Minches, and in 1894 went on to suggest similarly discordant ground-ice movement south of Skye in the Sea of the Hebrides (Fig. 6.11). Using charts of the sea floor to delineate the course of deep troughs, Geikie proposed that ground-ice moved from the Inner Sound and Loch Dunvegan across the Little and North Minches towards the steep eastern face of the Outer Hebridean chain, and was deflected to the north and south, eroding a deep channel close to the coast, while upper ice levels flowed overhead onto the islands. In support of his theory, Geikie introduced the evidence of the drift deposits of Lewis, which are, for the most part, a rubbly till composed of local rocks, to be found mainly on the western side of the island. Erratics are absent, apart from those of local origin, and for long there had been dispute even about the direction from which ice had flowed onto the island, the NW–SE trending striae providing only an ambivalent indication of ice movement. The Lewis drift, however, can be explained by Geikie's model, whereby only the cleaner, upper ice layers flowed over the islands, depositing no erratics, while the lower levels, containing all the ground-moraine picked up on the passage from sources in Skye and the mainland, was deflected to the north or south without reaching the coast. Only in the far north, near the Butt of Lewis, where a gently shelving coast allowed the ice to ride up onto the islands, is the ground-moraine to be seen. This deposit, already mentioned in regard to its supposed inter-glacial horizon, is in stark contrast to

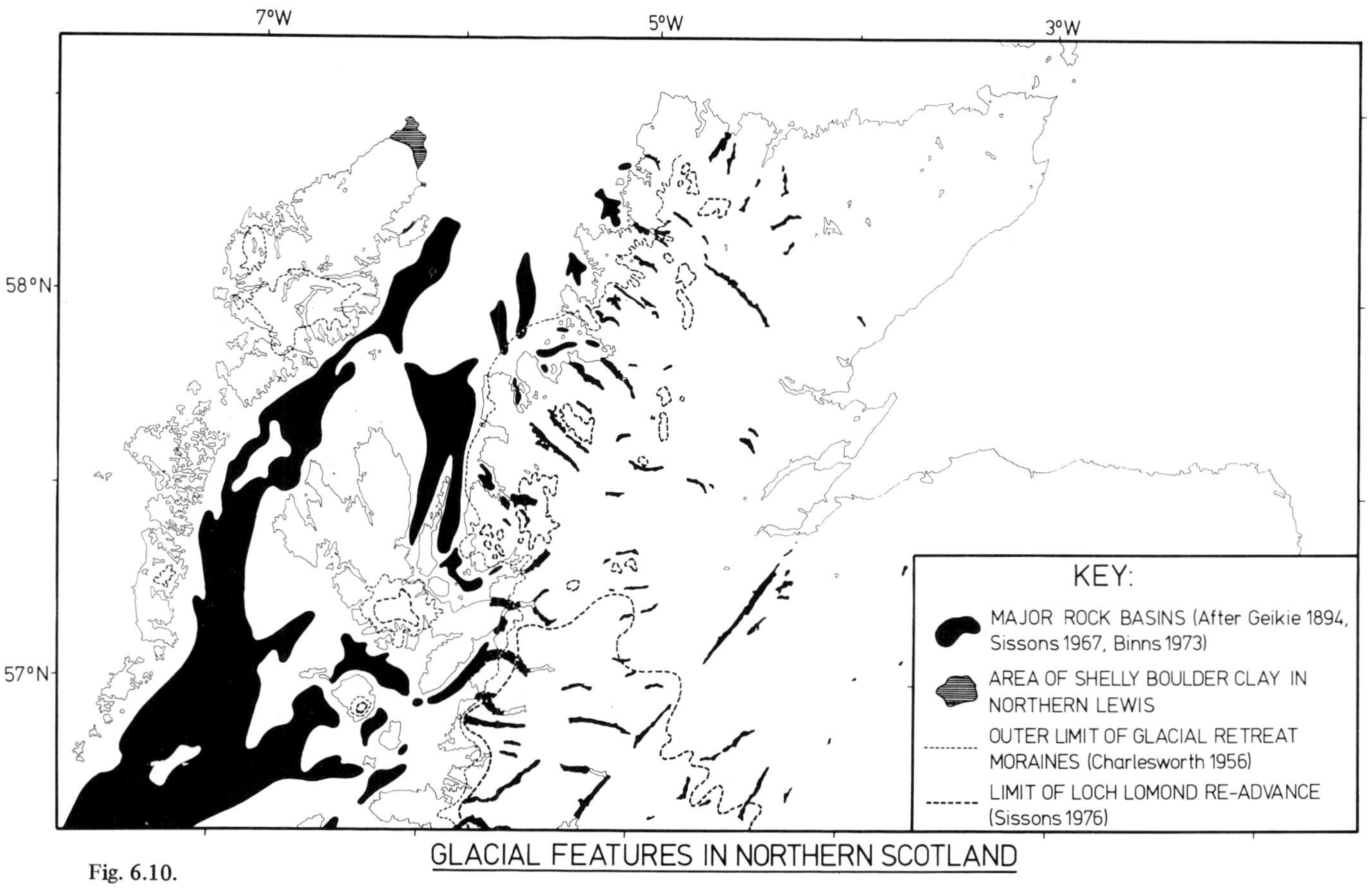

Fig. 6.10. GLACIAL FEATURES IN NORTHERN SCOTLAND

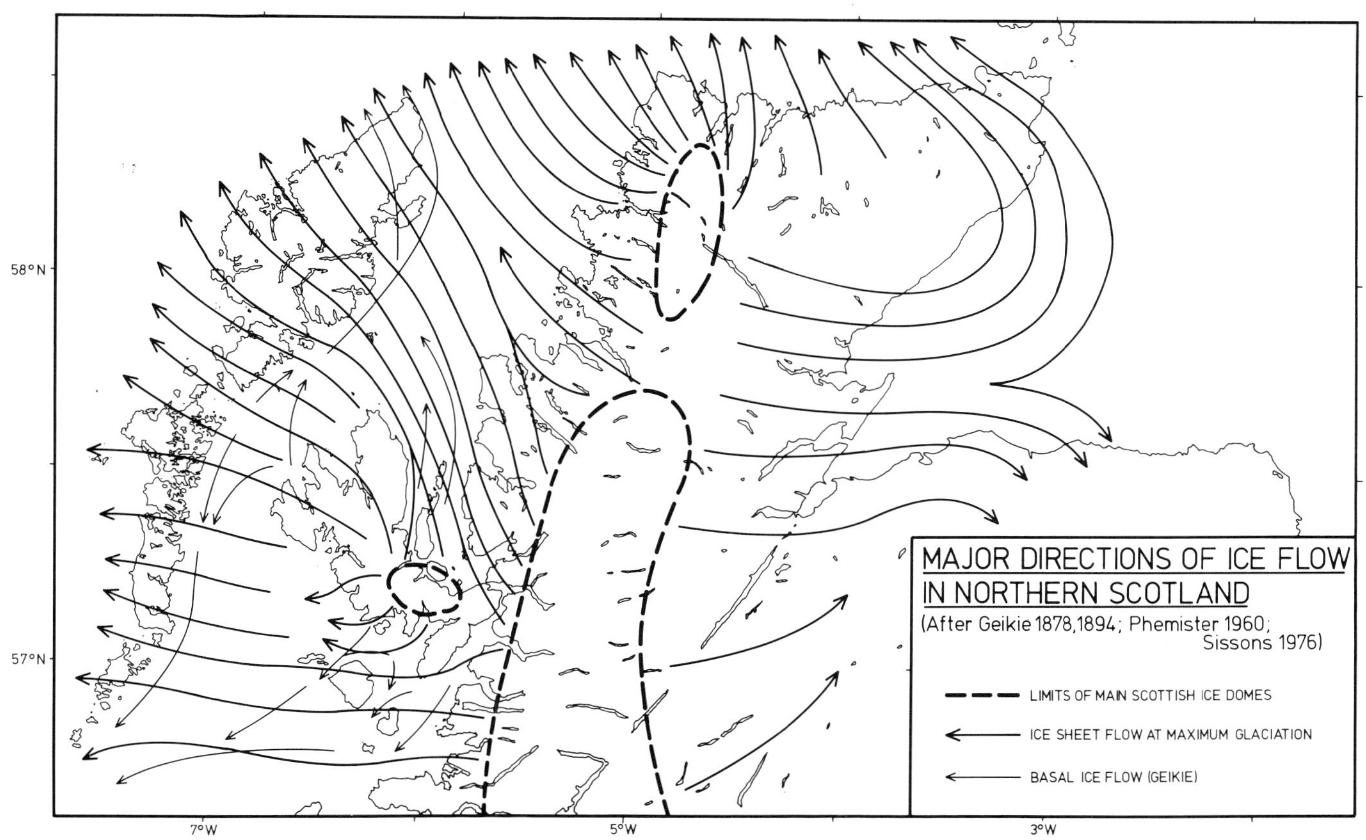

Fig. 6.11.

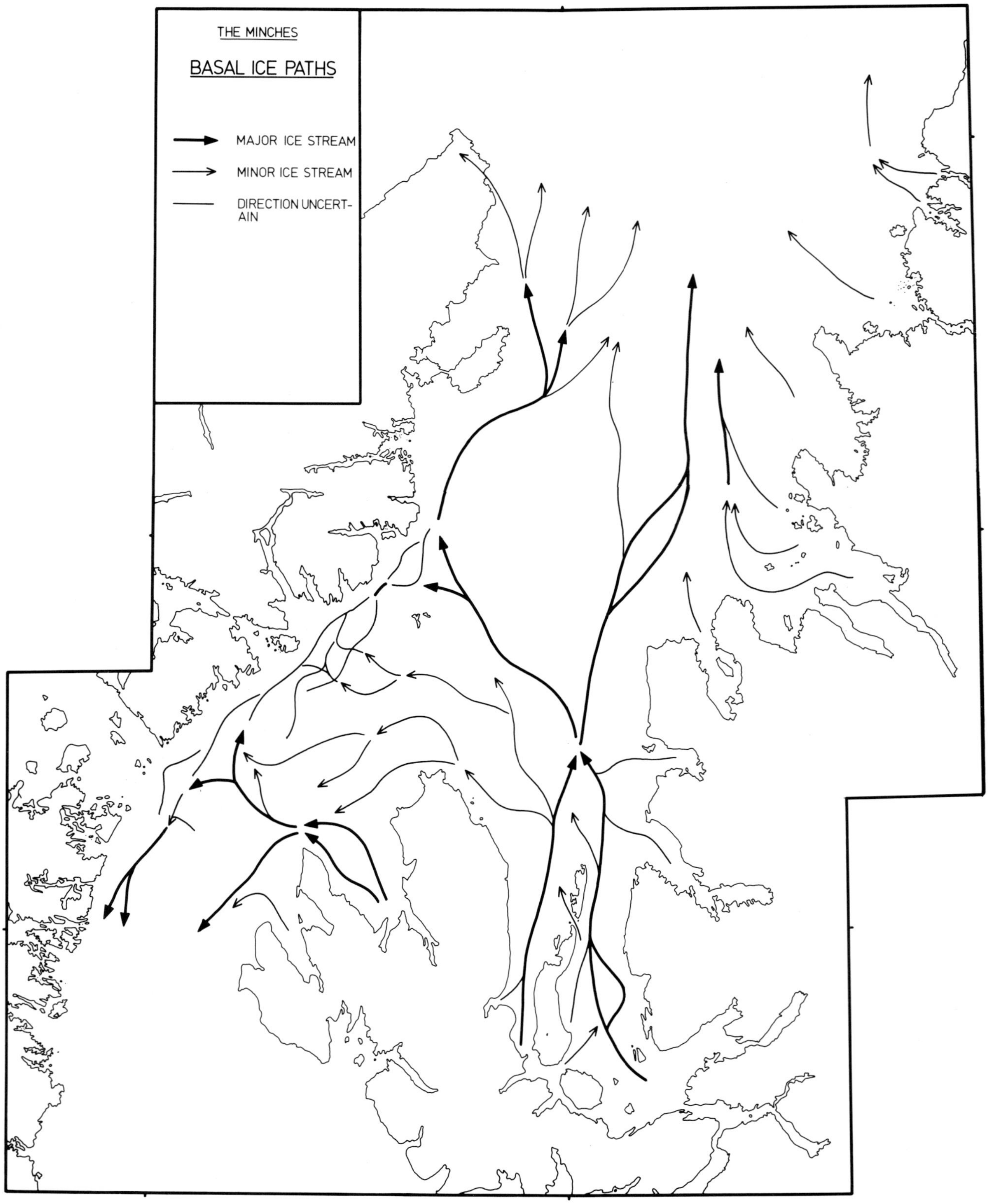
THE MINCHES
BASAL ICE PATHS
MAJOR ICE STREAM
MINOR ICE STREAM
DIRECTION UNCERT-AIN

Fig. 6.12.

the rubbly till found on the rest of the island. It is a stiff boulder clay, containing broken marine shells, and rock types included within it (as noted by Geikie, 1878) are gneisses, red sandstones, conglomerates, liver-coloured quartz pebbles (possibly from Cambrian conglomerates) and fragments of quartz porphyry which do not occur in situ on Lewis. The detailed bathymetric map now available allows us to postulate ground-ice movements with greater precision than Geikie (Fig. 6.12).

Not only did Geikie, in 1894, describe many glacially over-deepened rock basins found on land in Scotland, but he also discussed several major basins in the sea floor between the mainland and the Outer Hebrides. Some of these he termed *deflection basins*, for they were formed where ground-ice was deflected by topographic obstructions, resulting in locally increased erosive power. The form of some of the individual basins was subtly analysed; for instance, Geikie noticed that the deepest parts of the Inner Sound occur where the channel is narrowest, this constraint having caused an increase in the vertical erosive capability of the ice. The basin shallows as it broadens into the North Minch. Also, the Sound of Raasay is less deep than the Inner Sound because it was protected from the full pressure of the deflected mainland ice by the islands of Raasay and Rona, and carried only local ice from Skye. With the benefit of improved bathymetric maps and a greater knowledge of the geology of the sea floor, Sissons (1967, 1976) again examined the submerged rock basins off North-West Scotland and attributed their great size (relative to those on the adjacent land areas) to the exposure of soft Mesozoic rocks on the sea floor, and the existence of faults, for instance, in the Inner Sound, where much of the over-deepening can be attributed to erosion along fault lines. Deflection of lower ice levels along the Minches Fault, at the steeply rising Harris and Uist coasts, caused intense erosion, and resulted in the breaching of the main pre-glacial watershed at the Sound of Shiant. Off Lewis, the Minches Fault is further from the coast and late Palaeozoic–early Mesozoic sediments, the Stornoway Beds, occur to the west; consequently, it becomes a decreasingly important bathymetric control and, north of Stornoway, the Hebridean coastline is gently shelving. On the mainland side of the North Minch, the absence of a pronounced boundary between the Mesozoic sediments and the Precambrian basement has produced another gently shelving coastline with large embayments.

The over-deepened rock basins of the Minches are typical of the glacial features of northern Scotland in that they are products of erosion by powerful ice streams. Features resulting from deposition are less common than in other parts of Britain. Extensive deposits of ground moraine from the main ice cap have been recognised on the Scottish mainland only in Caithness. In the north-west, boulder clay is found solely in the valley floors (where it may be locally up to 30 metres thick) and on low plateaux, where it is of uncertain age. Scotland is an area where glacial erosion dominated and evidence of preceding glaciations has been removed by later ice sheets. Thus tills are rarely thick, at least where they have been examined on land.

Sissons (1967) suggested that much of the material removed from the Scottish Highlands has been deposited in the surrounding seas, and certainly, Geikie's model of ground-ice movement would imply the continuation of the north Lewis boulder clays into the North Minch. Up to 22 m of boulder clay is exposed on the north-west coast of Lewis at Swainbost. The basement rocks of Lewis give rise to a rubbly till that contains little silt or clay, and contrasts with the stiff clays of the north. Mesozoic rocks, because of their high shale and clay content, would generate a more clayey till and the passage of the ground-ice over the Mesozoic sediments of the Minches floor may have resulted in this distinctive Lewis boulder clay. However, no included rock fragments of Mesozoic age have been reported from these deposits.

Besides boulder clay, another variety of glacial deposit, found on land in North-West Scotland, is terminal and lateral moraine from valley glaciers. Charlesworth (1956) mapped strings of terminal moraines in the Scottish Highlands, but many of his attempted correlations are now doubted. However, Charlesworth's maps suggest that, in many valleys, moraines are continued beyond the present coast, particularly in Gruinard Bay, Gairloch and Loch Torridon. Sissons (1967) maintains that the bars that often occur at the mouths of the sea lochs are partly of solid rock and partly of moraine deposits. A large lateral moraine in the North Minch will be described later. Glacial (and post-glacial) sedimentation is the remaining factor that has contributed to the present bathymetry, having partially obscured the erosional surfaces.

A late glacial palaeogeographic reconstruction

During glacial and post-glacial times, the location of the Scottish shoreline reflected both eustatic changes of sea level and the isostatic movement of the land surface. The nature and extent of eustatic changes are still controversial, especially those of the post-glacial period, and workers in different parts of the world have arrived

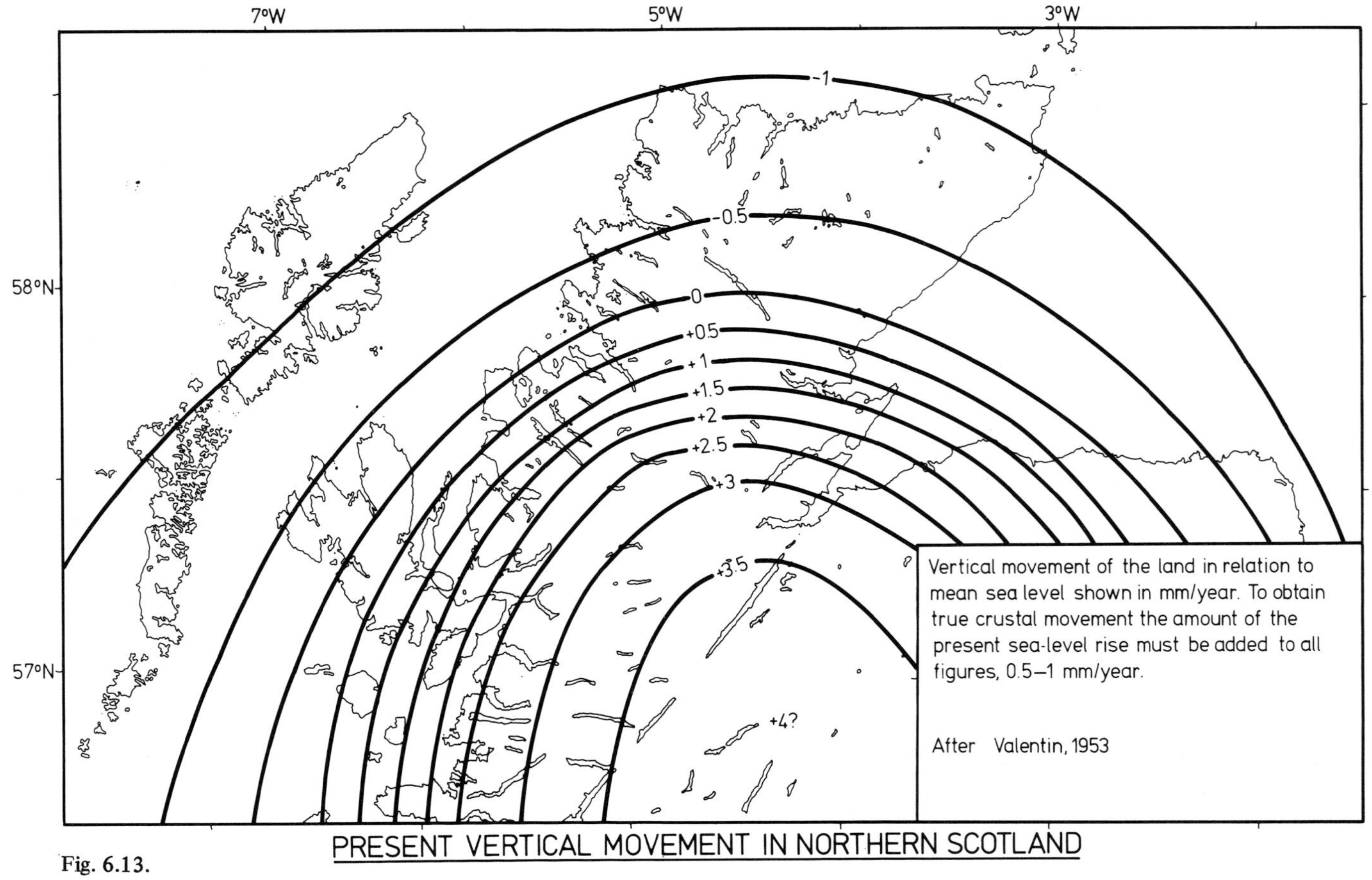

Fig. 6.13.

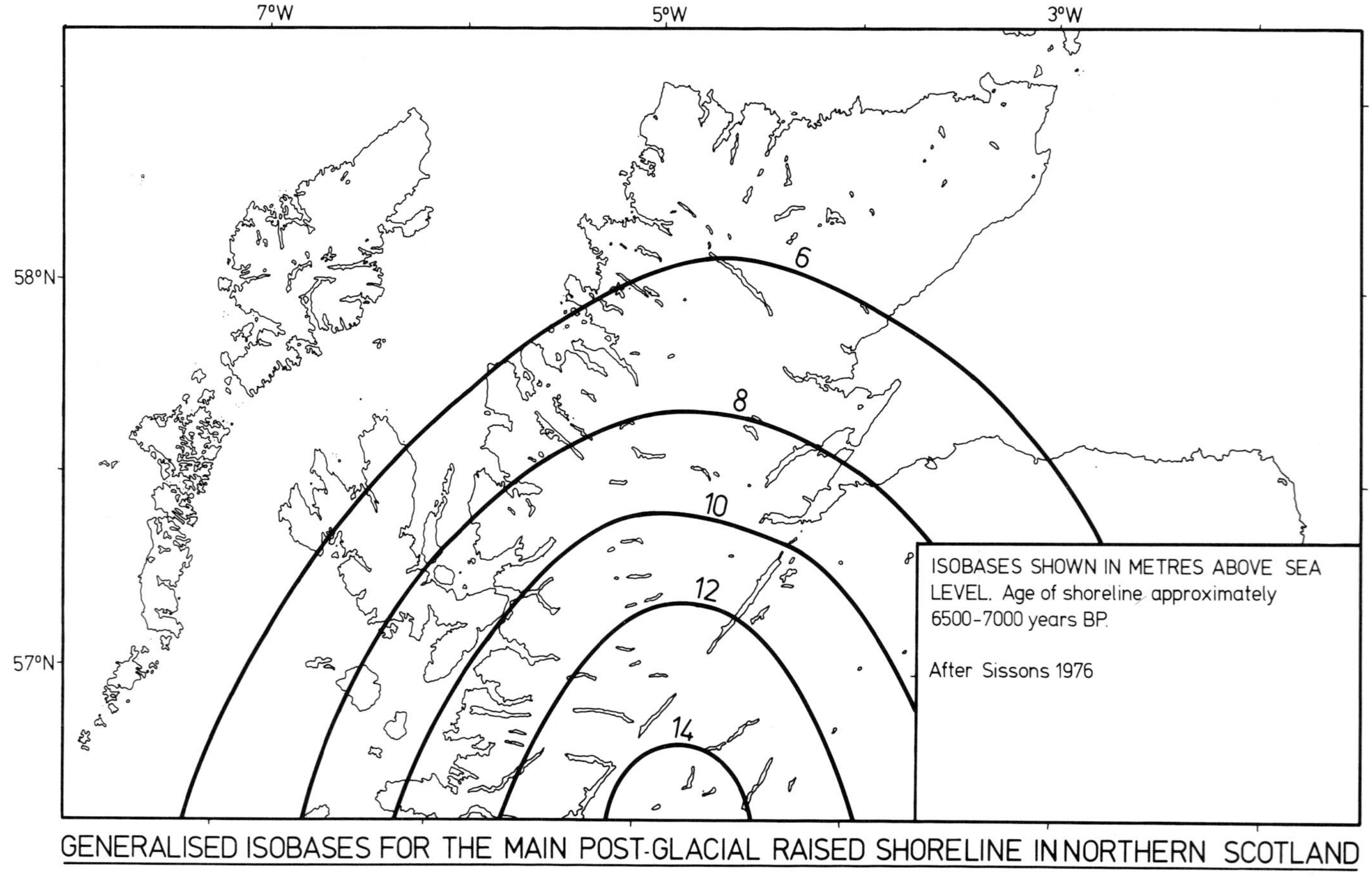

Fig. 6.14.

at varying conclusions. Godwin and Willis (1958) suggest that a fall in sea level of the order of 100 m occurred at maximum glaciation, with a steady rise since then to about −50 m at 11,000 years B.P. and present sea-level at 5500 years B.P. However, a maximum lowering of up to 150 m has been postulated by others (Emery, 1969).

Upon the problem of determining precise eustatic changes must be superimposed the complexity of isostatic adjustment in Scotland, following the removal of the Pleistocene ice sheet. Rates of relative sea-level change have been evaluated by Valentin (1953) from tide-gauge observations at various British ports, and used to construct a map (Fig. 6.13) showing the present vertical movement of the British Isles. The lines indicating movement for western Scotland are termed "conjectural" owing to a lack of local records. The map suggests that maximum rebound is taking place in the southern part of the Western Highlands, close to where ice thickness was at a maximum. Valentin's map is very similar to ones developed by Sissons (1967, 1970, 1976), on which the relative rise of the land surface during postglacial times has been expressed in terms of the measured uplift of raised beaches along the Scottish coasts (Fig. 6.14). Beaches closer to the southern parts of the Western Highlands have been uplifted to a greater extent than those more remote from the centre of ice accumulation. There are no well documented raised beaches in the Outer Hebrides. Both Sissons' and Valentin's maps suggest that, while most of the Scottish mainland and the Inner Hebridean islands have risen relative to the sea during post-glacial times, the Outer Hebrides and the Durness area in the extreme north-west of the mainland have suffered submergence. Glacial rebound is limited to a roughly elliptical dome, centred on the Western Highlands, the margin of which is marked by a hinge-line of zero vertical movement passing through the Little and North Minches. The sinking of out-lying areas may be due to the mass transfer of sub-crustal material inwards to compensate for the uplift of the central dome.

While the two maps suggest that there is a regular decrease in the rate of uplift away from the Western Highlands towards the Outer Hebrides, Stephens and Synge (1966) have pointed out that sharp changes in the rebound rate could occur across faults or other pronounced geological boundaries. The Minches would seem to provide ample opportunities for this phenomenon to occur, although abrupt changes in the rebound rate would be extremely difficult to detect as the faults and major geological boundaries in this region tend to be situated near to the coastline, beyond which isostatic movements cannot be as conveniently measured as on land. McCann and Chorley (1967), through trend surface analysis of post-glacial shorelines, have discovered that Skye has experienced greater isostatic depression than the mainland. In addition to short-term crustal movements caused by the glaciation must be added certain long term effects, for instance, Sissons' suggestion that the Minches Basin has been sinking relative to the Highlands since Tertiary times.

An attempted palaeogeographic reconstruction was made by Geikie as early as 1894, suggesting a late-glacial shoreline well to the west of the Outer Hebrides, and the existence of large fresh-water lochs in the rock basins of the Minches. However, Geikie assumed a greater lowering of sea level than present research allows, and the necessary adjustment of his coastline would turn his fresh-water lochs into salt-water bodies. A modern estimate of submergence in the Uists (Ritchie, 1966) gives a maximum of 120 m over the last 12,000 years. Valentin's map predicts a 120-m submergence over the same period. Whereas in outlying areas, a steady submergence has been caused by the combination of a rising sea-level and a falling land surface, within the rebound dome the position of the coast line during post-glacial times has, of course, been determined by the relative rates of eustatic and isostatic movement. With a history of this degree of complexity, and a lack of positive evidence of the age of raised beach deposits, wave cut benches, etc., the north-west Scottish area will have to wait for a more confident interpretation of its recent palaeogeography than can be achieved here. Figure 6.15 is a sketched reconstruction drawn for a date of 12,000 years B.P. (the approximate time of the melting of the main Scottish ice sheets), assuming a relative submergence of 120 m in the Outer Hebridean islands. The coastline west of the Outer Hebrides follows the 120-m submarine contour, but these outlying areas may have stood even higher relative to the Uists. Towards the mainland, within the dome of rebound, the generally steep coastline would have restricted the geographical migration of the shore, which must have stood at the levels of the raised beaches now exposed on land.

Figure 6.15 reveals that much of the Minches area was dry land at the time of the waning of the main ice sheets. However, the deep deflection basin running along the Outer Hebridean coast formed a long and narrow sea loch, extending northwards as far as the Eye Peninsula, and entering the Atlantic south-west of Barra Island. The rock basins of the Inner Sound and Sound of Raasay would similarly have been flooded, and tenuously connected to the main sea loch through a channel between the Shiant Isles and the Shiant East Bank, and to the

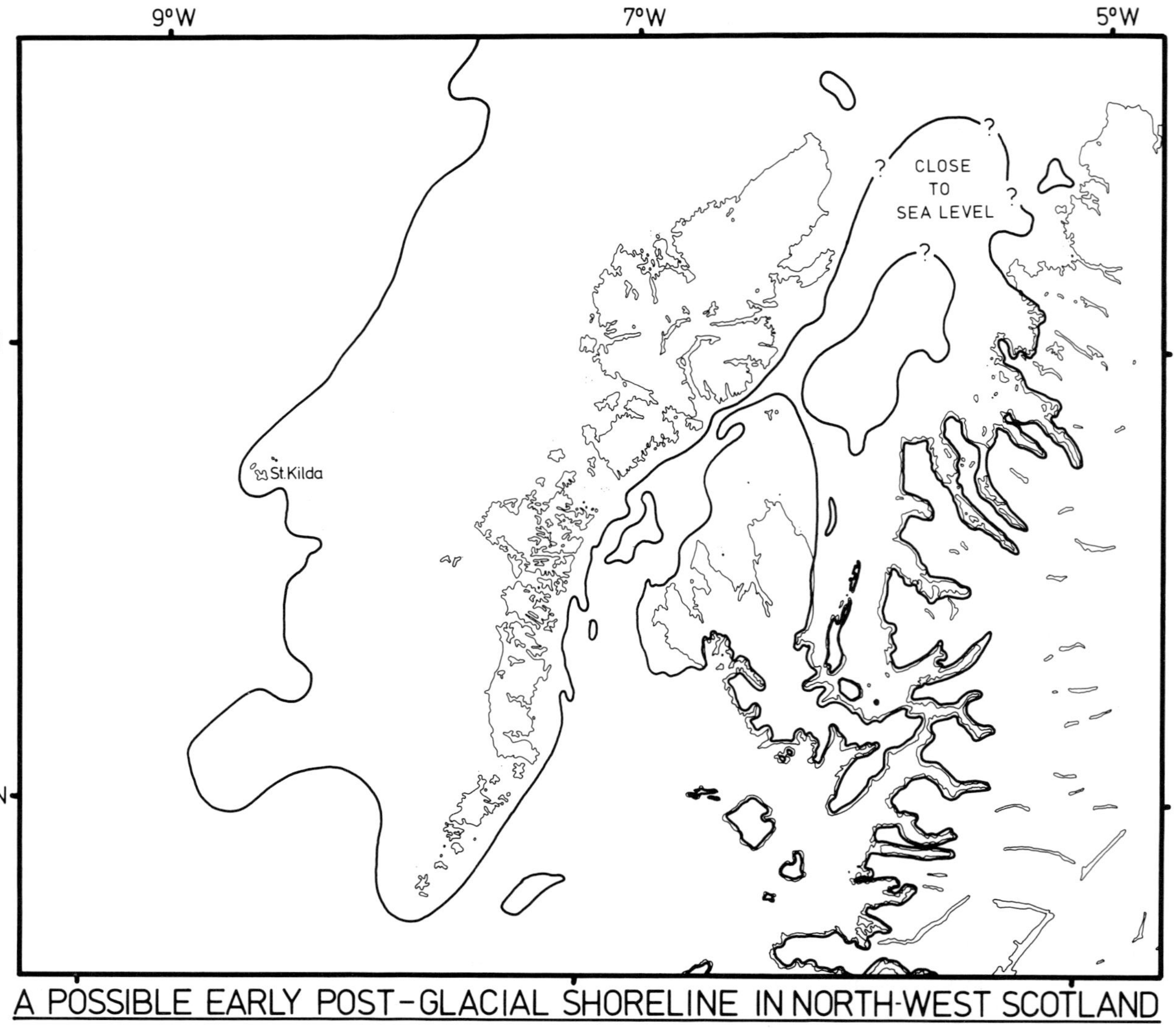

Fig. 6.15.

Loch Broom—North Minch area by a second channel off Rubh'Re. A 120-m submergence of the Uists, when interpolated around the margin of the rebound dome into the North Minch, would put its floor (with average depth today of about 120 m) very close to sea-level in late-glacial/early post-glacial times. With such an even topography, the precise location of the coast on the North Minch floor is hard to determine. As sea level quickly rose in the early post-glacial millenia, the major sea lochs must have grown to merge with scattered fresh-water bodies occupying smaller hollows on the Minches floor, the main Atlantic coastline, meanwhile, retreating eastwards towards the present Outer Hebrides, and southwards into the North Minch.

There is abundant evidence for the recent submergence of shallow seas around the Outer Hebrides. At low spring tides, peat and tree stumps are exposed on the foreshore in many parts of the islands, and have been dated as between 8000 and 9700 years old (Sissons, 1967). Peat has been brought up on ship's anchors many kilometres west of Lewis. The last parts of the Minches to have been flooded must have been the Shiant East Bank and the Skye—Shiant Platform, and many dangerous rocks are still exposed between Skye and the Shiant Isles.

Glacial sedimentation in the Minches

Identification of Quaternary sediment on seismic profiles

Several hundred kilometres of continuous seismic profiles have been recorded by both the I.G.S. and Uni-

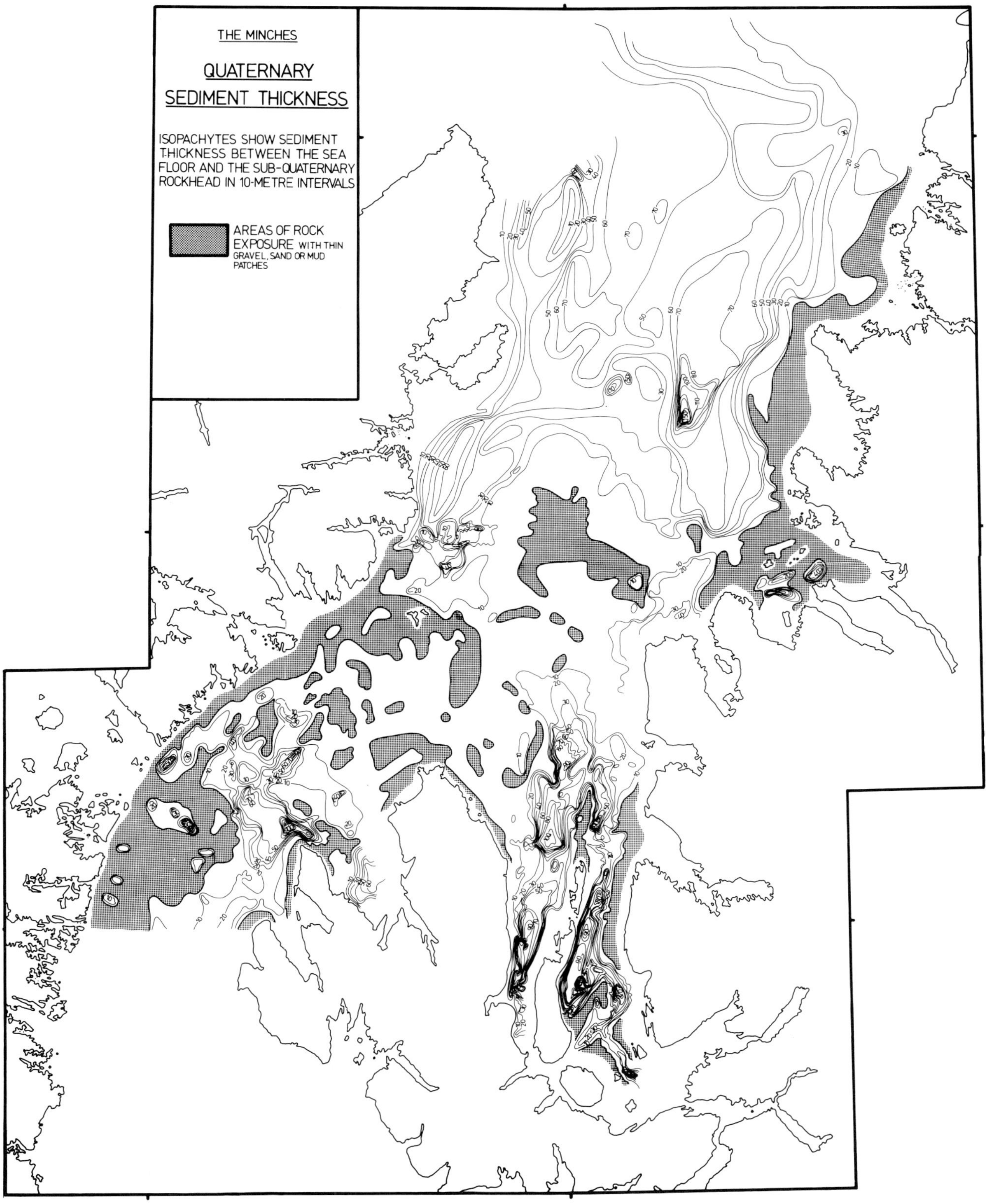
THE MINCHES
QUATERNARY
SEDIMENT THICKNESS
ISOPACHYTES SHOW SEDIMENT THICKNESS BETWEEN THE SEA FLOOR AND THE SUB-QUATERNARY ROCKHEAD IN 10-METRE INTERVALS
AREAS OF ROCK EXPOSURE WITH THIN GRAVEL, SAND OR MUD PATCHES

Fig. 6.16.

versity College London in the Minches (Fig. 6.3). Several models of profiling system were employed, a Sparker sound source being generally used. The Quaternary sediments (glacial and post-glacial) are generally easily discernible as a distinctive unit between the sea floor and a major seismic reflector, frequently associated with an angular unconformity, which can be interpreted as the sub-Quaternary rockhead. This reflector can be correlated with the location of solid rock at several I.G.S. borehole sites in the Minches (some of which are described below). Where exposed at the sea floor, it has been cored, yielding rock samples from Precambrian to Jurassic and Tertiary ages. The same reflector can be traced into the Sea of the Hebrides, where it has similarly been identified as the sub-Quaternary rockhead (Binns et al., 1974a). Three sections of typical continuous seismic profiles are presented in Fig. 6.18 and display the appearance of the rockhead reflector and the overlaying Quaternary sediments.

In order to calculate the thickness of the Quaternary sediment sequence a seismic velocity of 1.8 km/s was assumed (Fig. 6.16). Places where rockhead is exposed at the sea floor were determined with the additional examination of side-scan sonar records (Fig. 6.4) and also PDR/"Pinger" data, which were available for the majority of geophysical traverses. The isopachyte map in Fig. 6.16 reveals that Quaternary sediment occurs in three major bodies, separated by areas that possess less than 10 m of sediment and which have considerable rock exposure. The sediment bodies are, in places, up to 150 m in thickness.

Of the three major sediment bodies, that of the North Minch is easily the largest. It extends from the margins of the Shiant East Bank beyond the northern limits of the area of study. Maximum thicknesses within the body occur in two over-deepened and partially infilled troughs, which have been eroded by ice along the paths of important pre-glacial drainage channels, although other factors have also contributed to their development. The Cape Wrath Bank, like most of the mainland coastal margin, is covered by less than 10 metres of sediment, although no rock outcrop occurs. The 10-m isopachyte off the mainland coast closely follows the line of the Permo-Triassic–Precambrian contact. The Greenstone Ridge, extending from Greenstone Point northwards into the North Minch, has no expression on the rockhead surface, and its morainic nature will be established below. Local ice-flow paths strongly suggest that it formed as a lateral moraine between coalescing ice-flows from the Inner Sound and the Loch Broom–Gruinard Bay area. The Rubh'Re–Kebock Head basement ridge, which offered greater resistance to glacial erosion than the sediments either side, now stands exposed. Only a minor sediment body exists in the region of the ridge, around Rubh'Re itself. To the south, the second largest Quaternary sediment body in the Minches occupies the Inner Sound and the Sound of Raasay. Near the island of Rona, 80–90 m of sediment have been deposited over a rockhead eroded to 410 m below present sea-level. The resulting sea floor, at 320 m, is, nevertheless, the deepest part of the British continental shelf. The remaining sediment body is that of the Loch Snizort–Little Minch area, to the west of Skye.

History of glacial sedimentation

The study of continuous seismic profiles not only allows us to measure the thickness of Quaternary sediments but also to infer something of their nature. Several sediment types can be recognised on seismic records with a reasonable degree of certainty, a feature which Binns et al. (1974a) have used in an attempt to describe lithological variation in the Sea of the Hebrides. In North-West Scotland, boreholes that pass through appreciable thicknesses of Quaternary sediment, and which can be used to correlate observed lithology with seismic character, are limited in number, making interpretation of reflection records over a wide area only tentative. In the Minches, just four boreholes provide Quaternary sequences thick enough for the "calibration" of seismic profiles.

In the Sea of the Hebrides, Binns recognised four formations within the Quaternary sequence possessing contrasting textures on seismic profiles. At the base, Formation 1 was observed to rest directly upon rockhead and was identified in boreholes as till. It is distinguishable on seismic records by a fine internal texture of jumbled and hummocky reflectors and an uneven upper surface. Above, Formation 2 consists of firm, poorly sorted sandy muds, with scattered lithic pebbles, and is identifiable on seismic records by a generally structureless texture with a few widely spaced horizontal reflectors. The formation is most thickly developed in troughs along the courses of major ice streams, and while micropalaeontological assemblages show it to be of marine origin, it appears to have been deposited in close association with ice and may represent a flow-till as described by Carey and Ahmad (1961). Formation 3 is recognised on the reflection records by closely-spaced horizontal reflectors, which fade laterally and cannot be correlated with any lithological feature in the boreholes. Samples show it to consist of poorly-sorted sandy muds, somewhat similar to Formation 2, but with less pebbles. Micropalaeontological evidence in the Sea of the Hebrides suggests it is of late-glacial age, and Binns et al. hold it to be of peri-glacial origin, laid down a short distance from a retreating ice front. Formation 4 consists of a thin layer of modern sediment, only a few metres thick, the properties and origin of which will be studied separately in some detail in succeeding sections.

Seventeen selected continuous seismic profile lines are shown in the form of line drawings in Fig. 6.17. In addition, Fig. 6.18 presents three original records with particularly well developed glacial sequences. Fig. 6.18a displays typical Formation 1 reflectors over an indistinct rockhead reflector, and is the same section of record as shown on Line 8 (Fig. 6.17d), although it is there reversed. Figure 6.18b is part of Line 6 (Fig. 6.17c) and shows Formation 1 and 2 reflectors in the south of the Stornoway Basin. Finally, typical Formation 1 and 3 reflectors are shown on Fig. 6.18c, this record corresponding to the right half of Line 10 (Fig. 6.17c).

Seismic profile Line 1 (Fig. 6.17b) passes within 1 km of I.G.S. borehole 72/33. The borehole penetrated 37 m of Quaternary sediment before reaching rockhead; at the nearest point on the seismic line there are approximately 32 m of sediment. On the seismic record, the upper 11 m of the Quaternary succession has strong and continuous horizontal reflectors of Formation 3 type, and the remainder, a mass of jumbled and hummocky reflectors, is tentatively identified as Formation 1. The borehole log and micropalaeontological report for the site have been made available by the I.G.S. and can be summarised as in Table 6.1.

Allowing for the distance between the borehole and the seismic line, there is quite a reasonable correlation between the two. The late-glacial shelly boulder clays of the lower part of the sequence agree with the Formation 1 identification from the seismic data, while the upper muddy sands appear to be Formation 3 peri-glacial deposits, passing upwards into modern sands at the sea floor.

Borehole 71/77, at the mouth of Loch Snizort, is located at the northern end of seismic Line 9 (Fig. 6.17c) and proved 34 m of dark olive-grey boulder clay over the rockhead. Line 9 shows jumbled and hummocky reflectors of Formation 1 type throughout almost all of the Quaternary, again confirming the correlation of this seismic texture with till. Borehole 72/73, near Tolsta Head, Lewis, penetrated over 27 m of boulder clay under a thin veneer of modern sand, which again correlates well with reflectors of Formation 1 type on seismic Line 17 (Fig. 6.17f). Borehole SLN 55 was located at the junction of seismic Lines 15 and 16 (Fig. 6.17f), both of which display a mass of hummocky, but subhorizontal reflectors throughout the whole Quaternary sequence. At this site, the borehole penetrated over 65 m of dark grey, tenaceous, sandy and pebbly clay before reaching rockhead, and the character of the seismic profile appears to be somewhere between Formations 1 and 2. Geological data for this and the preceding boreholes has been kindly provided by Dr. N.G.T. Fannin of the Institute of Geological Sciences (personal communication, 1976).

The seismic profiles from the North Minch are, in general, not as easy to interpret as those from further south, and give rise to problems apparently not encountered by Binns et al. (1974a,b) in the Sea of the Hebrides. There, they were able to arrange the four formations in strict chronological order. Line 1 from the

TABLE 6.1

Lithology and micropalaeontology summarised from I.G.S. borehole 72/33

Lithology	Depth (m)	Micropalaeontology
Fine grey sand with shell fragments, becoming more muddy with increasing depth.	Sea-floor	modern boreal fauna fewer cold water elements
Stiff grey-brown boulder clay with shell fragments and small pebbles	25	climatic amelioration - more temperate forms
Sandy boulder clay near base with cobble size rock fragments	30	increased number of cold water forms (?Loch Lomond Re-advance) Late Glacial fauna of mixed cold and warm water forms (?Allerød Interstadial)
Rock	36.8	

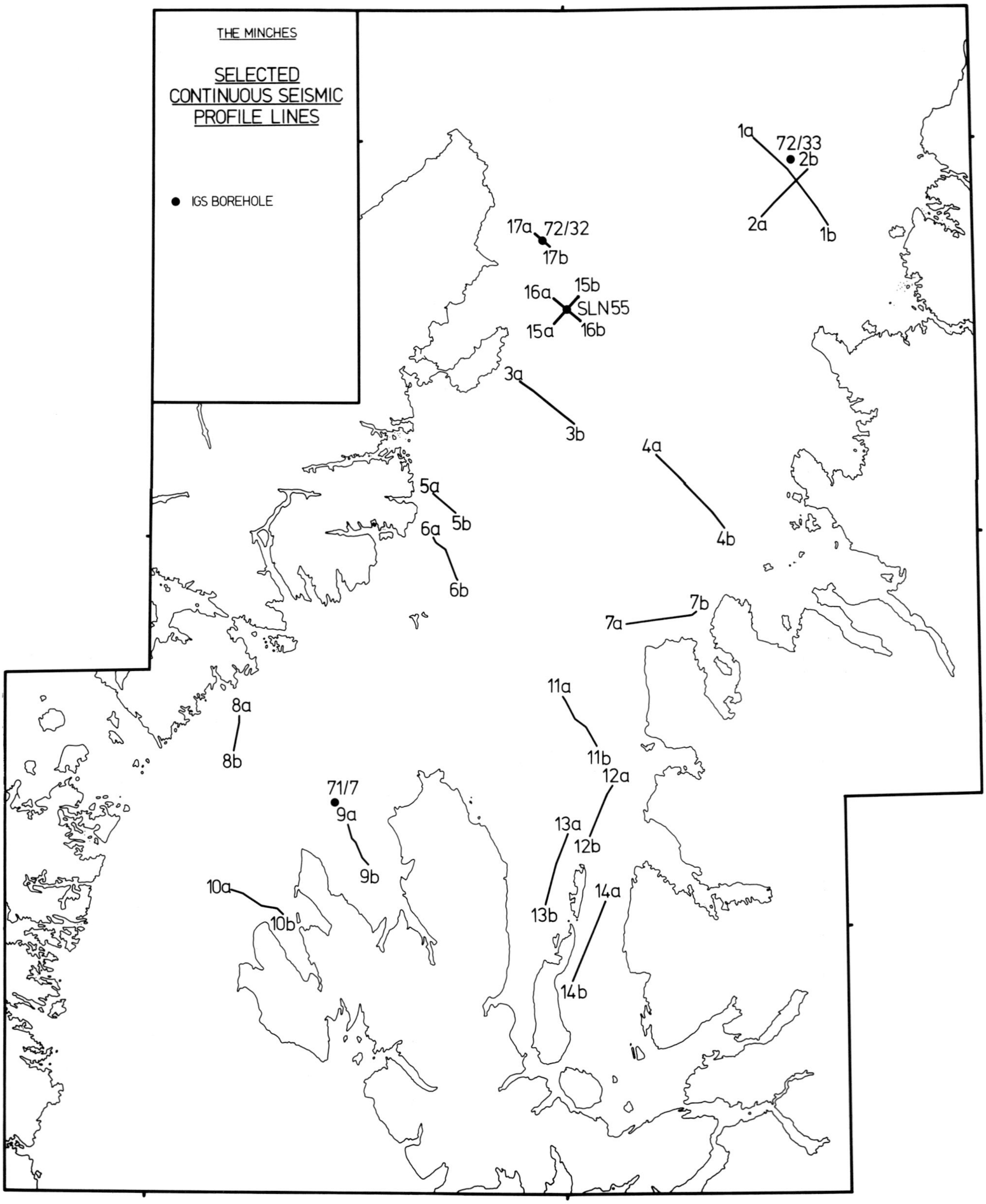
THE MINCHES
SELECTED
CONTINUOUS SEISMIC
PROFILE LINES
IGS BOREHOLE
1a
72/33
2b
2a
1b
17a
72/32
17b
16a
15b
SLN55
15a
16b
3a
3b
4a
4b
5a
5b
6a
6b
7a
7b
11a
11b
12a
8a
8b
71/7
9a
9b
13a
12b
10a
10b
13b
14a
14b

Fig. 6.17a.

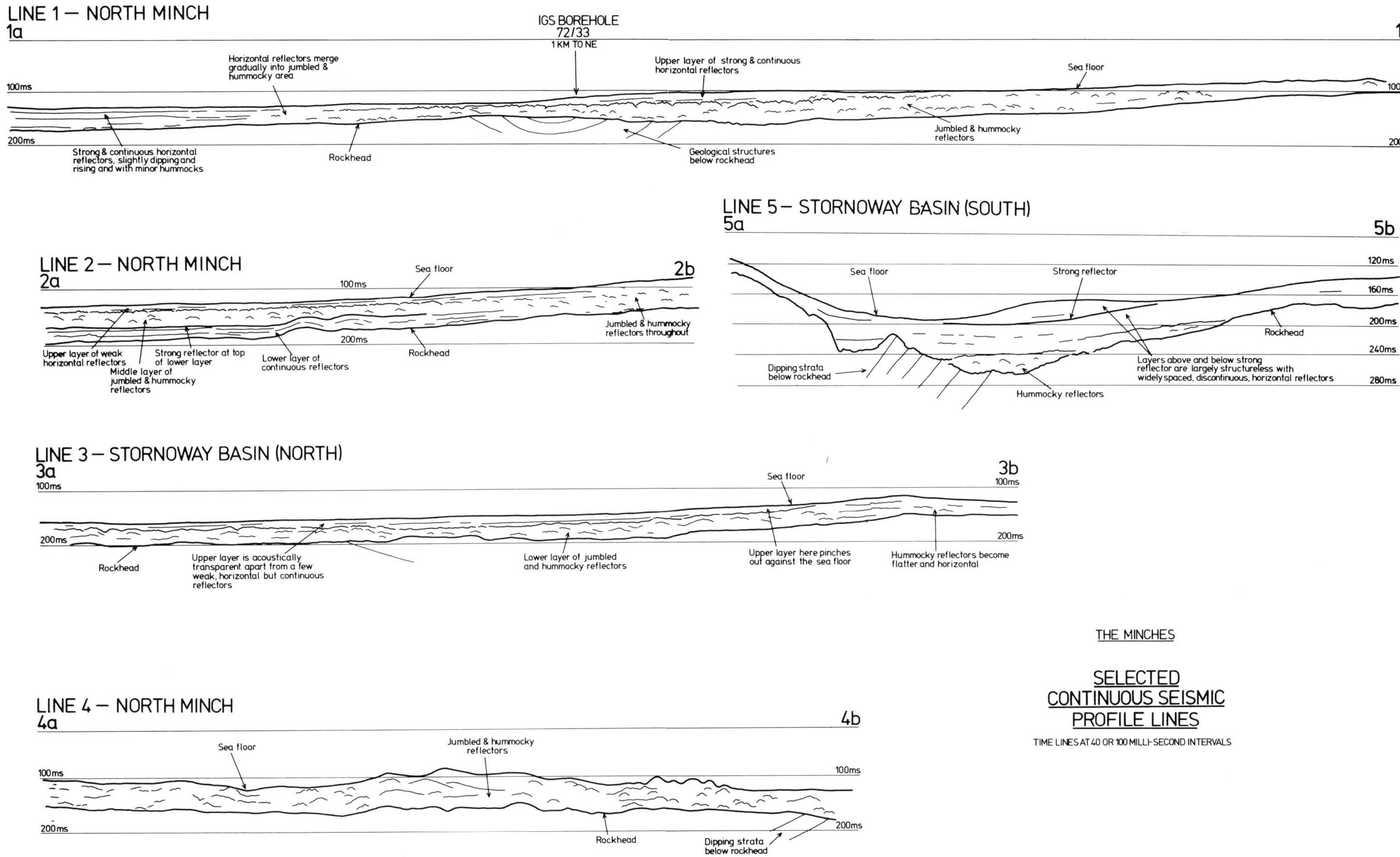
LINE 1 — NORTH MINCH
1a
IGS BOREHOLE 72/33
1 KM TO NE
1b
Horizontal reflectors merge gradually into jumbled & hummocky area
Upper layer of strong & continuous horizontal reflectors
Sea floor
100ms
200ms
Strong & continuous horizontal reflectors, slightly dipping and rising and with minor hummocks
Rockhead
Geological structures below rockhead
Jumbled & hummocky reflectors
LINE 5 — STORNOWAY BASIN (SOUTH)
5a
5b
120ms
160ms
200ms
240ms
280ms
Sea floor
Strong reflector
Rockhead
Dipping strata below rockhead
Layers above and below strong reflector are largely structureless with widely spaced, discontinuous, horizontal reflectors
Hummocky reflectors
LINE 2 — NORTH MINCH
2a
2b
100ms
200ms
Sea floor
Jumbled & hummocky reflectors throughout
Upper layer of weak horizontal reflectors
Strong reflector at top of lower layer
Middle layer of jumbled & hummocky reflectors
Lower layer of continuous reflectors
Rockhead
LINE 3 — STORNOWAY BASIN (NORTH)
3a
3b
100ms
200ms
Sea floor
Rockhead
Upper layer is acoustically transparent apart from a few weak, horizontal but continuous reflectors
Lower layer of jumbled and hummocky reflectors
Upper layer here pinches out against the sea floor
Hummocky reflectors become flatter and horizontal
THE MINCHES
SELECTED CONTINUOUS SEISMIC PROFILE LINES
TIME LINES AT 40 OR 100 MILLI-SECOND INTERVALS
LINE 4 — NORTH MINCH
4a
4b
Sea floor
Jumbled & hummocky reflectors
100ms
200ms
Rockhead
Dipping strata below rockhead

Fig. 6.17b.

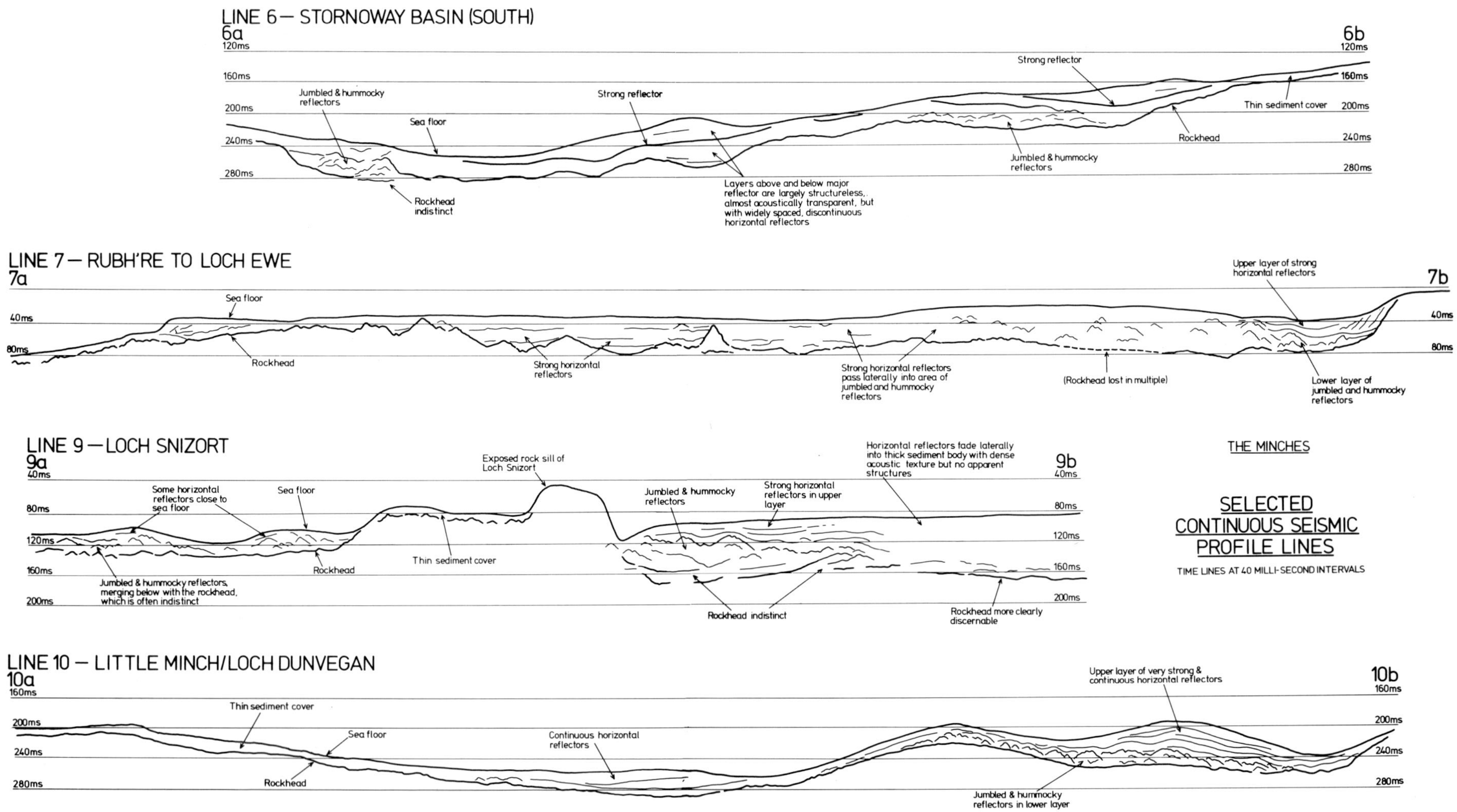

LINE 6 — STORNOWAY BASIN (SOUTH)
6a
6b
120ms
160ms
200ms
240ms
280ms
Jumbled & hummocky reflectors
Sea floor
Rockhead indistinct
Strong reflector
Layers above and below major reflector are largely structureless, almost acoustically transparent, but with widely spaced, discontinuous horizontal reflectors
Strong reflector
Jumbled & hummocky reflectors
Rockhead
Thin sediment cover
LINE 7 — RUBH'RE TO LOCH EWE
7a
7b
40ms
80ms
Sea floor
Rockhead
Strong horizontal reflectors
Strong horizontal reflectors pass laterally into area of jumbled and hummocky reflectors
(Rockhead lost in multiple)
Upper layer of strong horizontal reflectors
Lower layer of jumbled and hummocky reflectors
LINE 9 — LOCH SNIZORT
9a
9b
40ms
80ms
120ms
160ms
200ms
Some horizontal reflectors close to sea floor
Sea floor
Jumbled & hummocky reflectors, merging below with the rockhead, which is often indistinct
Rockhead
Thin sediment cover
Exposed rock sill of Loch Snizort
Jumbled & hummocky reflectors
Strong horizontal reflectors in upper layer
Rockhead indistinct
Horizontal reflectors fade laterally into thick sediment body with dense acoustic texture but no apparent structures
Rockhead more clearly discernable
THE MINCHES
SELECTED CONTINUOUS SEISMIC PROFILE LINES
TIME LINES AT 40 MILLI-SECOND INTERVALS
LINE 10 — LITTLE MINCH/LOCH DUNVEGAN
10a
10b
160ms
200ms
240ms
280ms
Thin sediment cover
Sea floor
Rockhead
Continuous horizontal reflectors
Upper layer of very strong & continuous horizontal reflectors
Jumbled & hummocky reflectors in lower layer

Fig. 6.17c.

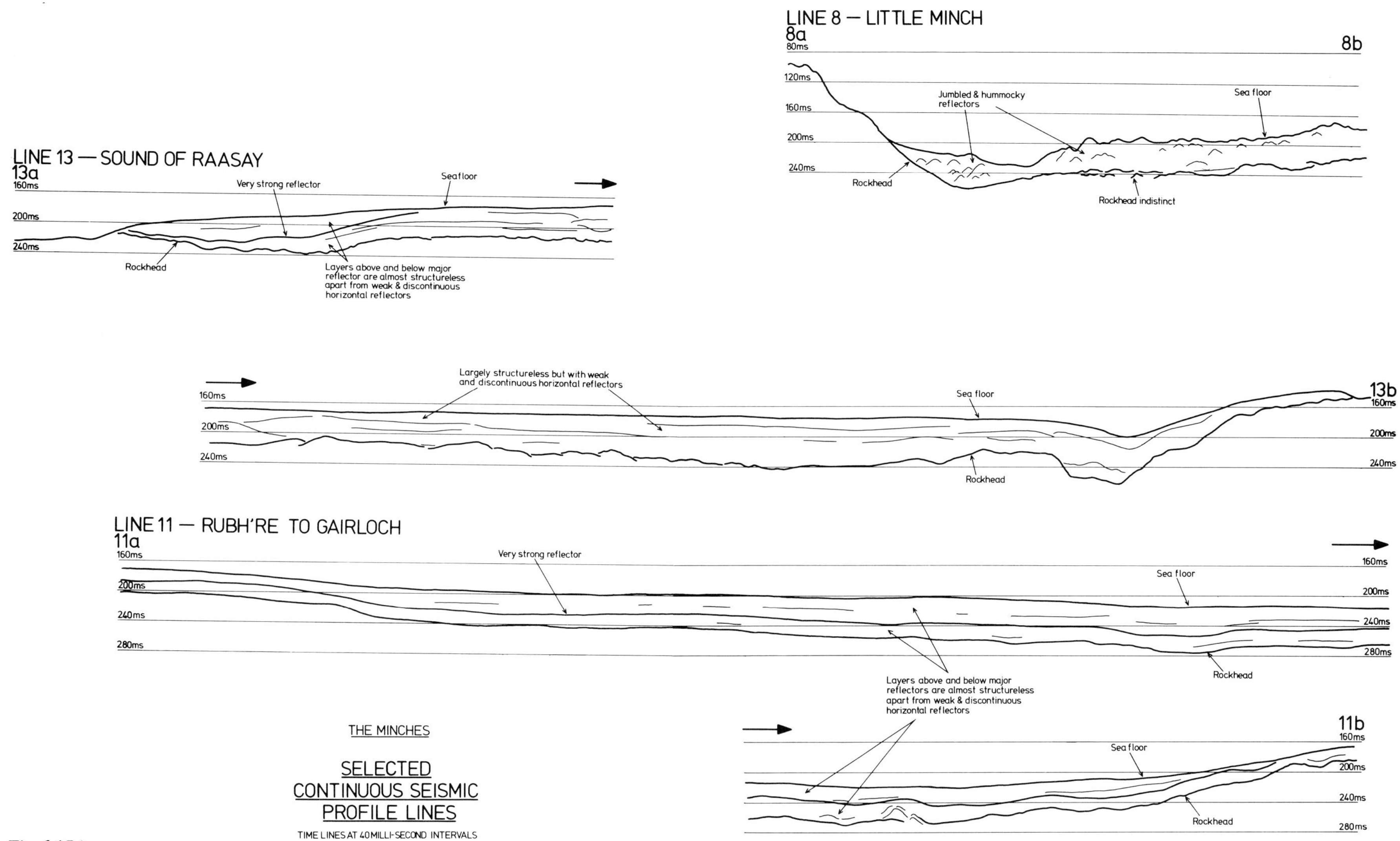
LINE 13 — SOUND OF RAASAY
13a
160ms
200ms
240ms
Very strong reflector
Seafloor
Rockhead
Layers above and below major reflector are almost structureless apart from weak & discontinuous horizontal reflectors
13b
LINE 8 — LITTLE MINCH
8a
80ms
120ms
160ms
200ms
240ms
8b
Jumbled & hummocky reflectors
Sea floor
Rockhead
Rockhead indistinct
Largely structureless but with weak and discontinuous horizontal reflectors
Sea floor
Rockhead
LINE 11 — RUBH'RE TO GAIRLOCH
11a
160ms
200ms
240ms
280ms
Very strong reflector
Sea floor
Rockhead
Layers above and below major reflectors are almost structureless apart from weak & discontinuous horizontal reflectors
11b
Sea floor
Rockhead
THE MINCHES
SELECTED CONTINUOUS SEISMIC PROFILE LINES
TIME LINES AT 40 MILLI-SECOND INTERVALS

Fig. 6.17d.

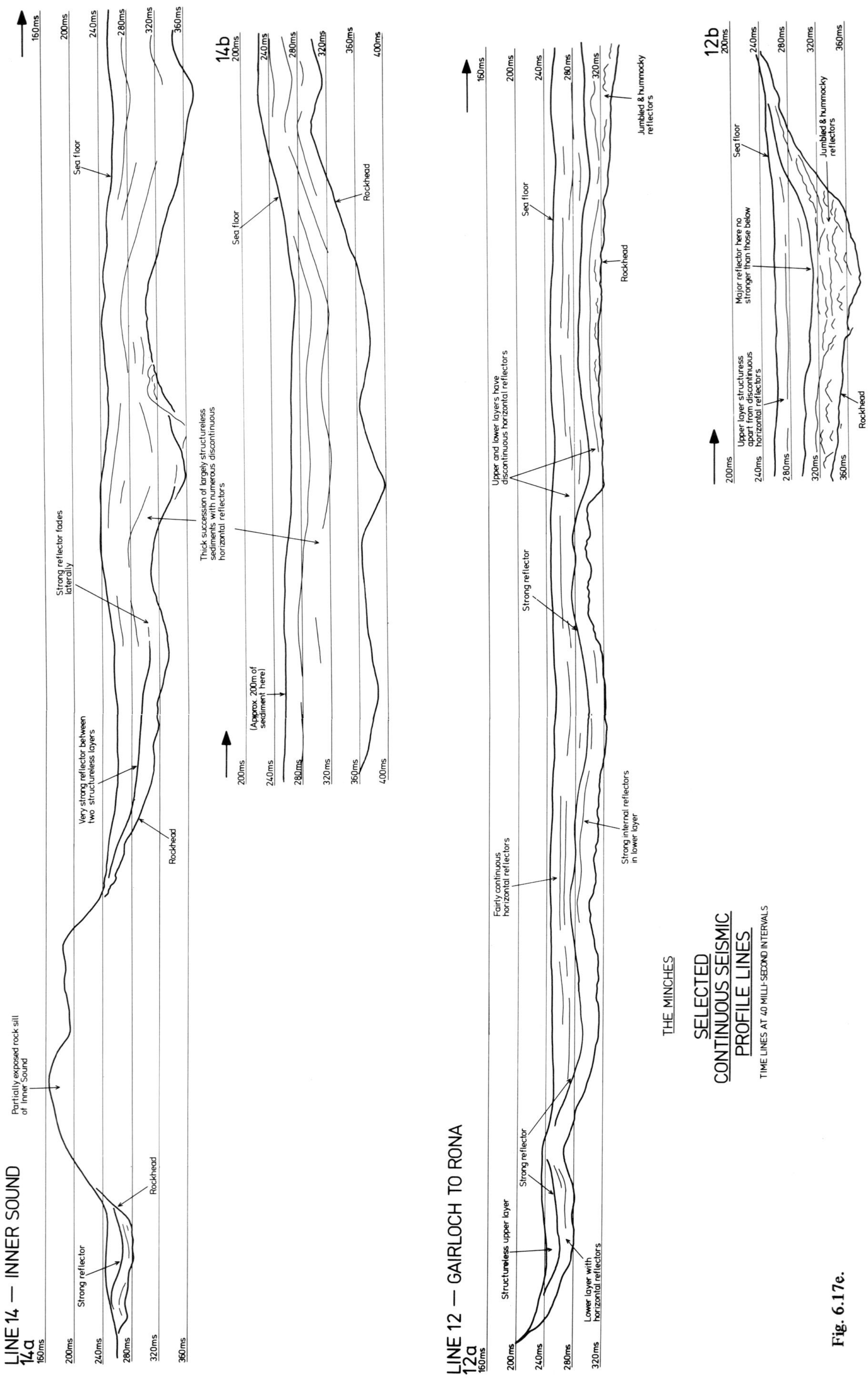
LINE 14 — INNER SOUND
14a
160ms
200ms
240ms
280ms
320ms
360ms
Strong reflector
Rockhead
Partially exposed rock sill of Inner Sound
Very strong reflector between two structureless layers
Rockhead
Strong reflector fades laterally
Sea floor
Thick succession of largely structureless sediments with numerous discontinuous horizontal reflectors
14b
(Approx. 200m of sediment here)
Sea floor
Rockhead
400ms
LINE 12 — GAIRLOCH TO RONA
12a
Structureless upper layer
Strong reflector
Lower layer with horizontal reflectors
Fairly continuous horizontal reflectors
Strong internal reflectors in lower layer
Strong reflector
Upper and lower layers have discontinuous horizontal reflectors
Sea floor
Rockhead
Jumbled & hummocky reflectors
12b
Upper layer structureless apart from discontinuous horizontal reflectors
Major reflector here no stronger than those below
Sea floor
Jumbled & hummocky reflectors
Rockhead
THE MINCHES
SELECTED CONTINUOUS SEISMIC PROFILE LINES
TIME LINES AT 40 MILLI-SECOND INTERVALS

Fig. 6.17e.

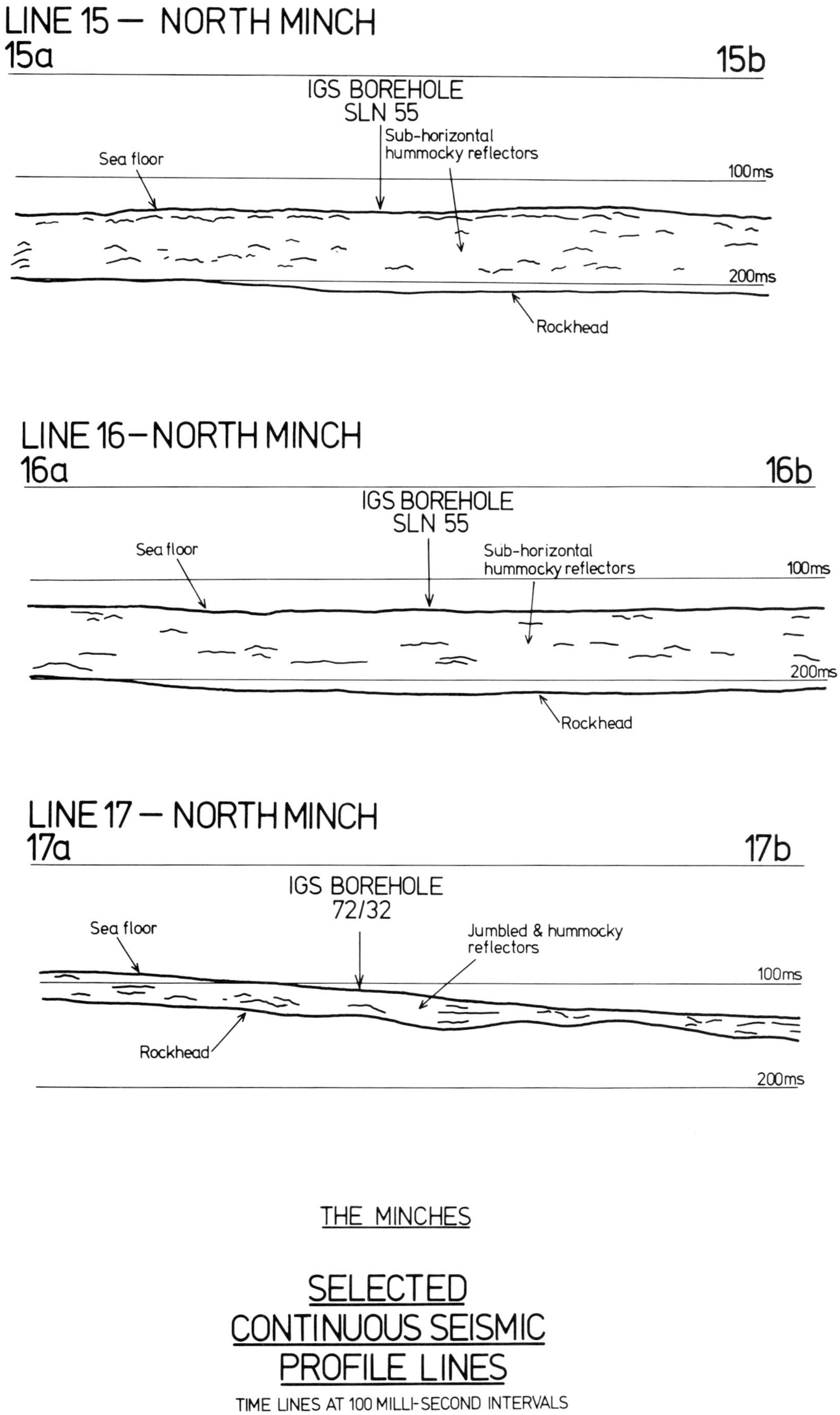
LINE 15 — NORTH MINCH
15a
15b
IGS BOREHOLE
SLN 55
Sub-horizontal
hummocky reflectors
Sea floor
100ms
200ms
Rockhead
LINE 16—NORTH MINCH
16a
16b
IGS BOREHOLE
SLN 55
Sea floor
Sub-horizontal
hummocky reflectors
100ms
200ms
Rockhead
LINE 17 — NORTH MINCH
17a
17b
IGS BOREHOLE
72/32
Sea floor
Jumbled & hummocky
reflectors
100ms
Rockhead
200ms
THE MINCHES
SELECTED
CONTINUOUS SEISMIC
PROFILE LINES
TIME LINES AT 100 MILLI-SECOND INTERVALS

Fig. 6.17f.

North Minch, however, reveals Formation 1 tills passing laterally, to the north-west, into Formation 3 sediments. Line 2 also shows lateral changes in this area, this time towards the south-west, with a Formation 1 sequence developing a distinguishable sub-unit of Formation 3 reflectors at the base, *overlain by* Formation 1 deposits. Although the lithological reality of these seismically-indicated facies changes cannot be confirmed without further borehole evidence, it appears that in the North Minch at least, we are not able to speak of the four formations as time-stratigraphic units.

Possible explanations for facies variations such as this within the glacial sediments come from a consideration of land–sea-level relationships in late glacial times (Fig. 6.15) and the theories and observations of modern Arctic sedimentation given by Carey and Ahmad (1961). The nature of the North Minch sediments is consistent with their formation below a wet-base glacier on a grounded shelf zone in late glacial times. When a wet-base glacier (one with basal ice at melting temperature and basal meltwater) passes from land onto an area lying below sea level, the glacier gradually decreases in weight until it loses contact with the sea bed at some distance from the shore. Strong horizontal shear beneath the ice pushes and rolls the till forwards towards the "buoyancy line", at which point a submarine slope is developed which advances with the deposition of till foresets (see Carey and Ahmad, 1961, fig. 1B). This has the effect of determining a base level for the till (somewhat analogous to the base-level concept for marine sediments in shallow water) dependent upon sea level and ice thickness. Till will build up rapidly to the base of the ice or will be cut down to that level while the foreset slope advances, forming a level-topped till platform that may be many kilometres in extent.

Line 7 (Fig. 6.17c) crosses the minor sediment body north of Rubh'Re and reveals that it possesses a remarkably flat surface over an undulating rockhead, and a steep scarp edge overlooking the deep water of the northern Inner Sound. Sediments are of Formation 1 and 3 types. Formation 1 tills occur towards the end of the line, opposite the mouth of Loch Ewe, a location at which ice probably attained a greater thickness than over the flanks of the sediment body, where Formation 3 stratified deposits were best developed under an increasingly more "buoyant" glacier. The level upper surface can be interpreted as the base level for till formation below the ice, and the steep scarp as the foreset slope at the "buoyancy line". The occurrence of Formation 3 reflectors within the sediment body suggests that they may not only represent peri-glacial deposits, as Binns' work in the Sea of the Hebrides shows, but can also indicate material of sub-glacial origin. Micropalaeontological evidence and radiocarbon dating of 8680 ± 250 years B.P. for Formation 3 deposits in the Sea of the Hebrides (Binns et al., 1974b), as well as the faunal assemblages from Minches borehole 72/33 described above, prove that at least some sediments associated with Formation 3-type reflectors have a peri-glacial origin, but it now seems clear that some may also be fluvially stratified deposits laid down below a wet-base glacier on a grounded shelf.

A reconsideration of Lines 1 and 2 (Fig. 6.17b) amplifies this point. On these tracks, Formation 1 reflectors occur throughout the glacial sequence on the higher ground towards the Cape Wrath Bank. The palaeogeographic reconstruction (Fig. 6.15) suggests that this area was above sea level in late-glacial times, while the central North Minch formed a shallow sea. The site of I.G.S. borehole 72/33 must have been very close to the coast at this time. It contains marine fossils at many horizons, indicating an age for the succession from the late-glacial to modern, spanning the Allerød Interstadial and Loch Lomond re-advance. Formation 3 reflectors appear at the top of the succession and their age indicates a peri-glacial origin. However, proceeding downslope on Lines 1 and 2, the Formation 1 deposits pass laterally into Formation 3, sometimes lying below Formation 1. The Formation 3 reflectors again must represent fluvial deposits laid down below a glacier becoming increasingly "buoyant" downslope. Likely conditions for the development of such a late-glacial sequence are described by Carey and Ahmad (1961):

> "Under a wet base glacier, eskers may develop at the glacier-till contact far below sea level, and if there is a rise of sea level owing to partial amelioration of climate, these river-washed channel sediments might be buried under firstly sub-marine outwash sands and silts with interbedded flow-tills; then under forset tills; then covered by a substantial thickness of normal till; all with a continuously warming climate and representing a single retreat phase. Yet such a section would probably be misinterpreted as a record of complex oscillating glacial and interglacial conditions."

The last remark calls for a re-examination of the supposedly inter-glacial marine sediment that occurs between two boulder clays at the Butt of Lewis. The age of this deposit has never been precisely determined. Thick boulder clays containing broken marine shells and erratics from the Scottish mainland are not restricted to the area near the Butt of Lewis, but also occur in much smaller isolated exposures at Tolsta Head and on the Eye Peninsula, where they have been attributed with the same age and origin (Baden - Powell, 1938), although direct correlation is not possible with the major exposure to the north. Von Weymarn and Edwards (1973) have obtained a radiocarbon date of

27,333 ± 240 years B.P. for a thin lens of silt occurring on Tolsta Head, between the boulder clay and Lewisian bedrock. This date corresponds with others recently obtained in Scotland and is attributed to an interstadial period, before the growth of the glaciers responsible for the Lewis boulder clays and their mainland-derived erratics. Although no direct correlation is possible between Tolsta Head and the Butt of Lewis, the evidence that the former boulder clay is entirely the product of the most recent glaciation throws even more doubt on the validity of the "inter-glacial" marine beds at the Butt. However, the present elevation of the Butt of Lewis marine beds, should they be of late glacial and not inter-glacial origin, is hard to explain on an island where subsidence is so well documented, and perhaps further research will shed more light on the raised beaches of Lewis and Uist, the former apparently of late-glacial age, fleetingly mentioned by Baden-Powell (1938).

To the south, interpretations of the seismic records show greater similarities to the Sea of the Hebrides. Although not proved lithologically by borehole data within the Minches, the deep glacial troughs appear to have thick Formation 2 deposits (Lines 5, 6, 11, 12, 13, 14) with underlying patches of Formation 1 till over the rockhead. As sea water must have penetrated under the ice in these troughs during late-glacial times, the conclusion of Binns et al. (1974) that, in the Sea of the Hebrides, Formation 2 is a glacial-marine sediment laid down in association with major ice streams, is also compatible with the evidence from the Minches. Carey and Ahmad (1961) have postulated that glacial-marine sediments such as these may consist largely of flow-tills, laid down after slumping of foreset tills on the steep sediment face under the glacier "buoyancy line". Binns et al. (1974a) have also suggested this mode of origin accounts for the firm till-like lithology of the marine Formation 2 sequence sampled in a Sea of the Hebrides borehole. In the Minches, Formation 2 is best developed near to the area where the major ice flows would have attained buoyancy in late glacial times, that is at the inner ends of troughs like the Stornoway Basin and the Inner Sound, and it thins in the downstream direction. Seismic profiles reveal that the thin Quaternary sediment of the Little Minch can be explained by an almost complete absence of Formation 2 deposits, even in the deep basins, although Formation 1 tills are well developed (Line 8, Fig. 6.17d). The Little Minch may have been beyond the range of the till flows originating in the north, while most of the sediment

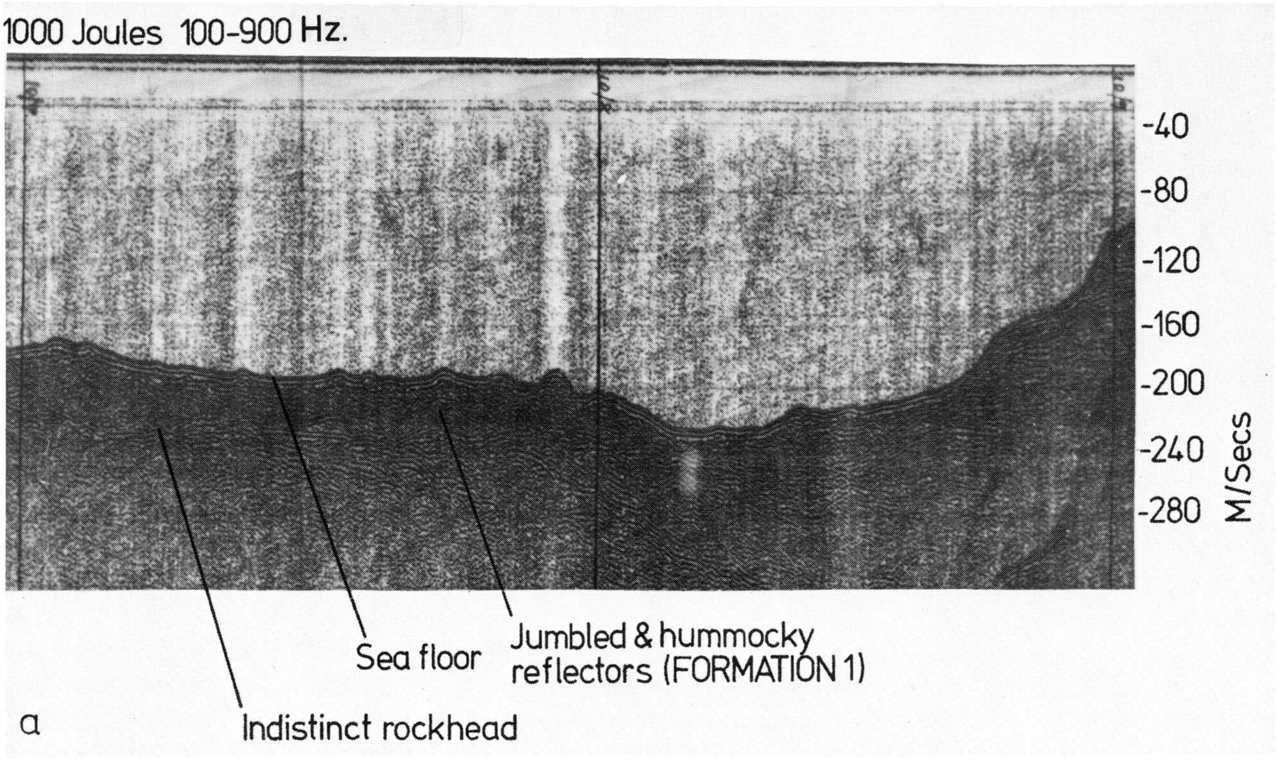

Fig. 6.18a (for legend see next page).

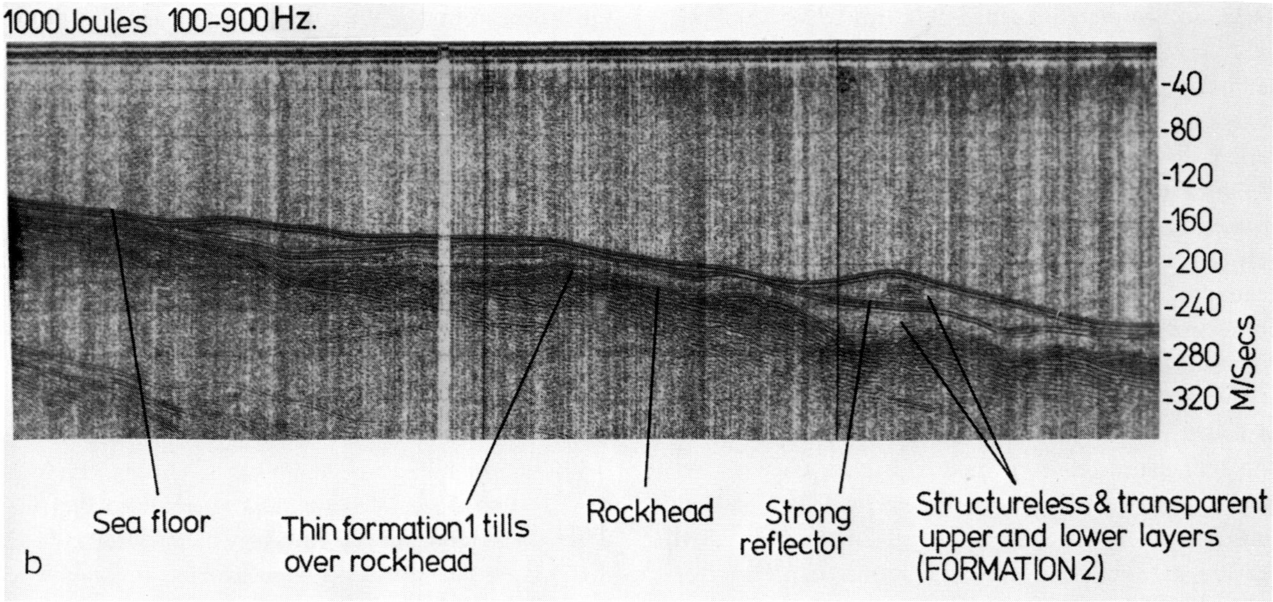

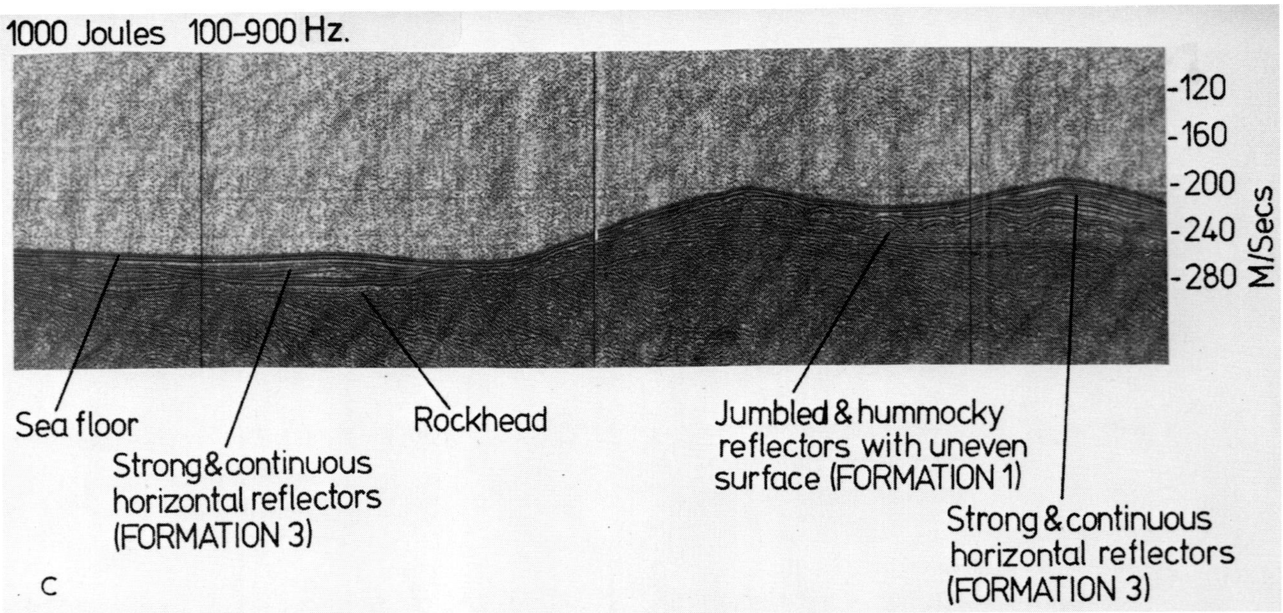

Fig. 6.18. Photographs of selected continuous seismic profiles.

emanating from the Loch Snizort glacier was trapped behind the large rock sill (Line 9, Fig. 6.17c). At the northern end of the Stornoway Basin, Lines 3, 15 and 16 (Figs. 6.17b and 6.17f) and borehole SLN 55 (described above) indicate a gradual passage from the Formation 2 deposits of the trough to the Formation 1 and 3 deposits characteristic of the North Minch. One feature of Formation 2 sediments in the Minches, not mentioned by Binns et al. (1974a), is the development of a strong reflector within the formation itself. Various relationships between the upper and lower parts of the formation are shown on Lines 5, 6, 12 and 13. The reflector can be followed from north of the Stornoway Basin to the mouth of the Inner Sound, where it eventually disappears (Line 14). Without lithological or palaeontological evidence, the significance of this seismic horizon is debatable.

The modern environment: the circulation of the Minches

The movement of water in the Minches can be attributed to a combination of three factors: tidal currents, non-tidal currents and waves. This section compiles and discusses published oceanographic data, and presents some new information, obtained during recent research cruises.

Tides and tidal currents

Detailed descriptions of tidal currents in the Minches are included in the Admiralty Pilot for the west coast of Scotland, and these are designed to complement the information published on the Admiralty chart coverage of the area. The Admiralty Tidal Stream Atlas gives data for current flow over the course of an entire tidal cycle, and information relating to the passage of the tidal wave through Scottish waters can be gathered from the Admiralty Tide Tables for Major and Secondary Ports. These four Admiralty sources were used to construct a series of maps (Figs. 6.19–6.23) which summarise the nature of the tide and tidal currents in the Minches.

The North-West Scottish tide is principally semi-diurnal and is controlled by a minor Atlantic amphidromic system between Scotland and Iceland, which sweeps co-tidal lines northwards along the western British coasts. The islands of North-West Scotland act only as a partial barrier to the passage of this progressive tidal wave over the continental shelf allowing it to move from the open Atlantic into the Sea of the Hebrides, then into the Little Minch, North Minch, and finally, past the Butt of Lewis, where it rejoins the main part of the wave, which has, meanwhile, moved similarly northwards west of the Outer Hebrides. The passage of the wave through these waters is reflected in the time of high tide at the various ports and harbours along the coasts. High tide arrives almost two hours earlier at Barra Head than it does at Cape Wrath. Using the tidal prediction data for major and secondary ports in the Admiralty Tables, it is possible to map the progress of the co-tidal line of high water through the Sea of the Hebrides and the Minches (Fig. 6.19). Unfortunately for our purposes, most of the Admiralty data are for harbours situated in lochs, bays and sounds adjoining the main water bodies (these offering sheltered anchorages) which have modified tides due to shallow water friction effects and the particular geometry of the subsidiary body. Some minor interpolations of the data have, therefore, been necessary to depict the passage of the wave through more open water.

The map reveals that the progress of the tidal wave through the Minches is not simple. First, there are indications that a standing wave oscillation exists in the Inner Sound. The co-tidal line that represents high water 20 min before Ullapool approaches the mouth of the Inner Sound, but no later lines exist within it and the time of high water is practically constant throughout the body. Uniform arrival of high water is one of the three principal diagnostic features of a standing wave in a short gulf (Doodson and Warburg, 1941). A second characteristic, similarly observed in the Inner Sound, is an increase of tidal range towards the head of the body. The Tide Tables reveal that a springs mean range of 4.5 m at Loch Ewe increases to 4.6 m at Gairloch and 4.8 m at Loch Carron, while a range of 4.3 m exists in more or less open water at the Summer Isles (see inset, Fig. 6.19). Thirdly, standing wave oscillations display maximum tidal current velocities at mid-tide, which again is a phenomenon that has been recorded in the Inner Sound, this time by a current meter located on the sea bed near the mouth of the Sound over a 28-day period; the new data will be presented below.

Outside the Inner Sound, maximum tidal current velocities are associated with high and low water, the normal condition for a passing progressive wave. The Inner Sound standing wave cannot be a simple feature, as this body itself has large subsidiary inlets, which in turn can be expected to have their own oscillations. The 4.9-m spring tidal range recorded in Loch Torridon may indicate the presence of at least one subsidiary system. In addition, the Inner Sound does not possess a closed end, as the link with the Sound of Sleat through the Kyle of Loch Alsh allows the passage of water at each tide. Here, two bodies of water having different shapes

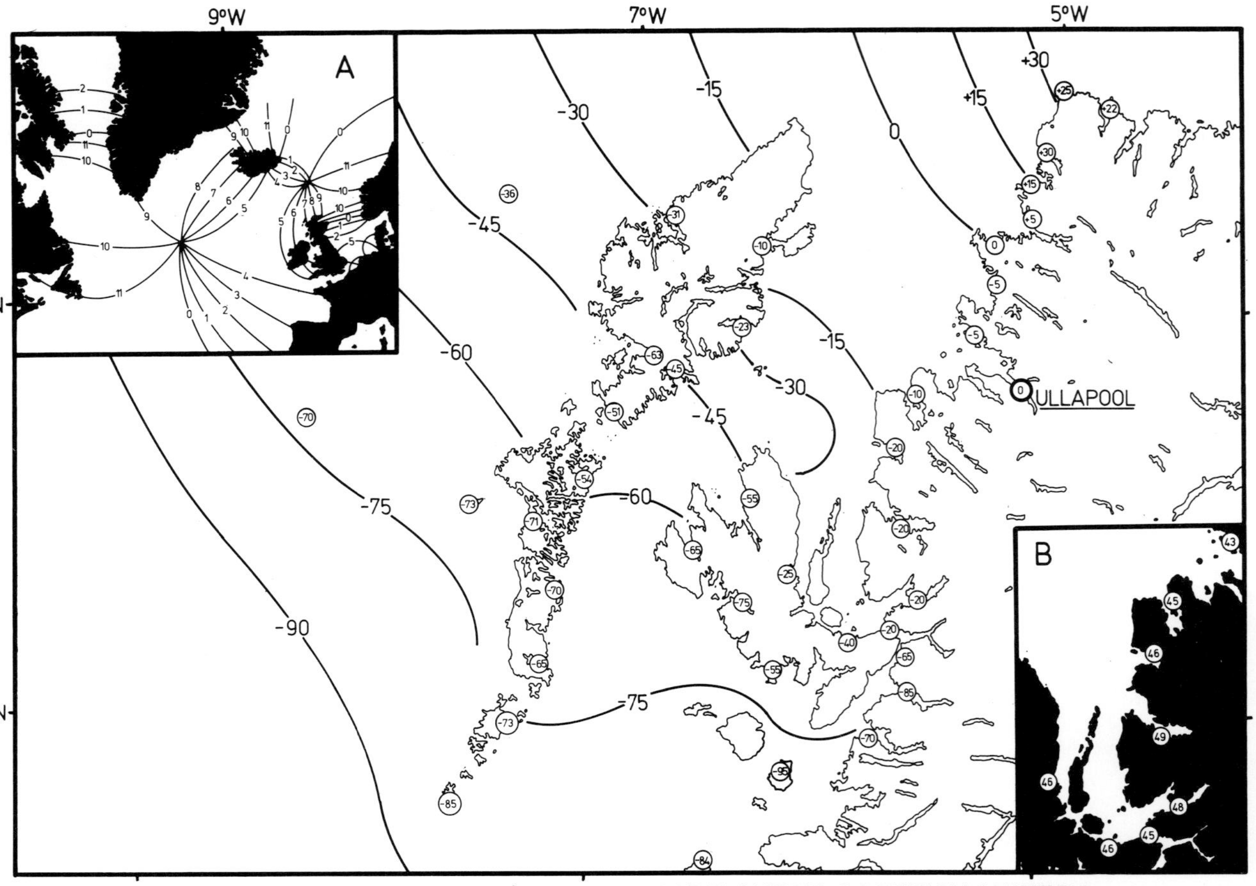

THE PASSAGE OF THE TIDAL WAVE THROUGH NORTHWEST SCOTTISH WATERS Estimated position of co-tidal lines in minutes relative to high water at Ullapool. Time of high water (in minutes relative to Ullapool) for secondary ports listed in Admiralty Tide Tables shown in circles. INSET A: CO-TIDAL LINES OF THE SEMI-DIURNAL TIDE IN THE NORTH ATLANTIC OCEAN, the major amphidromic systems controlling North West Scottish tides (from Sverdrup, Johnson & Fleming 1942). INSET B: MEAN SPRINGS TIDAL RANGE (IN CENTIMETRES) OF THE INNER SOUND AND ADJACENT AREAS (from Admiralty Tide Tables for secondary ports).

Fig. 6.19.

and orientations meet "back-to-back". Tidal gradients through the passage must be great, for the Admiralty Pilot reports dangerous turbulence in the area, associated with currents with speeds of up to 5 m/s. The tide experienced at Broadford Bay at the south of the Inner Sound is puzzling, and seems to bear little similarity to tides elsewhere in the main body. A satisfactory explanation of the complex tidal features of the area clearly needs much more data than is presently available.

In the Little and North Minches, maximum tidal current speed is associated with high and low water, and slack currents with mid-tide. The currents begin to flow sooner in the south than in the north, as there the tidal wave arrives earlier. The pattern of flow is simple, with flood streams (Fig. 6.20) heading south-west from Cape Wrath into the Little Minch, filling adjoining lochs and bays, and with ebb streams (Fig. 6.21) flowing in the reverse direction. Eddies occur in a few places around prominent headlands, or where rocks and islands stand in mid-current. Fig. 6.22 reveals that the strongest currents occur near shallow, shelving coasts and in constricted channels. There are three areas with particularly strong tidal currents; the mainland coastal waters from Rubh'Re to Cape Wrath; the Lewis coastal water from Tolsta Head to the Butt; and the northern part of the Little Minch, between Rubha Hunish and Harris. This latter area forms a restricted channel for the passage of tidal streams, being much shallower and narrower than the adjoining North Minch or southern Little Minch. Surface currents of 2 m/s are experienced during spring tides in the Sound of Shiant, and violent eddies and turbulence are noted in the Admiralty Pilot around the Shiant Isles and other obstructions in the path of the currents. Here speeds may be considerably greater than in the Sound of Shiant, but have not been estimated by the Admiralty. The shallow water off head-

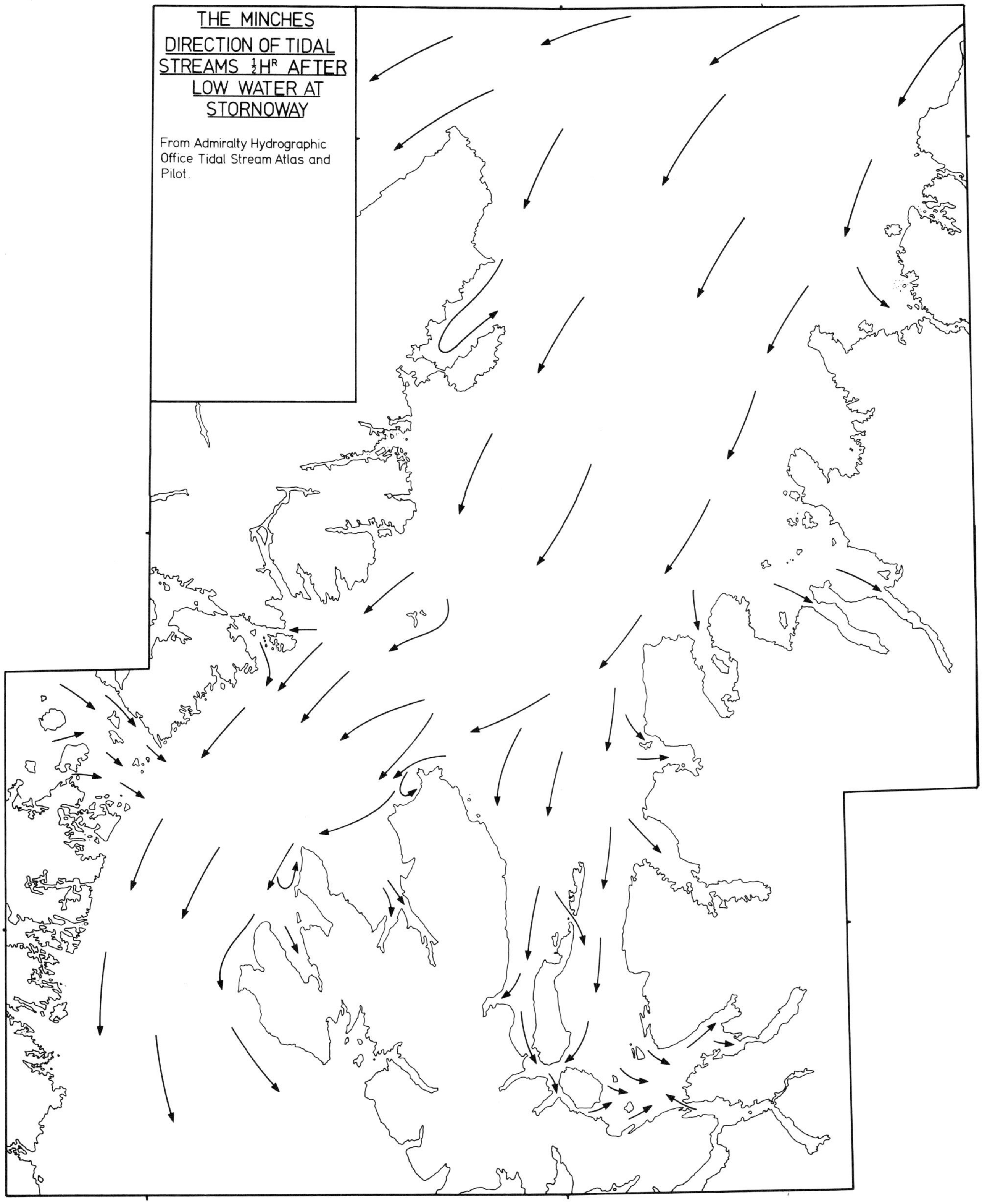
THE MINCHES
DIRECTION OF TIDAL STREAMS ½HR AFTER LOW WATER AT STORNOWAY
From Admiralty Hydrographic Office Tidal Stream Atlas and Pilot.

Fig. 6.20.

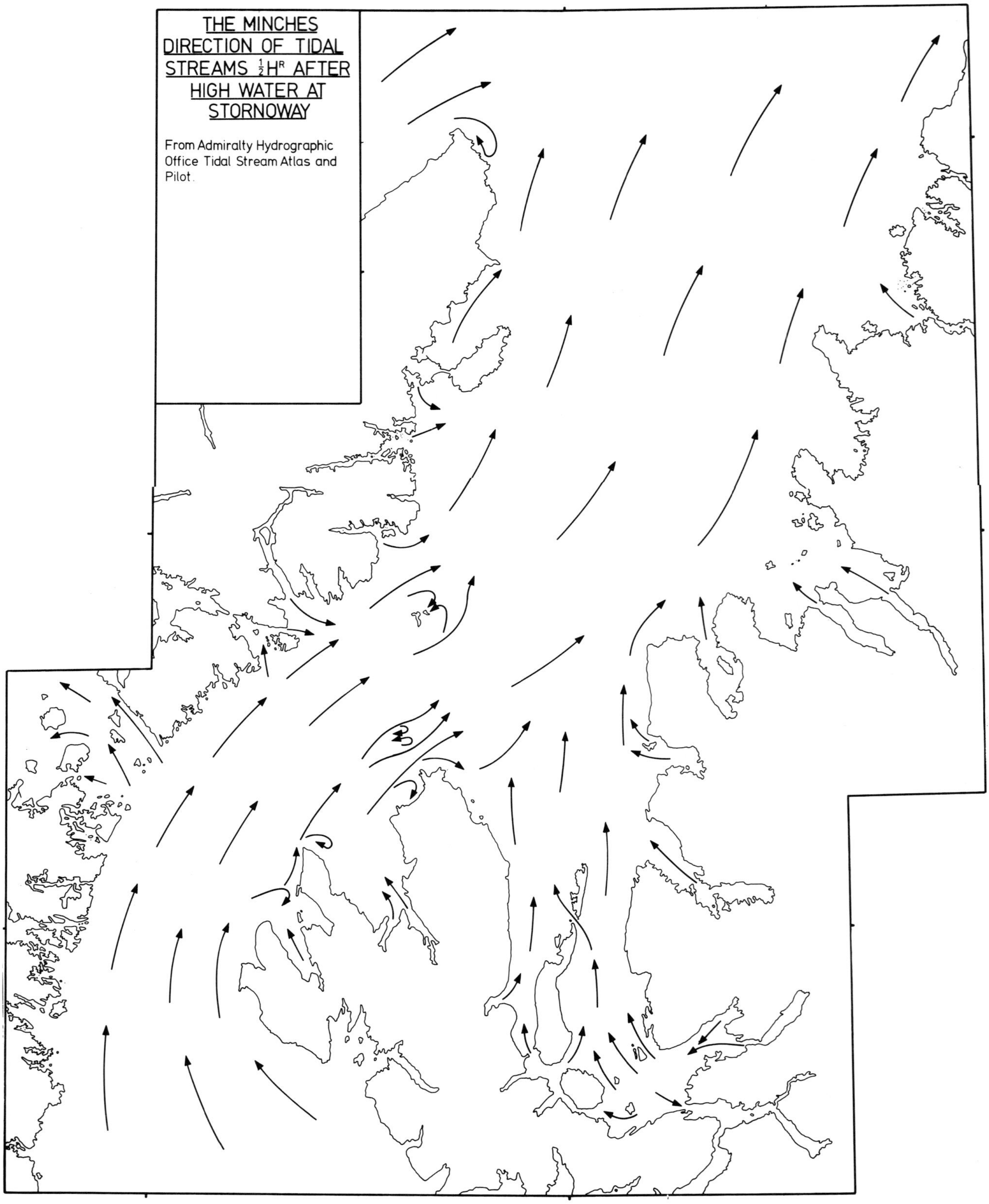
THE MINCHES
DIRECTION OF TIDAL
STREAMS ½Hr AFTER
HIGH WATER AT
STORNOWAY
From Admiralty Hydrographic Office Tidal Stream Atlas and Pilot.

Fig. 6.21.

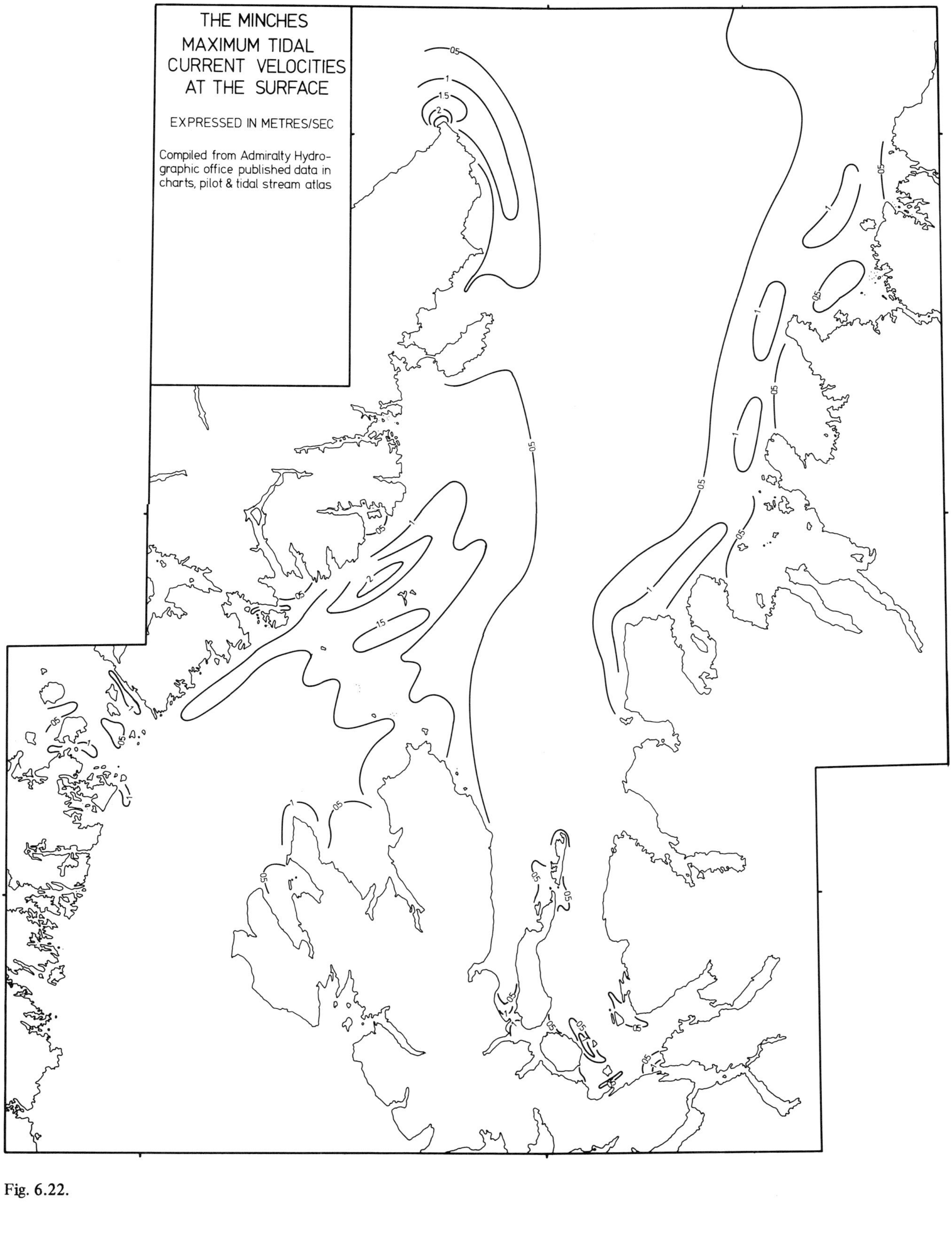
THE MINCHES
MAXIMUM TIDAL
CURRENT VELOCITIES
AT THE SURFACE
EXPRESSED IN METRES/SEC
Compiled from Admiralty Hydrographic office published data in charts, pilot & tidal stream atlas
05
1
1.5
2

Fig. 6.22.

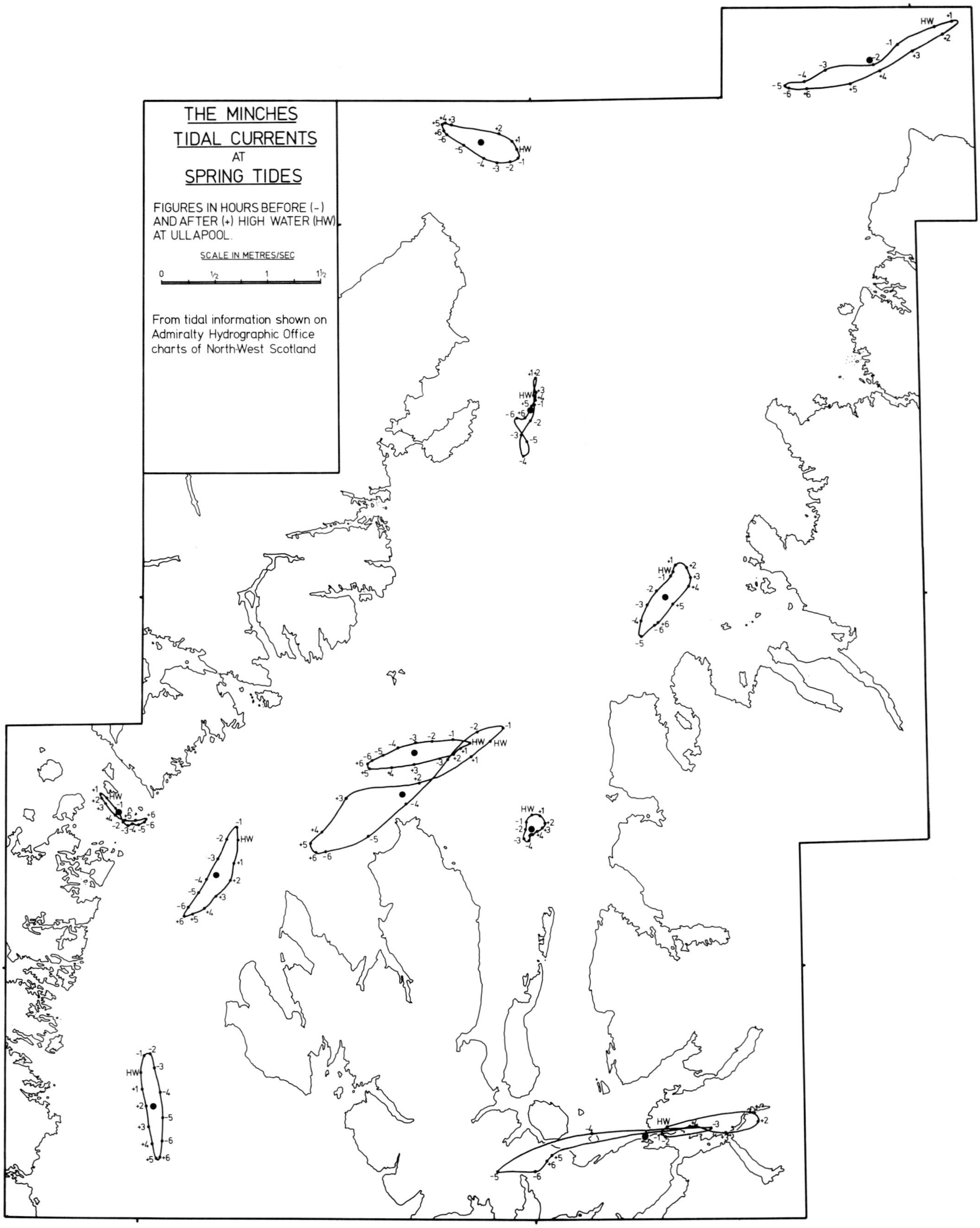
THE MINCHES
TIDAL CURRENTS
AT
SPRING TIDES
FIGURES IN HOURS BEFORE (-) AND AFTER (+) HIGH WATER (HW) AT ULLAPOOL.
SCALE IN METRES/SEC
0 1/2 1 1 1/2
From tidal information shown on Admiralty Hydrographic Office charts of North-West Scotland

Fig. 6.23.

lands along the mainland coast is marked by the development of locally strong tidal streams, while the bays between have more gentle currents. The strong flow around North-East Lewis partly reflects the shallow coastal waters and partly some unusual tidal circumstances. Figure 6.19 shows that the progress of the tidal wave through the Little Minch is slowed by its traverse of the shallow water of the Skye–Shiant Platform. Consequently, there is a difference of approximately 15 min between the arrival of high water on the east and west sides of the Butt of Lewis, which is responsible for the marked strengthening of tidal currents around the headland, in a reaction to the steep tidal gradient.

Non-tidal currents

Those currents not induced by tidal forces are revealed by the study of the physical and chemical properties of Minches water. An examination of the Institute of Oceanographic Sciences' data file for the Minches area (comprising almost 900 stations occupied up to 1974) showed that, at any one place, seasonal values of salinity and temperature can vary greatly; in fact, over as great a range as the geographical variations throughout the whole region at any one time. Large fluctuations of these parameters from year to year have already been pointed out by Craig (1951, 1954) and Craig and Slinn (1955). For this reason it has proved impossible to combine all the I.O.S. data gathered over many years of observations into maps of averaged salinity or temperature conditions, and the most reliable indicators of drift currents are, therefore, individual surveys made over as short a period of time as possible. Craig completed several such surveys during the 1950's for the Department of Agriculture and Fisheries, Aberdeen (Craig, 1951, 1954; Craig and Slinn, 1955) and in 1959 summarised his conclusions on the circulation of Scottish coastal waters. Although conditions varied considerably between his surveys, Craig achieved a good composite picture of non-tidal circulation in the Minches.

Non-tidal currents in the North-Western Approaches to the British Isles travel in a north-easterly direction close to the western coasts of the Outer Hebridean islands (Tulloch and Tait, 1959) in a direction consistent with the North Atlantic Drift current. Much of the water passing the Hebridean area has flowed along the shores

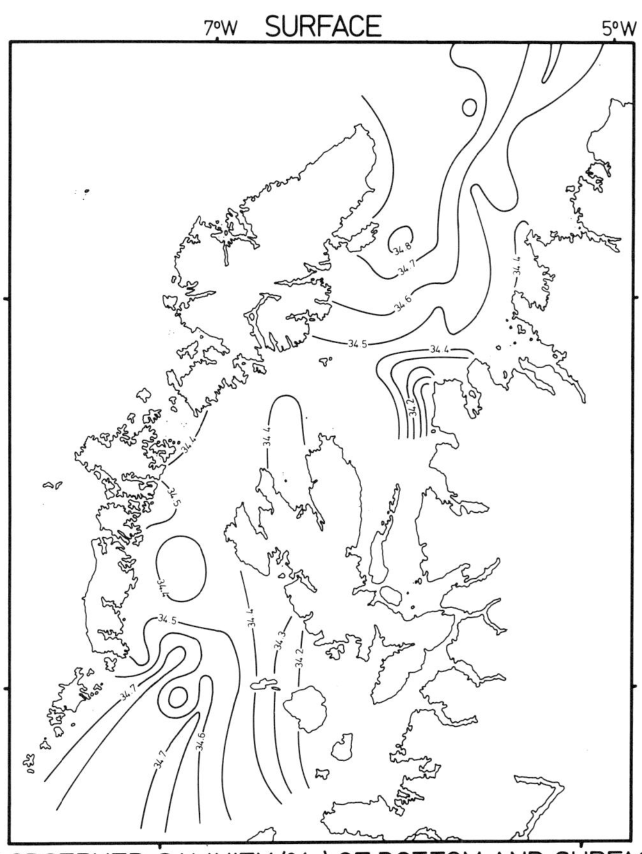

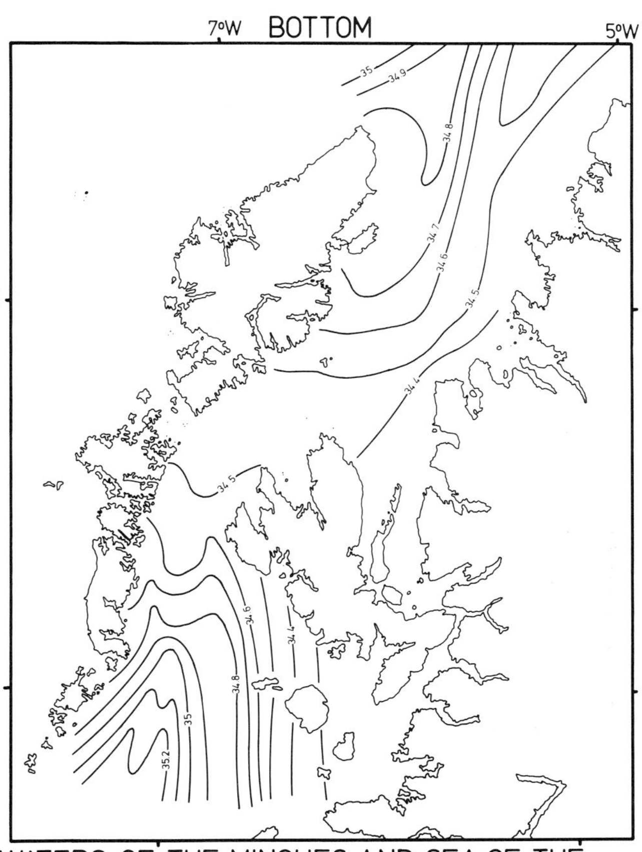

OBSERVED SALINITY (‰) OF BOTTOM AND SURFACE WATERS OF THE MINCHES AND SEA OF THE HEBRIDES, SEPTEMBER/OCTOBER 1951. (AFTER CRAIG, 1951)

Fig. 6.24.

Fig. 6.25.

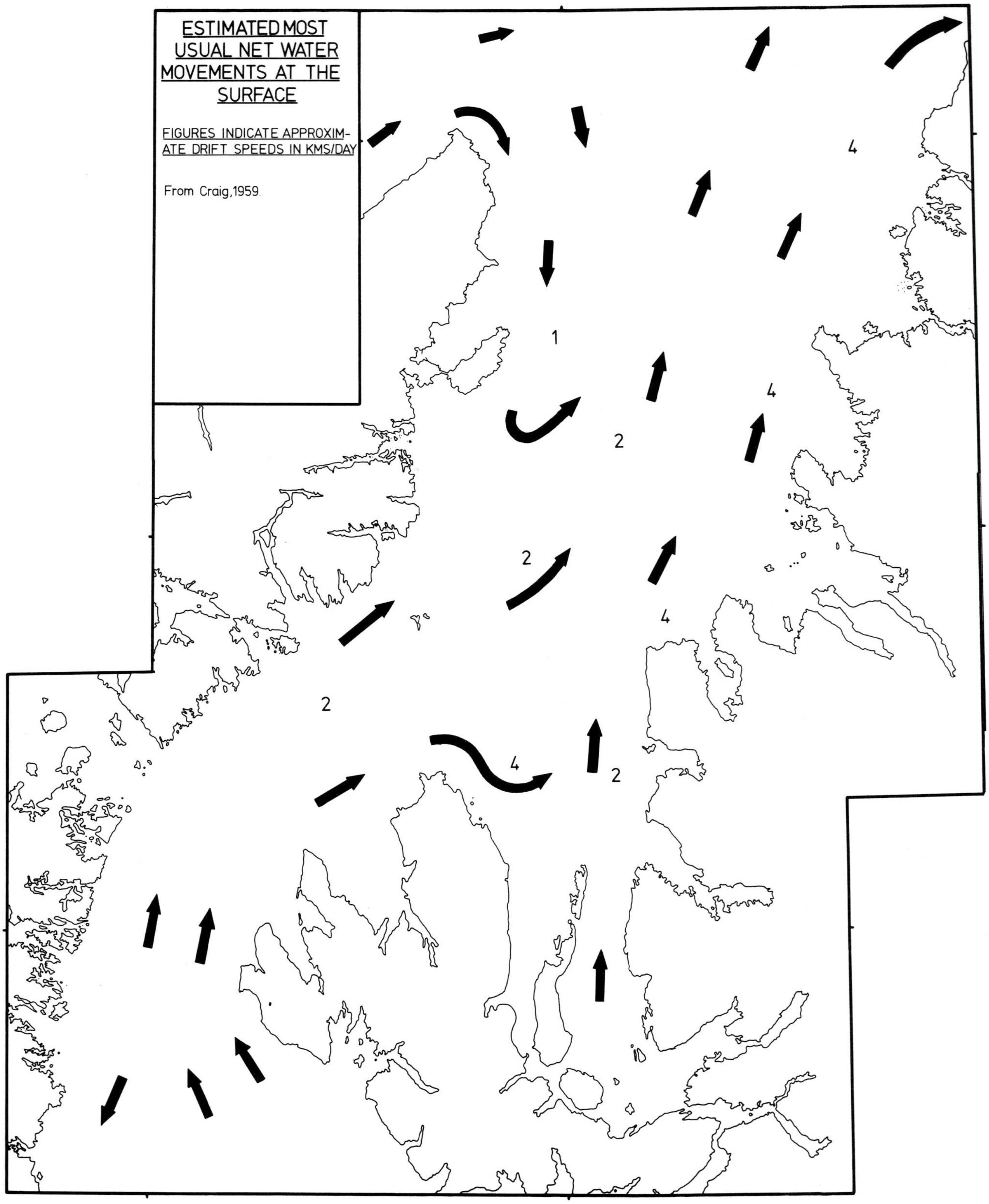
ESTIMATED MOST USUAL NET WATER MOVEMENTS AT THE SURFACE
FIGURES INDICATE APPROXIMATE DRIFT SPEEDS IN KMS/DAY
From Craig, 1959.
4
1
4
2
2
4
2
4
2

Fig. 6.26.

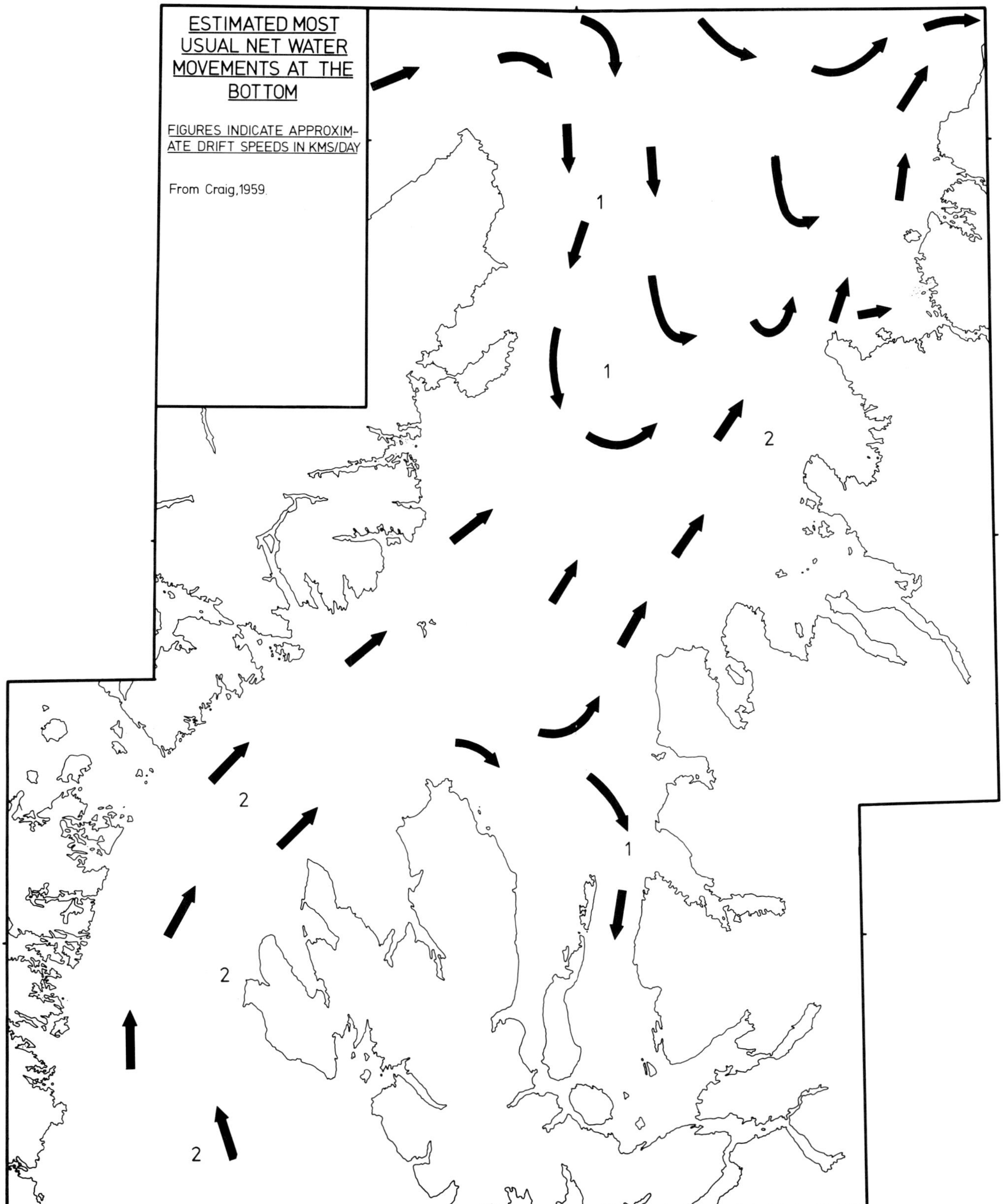
ESTIMATED MOST USUAL NET WATER MOVEMENTS AT THE BOTTOM
FIGURES INDICATE APPROXIMATE DRIFT SPEEDS IN KMS/DAY
From Craig, 1959.
1
1
2
2
1
2
2

Fig. 6.27.

of Ireland and Scotland, or through the Irish Sea, and is already considerably diluted by fresh water run-off. Salinities lower than 34.5 parts per thousand are reported even in open water around the Hebrides and, despite the proximity of the Atlantic, the water off North-West Scotland is considerably less "oceanic" than that of the northern North Sea. What may be described as truly oceanic water only penetrates into one part of the Hebridean area; a wedge of dense saline water is found at depth in the Sea of the Hebrides and southern Little Minch, overlain by more diluted and lighter continental shelf water, but it does not extend further north than the latitude of Loch Dunvegan (see Figs. 6.24 and 6.25). The rest of the Little Minch and the North Minch is characterised by vertically and horizontally homogeneous, diluted continental shelf water, with an apparent outflow to the north-east past Cape Wrath, while slightly more saline water from the outer shelf can be found rounding the Butt of Lewis and entering the North Minch.

Craig has estimated the usual drift current directions and speeds from the results of his salinity and temperature surveys (Figs. 6.26 and 6.27). The general south-west to north-east drift through the Minches corresponds with the Atlantic pattern for the North-Western Approaches. The Admiralty Pilot also recognises this drift and points out that its strength may be exaggerated by strong and continuous south-west and westerly winds up to a degree that it can form a hazard to navigation, while periods of strong and long continued north-easterly or easterly winds may see a *reversal* in the direction of flow of the whole system. Such an event might be expected to have a marked effect on the salinity of Minch waters, as water entering from the north would be leaving the open sea, and would not have the same degree of dilution as the normal south-westerly drift. The wide fluctuations of salinity seen in the Minches during many years of observations could well be a result of this phenomenon. Meteorological office statistics quoted in the Admiralty Pilot indicate that, on average, the wind at Stornoway blows from the south, south-west or west for 46% of the time, and from the north, north-east or east for 22% of the time. Comparisons of the maps showing tidal and non-tidal current directions (Figs. 6.20, 6.21 and 6.26, 6.27) reveal that, in most places, the two types of movement follow roughly the same trends. This means that tidal currents are either slightly accelerated or retarded by the non-tidal drift, according to the set of the tide and the temporary direction of the wind-controlled flow.

Waves and wave-induced currents

The depth to which wave-induced oscillations are capable of disturbing sediment on the sea floor depends, firstly, upon the dimensions of the sediment particles, and secondly, upon the size of the waves. As we shall see later, the grain size of the surface sediment generally varies in quite close relationship to the bathymetry, the high banks being covered with gravels, while in the deep basins muddy sediments dominate. The absence of fine-grained deposits in shallow water, except in certain more sheltered locations, suggests that agitation of the sediment by waves has liberated all but the coarser grades, enabling it to be carried away by tidal or other currents.

Wave disturbance is not the only factor that could prevent the accumulation of fine-grained material in shallow water; there is a tendency for tidal currents to be stronger in the shallows than in the deeper areas. At any one time, a spectrum of waves will occur in the Minches, containing waves of varying size, each capable of causing disturbances at the sea floor of differing strength. Additionally, between storms and periods of calm weather, there will be enormous variations in the influence of wave energy upon the sea floor. Observations made at lightships around the British Isles reveal that the periods of the largest waves recorded are 7.6 s in the Irish Sea, 8.4 s in the North Sea, and 6.4 s in the Straits of Dover (King, 1975). A wave with an 8-s period will begin to feel bottom at a depth of 50 m (Sverdrup et al., 1942). However, analysis of storm-wave spectra by Hadley (1964) shows that waves exist in British waters with even longer periods and analysis of force 10 gales in the Celtic Sea demonstrated that the longest period waves could produce bottom velocities of 52 cm/s at a depth of 183 m. Admittedly, the Minches are a much more sheltered area than the Celtic Sea, but there is exposure to violent storms from at least one quarter, the north, where comparable wave spectra might be expected. The Minches suffer northerly winds most frequently in the spring and autumn and storms occur for an average of 48 days/year (Admiralty Pilot statistics). Thus, it seems that we must not presume wave influence in the Minches to be limited to the shallower regions, and probably only the deepest and most sheltered basins never experience significant wave-induced motion.

Longshore and rip currents may be important in shallow coastal waters. Longshore currents of 50 cm/s have been measured by Shepard and Inman (1950) and rip currents of 250 cm/s by McKenzie (1958). The University College and I.G.S. sediment sampling and geophysical programmes have been limited, by the size of vessels used, to relatively deep water, outside the immediate coastal zone, but local currents there have to be taken into account to explain certain facets of sediment distribution in the Minches. On the beaches, longshore drift may be responsible for additional sediment transport and local longshore drift directions will be discussed later in connection with the distinctive sediments to be found between the Butt of Lewis and Broad Bay. Lastly, the ability of breaking internal waves to transport bottom sediment on the continental shelf has been investigated by Southard and Cacchione (1972). It has already been noted that the Minches are characterised by vertically-homogeneous, diluted water, and that there is only one area where a halocline exists to suggest an interface which might conceivably be subject to internal wave effects; this region lies in the southern part of the Little Minch, where a wedge of saline oceanic water penetrates at depth from the Sea of the Hebrides. No observations on internal waves here or anywhere in the Minches have been reported.

Non-tidal currents and organic phosphate distribution

While annual variations of the salinity and temperature of Minches waters are large, the monthly organic phosphate (PO_4-phosphorus) content appears to be a relatively conservative property from year to year. Observations from the Institute of Oceanographic Sciences' data file have been combined to produce a map of average organic phosphate concentration in the surface water for the month of October (Fig. 6.28). The map reveals that organic phosphate is generally less concentrated near the coast than in open water, almost certainly the result of fresh-water dilution from run-off, as salinity also tends to decrease landwards. In October, the Little Minch has a relatively low organic phosphate content, while the highest values are to be found in the North Minch, between the Sound of Raasay and Enard Bay.

A station just north of the Butt of Lewis has been repeatedly occupied for phosphate (and other) observations, enabling us to examine the seasonal variations in concentrations (Fig. 6.28, inset). A marked increase in organic phosphate in mid to late summer is revealed, the result of the utilisation of available nutrient phosphate by the greater organic activity of the summer months. The annual range of organic phosphate concentration at this station, about 0.5 μg atoms/l, is in fact, the same as that observed in the English Channel (Sverdrup et al., 1942; Armstrong and Butler, 1960, 1962).

In autumn, the organic phosphate level increases once more, owing to the decomposition of organic material after the summer bloom. Comparing the organic phosphate distribution map with that for net surface water movement (Fig. 6.26), it is apparent that the October organic phosphate maximum in the North Minch occurs on the down-stream side of the Shiant East Bank and the Skye–Shiant Platform. These banks, penetrating well into the photic zone, are noted in the Admiralty Pilot as productive fishing grounds. They are covered with abundant, fresh organic debris from a varied bottom fauna, and must represent areas of high organic productivity. It seems that the autumn decay of organic material supplies phosphorus to the sea water passing over the banks. In addition, observations clearly reveal that the seasonal variation of organic phosphate content is much greater in the October maximum zone than it is in the surrounding waters. In the Little Minch, organic phosphate concentration is lower, and seasonal changes almost non-existent, for here water is entering from more open parts of the continental shelf. The large seasonal organic phosphate variation north of the Butt of Lewis reflects the passage of water towards that point from off the shallow sea-floor west of the Outer Hebrides, another noted area of high organic productivity and active commercial fishing. Towards Cape Wrath, water enters the North Minch from a more northerly direction, from the open sea, and organic phosphate concentration is lower and seasonal variation less pronounced. The tongue of high organic phosphate values extending from the Butt of Lewis towards the central North Minch supports Craig's estimate of net surface water movement around the Butt in this direction.

The circulation in Lochs and Sounds

Milne (1972) has divided the Scottish west coast sea lochs and inlets into three categories, according to their form and the nature of their circulation. Type "A" lochs and inlets have deep water entrances and no sills, allowing full penetration of a relatively high salinity wedge of coastal water at the bottom, thus giving rise to an "estuarine" circulation, for inflowing saline bottom water complements the outflowing surface current of diluted water. Type "A" lochs in the Minches include Lochs Laxford, Maddy, Gairloch and Kishorn, as well as larger bodies like Loch Snizort, the Sound of Raasay and the Inner Sound. Estuarine-like circulation in these larger bodies has independently been suggested by Craig's maps of net water movement (Figs. 6.26 and 6.27). Type "B" lochs possess a sill, and thus are separated from the main body of coastal water. In the Minches, lochs of this category include Lochs Inchard, Broom, Ewe and Carron, and replenishment of water

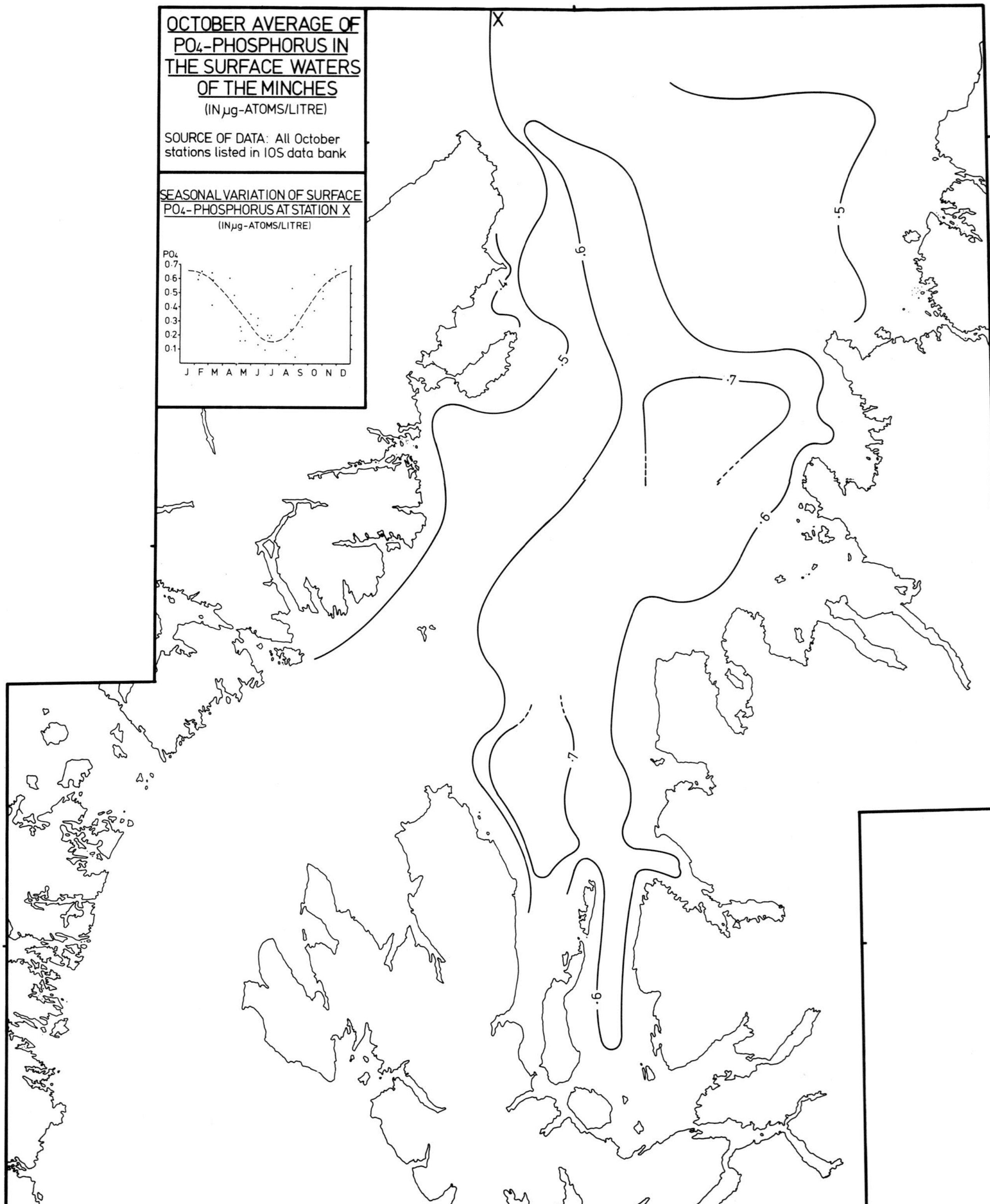
OCTOBER AVERAGE OF
PO4-PHOSPHORUS IN
THE SURFACE WATERS
OF THE MINCHES
(IN µg-ATOMS/LITRE)
SOURCE OF DATA: All October stations listed in IOS data bank
SEASONAL VARIATION OF SURFACE
PO4-PHOSPHORUS AT STATION X
(IN µg-ATOMS/LITRE)
PO4
0·7
0·6
0·5
0·4
0·3
0·2
0·1
J F M A M J J A S O N D
X
·5
·6
·7

Fig. 6.28.

depends on a combination of factors (sill depth, tidal range, fresh water inflow and storm surges) and may be of a regular or intermittent nature. Type "C" lochs are complex bodies, with more than one sill, for instance, Loch Torridon, the lower section of which is of type "A", while two sills divide upper sections. Due mainly to problems of accessibility, sample stations within lochs are few, and those that do exist are exclusively within type "A" lochs, where the absence of a sill gives easier entry. However, the restricted circulation of type "B" and "C" lochs means that sedimentological continuity with the system offshore must be more tenuous than in the case of category "A", but to what extent is not possible to estimate.

Bottom current observations

Although salinity–temperature surveys reveal net circulation, and there are abundant descriptions of tidal currents at the surface, it is the present sea-floor environment that concerns us most of all in our attempt to understand modern sediment distribution. In order to compare bottom currents with published surface currents, internally recording Plessey current meters suspended one metre above the sea bed have been deployed at four stations in the Minches, as close as possible to the locations of Admiralty surface tidal data shown on the charts (Fig. 6.23). Each current meter was suspended from a open rigid frame which rests on the sea bed. In July 1973, the I.G.S. attempted to drill a borehole north-west of Cape Wrath, about 1 km from an Admiralty tidal data station. While the vessel was anchored a current meter was lowered and ten hours of data obtained (Fig. 6.29) before the site had to be abandoned due to drilling difficulties. During that time, a generally south-westerly flow was recorded, becoming

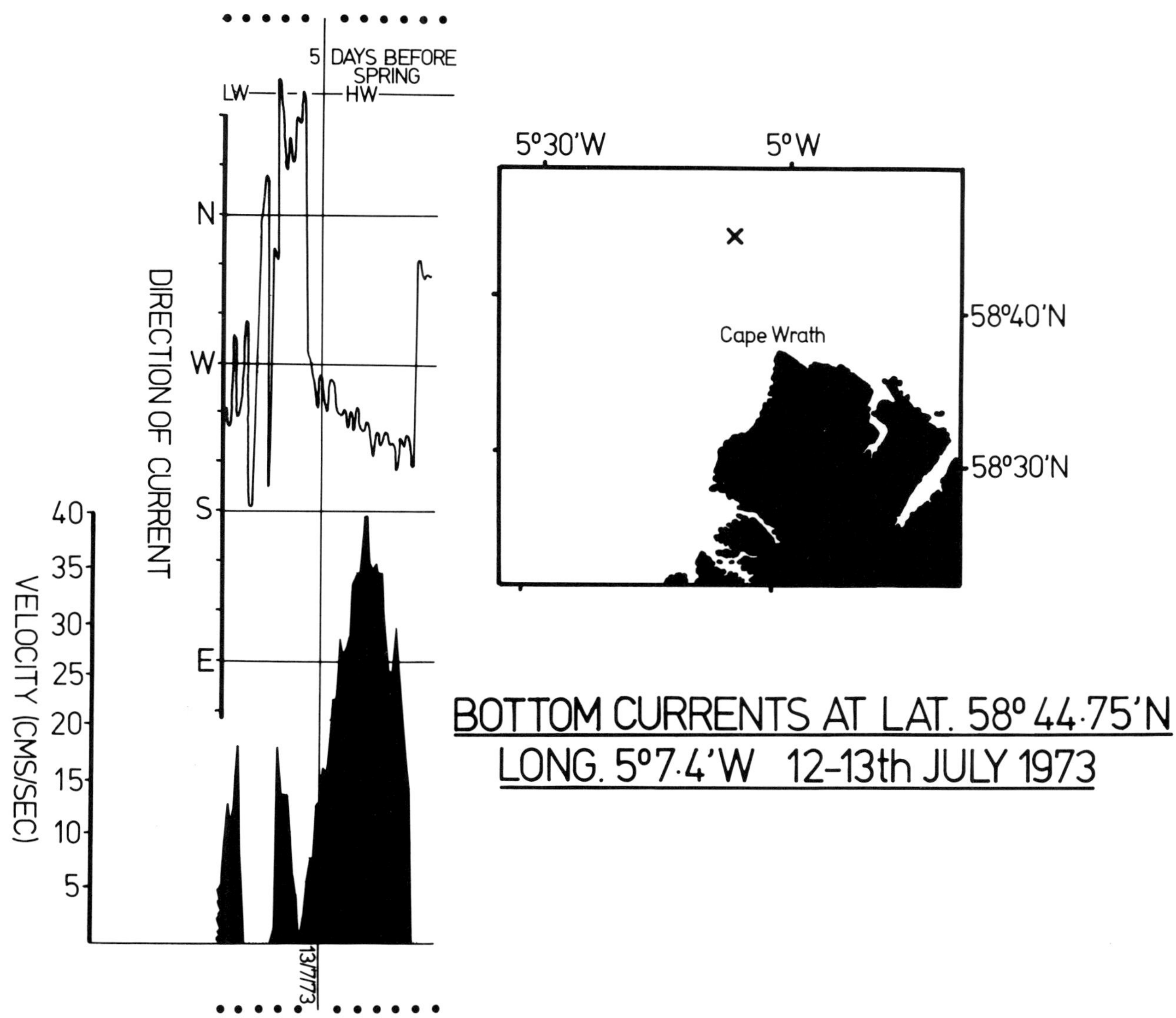

Fig. 6.29.

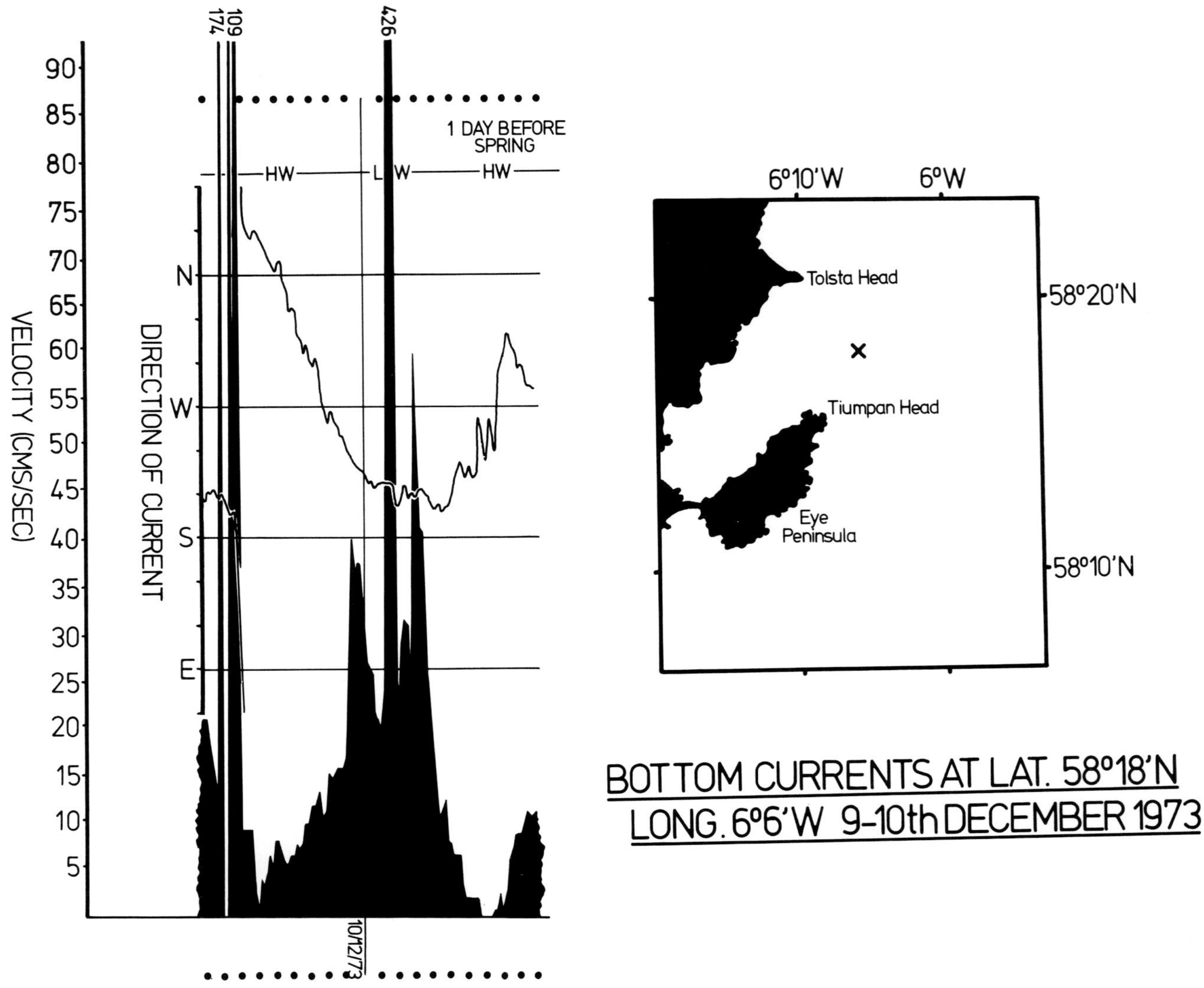

Fig. 6.30.

strongest at high water (40 cm/s) though it was also present, but weaker, at low water. At mid-flood, the flow temporarily reversed and a weak north-easterly current set in. At the surface (Figure 6.23) the Admiralty data indicate a simpler situation, with a west-south-westerly flow at low water alternating with a high water stream to the north-east, both flows, at spring tides, attaining velocities in excess of 75 cm/s. It is unfortunate that current meter observations had to be abandoned before more data were obtained.

The long-term deployment of moored current meters formed part of the planned programme of a 1973 University College cruise to the Minches, but bad weather and instrument malfunction resulted in the short term occupation of only two stations. The first was off the Lewis coast between Tolsta and Tiumpan Heads, where, while sheltering from heavy weather, a meter was deployed for a 19-h period (Fig. 6.30). According to Admiralty data, maximum tidal currents to be expected at the surface at this location are of the order of 50 cm/s. On the sea floor a south-westerly current developed at low water, with peaks of 40 and 60 cm/s, although higher "gusts" from the south-west were recorded, as high as 426 cm/s. Although these velocities seem extremely high they appear to indicate real events, possibly resulting from the storm that was running at the time and are due to wave induced rather than tidal currents. At high water, a northerly or north-easterly flow occurred, but was not so well defined as the south-westerly flow into Broad Bay.

The next day, at spring tide, the weather had improved enough to place a current meter on the sea bed north of Rubha Hunish, Skye (Fig. 6.31) on the site of the more northerly of two closely spaced Admiralty tidal stations. Here a slightly longer run, encompassing two tidal cycles, was achieved. The surface flow in this

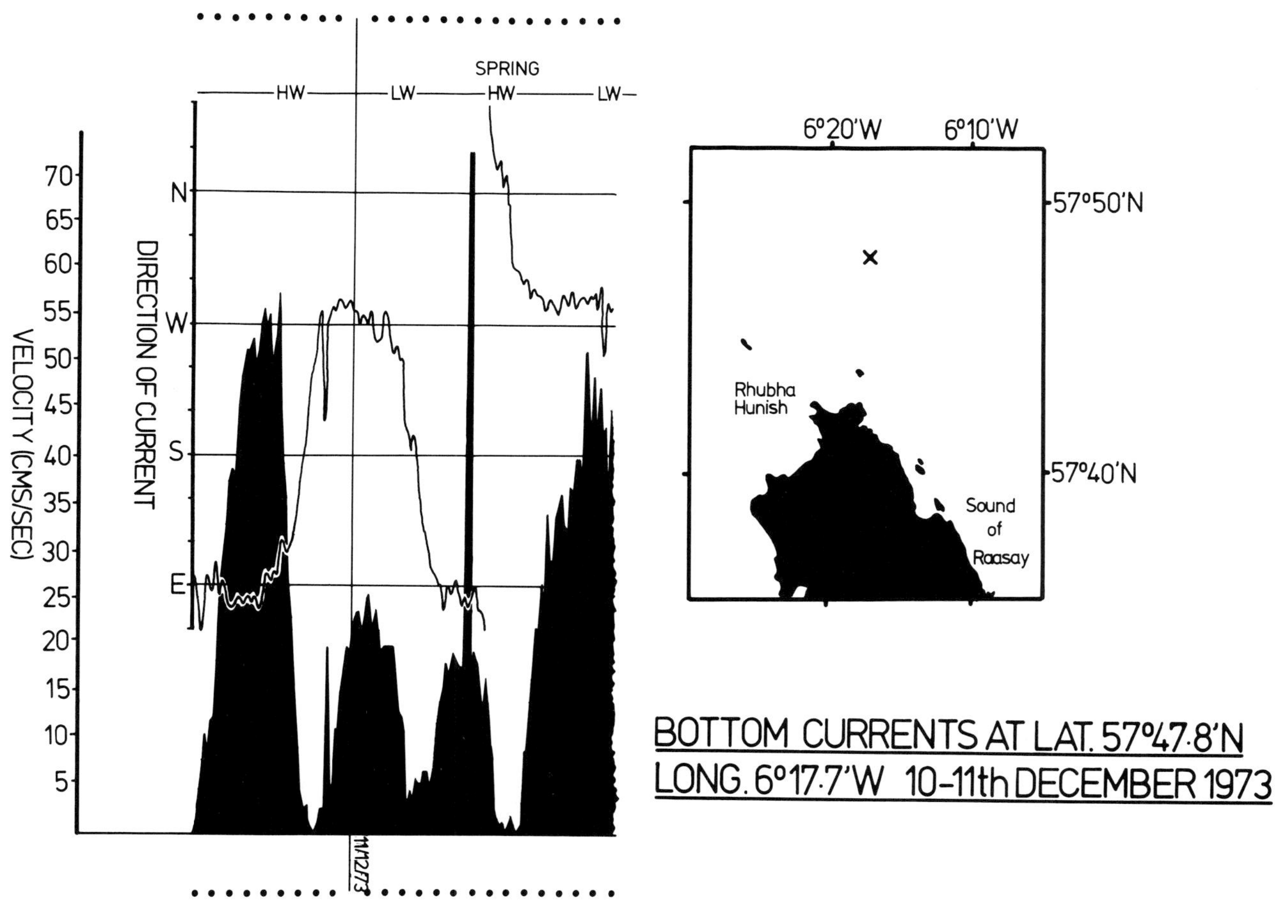

Fig. 6.31.

area is very simple, with a westerly stream at low water and an easterly current at high water, both with a velocity of about 50 cm/s. On the sea floor, however, a 50-cm/s easterly flow at high water was followed by a westerly flow of only half that strength at the ensuing low tide. A similarly weak flow to the east (with one stronger "gust") occurred at the next high water, and then, on the following low tide, a strong 50 cm/s flow to the west.

In May 1974, the Hydrographic Department (Ministry of Defence) undertook to deploy a current meter on the sea floor for a 28-day period within the Inner Sound, where one of their vessels was completing a survey. Unfortunately, the meter was moored some distance north of its intended location in the mouth of the Inner Sound, in a passage between the island of Rona and the feature known locally as Hinds Shoal. The weak surface currents of this area were proved also to be characteristic of the sea-floor environment, and even at spring tides (Fig. 6.32) the peak flow was only about 15 cm/s. At neap tides flow was negligible for days on end. Currents were strongest at mid-tide, which, as we have already seen, indicates the presence of a standing wave oscillation. The most surprising result from this set of observations is that the northerly current regularly attained a greater strength and duration than that to the south, the reverse of what may have been expected from Craig's estimate of net bottom water circulation. The location of the meter, close to the Rona/Hinds Shoal channel, may have resulted in its recording a flow not typical of the rest of the entrance to the Inner Sound. There seems to be no ready explanation for the contradiction, other than that bottom currents may be more complex than net drift patterns suggest.

The results from the four current meter stations imply that, while we may assume a general similarity between maximum current strength at the surface and the sea floor, the sea floor currents have aspects that cannot be easily explained or related to surface flow and net drift directions, and we must not consider surface tidal conditions to be necessarily indicative of those at depth. The only way in which we can begin to speak with confidence of bottom current conditions would be with the possession of data collected over longer periods. The

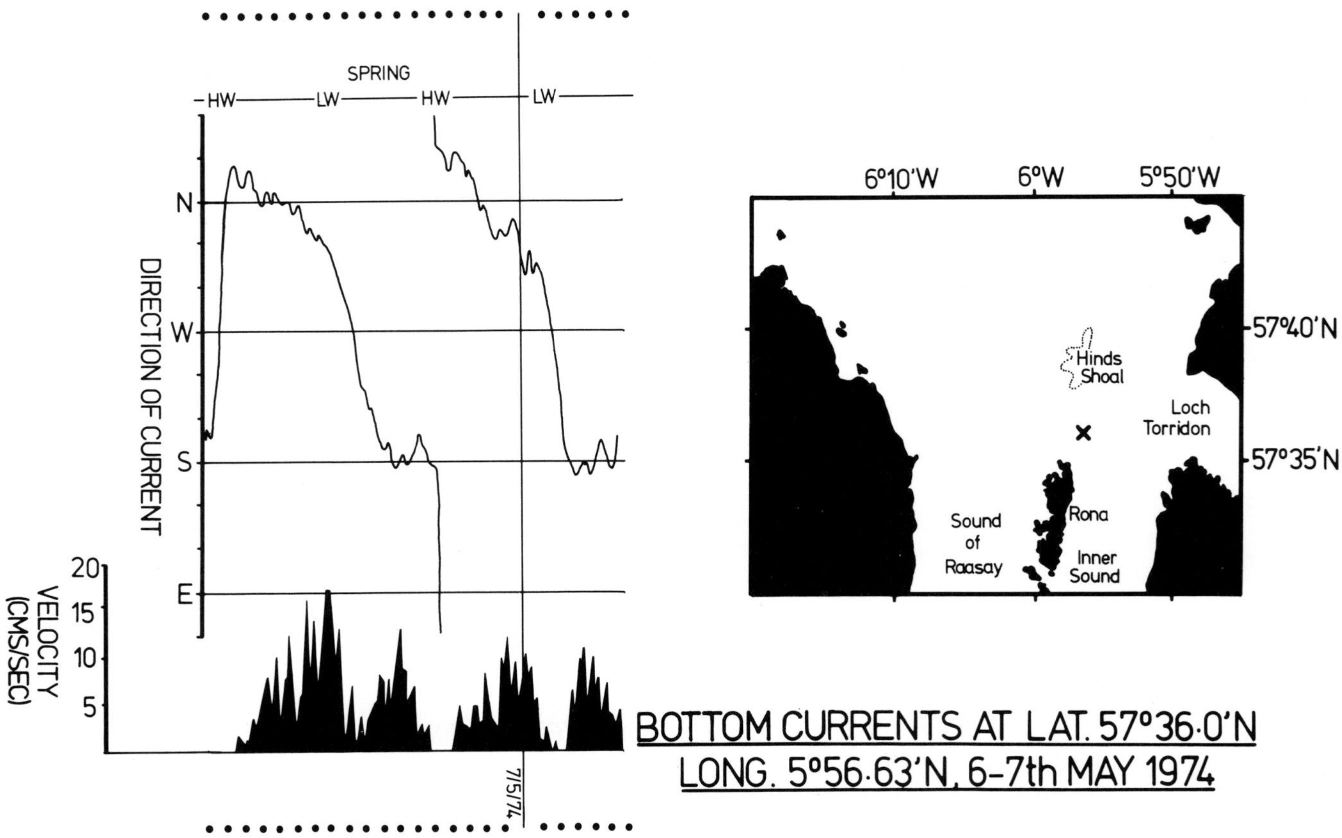

Fig. 6.32.

stations discussed here cast only a little light on the problem of bottom currents, being generally too short to distinguish normal flow from any abnormal conditions which may have been encountered. Even during the 28 days of the Inner Sound station, though encompassing both spring and neap tidal ranges, the Minches did not experience any more than moderate winds, and a gale could have produced strikingly different currents than those observed. Nevertheless, the new data obtained certainly do not contradict the concept of the shallow banks being swept by strong tidal currents while much weaker currents occur in the deep, mud-floored hollows.

Surface sediments

Facies variations

In order to make reasonably confident comparisons of the various properties of the surface sediment in different parts of the Minches, only Shipek Grab samples have been used in this part of the study. To separate the gravel, sand and mud fractions the samples were wet sieved, the dried fractions were weighed and the results checked against the original dry weight. All oven drying was performed at a low temperature (less than 50° C) to avoid baking the mud and hampering later particle size analysis. Each of the samples was classified according to the system of sediment nomenclature suggested by Folk (1965). The appropriate sediment categories were plotted onto a base map, and the location of the boundaries between them was decided with reference to bathymetry (for, as we shall see, sediment facies variations are closely paralleled by changes in water depth) and descriptions of the bottom sediments given on Admiralty charts (a very rough guide only, but of some interpretative value). The resulting sediment facies distribution map (Fig. 6.33) depicts five facies groups, which are combinations of Folk's original fifteen categories (Table 6.2).

The *gravel* facies group is associated with shallow areas, particularly the Skye–Shiant Platform and the Shiant East Bank, and is also found off many headlands on the mainland and Hebridean coasts. No significant gravel deposits were found anywhere in deep water. By definition, all samples placed in this category contain more than 30% (by weight) of gravel-sized material. With less than 30% gravel, and more than a 1 : 1 ratio with sand, the *muddy* sediment group includes four of Folk's categories, and comprises gravelly or sandy muds.

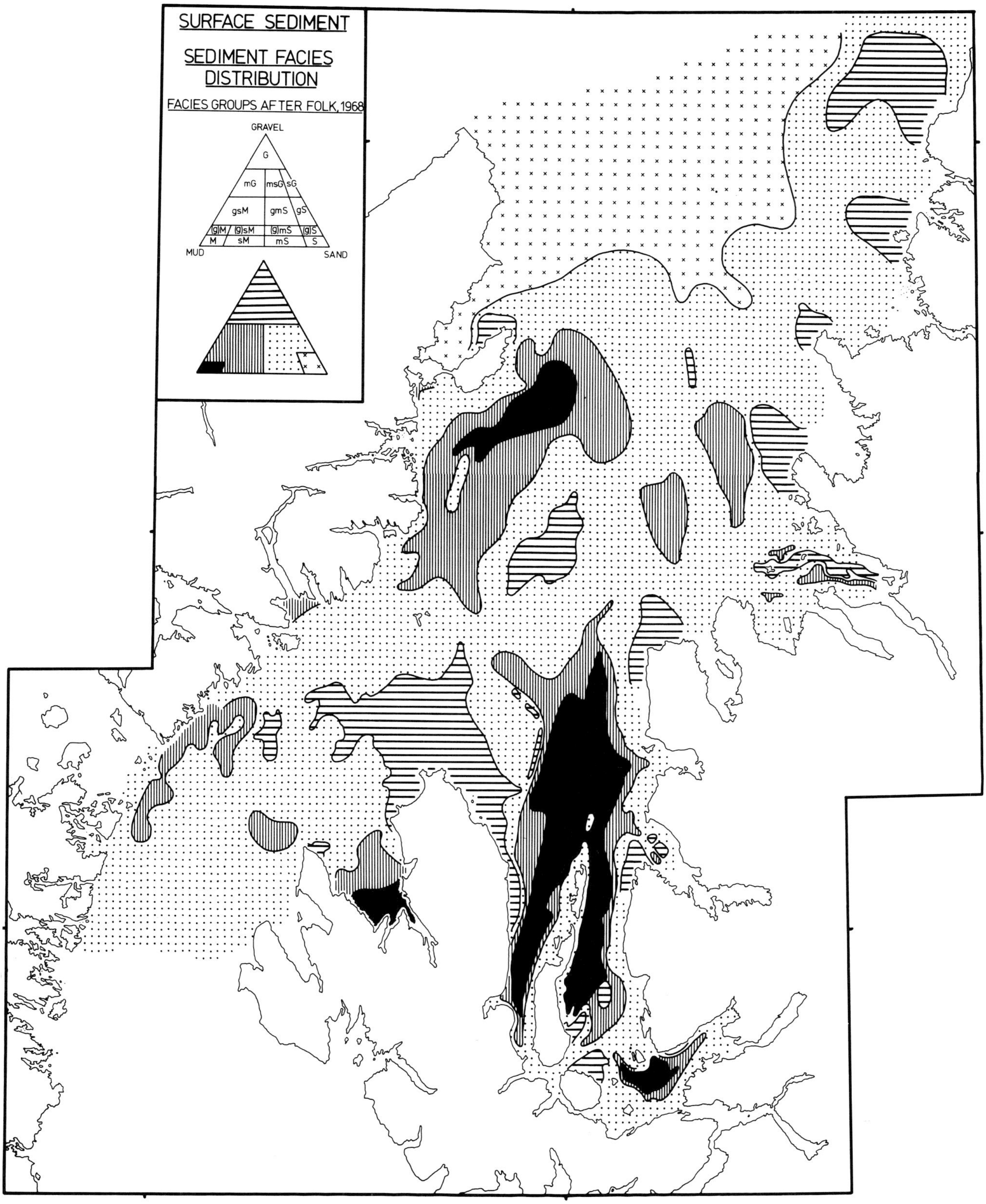
SURFACE SEDIMENT
SEDIMENT FACIES DISTRIBUTION
FACIES GROUPS AFTER FOLK, 1968
GRAVEL
G
mG
msG
sG
gsM
gmS
gS
(g)M
(g)sM
(g)mS
(g)S
M
sM
mS
S
MUD
SAND

Fig. 6.33.

TABLE 6.2

Sediment facies groups after Folk (1965)

Facies group	Folk categories
Gravel	gravel
	muddy gravel
	muddy sandy gravel
	sandy gravel
Muddy	gravelly mud
	slightly gravelly mud
	slightly gravelly sandy mud
	sandy mud
Mud	mud
Sandy	gravelly muddy sand
	gravelly sand
	slightly gravelly muddy sand
	muddy sand
Sand	sand
	slightly gravelly sand

Muddy sediment occurs in deeper and more sheltered areas, the largest of which stretches from the southern flanks of the Shiant East Bank into the Sound of Raasay and the Inner Sound. The northern side of the Bank is similarly encircled by muddy sediments, particularly in the deep Stornoway Basin to the north-west. On the north-east side, a broad muddy hollow between the Bank and Rhu Coigach is split into two sections by the sand and gravel covered Greenstone Ridge. Muddy sediments also occur in many of the lochs, particularly Loch Snizort. Folk's single category *mud* has here been given the status of a separate "group" due to the large area it alone covers within the muddy facies. With more than 9 : 1 ratio against sand and less than 0.01% gravel, these better sorted muds occur in the most protected and deepest parts of the Inner Sound, Sound of Raasay, Loch Snizort and the Stornoway Basin. The rest of the Minches is covered by sandy sediments. In the north-west of the North Minch, well sorted *sand* is found, with more than a 9 : 1 sand/mud ratio and less than 5% gravel. The *sandy* sediment found elsewhere has a higher mud or gravel content. The sediment facies map, although simple in conception, conveniently groups the sediment types in an easily comprehensible manner, and may be regarded as a fundamental guide to the nature of the surface sediment.

Thickness

In order to determine the thickness of the surface sediment layer, that is, the post-glacial re-worked sediments of Formation 4, an examination has been made of the logged descriptions of several hundred sediment cores collected by the I.G.S. Core recovery varied from over 250 cm in soft muds to about 50 cm in muddy sands and as little as 10 cm in gravels or well sorted sands, which proved the most difficult sediments to penetrate and recover. In most cases, the upper section of the core possessed a similar lithology to the grab samples collected at the same station, but on occasion, considerable differences were noted. This may be due to the variability of bottom sediment over short distances, coupled with drifting of the survey vessel during sample operations, or it may reflect the contrasting modes of operation of grab and core samplers. In many areas, the lithology of the surface sediment changes quickly and radically with depth, coarse sediment invariably passing downwards into finer material. The character of these variations displays distinct geographical patterns that can be summarised with the aid of a number of selected grab and core descriptions arranged into three traverses over major topographic features (Fig. 6.34).

To the north of the Shiant East Bank is a belt of muddy sands and sandy muds extending from Enard Bay towards the northern Lewis coast, which pass downwards within a few centimetres of the surface into a notably muddier sediment. Almost all cores of reasonable length in this area show a similar transition (Fig. 6.34, Traverse *A*) and some reach down to a dark, plastic and occasionally pebbly clay. These dark clays, of a distinctly glacial aspect, are encountered at other localities in the North Minch. They are thinly covered by pebbly and shelly sand near Tiumpan Head, north of the Eye Peninsula, and, on Greenstone Ridge, are overlain by poorly sorted shelly gravel (Traverse *A*). It is likely that the clays are Formation 1 tills, whereas the muddy sands represent Formation 3 peri-glacial deposits, slightly resorted and deprived of the finer fractions towards the surface. The shelly sands and gravels on the shallow seabed of the Greenstone Ridge and Eye Peninsula could represent a more strongly winnowed lag deposit derived from the underlying boulder clay. Such an interpretation would be consistent with our knowledge of both the glacial sediments and the marine environment of the Minches. The Formation 1 tills that characterise the glacial sediment of shallower areas in the North Minch would now be subjected to greater wave energy and stronger tidal currents than the sandy peri-glacial deposits of the deeper and less agitated parts. Near both coasts of the North Minch, cores have penetrated thicker sand deposits that show no change of lithology with depth. A vibrocore sample near Tolsta Head recovered 175 cm of shell sand (Traverse *A*). These distinctive sand deposits, which will be discussed more fully later, appear to be a coastal development, although it is not possible to compare their thickness directly with

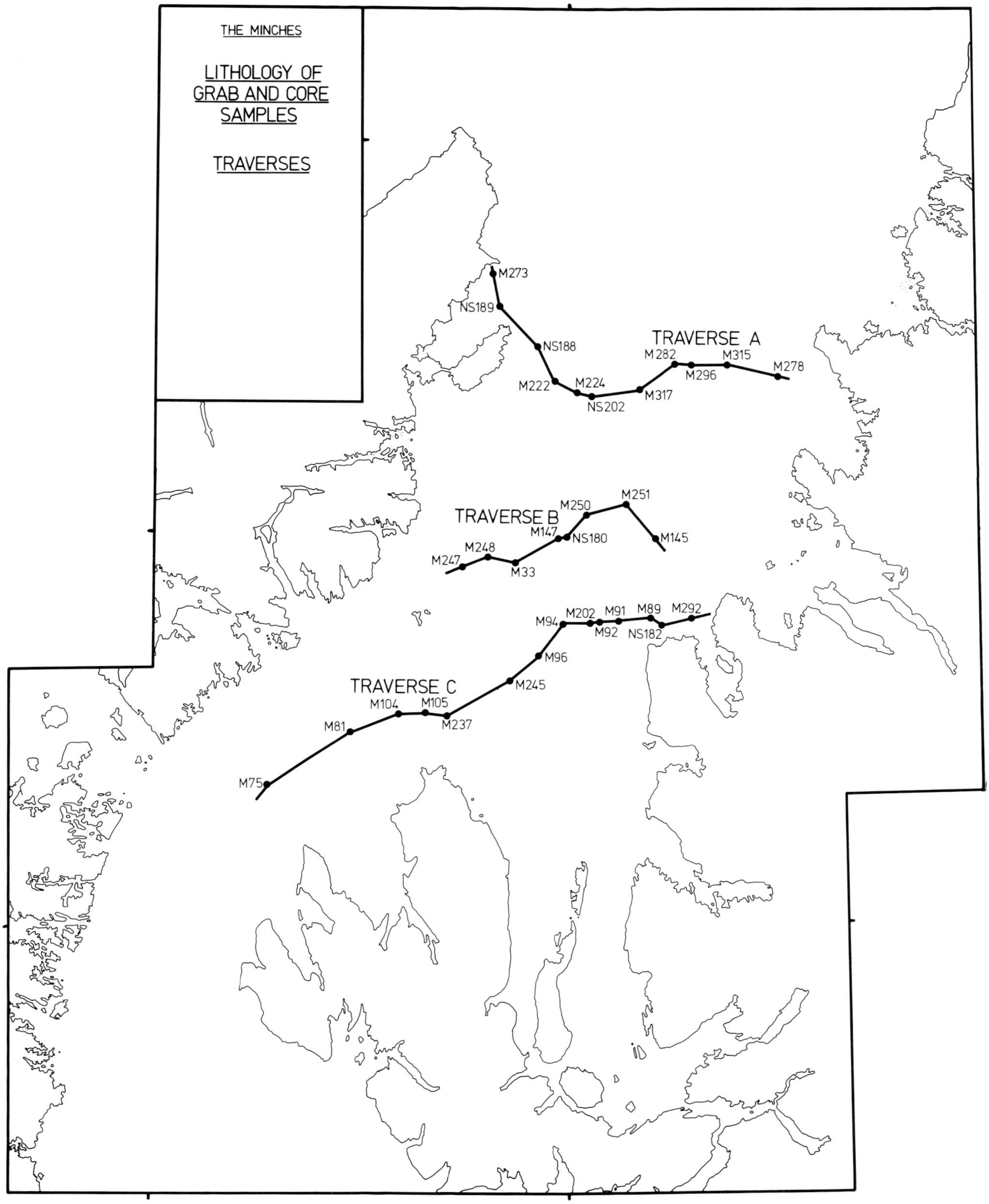
THE MINCHES
LITHOLOGY OF GRAB AND CORE SAMPLES
TRAVERSES
TRAVERSE A
M273
NS189
NS188
M222
M224
NS202
M317
M282
M296
M315
M278
TRAVERSE B
M247
M248
M33
M147
NS180
M250
M251
M145
TRAVERSE C
M75
M81
M104
M105
M237
M245
M96
M94
M202
M92
M91
M89
NS182
M292

Fig. 6.34a.

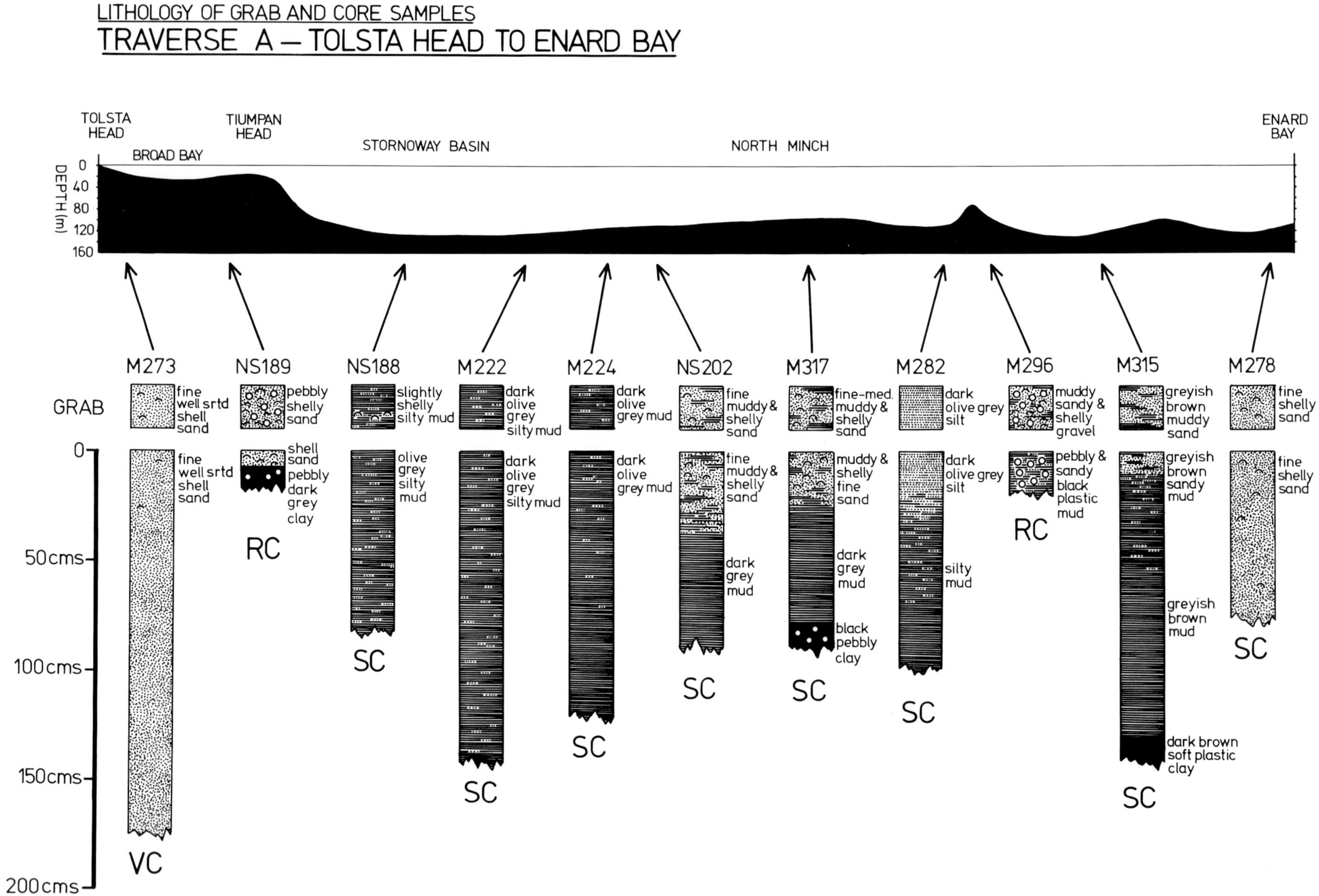
LITHOLOGY OF GRAB AND CORE SAMPLES
TRAVERSE A – TOLSTA HEAD TO ENARD BAY
TOLSTA HEAD
BROAD BAY
TIUMPAN HEAD
STORNOWAY BASIN
NORTH MINCH
ENARD BAY
DEPTH (m)
0
40
80
120
160
M273
NS189
NS188
M222
M224
NS202
M317
M282
M296
M315
M278
GRAB
fine well srtd shell sand
pebbly shelly sand
slightly shelly silty mud
dark olive grey silty mud
dark olive grey mud
fine muddy & shelly sand
fine-med. muddy & shelly sand
dark olive grey silt
muddy sandy & shelly gravel
greyish brown muddy sand
fine shelly sand
0
50cms
100cms
150cms
200cms
fine well srtd shell sand
VC
shell sand
pebbly dark grey clay
RC
olive grey silty mud
SC
dark olive grey silty mud
SC
dark olive grey mud
SC
fine muddy & shelly sand
dark grey mud
SC
muddy & shelly fine sand
dark grey mud
black pebbly clay
SC
dark olive grey silt
silty mud
SC
pebbly & sandy black plastic mud
RC
greyish brown sandy mud
greyish brown mud
dark brown soft plastic clay
SC
fine shelly sand
SC

Fig. 6.34b.

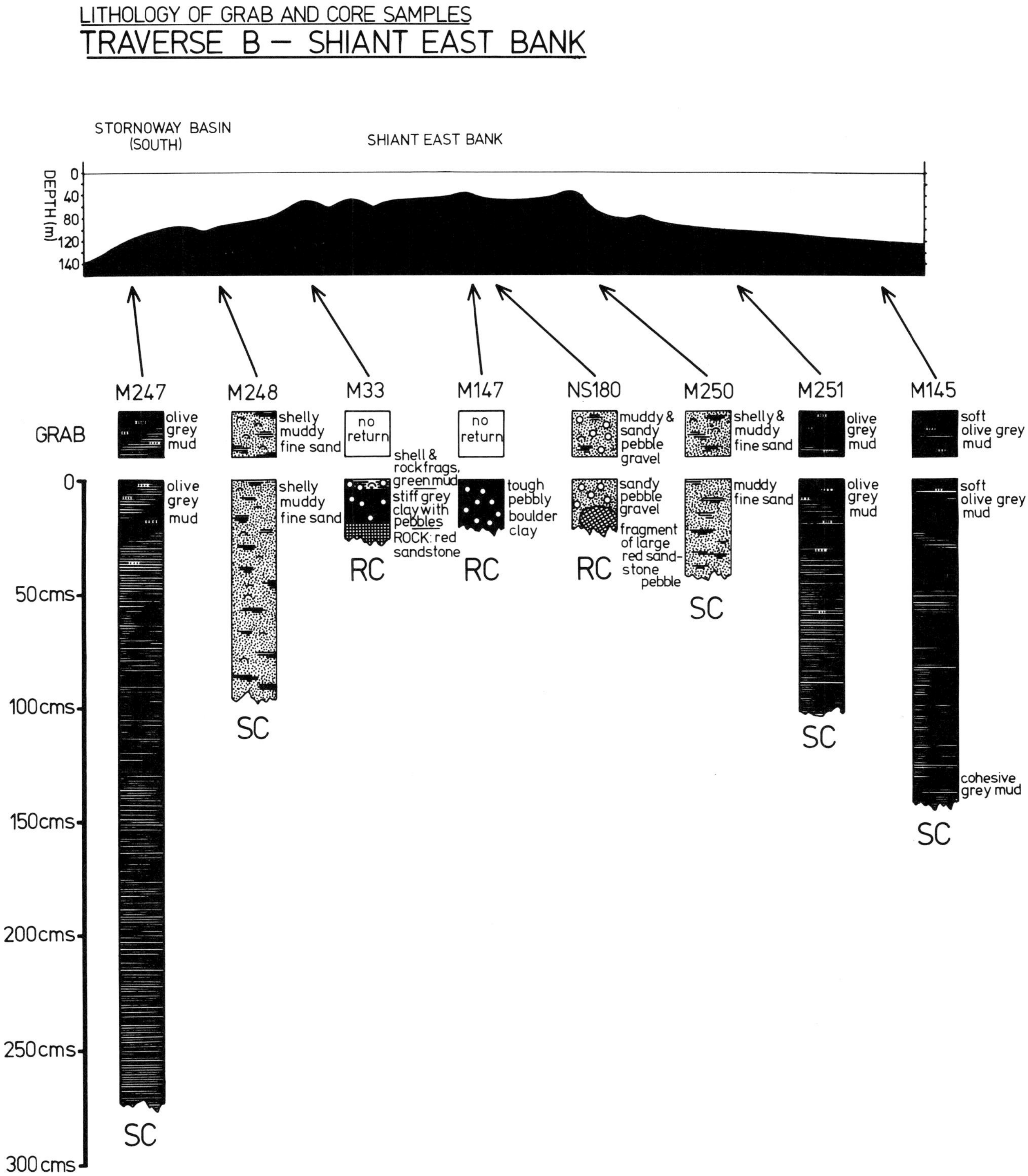
LITHOLOGY OF GRAB AND CORE SAMPLES
TRAVERSE B – SHIANT EAST BANK
STORNOWAY BASIN (SOUTH)
SHIANT EAST BANK
DEPTH (m)
0
40
80
120
140
M247
M248
M33
M147
NS180
M250
M251
M145
GRAB
olive grey mud
shelly muddy fine sand
no return
no return
muddy & sandy pebble gravel
shelly & muddy fine sand
olive grey mud
soft olive grey mud
0
50cms
100cms
150cms
200cms
250cms
300cms
olive grey mud
shelly muddy fine sand
shell & rock frags, green mud
stiff grey clay with pebbles
ROCK: red sandstone
tough pebbly boulder clay
sandy pebble gravel
fragment of large red sandstone pebble
muddy fine sand
olive grey mud
soft olive grey mud
cohesive grey mud
SC
SC
RC
RC
RC
SC
SC
SC

Fig. 6.34c.

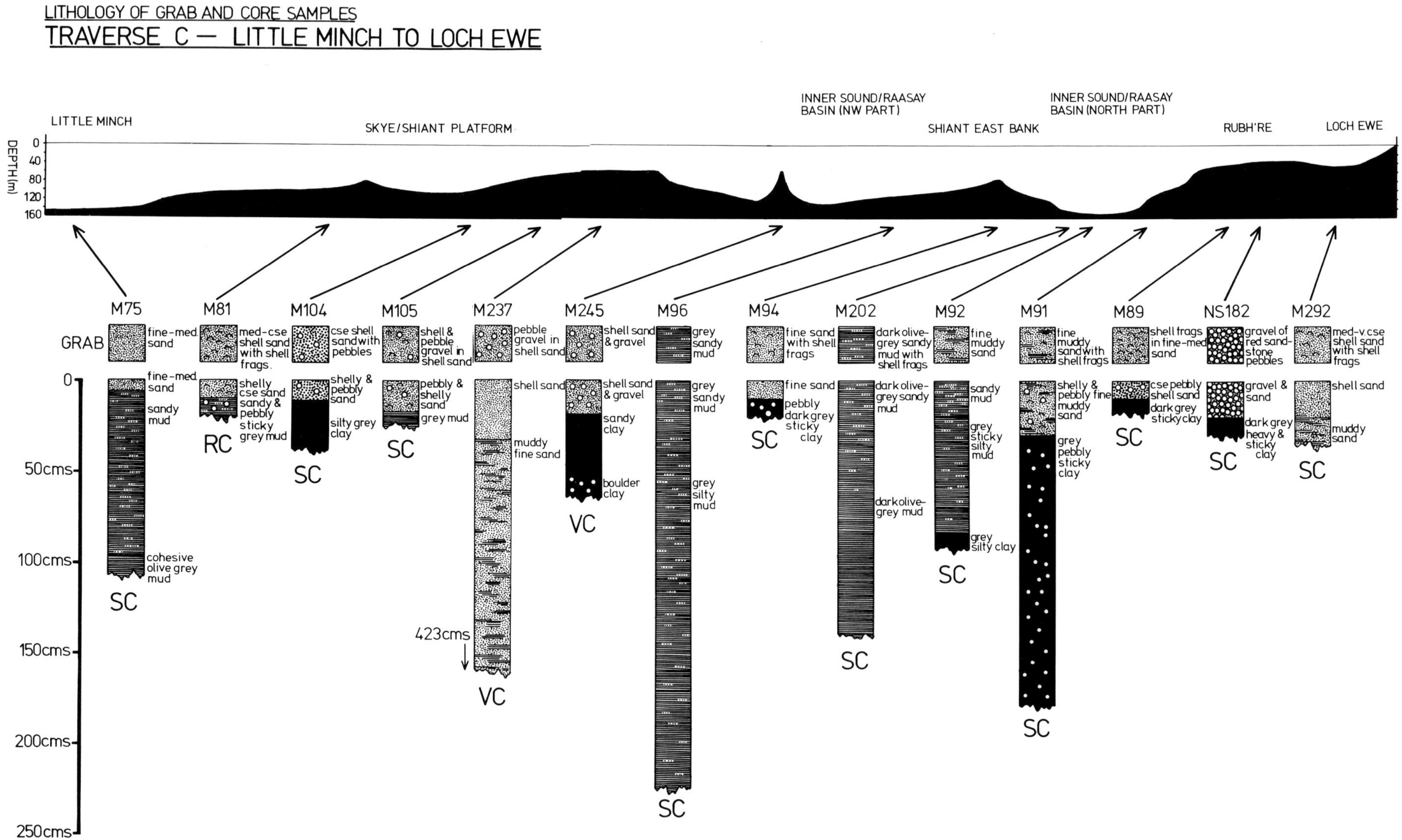
LITHOLOGY OF GRAB AND CORE SAMPLES
TRAVERSE C — LITTLE MINCH TO LOCH EWE
LITTLE MINCH
SKYE/SHIANT PLATFORM
INNER SOUND/RAASAY BASIN (NW PART)
SHIANT EAST BANK
INNER SOUND/RAASAY BASIN (NORTH PART)
RUBH'RE
LOCH EWE
DEPTH (m)
0
40
80
120
160
GRAB
0
50cms
100cms
150cms
200cms
250cms
M75
fine-med sand
fine-med sand
sandy mud
cohesive olive grey mud
SC
M81
med-cse shell sand with shell frags.
shelly cse sand sandy & pebbly sticky grey mud
RC
M104
cse shell sand with pebbles
shelly & pebbly sand
silty grey clay
SC
M105
shell & pebble gravel in shell sand
pebbly & shelly sand
grey mud
SC
M237
pebble gravel in shell sand
shell sand
muddy fine sand
423cms
VC
M245
shell sand & gravel
shell sand & gravel
sandy clay
boulder clay
VC
M96
grey sandy mud
grey sandy mud
grey silty mud
SC
M94
fine sand with shell frags
fine sand
pebbly dark grey sticky clay
SC
M202
dark olive-grey sandy mud with shell frags
dark olive-grey sandy mud
dark olive-grey mud
SC
M92
fine muddy sand
sandy mud
grey sticky silty mud
grey silty clay
SC
M91
fine muddy sand with shell frags
shelly & pebbly fine muddy sand
grey pebbly sticky clay
SC
M89
shell frags in fine-med sand
cse pebbly shell sand
dark grey sticky clay
SC
NS182
gravel of red sand-stone pebbles
gravel & sand
dark grey heavy & sticky clay
SC
M292
med-v.cse shell sand with shell frags
shell sand
muddy sand
SC

Fig. 6.34d.

the North Minch sands further offshore, which were rarely cored by the I.G.S. owing to poor recovery. Only 20 cm of shelly sand were encountered over boulder clay in Borehole 72/32, 5 km east of Tolsta Head.

The small Quaternary sediment body around Rubh'Re has yielded cores of dark grey, heavy and sticky, pebbly boulder clay, covered by what we may again interpret as a thin, winnowed, lag deposit of pebbly and shelly sand and gravel (Traverse *C*). Cores on the Shiant East Bank often penetrate to rock and are seldom of any great length. Frequently, the rockhead is overlain by a tough and pebbly boulder clay (Traverses *B* and *C*) and the pebbly and shelly gravels found at the surface form only a very thin layer. The sides of the bank produce cores of muddy fine sand, which show no rapid change in lithology downwards, and the surrounding hollows have thick deposits of modern mud.

The thickest surface sediments occur in the mud-floored hollows of the Inner Sound, Sound of Raasay, Loch Snizort, Stornoway Basin and other smaller areas. In many cases core penetration of two or more metres failed to encounter a significantly different sediment from that at the surface. However, sedimentation rates calculated from palaeomagnetic studies in the Inner Sound (Bishop, 1975), if typical of the other muddy hollows, indicate that there may be no more than about 10–12 m of post-glacial muds present. There is a tendency for the surface mud to be slightly more sandy than that below, and for there to be minor colour changes with depth, often from olive-grey to grey. The more deeply buried mud is generally a little more cohesive than the surface deposit, but nowhere did cores penetrate a contrasting lower sediment. It is likely that the modern muds pass downwards into the Formation 2 glacial-marine deposits that infill much of these deeply excavated hollows.

Although the Inner Sound has yielded many long mud cores, the shallower areas, particularly between Applecross Bay and Raasay, have only a thin sandy or muddy sediment cover over rock. The cores from the Skye–Shiant Platform resemble those from the Shiant East Bank; thin, pebbly sands and gravels over boulder clay or rock (Traverse *C*), and although slightly thicker sand sequences are occasionally to be found, these become more fine grained with depth. The fine sand typical of much of the Little Minch is seldom more than a few centimetres in thickness, and passes gradually downwards into sandy mud or a more cohesive grey mud, in a way similar to the sandy muds and muddy sands of the North Minch. The deep hollows of the western Little Minch, close to the Outer Hebridean coast, frequently reveal a thin mud or fine sand cover over rock or pebbly clay.

The examination of the lithological properties of the surface sediments expressed here in the simplest terms, seems to establish a pattern of marine re-working of glacial sediments in the less deep parts of the Minches, the finer fractions being removed from the boulder clays in shallow areas to leave a lag deposit of shelly and sandy gravel, while thick modern muds have been deposited in the neighbouring hollows. To investigate further sediment redistribution we shall make a closer examination of the textural properties of the surface sediment, and then discuss the gravel, sand and mud fractions.

Textural analyses

Attempts at relating the textural characteristics of a marine sediment to its history of transport or its environment of deposition have been a common feature of recent sedimentological studies on the continental shelf. Subtle techniques of grain-size analysis have been developed, and some of the more purely statistical treatments of the data have yet to be conclusively linked to specific environmental controls. In the light of such a rapid increase in knowledge and technique as has occurred in recent decades, we shall first review the methods of analysis used and their application to the problem of the origin of the Minches surface sediments.

Even the fundamental concept of grain size is far from simple to define. Pettijohn (1975) lists no fewer than eight definitions, varying from the measurement of the length and breadth of a particle to a definition dependent on the particle's settling velocity which is, ideally, a simple function of its dimensions. Each concept of size is related to a particular technique of grain-size analysis, and, unfortunately, no technique presently exists that is capable of handling the extreme range of grain sizes frequently to be found in a single Minches sample. This leaves no option but to construct the grain-size distribution curve of a sample from the combined results of two separate analytical techniques, *each dependent on fundamentally different concepts of grain size.* The gravel and sand fractions were analysed by dry sieving (the procedure outlined by Folk, 1965) at ½ phi intervals, whereas the grain size of the mud fraction was determined by pipette analysis (also according to Folk's procedure). Sieving measures the "sieve diameter" of a particle, that is, the minimal or least area of cross-section (Pettijohn, 1975). It is quite possible in theory, and was frequently observed in practice, for a single sieved fraction to contain a variety of grains of contrasting shape and volume. This was particularly noticeable in the case of sand of mixed biogenic and detrital origin, where platiform shell fragments often collected in the same fraction as rounded lithic grains which had a much greater volume per particle. Conversely, needle-like echinoderm spines, with a relatively large volume but a small cross-sectional area in one plane, passed through fine meshed sieves to collect with round detrital grains of much smaller volume. The sieve tends to sort, therefore, according to shape as well as grain size. When trying to relate the sieve-derived grain-size distribution of a sediment to those environmental processes that may have controlled its transport or deposition, shape sorting will obviously hinder accurate interpretation of the data. In addition, a platiform shell fragment may be hydrodynamically equivalent to a rounded detrital grain of quite different size, while platiform and rounded grains of the same sieve diameter may behave quite differently in identical environmental circumstances.

The sedimentation tower is often seen as an alternative method of grain-size analysis in the sand-size range, but it should be remembered that the reduction of settling velocities to grain-size equivalent is based on velocities calculated for *spherical* particles of a standard density, while a sample may contain a variety of both particle shapes and densities. Even the use of "raw" settling velocities for environmental comparisons is confounded by the problem of relating *vertical* settling velocities to current speeds needed for the *horizontal* transport, initial movement or deposition of particles of various shapes. This problem, together with the practical difficulties still encountered in the operation of the sedimentation tower, would make the results so obtained of no more real value to environmental analysis than those gained through sieving, provided that the sieve's tendency towards shape sorting is constantly held in mind, and the results treated with due caution.

Pipette analysis of silt and clay again depends on the measurement of settling velocities, the reduction of which to grain size, as in the case of the sedimentation tower, assumes that the particles are spheres of constant density. This assumption can be regarded as unrealistic for probably all natural sediments. Pettijohn (1975) has pointed out that the care taken to disperse clays prior to analysis in the laboratory, by both physical and chemical methods, destroys any flocculation structures that may have been present and which would have originally controlled deposition in the sea (see also Chapter 8). Also, faecal and other pellets that could have influenced mud transport are destroyed. Pelleted mud was observed on the retrieval of grab samples at many stations in the Inner Sound and North Minch. For these reasons, analysis of the finer fractions will probably be the least realistic aspect of the study of the relationship between the sediment and its environment, but pipette analysis remains the most commonly used and most accurate method of analysis available in this size range (Folk, 1965). Finally, we can only speculate about the importance of further unseen but natural controls on sediment distribution that are not revealed in laboratory analysis, particularly biological factors like bioturbation or the binding of sediment by algae or plants. During a submersible dive in the Little Minch (Eden et al., 1973) algae and sea-ferns were observed to be trapping sediment.

For the purpose of expressing grain size, phi units have been chosen rather than millimetres or microns, for the ease with which this logarithmic scale can be directly applied in various statistical treatments of the data.

Mean grain size

The mean grain size of the Minches sediment samples was calculated using the graphic mean formula of Folk (1965). A wide range of mean sizes occurs, from pebble to clay values. The mean grain size (Fig. 6.35) is closely related to water depth, the finest sediments being found in the deepest water and the coarsest on the shallow banks. In Fig. 6.36 mean grain size is plotted against water depth for all of the grab samples used in the study. A clear trend from shallow and coarse to deep and fine sediments exists, with a broad scatter of points. This trend reflects the greater energy level of the shallow marine environment, where waves and tidal currents are most influential, compared to the deeper and quieter areas. The scatter of points may be partly due to the shape-sorting tendency of sieves and the inherent problems of pipette analysis, which as we have seen, can be expected to cause distortion of the hydrodynamic significance of the grain-size values. Significantly, no gravel means occur in deep water, yet a number of samples do display unusually fine sediment in shallow areas. Zone *A* on Fig. 6.36 arbitrarily encloses 26 samples with mean grain sizes much lower than that normally expected for their depth. The geographical distribution of these samples is as follows:

Sound of Raasay	7	Loch Maddy	2
Loch Snizort	6	Gruinard Bay	1
Inner Sound	5	Stornoway Bay	1
Broad Bay/Tolsta Head	3	Loch Seaforth	1

The samples all come from lochs, bays and sounds, and none is from the open water of the North or Little Minches. This would seem to indicate that, while bathymetry is the main influence on grain size in open water, the shelter offered by the adjoining sounds and lochs encourages mud accumulation at shallower depths. How the mud is transported to the adjoining bodies

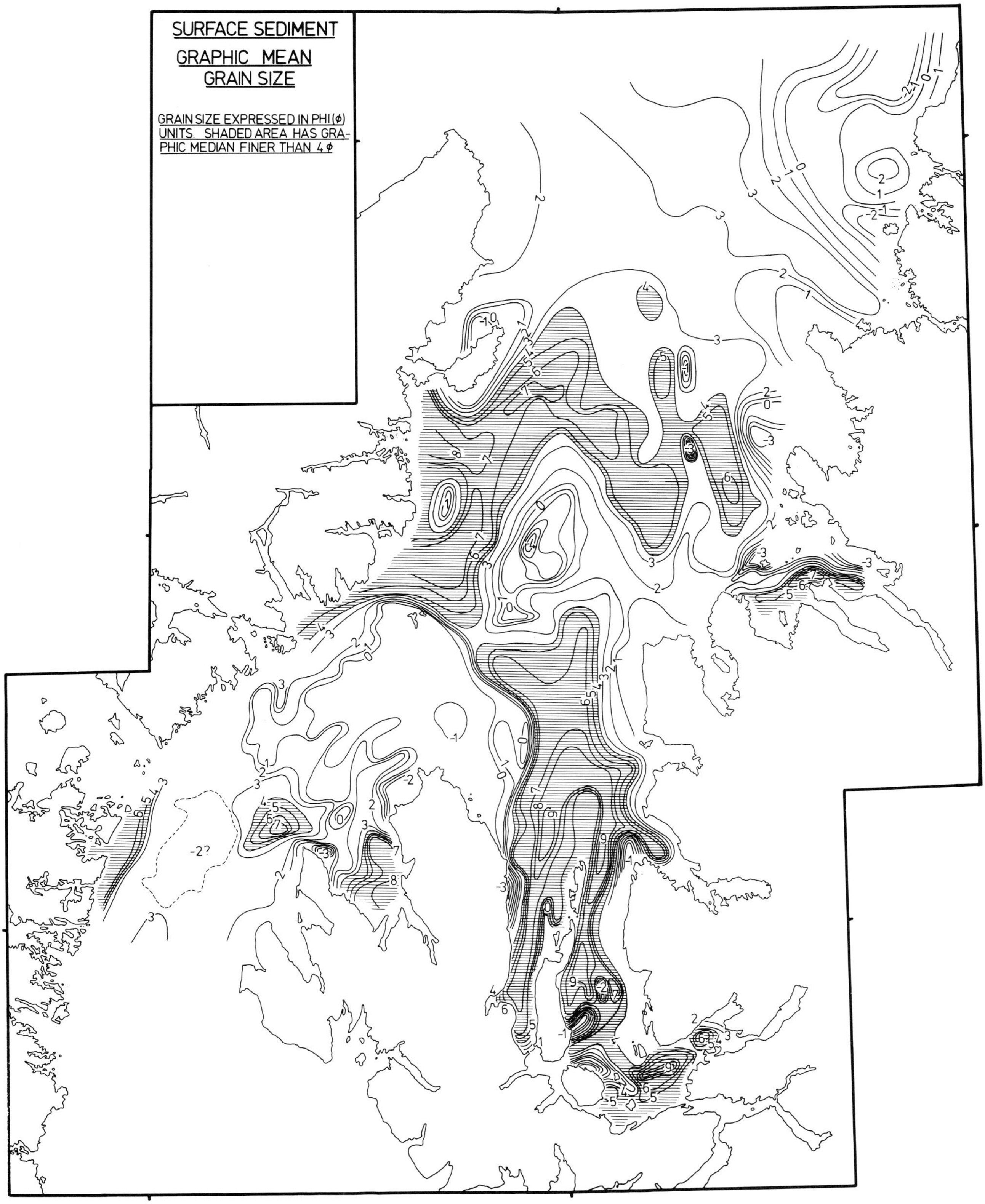
SURFACE SEDIMENT
GRAPHIC MEAN GRAIN SIZE
GRAIN SIZE EXPRESSED IN PHI (ϕ) UNITS. SHADED AREA HAS GRAPHIC MEDIAN FINER THAN 4ϕ

Fig. 6.35.

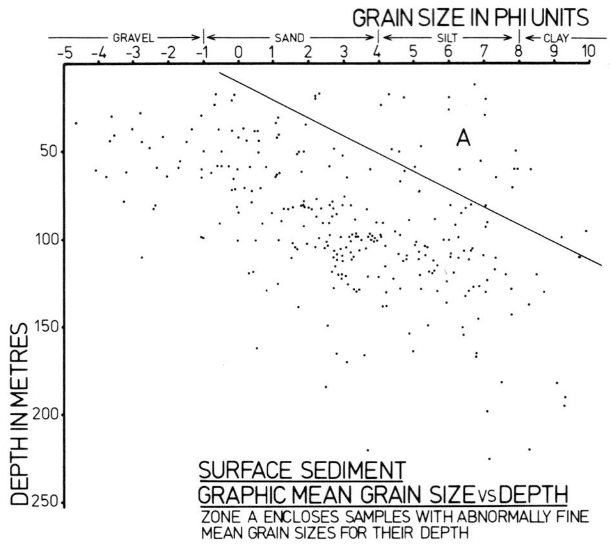

Fig. 6.36.

will be considered in due course. The three samples from the Broad Bay–Tolsta Head area represent a different set of circumstances, for these are the only ones in zone *A* to fall within the sand size range, and probably result from the infilling of Broad Bay with a distinctive shell sand which is in transit along the Lewis coast from the Butt.

The C-M diagram

C-M diagrams, on which the median grain size of a sample is plotted against the coarsest grain size present (in fact, the 1% value on the cumulative curve) were introduced by Passega (1957, 1964) in an attempt to relate grain-size distribution directly to depositional process. As median grain size has not been used at all in this study – for reasons given by Folk (1965) – graphic mean grain size becomes our parameter for expressing the average grain size of the sediment. C-M diagrams have been used by the U.S. Geological Survey (Schlee, 1973) in the analysis of sediment samples from the north-eastern continental shelf of the U.S.A., an area that includes the Gulf of Maine, which is floored by glacial sediments now being re-worked by the sea, and so very similar to the Minches. Schlee extended Passega's originally simpler diagrams by the addition of several hundred reference samples from known environments, so that samples with transport histories as diverse as glacial till, beach gravel, suspension deposits and current-laid sands, were shown to occupy distinct clusters when plotted on a C-M diagram. The fields used in this present study are based on Schlee's but in a slightly modified form in order to suit the spread of Minches data (Fig. 6.37). Each point on the diagram represents a single sample. Tractive current deposits occupy much of the area close to the C=M line and range in grain size from sand to gravel. Zone *F* consists of current-laid sands, and zone *C* of beach or "wave zone" gravel, which are both characterised by good sorting, and therefore, possess coarsest grains similar in size to the mean. These two fields are as determined by Schlee (1973). Fine-grained deposits possessing similar coarsest and mean grain sizes can be regarded as having been transported by suspension, and fall within zone *G*, another zone taken directly from Schlee's results. Zone *A*, located furthest from the C=M line, encloses sediments with mean grain sizes much finer than the coarsest particles present, and therefore, includes the poorest sorted sediments. In this zone, Schlee plotted known glacial tills, for ice has the ability to transport the widest range of grain sizes (although the texture of the till will always ultimately depend on the grain size of the source material). A large number of Minches samples falls into zone *A*, indicating that unmodified or little altered glacial deposits are probably now exposed in large areas of the sea floor. Zones *D*, *E* and *B* have not been used by previous workers and represent transition zones between the zone *A* tills and zones *G*, *F* and *C*. Zone *B* happens to fall roughly within the same area as a number of river terrace gravels plotted by Schlee, but this does not necessarily imply that the sediments from the Minches are of this origin. That a particular sample plots in the same field as a known sediment type is not absolute confirmation that it shares the

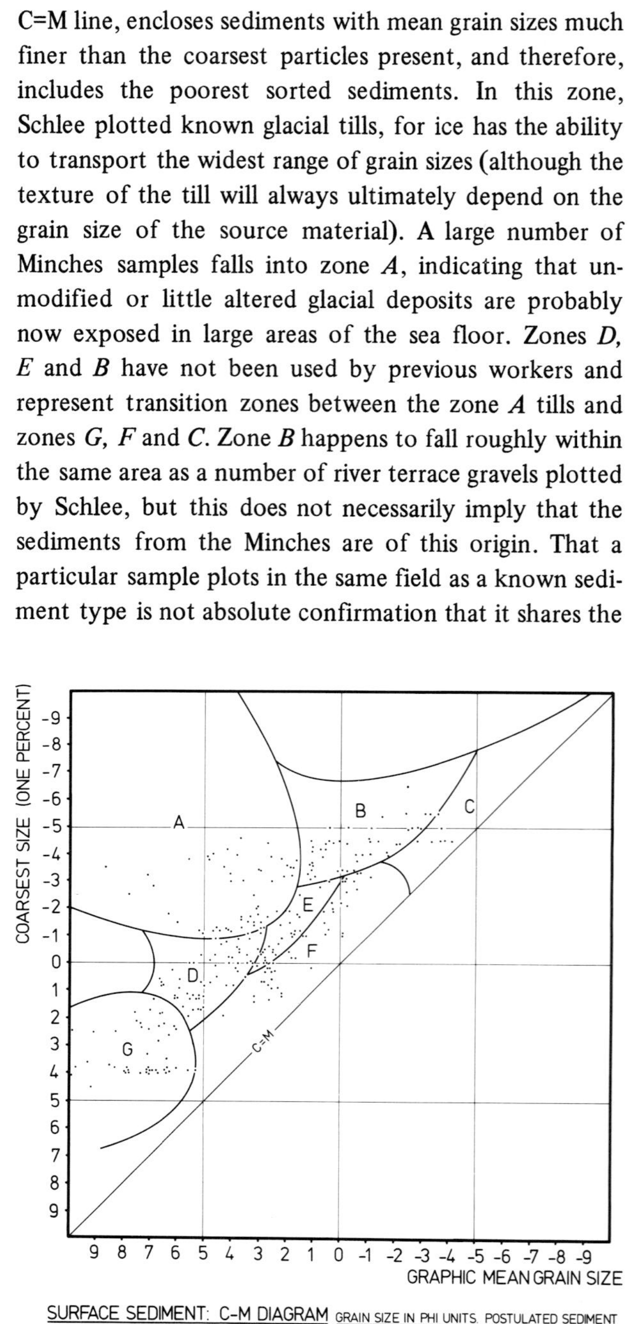

Fig. 6.37.

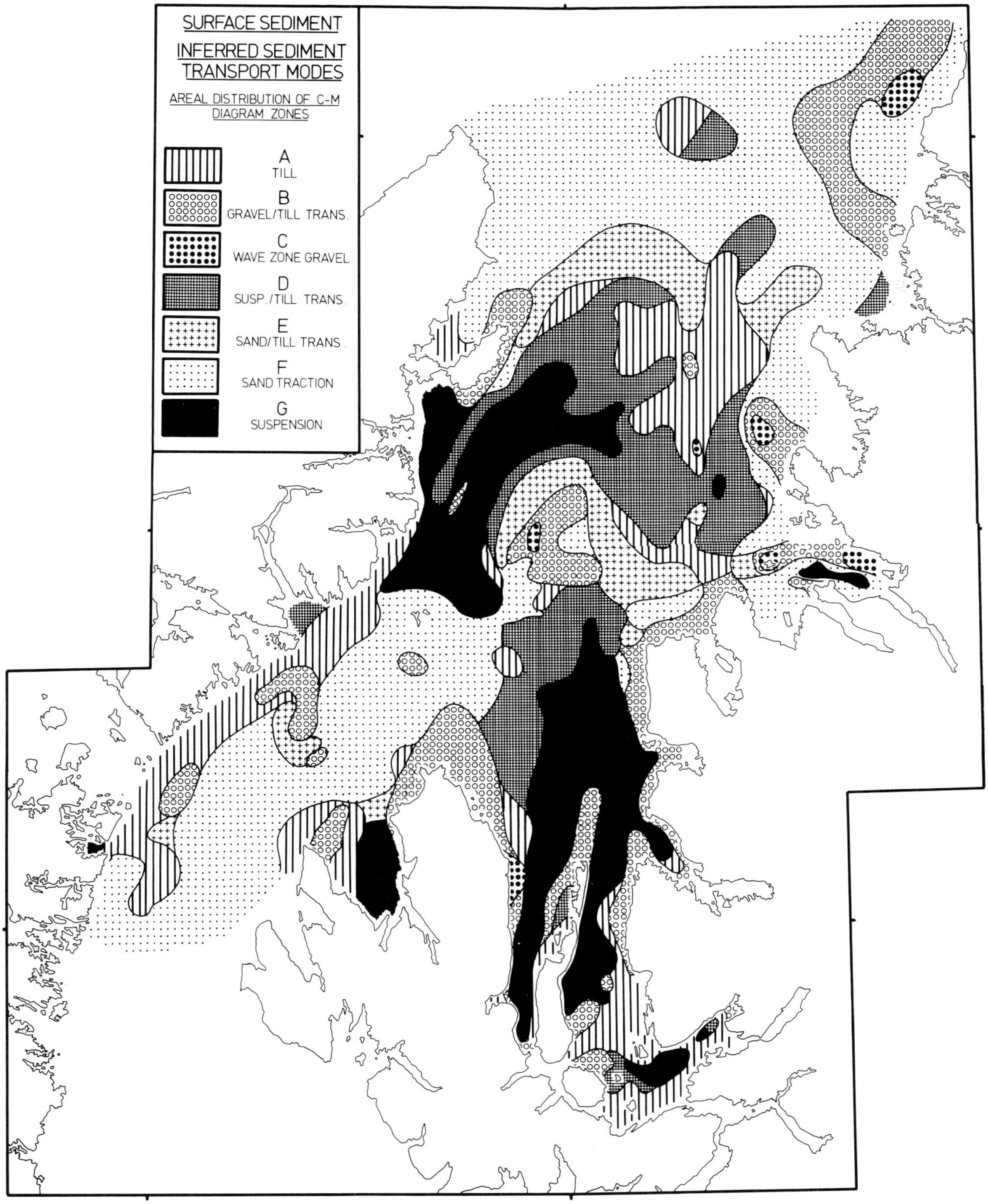
SURFACE SEDIMENT
INFERRED SEDIMENT TRANSPORT MODES
AREAL DISTRIBUTION OF C-M DIAGRAM ZONES
A
TILL
B
GRAVEL/TILL TRANS.
C
WAVE ZONE GRAVEL
D
SUSP./TILL TRANS.
E
SAND/TILL TRANS.
F
SAND TRACTION
G
SUSPENSION

Fig. 6.38.

same mode of origin, but only that the two sediments have certain diagnostic properties in common. All the transition zones introduced on the C-M diagram for the purposes of this study have a marked scattering of points, continuous spreads of which link zone *A* tills with gravels, sands and suspension deposited muds. At the same time, the gravel, sand and mud zones do not themselves possess as well defined a continuity of points, which would suggest that while these three sediment end-members are each closely related to the tills, they are not genetically inter-related to each other and tend to form discrete sedimentary units.

The geographical distribution of samples falling into each of the seven zones on the C-M diagram is shown on Fig. 6.38. Tills occur in the central North Minch, in the deep trough along the Outer Hebridean coast in the Little Minch, around Vaternish Point, in the Inner Sound and the Sound of Raasay, and in patches on the flanks of the Skye–Shiant Platform and Shiant East Bank. On the tops of the banks the tills pass into till-gravel transition deposits, and then into better sorted, wave-zone gravels in the very shallowest parts, suggesting again that the thin gravel cover on the banks is a winnowed lag deposit derived from boulder clay. The southern part of the Shiant East Bank and the northern part of the Skye–Shiant Platform are covered with traction-laid sands. This seems to be a result of the availability in these areas of abundant shell debris, for, as we shall see below, the sands consist almost entirely of locally derived broken and abraded shell fragments, forming a sediment blanket over the glacial deposits and bedrock below. The muds in the surrounding hollows have been deposited from suspension, and at the margins grade into sediments with textural properties akin to the tills. In the North Minch, tills pass northwards through a belt of transition deposits into a wide zone of traction-laid sand that extends northwards beyond the Butt of Lewis, and similar sand occurs in the southern Little Minch. In both of these areas we have already noted the presence of Formation 3 peri-glacial deposits at the top of the glacial sequence. The C-M diagram appears to reveal here the presence of traction-laid, peri-glacial sands on the sea floor, partially re-worked by the sea in the upper few centimetres. Later, we shall consider evidence to suggest that it is unlikely that the sands have gained their traction-characteristics under the modern regime, except perhaps, towards the extreme north, which is much more exposed and where sea-floor conditions may begin to resemble those of the outer continental shelf, and also between the Butt of Lewis and Broad Bay where a distinctive shell sand is in transit close to the coast. The only other deposits of traction-laid sand occur in a discontinuous belt close to the mainland coast, from Cape Wrath to Rubh'Re, in an area where strong tidal currents are experienced.

Modal grain size and polymodality

Modal grain size is held to be a more reliable parameter than mean grain size for environmental studies (Folk, 1965). The distribution of the primary modal grain size is shown in Fig. 6.39. The pattern is similar to the distribution of graphic mean grain size (Fig. 6.35) but the trends are somewhat simplified. Pronounced modal gradients, that is, several modal intervals forming a consecutive sequence over a relatively limited distance, have been taken to imply potential sediment transport directions (Schlee, 1973). In the Minches, modal gradients occur in several places, especially between the banks and the surrounding mud-filled hollows. Lesser modal gradients occur locally. Much of the northern North Minch and the top of the Skye–Shiant Platform have no obvious direction of modal gradient, which implies that sediment transport is not taking place, although we cannot rule out the possibility that the shape sorting tendency of sieve analysis has concealed important hydrodynamic properties of the sediment. Ferentinos (1976) using mean grain size rather than mode, has correlated grain size contours with a south-westerly swell and tidal currents in order to map sand transport paths over the continental shelf west of the Outer Hebrides.

It is likely that the modal gradients at the margins of the banks are largely the result of a progressive decrease in sorting capability between the shallow high energy environments and the surrounding hollows, and may only indirectly be indicative of transport paths. However, an increase in the order of polymodality (Fig. 6.40) is observed to accompany the modal gradients towards the deep mud-filled hollows, which suggests the influx and mixing of sediment. Use of the polymodality map in environmental interpretation is not easy. Most of the Minches floor is covered by polymodal sediment. Whereas polymodality is usually attributed to sediment mixing, in the case of the Minches, where typically polymodal glacial sediments are in abundant supply for marine reworking, an unmixed modern sediment may derive directly from a polymodal source. Although interpretation is difficult, the trends do suggest sediment influx at the foot of the modal gradients that encircle the banks. These fine polymodal sediments contrast with the sands of the Skye–Shiant Platform and the North and Little Minches, where unimodality accompanies the absence of modal gradient. On the banks, gravels are typically bimodal, owing to the admixture of lithic

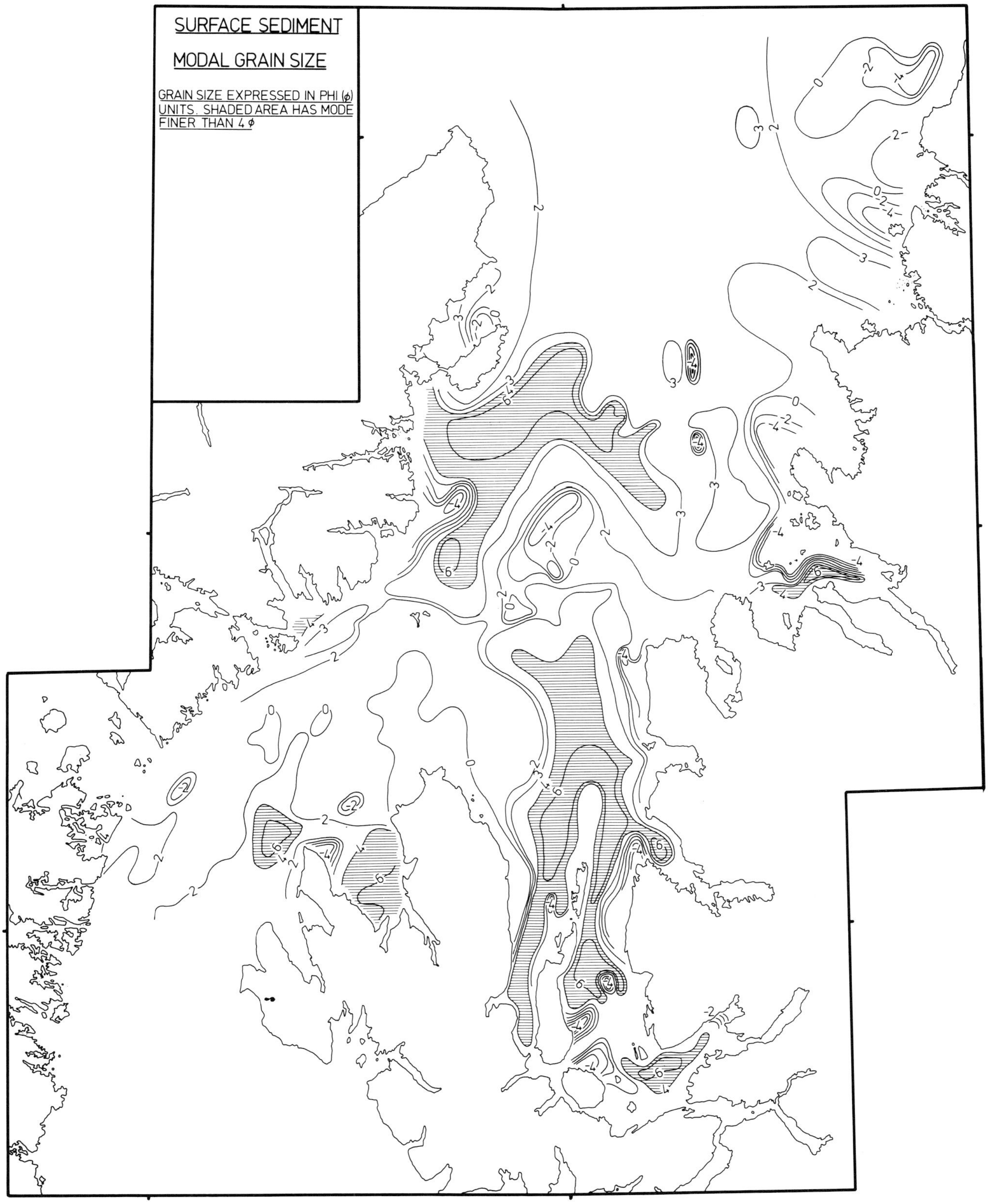
SURFACE SEDIMENT
MODAL GRAIN SIZE
GRAIN SIZE EXPRESSED IN PHI (ϕ) UNITS. SHADED AREA HAS MODE FINER THAN 4 ϕ

Fig. 6.39.

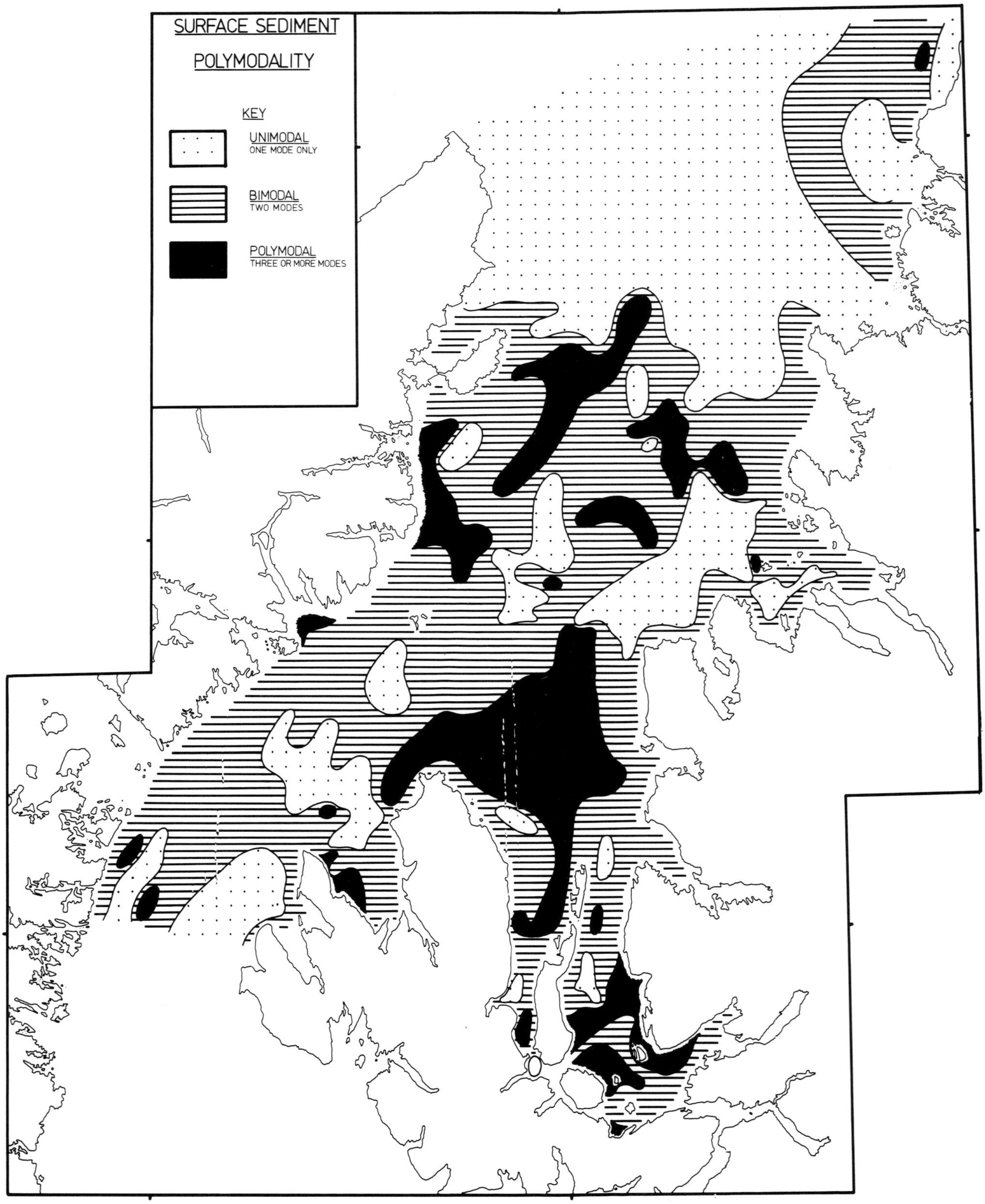

Fig. 6.40.

gravel and interstitial shell debris, but some unimodal gravel occurs on the Shiant East Bank.

Sorting

The sorting coefficient used in the preparation of Fig. 6.41 was the Graphic Standard Deviation of Folk (1965) and the division of sorting classes used was that also suggested by Folk. An extreme range of sorting values occurs in the Minches sediments, and examples occupying all of Folk's classes can be found. The most poorly-sorted sediments are the polymodal muds of the deep hollows, and the tills of the central North Minch and Inner Sound. Sandy and gravelly areas are generally better sorted. Some gravels on the Shiant East Bank are particularly well sorted, and on the Skye–Shiant Platform shell sands locally attain the highest class of sorting, indicating the reworking of biogenic debris by bottom currents. The sands of the northern North Minch are not as well sorted, except towards the Lewis coast, and to the south-west of the Cape Wrath Bank. Sands in the south-eastern part of the Little Minch are distinctive in their good sorting.

Skewness

The environmental implications of skewness distribution (Fig. 6.42) are not very certain. The skewness characteristic of sediment, like other textural (and compositional) properties, must primarily depend on its source material. Folk (1965) suggests that extremes of skewness occur in areas of sediment mixing, and that symmetry of the grain-size distribution curve suggests derivation from single source. However, our working hypothesis shall be that positive skewness, that is, an excess of fine material, indicates an influx of mud in suspension to an area of deposition, while negative skewness, or an excess of coarse material, indicates an area from which finer grades have been winnowed. This interpretation of skewness has been used previously by Hough (1942), Spencer (1963), Friedman (1967), Swift (1969) and Schlee (1973). It is possible that other factors could influence the skewness distribution. The Minches possess a complicated distribution of skewness, but, in general, the Inner Sound and the Sound of Raasay have positive-skewed sediment, as do Loch Snizort, Loch Maddy and Stornoway Bay – all places where the influx and deposition of mud from suspension has already been inferred from other textural properties. Much of the central North Minch also has positive-skewed surface sediment, including the area of till exposure indicated on the C-M diagram, and while the Stornoway Basin falls within the "nearly symmetrical" category, values in the area are, in fact, slighly positive. The sands of the northern North Minch, the Skye–Shiant Platform and the southern Little Minch are not markedly skewed, but significantly coarse-skewed sediment, indicating strong winnowing, occurs on gravel-covered parts of the Skye–Shiant Platform, the Shiant East Bank and near the mainland and the Hebridean coasts. The skewness pattern on the gravel-covered banks is complicated by small areas of fine-skewness, caused by the presence of sand-sized material, usually shell debris, among the pebbles. Skewness values, therefore, give further confirmation of a pattern of winnowing and redistribution of fine sediment from shallow areas, and deposition of mud from suspension in deeper hollows and adjoining lochs and sounds.

The gravel fraction

The distribution of gravel (by weight) within the surface sediment is shown in Fig. 6.43. Because of the larger particle size, a statistically representative sample of gravel must, of course, be of a greater volume than one of sand or mud. The volume of grab samples from the Minches varied widely, not all being up to the 1-kg minimum suggested by Folk for gravel (1965). Krumbein and Pettijohn (1938) suggest a 2-kg minimum. Between 50 and 100 g of sand, and 20–50 g of mud are required for confident analysis according to Folk, and almost all Minch samples were sufficient in these grades. Unfortunately, the surveying programme did not take into account the need for large gravel samples. The Shipek sampler has more difficulty in functioning on a hard gravel bottom than on softer mud or sand, and often does not provide as much as 1 or 2 kg of sediment. Rather than discard most of the samples from gravel-floored areas, we decided to use all gravel samples, bearing in mind that distribution maps for gravel properties cannot achieve the same order of accuracy as those for the other fractions. Fortunately, the wide variations in the texture and composition of Minches sediments allow us to discuss very broad patterns with reasonable confidence, but more detailed examination of the gravel fraction would require a specially designed sampling programme.

The distribution of gravel is closely related to bathymetry. Large gravel-free areas occur in the deep mud-floored hollows of the Inner Sound, Sound of Raasay, North Minch and Loch Snizort. The sand-covered areas of the Little Minch and North Minch have low gravel contents, and we may note that a lack of pebbles is a distinguishing feature of Formation 3 peri-glacial deposits, which underly both these areas. Gravel maxima occur on the Shiant East Bank, the Skye–Shiant Platform and in many coastal areas, particularly off the mainland headlands. The southern part of the Skye–Shiant Platform has a higher gravel content than the north, where biogenic sediments dominate, and locally increased gravel values are associated with the morainic Greenstone Ridge. The Minches gravels typically display an admixture of rounded and sub-angular shapes. However, on the Shiant East Bank the better sorted gravels of the central part consist of remarkably well rounded and uniform Torridonian pebbles. In areas with increasing biogenic gravel components, lithic pebbles may display considerable amounts of encrusting.

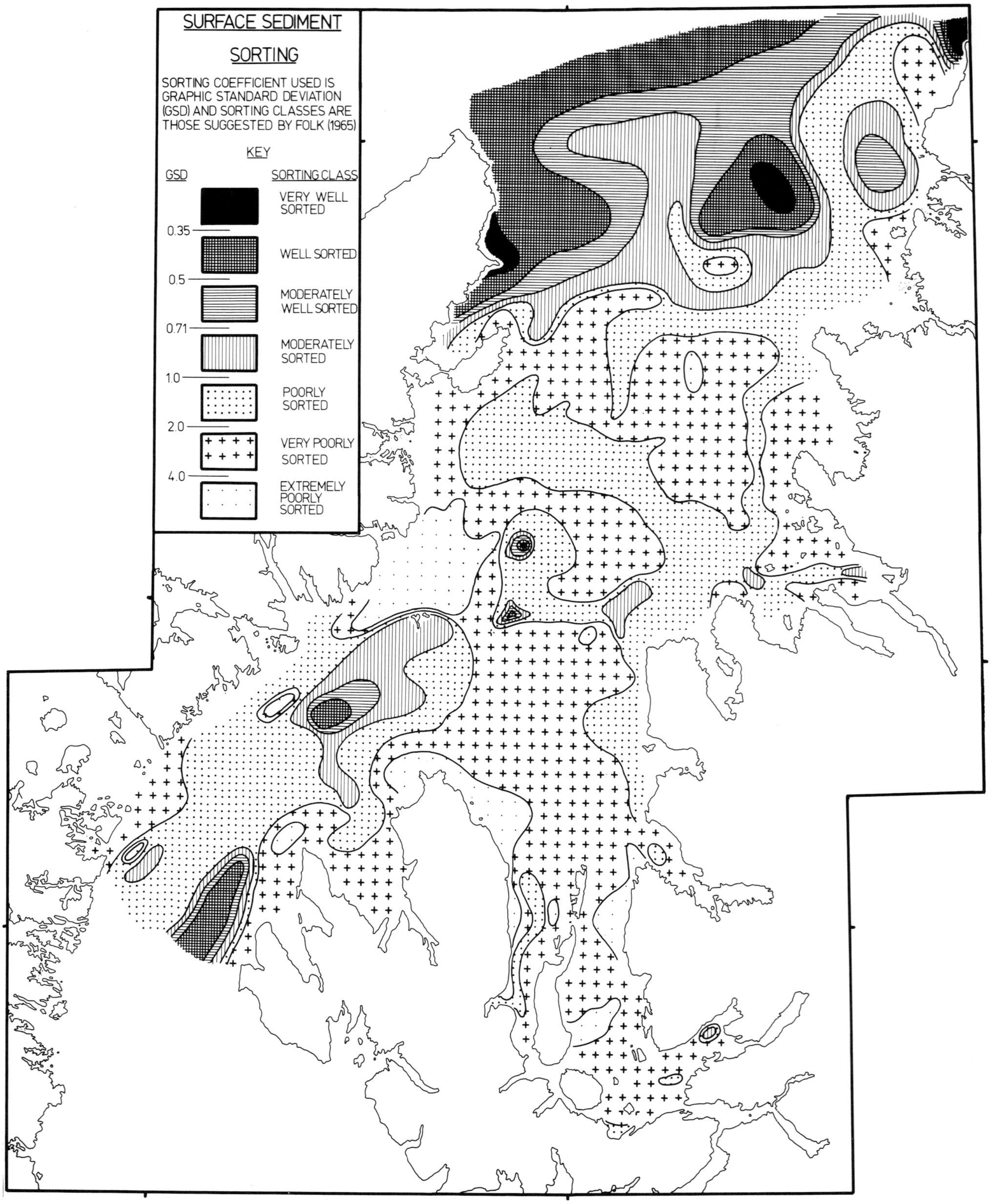
SURFACE SEDIMENT
SORTING
SORTING COEFFICIENT USED IS GRAPHIC STANDARD DEVIATION (GSD) AND SORTING CLASSES ARE THOSE SUGGESTED BY FOLK (1965)
KEY
GSD
SORTING CLASS
VERY WELL SORTED
0.35
WELL SORTED
0.5
MODERATELY WELL SORTED
0.71
MODERATELY SORTED
1.0
POORLY SORTED
2.0
VERY POORLY SORTED
4.0
EXTREMELY POORLY SORTED

Fig. 6.41.

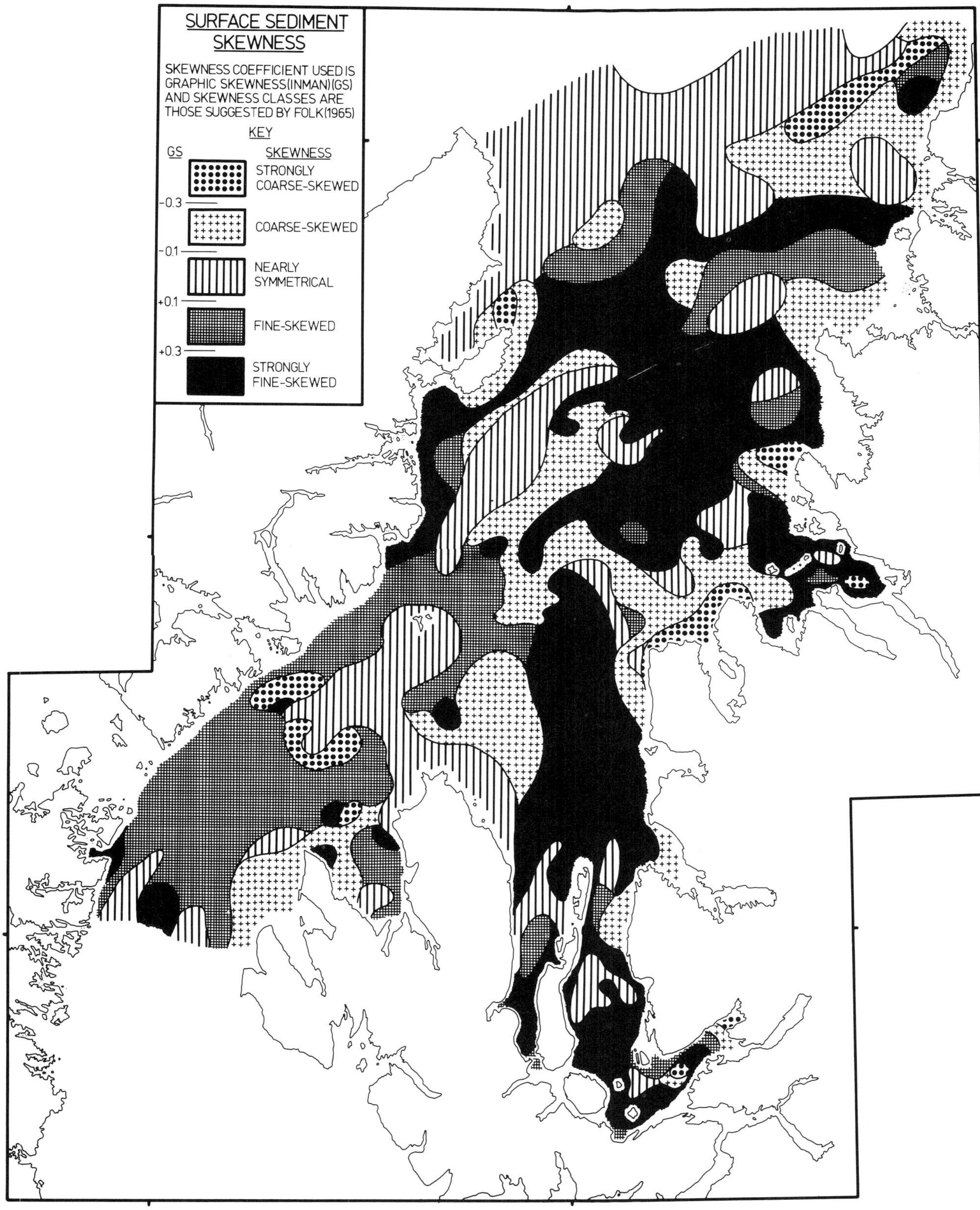
SURFACE SEDIMENT SKEWNESS
SKEWNESS COEFFICIENT USED IS GRAPHIC SKEWNESS(INMAN)(GS) AND SKEWNESS CLASSES ARE THOSE SUGGESTED BY FOLK(1965)
KEY
GS
SKEWNESS
STRONGLY COARSE-SKEWED
-0.3
COARSE-SKEWED
-0.1
NEARLY SYMMETRICAL
+0.1
FINE-SKEWED
+0.3
STRONGLY FINE-SKEWED

Fig. 6.42.

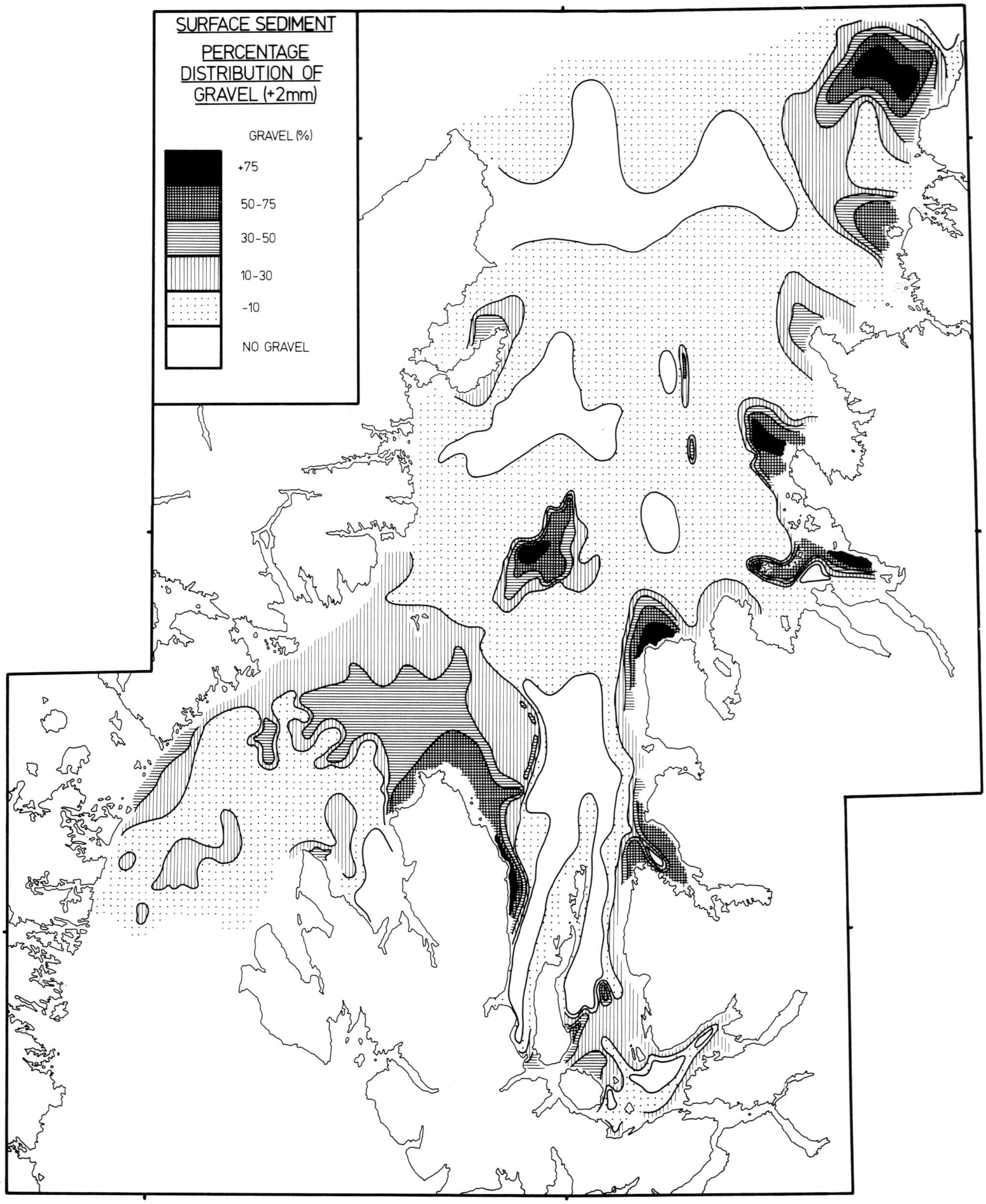
SURFACE SEDIMENT
PERCENTAGE
DISTRIBUTION OF
GRAVEL (+2mm)
GRAVEL (%)
+75
50-75
30-50
10-30
-10
NO GRAVEL

Fig. 6.43.

The lithic detrital gravel component

The percentages (by volume) of lithic and biogenic material within the gravel fraction of each sample were estimated by eye to the nearest 10% and the results mapped on Fig. 6.44. The detrital gravel component dominates in coastal areas, in a broad belt in the centre of the North Minch, extending north-eastwards from the Shiant East Bank, and in the northern Little Minch from the southern part of the Skye–Shiant Platform. Most of the areas with a high total gravel content are associated with high detrital gravel percentages. Abundant detrital gravel in areas where total gravel content is low occurs in the southern part of the Inner Sound and in the central North Minch, both areas where the C-M diagram (Fig. 6.38) indicates glacial tills, as well as in the Little Minch. Low detrital gravel content is usually associated with low overall gravel abundance and occurs in the deep, mud-floored hollows and in the sands of the northern North Minch. Large quantities of biogenic gravel occur only on the northern part of the Skye–Shiant Platform and parts of the Shiant East Bank.

Detrital gravel may be derived either from the land or from submarine exposure of bedrock or glacial sediment. The high detrital gravel content near the coasts suggests that direct derivation from the land is locally important. High lithic gravel content in the central parts of the North and Little Minches suggests that submarine sources are also significant. The rock types present within the detrital component may be an indication of its origin. In Fig. 6.45 the rock types found in an inspection of the gravel fraction have been plotted on a presence/absence basis. Four most easily distinguishable geological categories were used: Torridonian, Lewisian, Mesozoic (sediments) and basalt or gabbro. These groups were sufficient to include almost all the rock types encountered. Some rare, probably Palaeozoic pebbles and igneous rocks of indeterminable origin were the only exceptions and were not mapped. Some difficulty was encountered in the distinction of red sandstones, which could have been either of Torridonian or Mesozoic age. Comparison of Fig. 6.45 with the geological map of the Minches (Fig. 6.5) reveals that there are large areas with little similarity between the distribution of detrital gravel types and solid geology. North of the Shiant East Bank the rockhead is deeply buried by Quaternary sediments and components within the gravel fraction are entirely of Torridonian and Lewisian origin. There is no obvious relationship between the distribution of Torridonian and Lewisian pebbles in the North Minch and the outcrops of these formations on the mainland, although some lineation of the distribution pattern along a SE–NW trend is apparent. This lineation is interesting in that it conforms with the overall trend of glacial movement away from the mainland in the extreme north of the Minches.

The Shiant East Bank is an area with thin Quaternary cover and exposed rockhead, and consequently, the distribution of detrital pebbles bears more similarity to the solid geology than it does to the north. However, basalt and gabbro sills are exposed only on the south of the bank, yet basic pebbles are distributed as far as its northern limits. Similarly, Torridonian pebbles cease south of the outcrop of the basement rocks of the Rubh'Re–Kebock Head Ridge but can be found for some distance further north. The occurrence of Torridonian and Lewisian pebbles on the southern and eastern flanks of the bank may reflect glacial transport from the Inner Sound, where the same assemblage is found. In all, the distribution of lithic gravel on the Shiant East Bank suggests a northwards transport of locally derived material by an ice flow from the south, which was also responsible for emplacing some mainland-derived material on the south and east flanks.

The absence of Lewisian rocks on the coast from Greenstone Head to Loch Torridon is paralleled offshore by a detrital gravel component made up solely of Torridonian pebbles, while a minor occurrence of Mesozoic pebbles off Rubh'Re seems to be associated with local Permo-Triassic outcrops. Lewisian exposures in Loch Torridon give rise once more to a mixed Lewisian and Torridonian assemblage, which also characterises much of the Inner Sound. South of Applecross Bay, Mesozoic sediments and gabbro and basalt pebbles also contribute to the gravel fraction, in close association with the partially exposed Mesozoic rocks of the area. The Sound of Raasay similarly has a gravel assemblage rich in basalt and gabbro and Mesozoic material. The thinly covered Jurassic sediments and sills of the Skye–Shiant Platform have produced a detrital gravel component consisting mainly of these same rock types, but also with traces of Torridonian and Lewisian material, which may well have been emplaced by basal ice moving up over the platform from the Inner Sound and Sound of Raasay (see Fig. 6.12). That the thick glacial deposits of the northern Little Minch were derived from the Skye–Shiant Platform and the island of Skye itself is indicated by the abundance of gabbro and basalt pebbles and the paucity of Torridonian fragments in the gravel component. The Little Minch has a strikingly different gravel assemblage from the northern North Minch, which is derived from the Scottish mainland. The Outer Hebridean coastal area from the Uists to the Eye Peninsula is dominated by an almost uncontaminated gravel

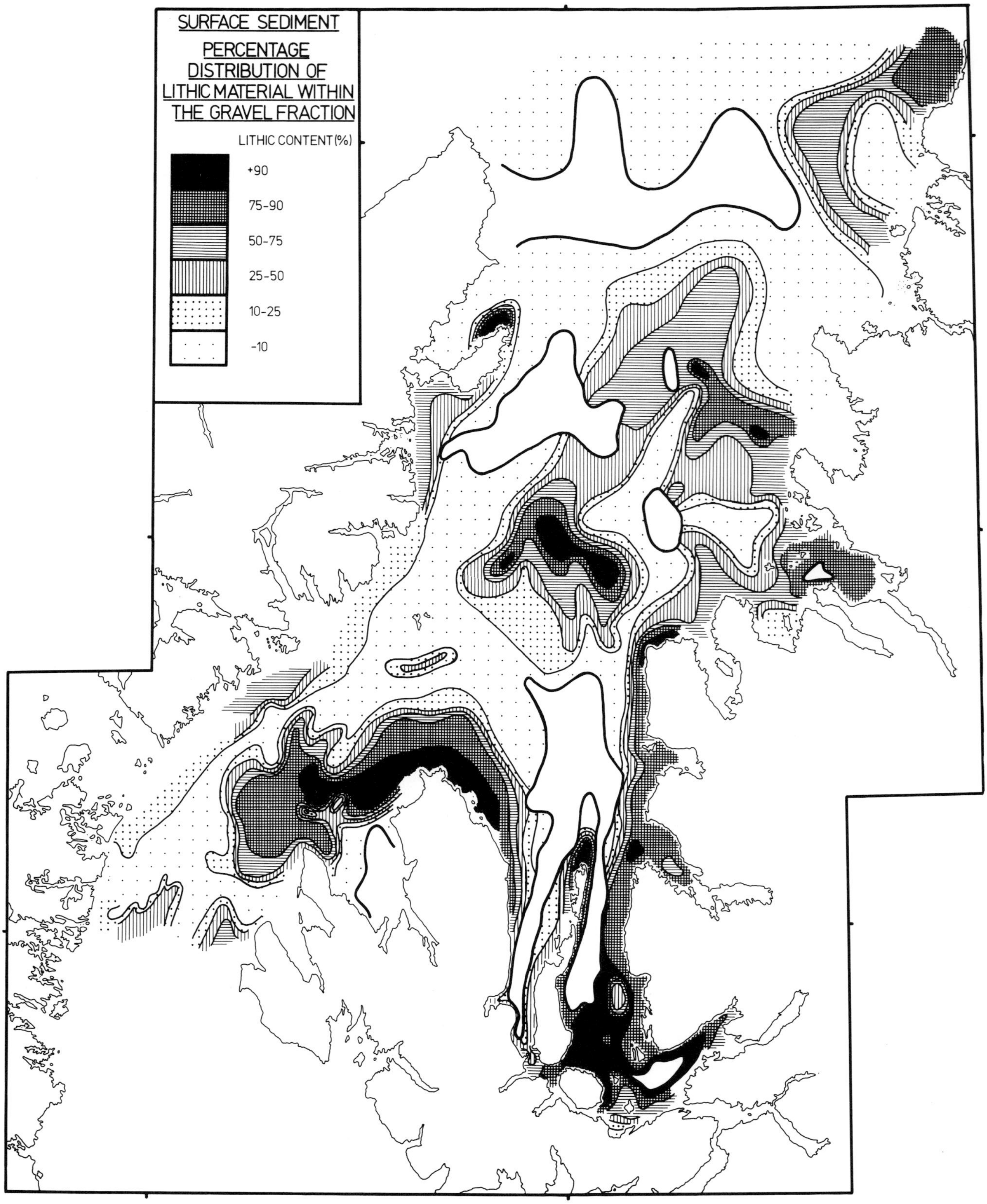
SURFACE SEDIMENT
PERCENTAGE DISTRIBUTION OF LITHIC MATERIAL WITHIN THE GRAVEL FRACTION
LITHIC CONTENT (%)
+90
75-90
50-75
25-50
10-25
-10

Fig. 6.44.

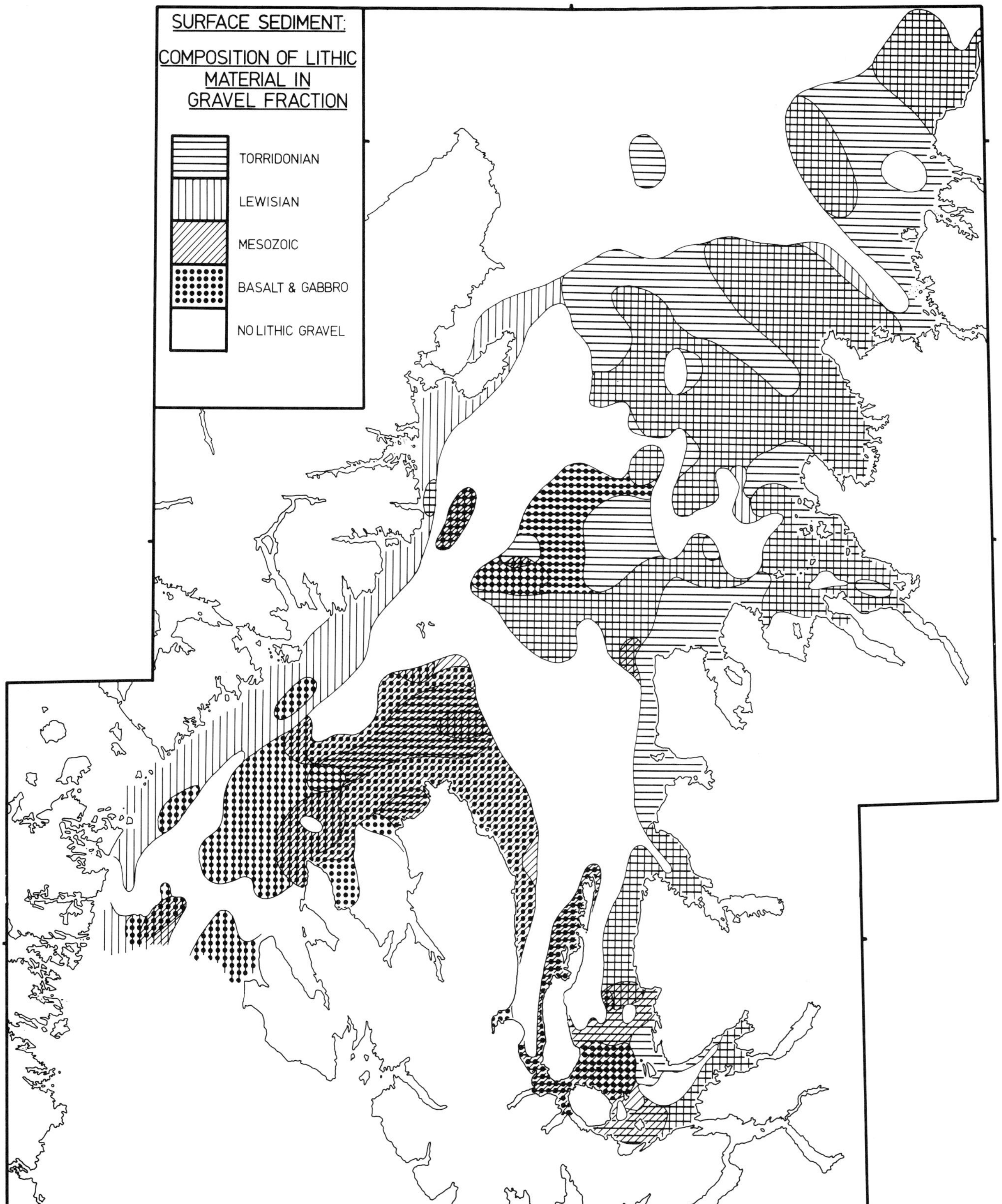
SURFACE SEDIMENT:
COMPOSITION OF LITHIC MATERIAL IN GRAVEL FRACTION
TORRIDONIAN
LEWISIAN
MESOZOIC
BASALT & GABBRO
NO LITHIC GRAVEL

Fig. 6.45.

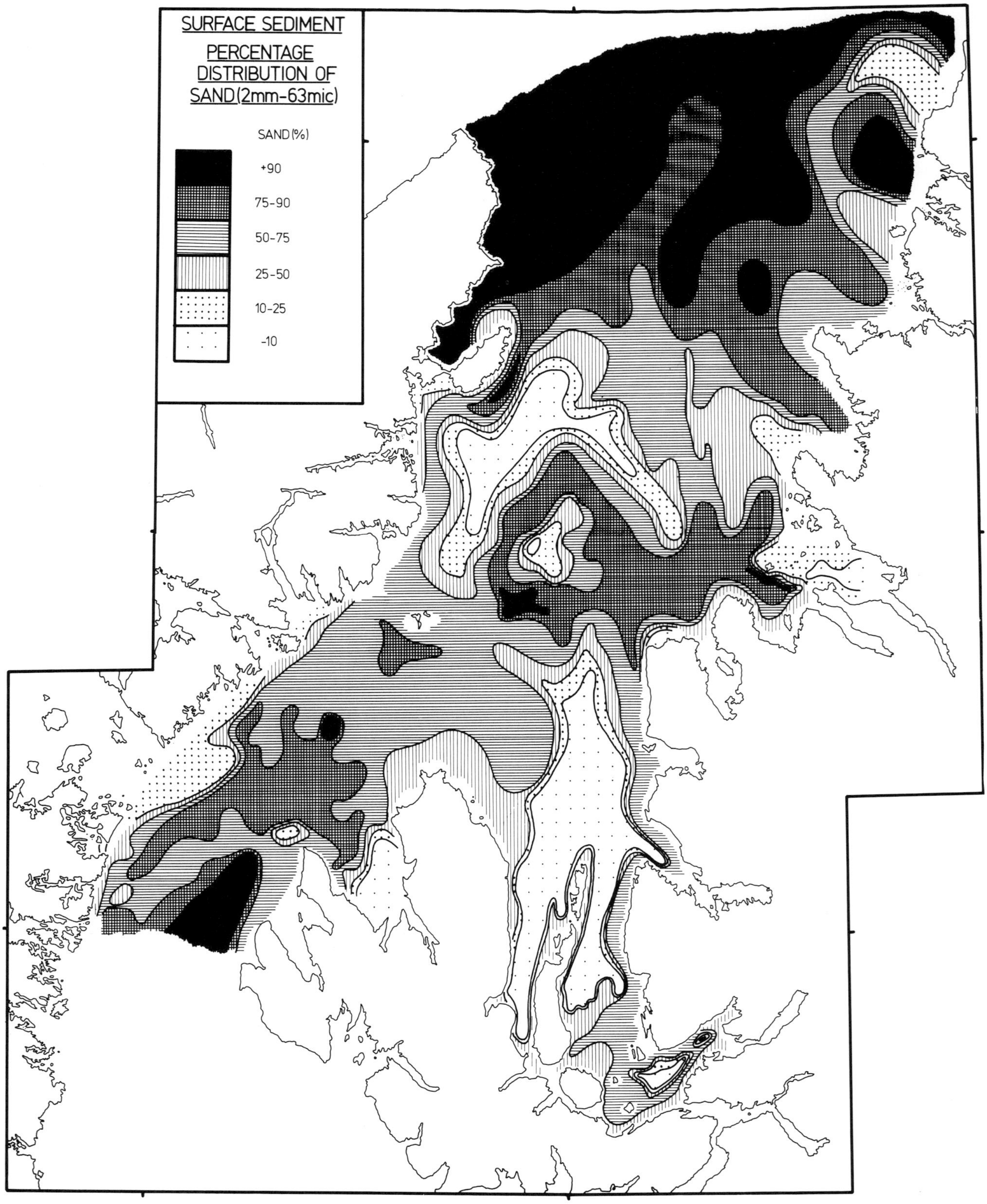
SURFACE SEDIMENT
PERCENTAGE
DISTRIBUTION OF
SAND (2mm-63mic)
SAND (%)
+90
75-90
50-75
25-50
10-25
-10

Fig. 6.46.

fraction of Lewisian rock types, again indicating the direct contribution of gravel from the land near the coasts. In summary, the distribution of rock types within the detrital gravel component can be adequately explained in terms of local rock exposure and glacial transport, and there are no indications of any significant amounts of transport by modern processes.

The biogenic gravel component

Gravels with up to 100% biogenic material occur in the Minches but are not widely developed, apart from on the northern part of the Skye–Shiant Platform and parts of the Shiant East Bank. These deposits are composed of the skeletal remains of a number of modern organisms, of which bivalves are the most important. Common bivalve genera include *Venus, Cardium, Glossus, Chlamys, Pecten, Modiolus* and *Ostrea.* Many types of gastropods occur but do not contribute so abundantly. Limpets are common and worm tubes, bryozoa and echinoderm plates and spines are also present. Shells occur in all states of freshness. Some are little worn, a few pairs of attached bivalve shells having been recovered, while at the other extreme, the gravel fraction grades into heavily abraded shell fragments of sand size. In more muddy or sandy sediments, the smaller gravel fraction may include a few of the species found on the banks, but some contrasting forms occur, especially in the muddy sands, where the gastropod *Turritella communis* is very common. Broken shells of thin-walled *Pecten* often make up the small biogenic gravel fraction of muddy sediments from deeper hollows. Two samples in the south of the Little Minch yielded small solitary corals, and Scaphopoda were found in the sands of the northern North Minch.

The sand fraction

The distribution of sand within the surface sediment is shown in Fig. 6.46. It can be seen that sand is mainly concentrated in the northern North Minch and in the Little Minch. A collar of sandy sediment encircles the Shiant East Bank at an intermediate depth between the gravelly summit and the muddy sediments below. Little sand is to be found in the mud basins of Loch Snizort, the Sound of Raasay, the Inner Sound and in the Stornoway Basin or on shallow areas such as the Shiant East Bank and the regions around the mainland headlands, which are floored by gravels.

Component analysis of sands

Analysis of the distribution of mineral components within the sand fraction was made by the following point-counting procedure. A suitably sized quartile split was prepared and scattered onto a glass slide, lightly greased to prevent slipping, so that the grains were as closely spaced as possible without overlapping. The slide was clipped onto a moving stage on a binocular microscope with a cross-hair fitted into one eye-piece, the stage being linked to a multi-channel counter. Successive grains falling under the cross-hair were identified and counted on separate channels, following a grid of 1 mm by 0.5 mm, early trials suggesting that this was the most informative spacing for the Minches samples. The grid was traversed until 500 identifications had been made. This total gives a 95% degree of confidence of an estimated component percentage as low as 5% (Van der Plas and Tobi, 1965; Pettijohn et al., 1973). The categories used for the analysis were:

1. Rock fragments	
2. Clear quartz	Detrital
3. Red-stained quartz	Detrital
4. Mica	Detrital
5. Dark/heavy minerals	Detrital
6. Other detrital fragments	Detrital
7. Shell fragments	Biogenic
8. Foraminifera	Biogenic
9. Ostracoda	Biogenic
10. Other calcareous material	Biogenic
11. Unidentifiable and others	

Categories 6 and 10 generally covered those grains of obscure composition, but occasionally grains were encountered that were totally unidentifiable. In these cases the 11th category was used, and this also served for certain spurious material, like industrial slag and coal in Portree Bay. Very rarely, authigenic glauconite was encountered, but not in sufficient quantity to merit its own category. This was counted in category 11. When the point counting programme was completed, percentage values were calculated and the results plotted (Figs. 6.47–6.53). The commonly made assumption that point counting gives a volumetric analysis only is true for a sample of identically shaped and orientated grains. In the present case, where platiform micas, for instance, are found along with rounded quartz grains, the same number of counts may be recorded for each type of grain although the volume is greatly different. This should be remembered when dealing with the distribution of platiform particles like mica or shell fragments, for their "volumetric" abundance may be exaggerated.

The biogenic sand component

A high concentration of biogenic sand within the sand fraction (Fig. 6.47) occurs on the Skye–Shiant Platform, and also in several other localities, such as

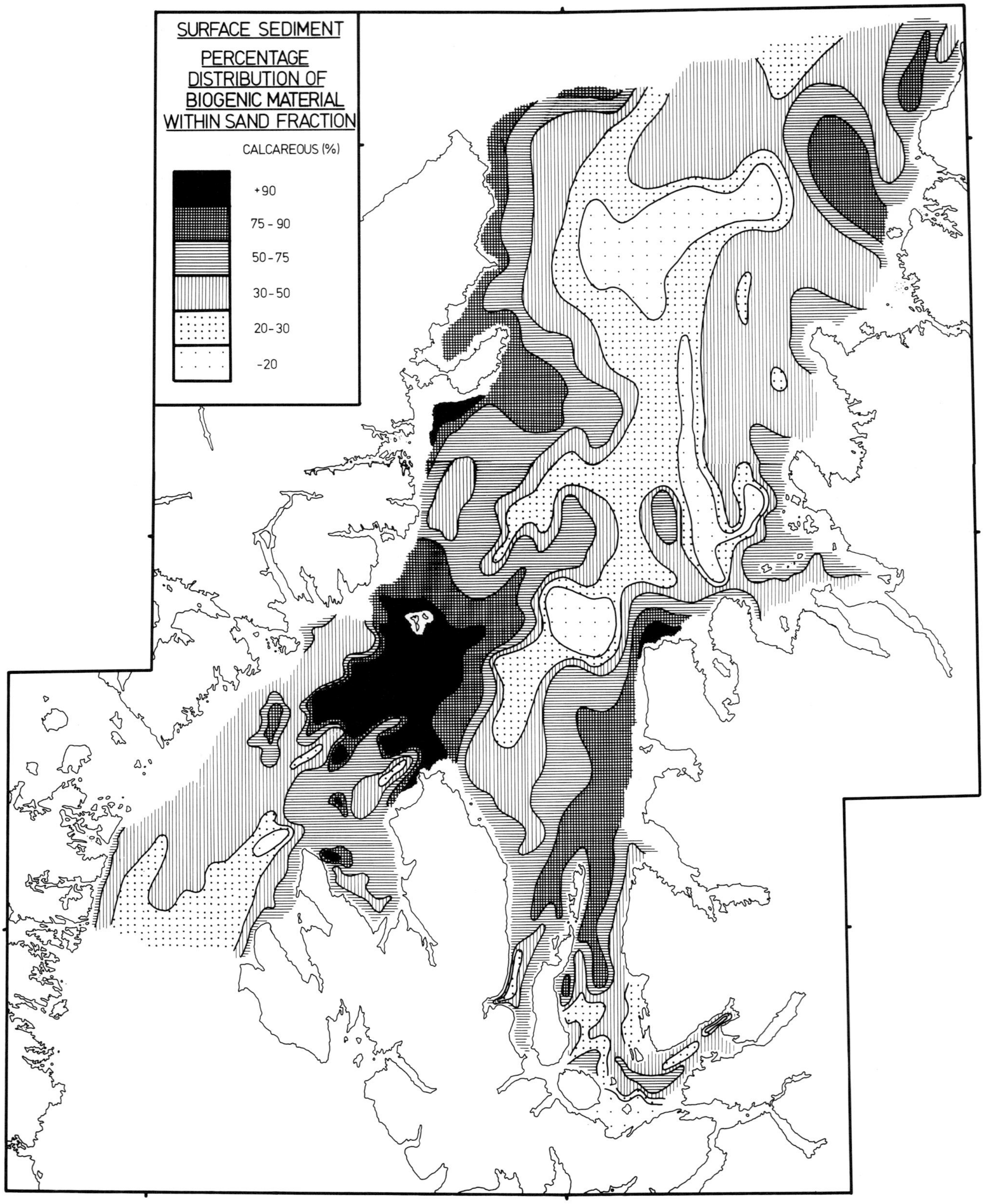
SURFACE SEDIMENT
PERCENTAGE
DISTRIBUTION OF
BIOGENIC MATERIAL
WITHIN SAND FRACTION
CALCAREOUS (%)
+90
75 - 90
50 - 75
30 - 50
20 - 30
-20

Fig. 6.47.

along the coast of Lewis and near the mainland headlands. The muddy basins of the Inner Sound, Sound of Raasay, Loch Snizort and Stornoway have high biogenic sand values, but in these places sand represents a low percentage of the sediment overall. In the North Minch, where total sand content is greater, biogenic material decreases in abundance away from the coasts to a minimum in the central area. A conspicuous biogenic sand minimum also occurs between the Shiant East Bank and the Rubh'Re, on the axis of the basement ridge. The Greenstone Ridge has a lower biogenic sand component than the surrounding sea-floor. As in the North Minch, there is in the Little Minch a minimum biogenic sand content in the central parts distant from the coasts.

Shell fragments

Shell fragments (Fig. 6.48) comprise the bulk of the biogenic sand found in the Minches, and where they are recognisable, are derived from bivalves, gastropods and echinoderms. The distribution resembles that of biogenic sand overall, with high values on the Skye–Shiant Platform, the Shiant East Bank and near the mainland and Lewis coasts. However, the mud basins here appear as areas of minimum shell fragment accumulation, most of their biogenic sand fraction being, in fact, composed of foraminifera tests, although the Stornoway Basin has a slightly higher shell fragment content than the others. The line of the Rubh'Re–Kebock Head basement ridge is marked, on either side of the Shiant East Bank, by a shell fragment minimum. Comparing the distribution of sand-sized shell fragments with that of gravel-size biogenic material it becomes apparent that three associations are present:

(1) Along the mainland headlands, in the area south of the Skye–Shiant Platform close to the Skye coast, and in the central part of the Shiant East Bank, shell sands are associated with dominantly detrital gravels.

(2) The area north of the Skye–Shiant Platform and the southern part of the Shiant East Bank both have highly biogenic gravel and abundant shell sand.

(3) Around the Lewis coast from Broad Bay to the Butt, shell sand is in abundance but there is little or no biogenic or detrital gravel present.

The first two cases can be explained by a change in the balance between detrital and biogenic sources in the creation of the surface sediment. In the former case, detrital supply must dominate, but in the second, organic production is great enough to mask the detrital sediment contribution. The first association occurs in shallow coastal areas where detrital supply of coarse sediment is at a maximum, and on the Shiant East Bank, which has a core of exposed resistant Precambrian rocks. The southern part of the bank, which has a foundation of Mesozoic sediments and sills, shares the same biogenic surface sediment as the Skye–Shiant Platform, which has an identical geological structure, apparently giving rise to a topography well suited to an abundant bottom fauna. Shell fragment occurrence declines rapidly at the edges of the Banks, and there is no evidence that any of the biogenic debris is carried into deeper water. Ferentinos (1976) was able to trace a trail of calcareous sand away from a similar source on a bank south-west of the Outer Hebrides, in a direction concordant with tidal currents and the dominant swell direction.

On the banks, the presence of a spectrum of shell-fragment sizes and states of abrasion indicates that the shell sands are being produced locally from modern marine organic production in these favourable habitats. There are no gravel-sized shell fragments found in association with the Lewis coastal shell sands, which consist of remarkably uniform, well sorted and highly abraded fragments. These properties indicate that it has not formed in situ and has been transported for some distance. The sands are located in an area of strong tidal currents (see Fig. 6.22) that stretches from Broad Bay to the Butt of Lewis, where particularly powerful currents are attributable to unique tidal conditions. Shell-sand distribution increases towards the Lewis coast, which has been described by Steers (1973). On the western coast of Lewis, shell sands occur on the beaches, presumably originating in the shallow and rocky sea-floor offshore, and drifting northwards towards the Butt of Lewis. On the east coast, the longshore drift of shell sand is to the south, in accordance with the long northerly fetch directions and wave diffraction around the Butt. The large north-east facing beach, Traigh Mhor, immediately north of Tolsta Head, is formed from shell sand and Steers reports that the "amount of sand on the beach is considerable and its main source appears to be offshore. The general direction of drift along this beach is to the south and Tolsta Head acts as a great groyne". South of Tolsta Head there are indications that beach sands are being swept out to sea.

In summary, there is strong evidence that the Lewis coastal shell sands have been and are being carried into the North Minch from sources on the outer continental shelf west of the Outer Hebrides by a combination of tidal currents and longshore drift, and are impounded in the sand flats of Broad Bay. We may recall here the discovery of a sand accumulation in Broad Bay at an abnormally shallow depth. The shell-fragment component decreases rapidly away from the coast and there are no indications that it is being transported further south than the Eye Peninsula, or eastwards for any con-

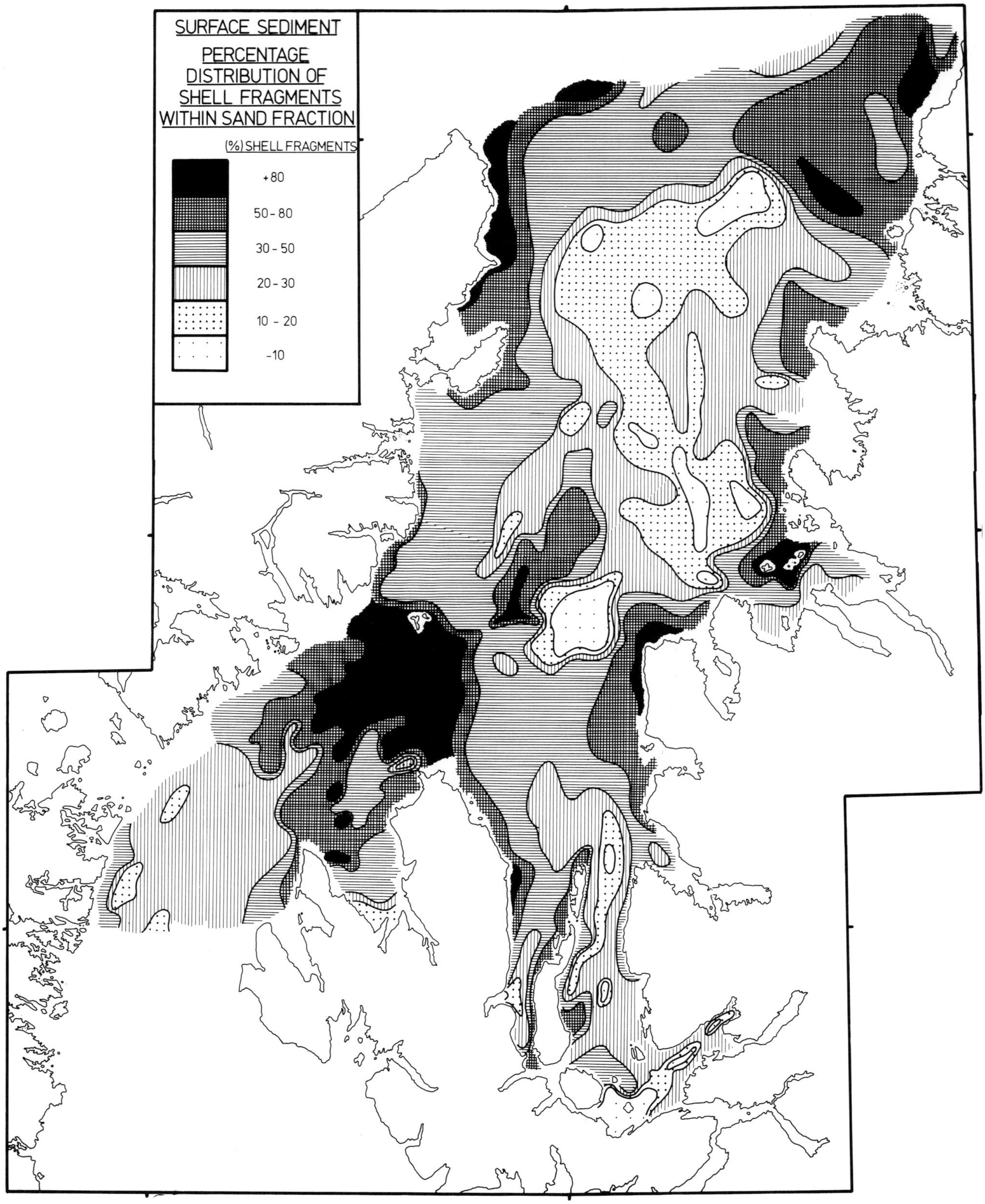
SURFACE SEDIMENT
PERCENTAGE
DISTRIBUTION OF
SHELL FRAGMENTS
WITHIN SAND FRACTION
(%) SHELL FRAGMENTS
+80
50-80
30-50
20-30
10 - 20
-10

Fig. 6.48.

siderable distance into the northern North Minch.

Foraminifera and Ostracoda tests

The percentage distribution of Foraminifera and Ostracoda tests is shown in Fig. 6.49. The abundance of Ostracoda is everywhere less than 2% of the total sand fraction and seems to vary closely with the content of Foraminifera. The total content of Foraminifera and Ostracoda within the sand fraction reaches more than 70% in Loch Snizort, 65% in the Inner Sound, and 50% in the Sound of Raasay and the Stornoway Basin. Away from these mud basins, the total abundance decreases to less than 5% in areas of sand and gravel, in many cases the point-counting data indicating less than 1%, although such a low value is statistically of little significance. The Foraminifera of the north-western quadrant of the Minches have been studied at University College by Edwards (1975). Distinct assemblages are present. The Shiant East Bank and the shallows north of Tiumpan Head have a benthonic fauna of thick-shelled Foraminifera, predominantly of attached forms, among which *Cibicides, Eponides,* the arenaceous *Gaudryina* and Miliolids are common. In the deeper water of the Stornoway Basin, a contrasting fauna occurs, characterised by smaller, thin-walled and delicate forms, like *Bulimina,* while *Hyalinea* is particularly common in the muddy sands of the central North Minch, although also occurring in the Stornoway Basin muds, where the occasional occurrence of blue-black sulphide deposits in the tests suggests reducing conditions within the sediment. The distribution of the fauna could depend on substrate variations or the energy of the environment; as each control varies with the other it is difficult to determine which is the dominant. The presence of small individuals of *Cibicides* in the Stornoway Basin muds may indicate winnowing and transport of the finer size grades away from the surrounding banks. A more reliable indication of this is the presence of fragmented and abraded Miliolid tests in the basin, while whole individuals are confined solely to the shallow banks.

In conclusion, the distribution of biogenic sand in the Minches seems to be typical of that described by Milliman (1974) for the Inner Shelf Zone, where the deposition of fine-grained detrital sediment in areas of low current activity reduces the abundance of biogenic sand, with the exception of Foraminifera tests. The sediments of outer continent shelves which are not close to major river mouths generally have high carbonate contents. It seems likely that the influx of shell sand into the North Minch around the Butt of Lewis represents the encroachment onto the Inner Shelf of a biogenic sediment more typical of the Outer Shelf Zone.

The detrital sand component

For the distribution maps (Figs. 6.50–6.53), detrital lithogenic components have been expressed as percentages of the total detrital fraction, rather than as proportions of the whole sand fraction, so that the distributions reflect only variations of the properties of the non-biogenic sediment.

(1) *Rock fragments.* Wide variations in the abundance of rock fragments are present within the detrital sand fraction (Fig. 6.50). Low values occur in the muddy hollows around the Shiant East Bank and in the lochs and sounds, but more than 70% of the sand in some parts of the Shiant East Bank and the Skye–Shiant Platform is made up of rock fragments. The proportion of rock fragments is often used as a measure of sediment maturity (Pettijohn, 1975); the greater the abundance, the more immature the sediment. However, in the case of the Minches, this may be too simple a concept. Glacial erosion, due to its association with intensely physical rather than chemical weathering action, produces large quantities of rock fragments. Where glacial deposits are exposed on banks and the finer fractions are winnowed away, the remaining sediment will be enriched in rock fragments which are larger and less easily removed than mud or smaller sand grains. This interpretation has been used by Trumbull (1972) to explain rock-fragment distribution on the north-eastern continental shelf of the U.S.A. In the Minches, rock-fragment distribution is closely related to bathymetry, maxima occurring around the coasts and on the banks. Although this points to winnowing as the cause of rock-fragment concentration, we cannot dismiss the possibility that some of the fragments are, in fact, derived from sea-floor exposures of rock on the banks, or from coastal cliffs.

The sediments of the northern North Minch are low in rock fragments, except for the Broad Bay to Butt of Lewis area, although the total non-biogenic component of the sand in this latter area is low. The Little Minch sediments are richer in rock fragments than those of the North Minch.

(2) *Red-stained quartz.* Much of the quartz present in the Minches samples possesses conspicuous red iron-staining, which tends to be accompanied by a highly rounded shape and surface frosting. The distribution of red-stained quartz is shown on Fig. 6.51, expressed as a percentage of the total quartz content. The existence of large amounts of stained quartz at certain localities may either reflect the derivation of sand from sedimentary rocks rich in such grains, or has been caused by staining during sub-aerial weathering in a Quaternary period of

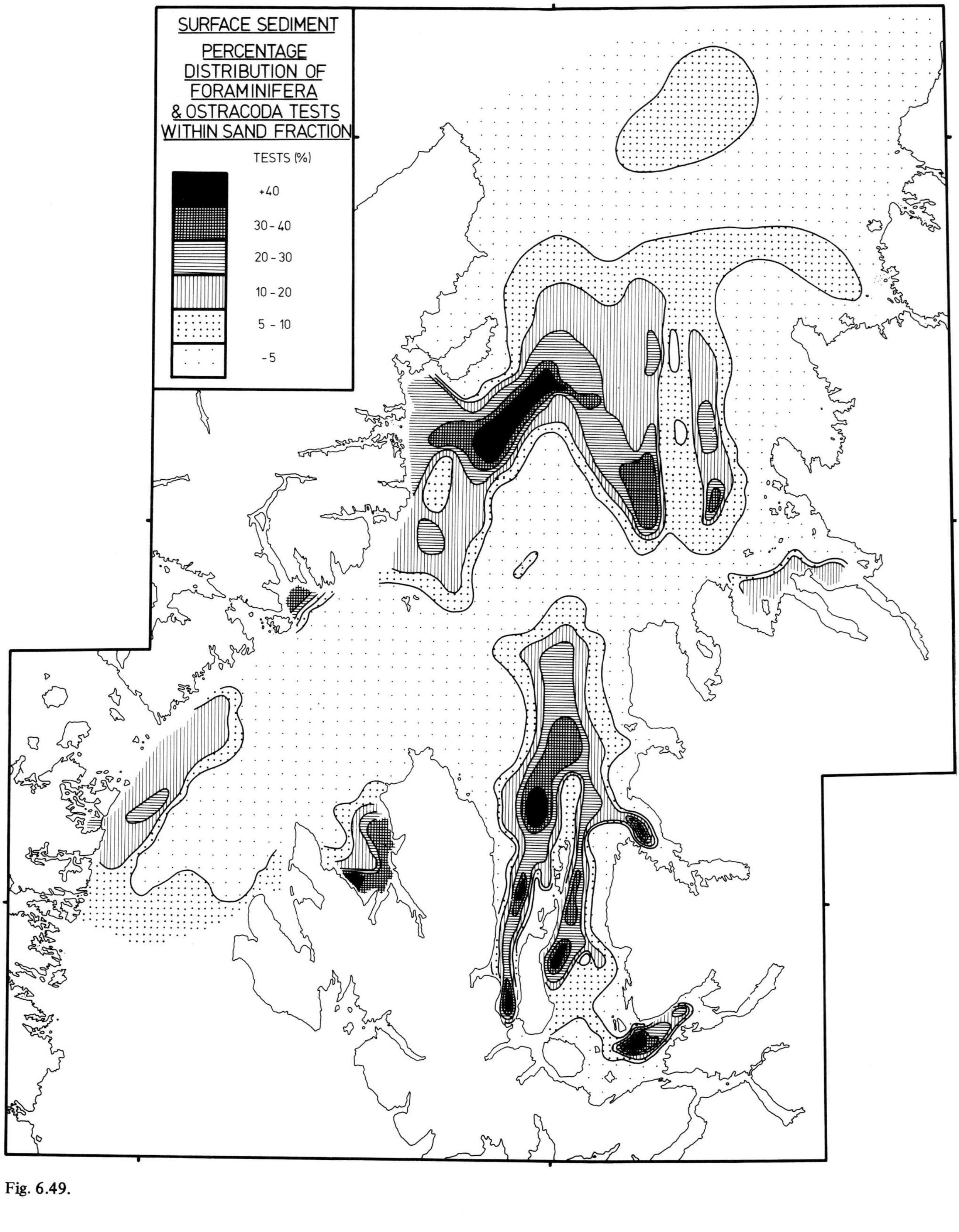
SURFACE SEDIMENT
PERCENTAGE DISTRIBUTION OF FORAMINIFERA & OSTRACODA TESTS WITHIN SAND FRACTION
TESTS (%)
+40
30 - 40
20 - 30
10 - 20
5 - 10
-5

Fig. 6.49.

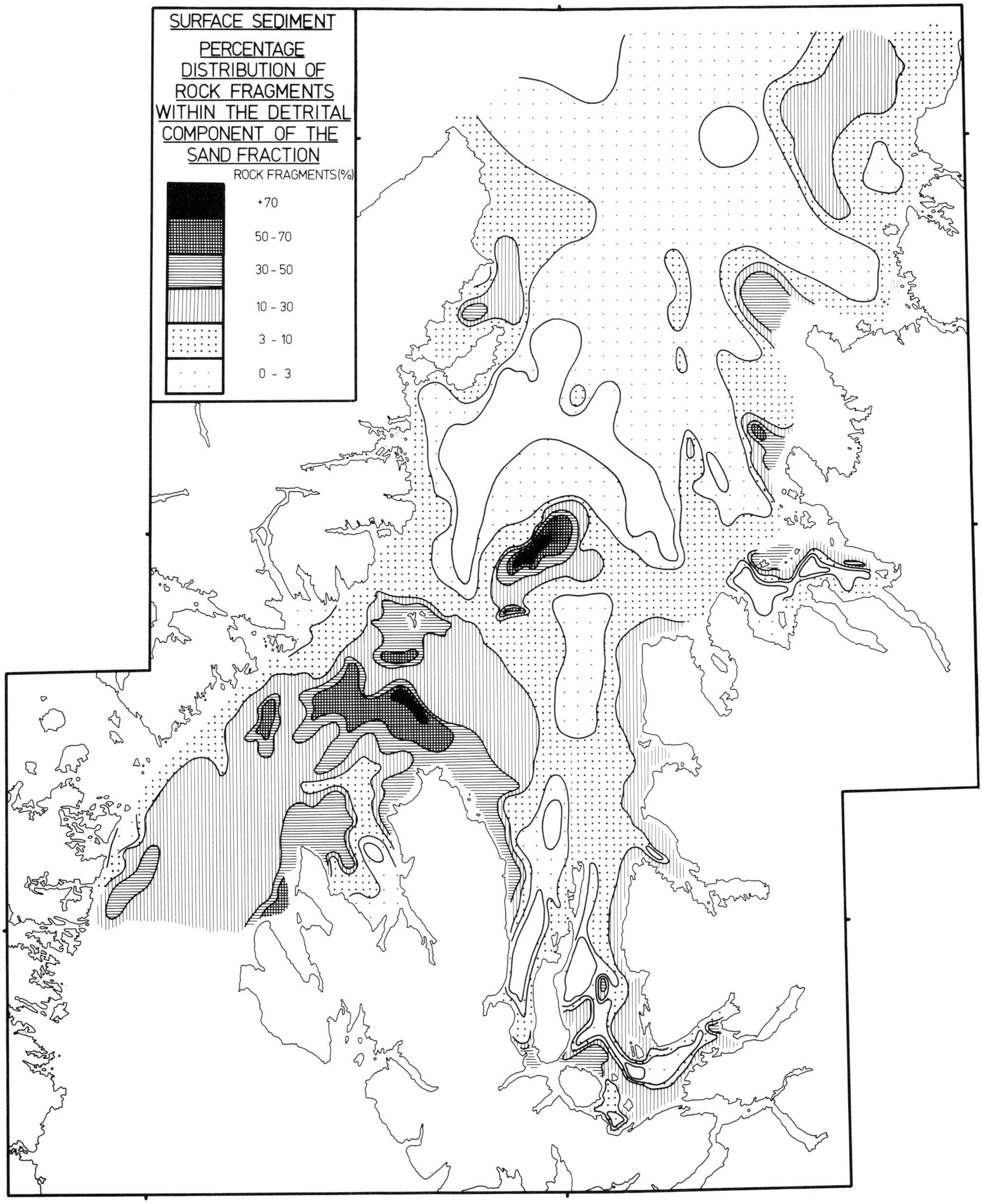
SURFACE SEDIMENT
PERCENTAGE DISTRIBUTION OF ROCK FRAGMENTS WITHIN THE DETRITAL COMPONENT OF THE SAND FRACTION
ROCK FRAGMENTS (%)
+70
50 - 70
30 - 50
10 - 30
3 - 10
0 - 3

Fig. 6.50.

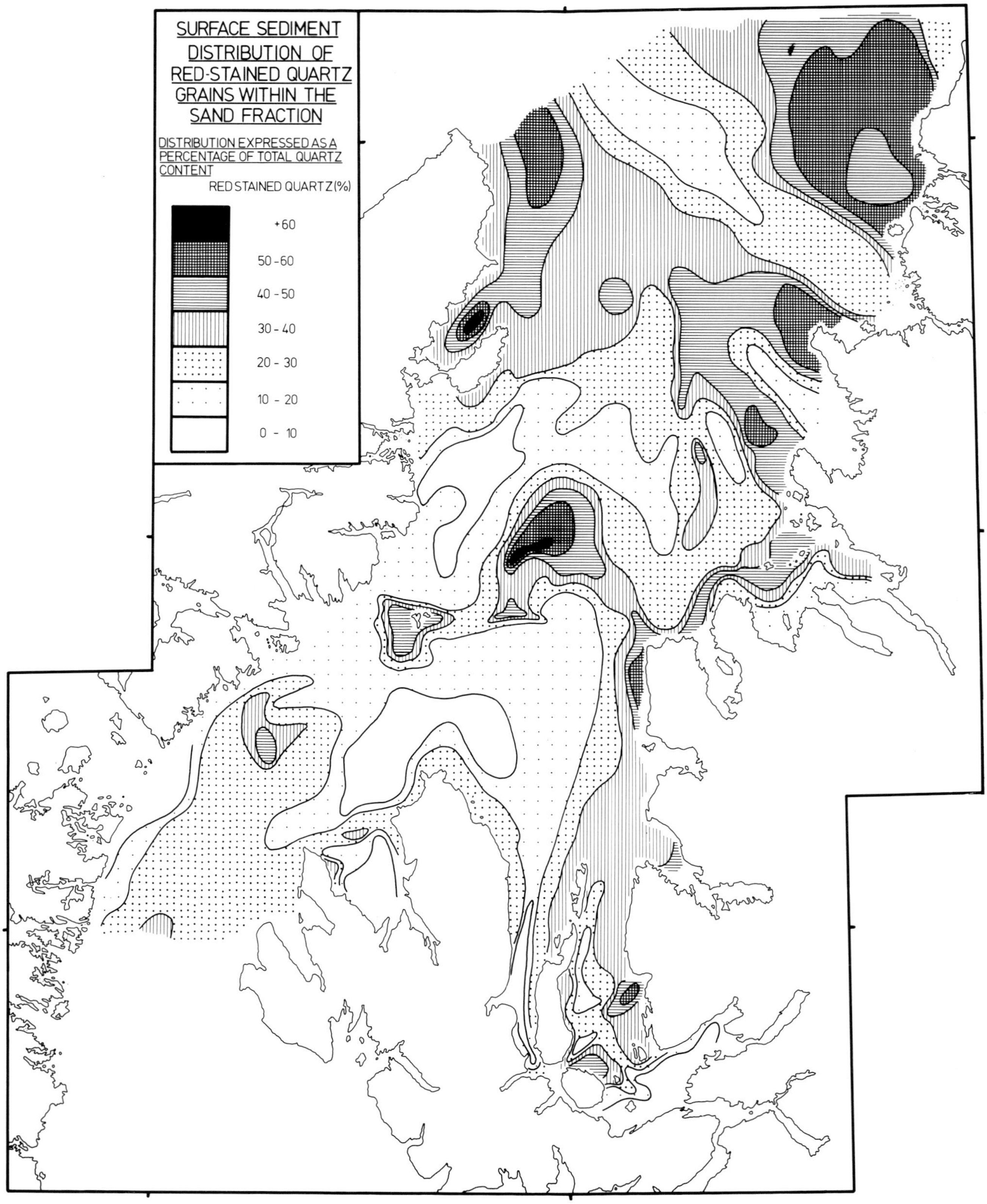
SURFACE SEDIMENT
DISTRIBUTION OF
RED-STAINED QUARTZ
GRAINS WITHIN THE
SAND FRACTION
DISTRIBUTION EXPRESSED AS A
PERCENTAGE OF TOTAL QUARTZ
CONTENT
RED STAINED QUARTZ (%)
+60
50 - 60
40 - 50
30 - 40
20 - 30
10 - 20
0 - 10

Fig. 6.51.

lower sea-level. The latter process has been held responsible for staining of quartz grains on the outer Nova Scotia shelf of eastern Canada (James and Stanley, 1968). In the Gulf of Maine, however, Trumbull (1972) has reported the presence of red-stained grains inherited from red Triassic sediments. Iron-staining, surface frosting and roundness are features associated with quartz grains deposited in a desert environment. Unfortunately, there is more than one iron-stained desert sandstone in the Minches area that could represent a source of red quartz. The Torridonian and Permo-Triassic sandstones are the obvious candidates, but ferruginous sediments also occur at horizons within the Jurassic.

On the Shiant East Bank, maximum red-quartz values are obtained over the central Torridonian outcrop of the basement ridge. The slightly lower parts of the bank to the north and south are underlain by softer, Permo-Triassic red sandstones and are covered by a sediment with less red quartz. In the extreme south of the bank, in an area of Jurassic sediment and basic igneous rocks, higher red-stained quartz values are again obtained, similar values also occurring on the nearby northern parts of the Skye–Shiant Platform around the Shiant Isles, although the rest of the Jurassic-floored Platform has a very small stained quartz content.

In the North Minch, sand, measurably poorer in red quartz, occupies a long depression extending north-westwards from Eddrachillis Bay, which has slightly finer sediment than the surrounding sea-floor (see Figs. 6.6 and 6.35). Finally, the quartz sand of the southern extremities of the Inner Sound has a very low red quartz content, and consists of well sorted, fine and very angular clear grains, quite different in character from any concentrations of sand found elsewhere. As Loch Kishorn, in which these sands are best developed, cuts into the Lower Palaeozoic sediments of the Moine Thrust Zone, it is likely that they are of very local derivation.

(3) *Dark and heavy minerals.* Dark and heavy minerals can sometimes represent more than 20% of the detrital sand component of Minches sediments (Fig. 6.52). As with rock fragments, Trumbull (1972) has found evidence that on the north-eastern continental shelf of the U.S.A. a high mafic mineral content can reveal a sediment closely related to glacial deposits, but may also reflect nearby exposures of pre-Pleistocene material. In the shallow parts of the Minches, where the gravel fraction shows the glacial sediments to be closely related to local rock exposures, interpretation becomes somewhat confusing. The distribution pattern of mafic minerals in the Minches would seem to indicate, however, that local geology is the dominant control, either directly or through glacial sediments. The Lewis coast, as far north as the Eye Peninsula, has a high dark mineral content, while the mainland coast is associated with a dark mineral minimum, especially the Torridonian headlands of Rubh'Re and Rhu Coigach, which suggests a greater mafic mineral derivation from the igneous and metamorphic Lewisian formation than from the Torridonian sediments. The Jurassic rocks exposed on the Skye–Shiant Platform also give rise to sediments poor in mafic components. Maximum values occur in the Sound of Raasay and Loch Snizort, leading away from Skye and possibly allowing us to trace glacial transport paths away from the island's Tertiary volcanics. In the North Minch, a dark mineral minimum over the Shiant East Bank is continued north-eastwards as far as the Greenstone Ridge, which also has low mafic mineral values. The minimum in the northern part of Broad Bay reflects again a local sediment source in the Stornoway Beds. The Lewis coastal area, from Broad Bay northwards to the Butt of Lewis, has a much lower proportion of mafic components than to the south, where Lewisian rocks are exposed. The extreme north of the North Minch is marked by a strong dark mineral maximum, but its cause is not clear, and may result from a source outside the Minches area.

(4) *Mica.* Because of their platform shape, mica particles in the sand fraction can be used as a measure of energy conditions at the sea-floor, and their presence within the sand fraction will indicate low energy environments and the contemporaneous deposition of mud (Doyle et al., 1968). Mica is abundant in all of the mud-floored hollows, lochs and sounds, but is absent from the shallower sandy floors and gravel banks, which Doyle takes as being an indication of winnowing, or, of course, an initial absence of mica. The highest mica contents are found in the sand fraction of sediments deposited near the Outer Hebridean coast (Fig. 6.53). This suggests a second control over mica distribution, the proximity of a source area, for the Lewisian metamorphics must provide a greater supply of mica than sediments of Torridonian or Mesozoic age, or Tertiary basic igneous rocks. A very high mica content is found in sediments from Loch Maddy. The Stornoway Basin similarly possesses high values.

Lower percentages of mica typify the sediments of the central North Minch. Mica has accumulated in Loch Snizort to only a minor degree, perhaps reflecting the paucity of local supply, for almost all of the Little Minch clear of the Uist coast has little mica. The Sound

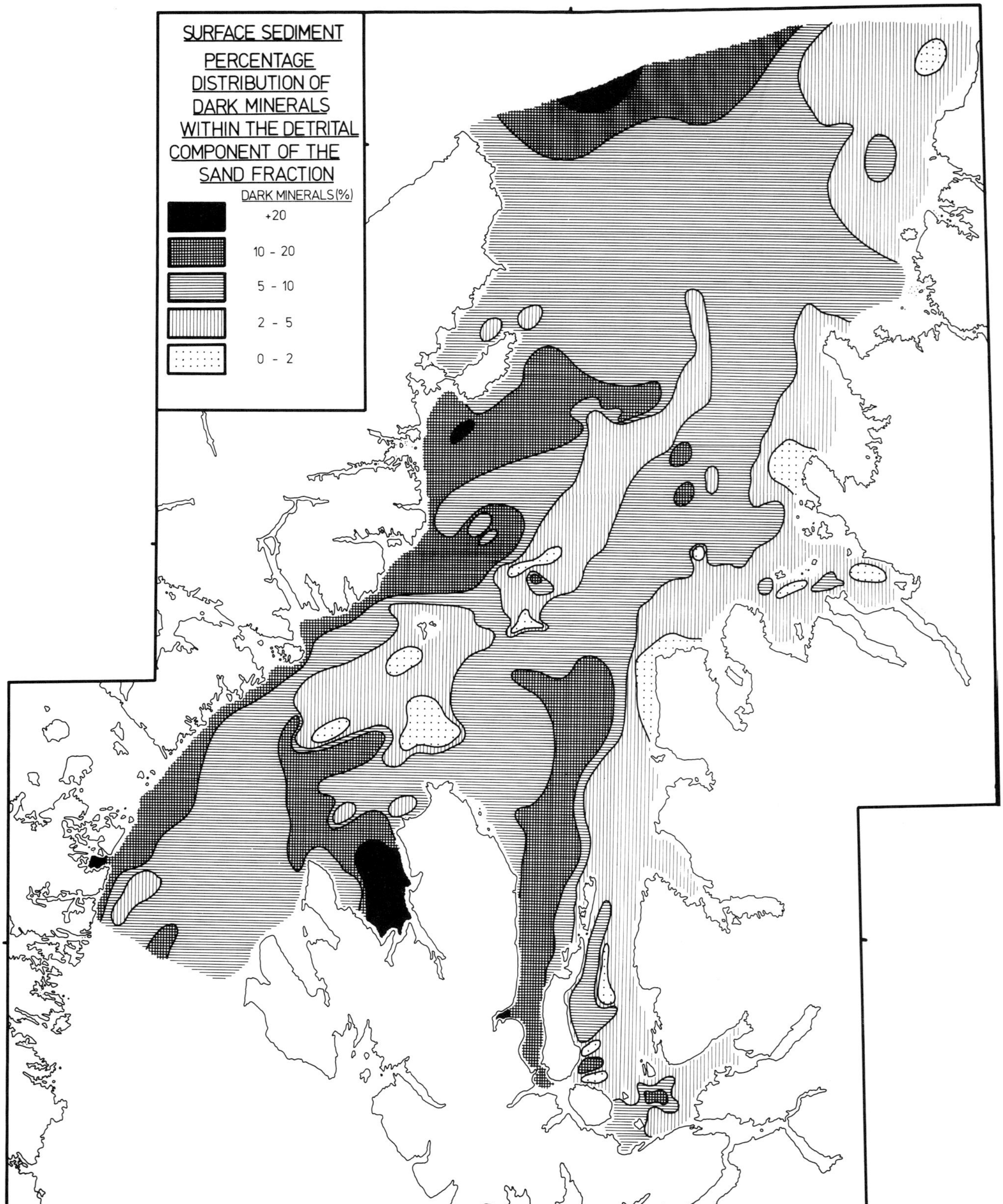
SURFACE SEDIMENT
PERCENTAGE
DISTRIBUTION OF
DARK MINERALS
WITHIN THE DETRITAL
COMPONENT OF THE
SAND FRACTION
DARK MINERALS (%)
+20
10 - 20
5 - 10
2 - 5
0 - 2

Fig. 6.52.

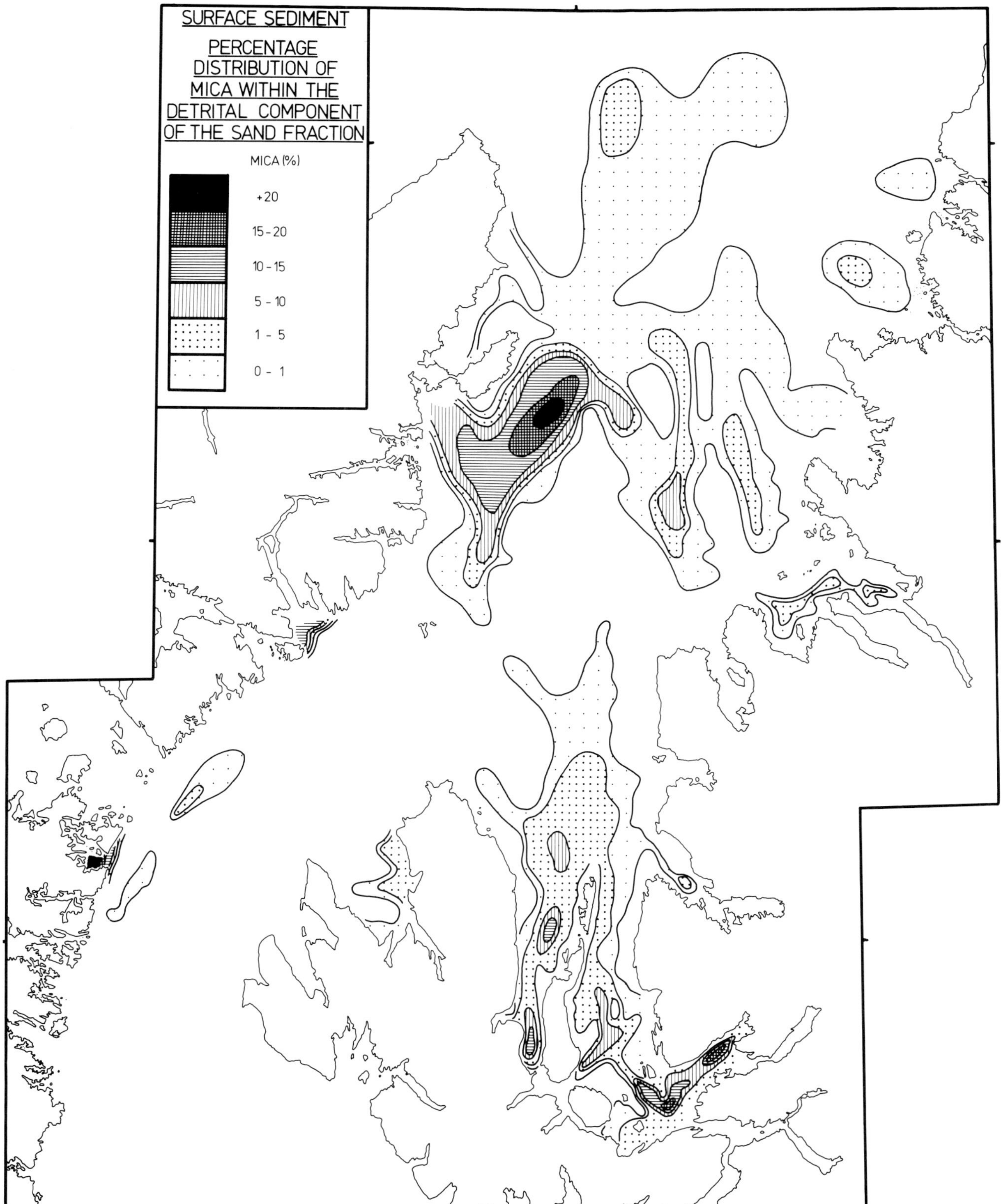
SURFACE SEDIMENT
PERCENTAGE DISTRIBUTION OF MICA WITHIN THE DETRITAL COMPONENT OF THE SAND FRACTION
MICA (%)
+20
15-20
10-15
5 - 10
1 - 5
0 - 1

Fig. 6.53.

of Raasay and the Inner Sound have large mica components, especially in the extreme south of the Inner Sound, reflecting the exposure of metamorphic rocks of the Moine Thrust Zone around Lochs Carron and Kishorn. The Shiant East Bank, the Skye–Shiant Platform and the mainland headlands lack a mica component in the sand fraction, which could be taken as firm indication of winnowing were it not for the fact that the rocks exposed in these areas are poor sources of mica.

In summary, the distribution of sand components does not provide any evidence of modern transport outside of the fine-sand grade. Although some smaller and broken Foraminifera tests have made their way into the Stornoway Basin from the surrounding shallows, along with small, clear, angular quartz grains and mica, larger red-stained quartz grains or rock fragments have not been transported. In shallower areas, the distribution of red-stained quartz, rock fragments and dark and heavy minerals all point to a local derivation from exposed rock, either on the sea floor or in the coastal cliffs, and even the distribution of the more easily transported mica component can largely be explained by proximity to mica-bearing source rocks (although this may also be a guide to areas of winnowing and mud deposition). Some features of the component distributions may partly reflect variations in the nature of the underlying glacial sediments, but as these are also determined by solid geology it is not easy to decide whether the geological control is acting directly or indirectly upon the surface sediment, especially in areas where the Quaternary sediment cover is thin. Only in one area, from the Butt of Lewis to Broad Bay, is there good mineralogical evidence for the large scale transport of sand, in a way concordant with longshore drift and tidal currents. Although some of the biogenic debris on the Skye–Shiant Platform may have been reworked locally into a well sorted shell sand deposit, there is no evidence that any of this material has been carried away from the bank.

The mud fraction

The distribution of mud in the surface sediments (Fig. 6.54) possesses a simple relationship with bathymetry. The banks and shallow headlands have low mud values, and maxima occur in the deep hollows, lochs and sounds. Low mud content characterises the Lewis coast north of Broad Bay, although south of Stornoway, mud content increases towards the steep coast. A comparison of Fig. 6.54 with the distribution map for clay grade sediment (Fig. 6.55) shows that much of the mud at intermediate depths is of silt size. Clays are confined solely to the deep hollows, lochs and sounds, only minimal values occurring elsewhere.

Clay-mineral composition

Although there has been much controversy in recent decades, authorities generally agree today that the ultimate origin of most clay minerals found in marine sediments lies on land. However, different interpretations still arise in assessing the degree of alteration which clay minerals undergo on leaving the continents and entering the marine environment. Some claim that major changes do occur at this transition, with clay minerals being altered to quite different varieties, while others maintain that such transformations, although they could occur, are rare, clays in the oceans being formed of essentially unaltered continental detritus. In the study of provenance of a sediment, alteration would clearly make interpretations hazardous.

According to the school of thought represented by Grim (1939, 1953) the unequal distributions of clay minerals in ocean-floor sediments are a result of environmentally induced mineralogical changes, geographical distributions representing little more than final, or arrested, stages of such transformation. Only a minor amount of provenance indication is admitted. However, others claim that there are no significant contrasts between marine and non-marine clays (Pettijohn, 1975) and have demonstrated connections between world-wide clay-mineral distributions and processes on adjacent continents (Biscaye, 1965; Griffin et al., 1968; Venkatarathnam and Biscaye, 1973). In the study of the Minches clays, two environmental relationships described by these workers are of particular importance. Chlorite is easily destroyed by chemical weathering, and is, therefore, not common in the tropics, but is found in higher concentrations in marine muds in polar regions, where the physical weathering of chlorite-rich igneous or metamorphic rocks enables it to reach the sea, either in the sand grade or finely pulverised as rock flour in boulder clay. The possibility of authigenic formation of chlorite in the sea is admitted, but only in very rare cases. However, no evidence for the authigenic formation of illite is recognised, and, on the contrary, the concentration of illite in a mud is taken to be an indication of the importance of river-borne solids in its accumulation. It must be noted that while this latter group of workers generally discounts the authigenic formation of clays, diagenetic alteration may take place, but only over a very long period of time.

On a much smaller scale than these world-wide environmental associations, work in such places as the Gulf of Mexico (Griffin, 1962), the Barents Sea (Wright,

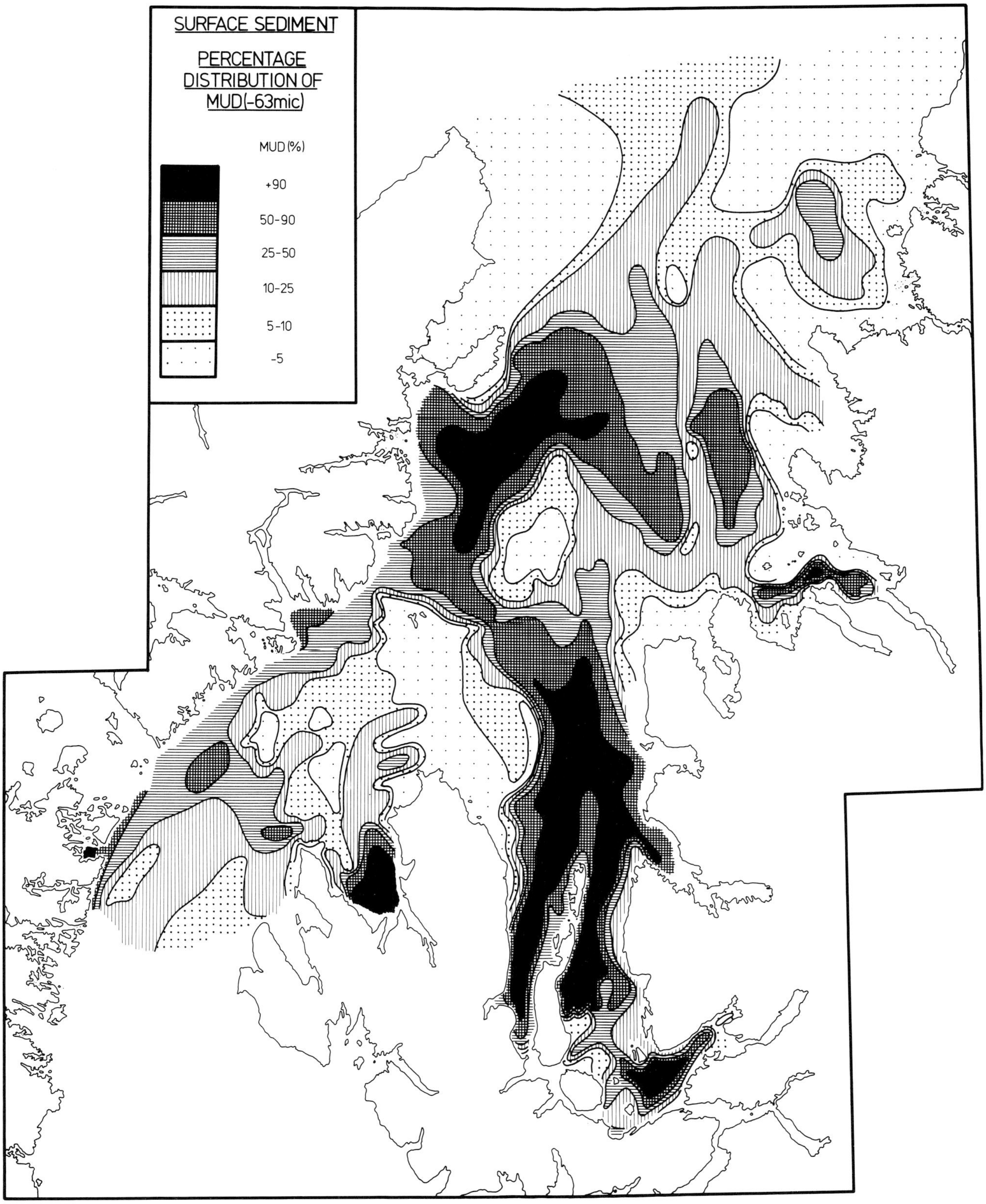
SURFACE SEDIMENT
PERCENTAGE DISTRIBUTION OF MUD(-63mic)
MUD (%)
+90
50-90
25-50
10-25
5-10
-5

Fig. 6.54.

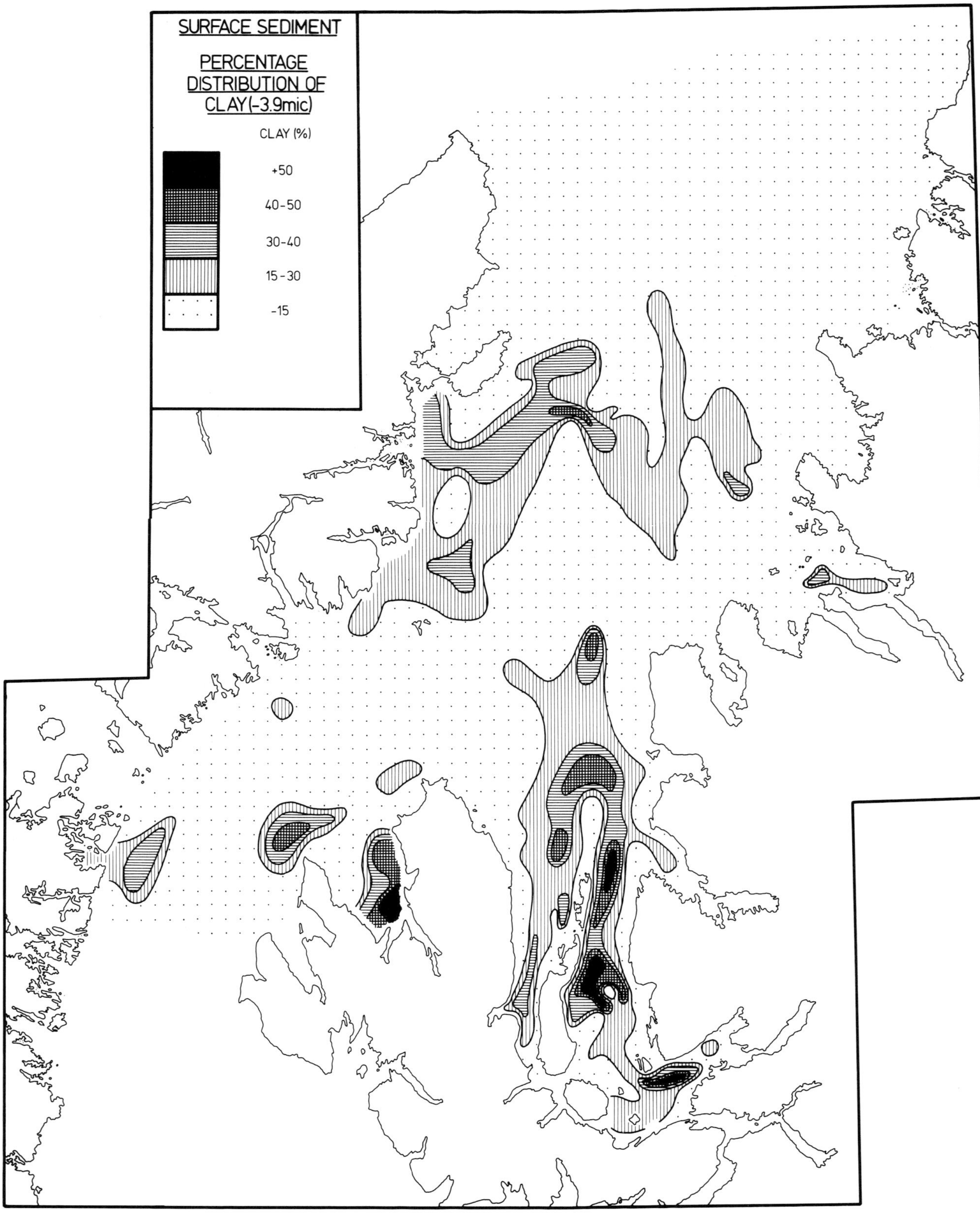
SURFACE SEDIMENT
PERCENTAGE DISTRIBUTION OF CLAY(-3.9mic)
CLAY (%)
+50
40-50
30-40
15-30
-15

Fig. 6.55.

1974) and the Sea of Japan (Saburo et al., 1974) has demonstrated correlations between marine clay mineral distribution and specific currents and river discharges. Although this recent work seems to confirm the validity of provenance studies based on clay-mineral distributions, the possibility and degree of mineralogical changes that the sediments could undergo in the ocean have still not been adequately determined and leaves some room for doubt.

Over 50 suitable clay-bearing grab samples from the Minches were selected at random, and, after disaggregation by a mechanical stirrer in a calgon solution, the clay fractions (less than 2 μ in size) were separated by a simple settling procedure. After being left to stand for several days, the clear calgon solution could be decanted, and the clay suspensions reduced to a thick clay paste, which was then dropped onto ceramic tiles and dried quickly in a warm room. This technique for preparing X-ray diffraction mounts of clays is known to be the most reliable of all available mounting procedures (Gibbs, 1965), although an important and previously undescribed hazard will be discussed below. The samples were placed in a dessicator at room temperature for a day before the diffractograms were made.

Diffractograms were made at a standard running speed for each of the clay samples. The intensity of the deflection from base line to the diffractogram peak is directly proportional to the concentration of the mineral producing it, when allowance has been made for absorption effects (Klug and Alexander, 1954). The problems in quantitative analysis from X-ray diffraction data primarily arise from the difficulty in determining mass absorption coefficients and effects of instrument geometry, which lead to relatively low degrees of confidence in the results. In this study, as we are only interested in changes in the *relative* proportions of the minerals present, we deal solely with the variations of diffractogram peak dimensions. All samples were prepared and analysed in identical fashion. The dimensions of the peaks were measured by the method suggested by Norrish and Taylor (1962), peak height being multiplied by the width at half the height. This reduced errors due to small variations in crystallite size, which can affect the height/width ratio of the deflection, yet is simpler than measuring the area under the peak, where problems of base line interpretation are severe. All samples revealed the same broad mineralogical composition of 2*M* mica/illite, la chlorite, quartz and calcite. These four minerals were all that were detected on the diffractograms, apart from minor amounts of feldspar at 3.22 Å, which were nowhere of measurable quantity. A typical diffractogram is shown on Fig. 6.56. The particular peaks measured were:

3.35 Å mica/illite and quartz
4.26 Å quartz
3.55 Å chlorite
3.04 Å calcite

These were chosen for their location close to each other on a level and easily recognisable section of baseline. The mica/illite value was calculated by removing the quartz components combined within the 3.35 Å peak, its size being calculated by measurement of the separate 4.26 Å quartz peak. The variability of the clay mineralogy of the Minches sediments is expressed in a triangular diagram (Fig. 6.57) in terms of relative percentages, dependent on simple diffractogram deflections.

Although the use of ceramic tile mounts for clay-mineral analysis is a standard procedure (Carrol, 1970) and has been used for many recent publications, we have been unable to find a discussion of the possibility of clay within the ceramic substrate contaminating the results from the clay sample. During the production of the dif-

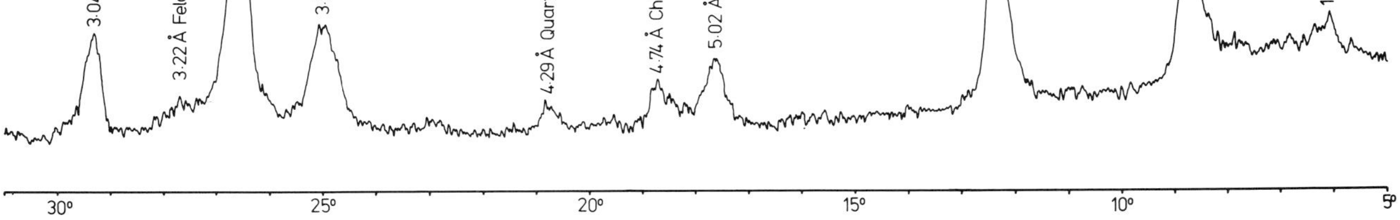

Fig. 6.56.

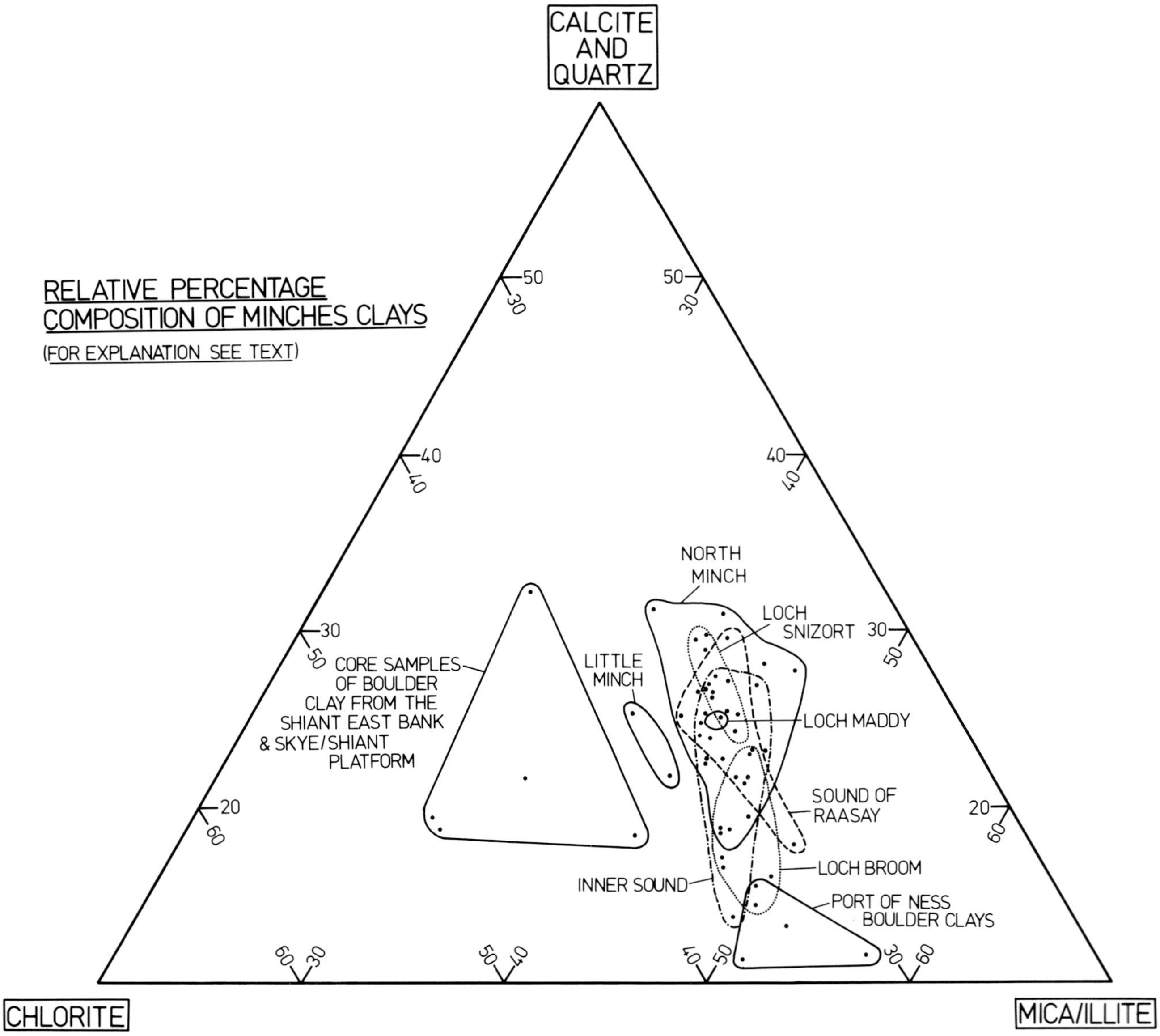

Fig. 6.57.

fractograms described below, we made a diffractogram of a blank ceramic tile, and obtained a set of peaks for illite (one of the minerals present in the Minches clays) about one third of the height of typical sample peaks. Whether or not peaks from the tiles had contributed to the final result depended upon to what depth the X-ray beam penetrates the sample target, and in order to check this point two series of test mounts were prepared. The first used a typical Minches clay, the second crushed chalk. For each material four mounts were made of varying thickness, ranging from an extremely thin mount, barely covering the ceramic tile, to a thick mount, of comparable thickness to those used in the study. With the chalk series, the thinnest mount almost obscured the illite peaks from the tile, which were visible as small baseline fluctuations. The three thicker mounts revealed no trace of peaks other than those of $CaCO_3$. With the Minches clay, the three thickest mounts yielded similar chlorite/illite ratios, but illite was exaggerated in the thinnest. We may conclude that the X-rays do not penetrate the sample layer in any but the thinnest mounts, and peaks from the ceramic tile have not contaminated the diffractograms of the 50 or more Minches clay samples. However, this problem has brought to the attention of the authors a lack of knowledge of the effects of mounting characteristics on quantitative X-ray analysis.

In addition to the grab samples, Fig. 6.57 also

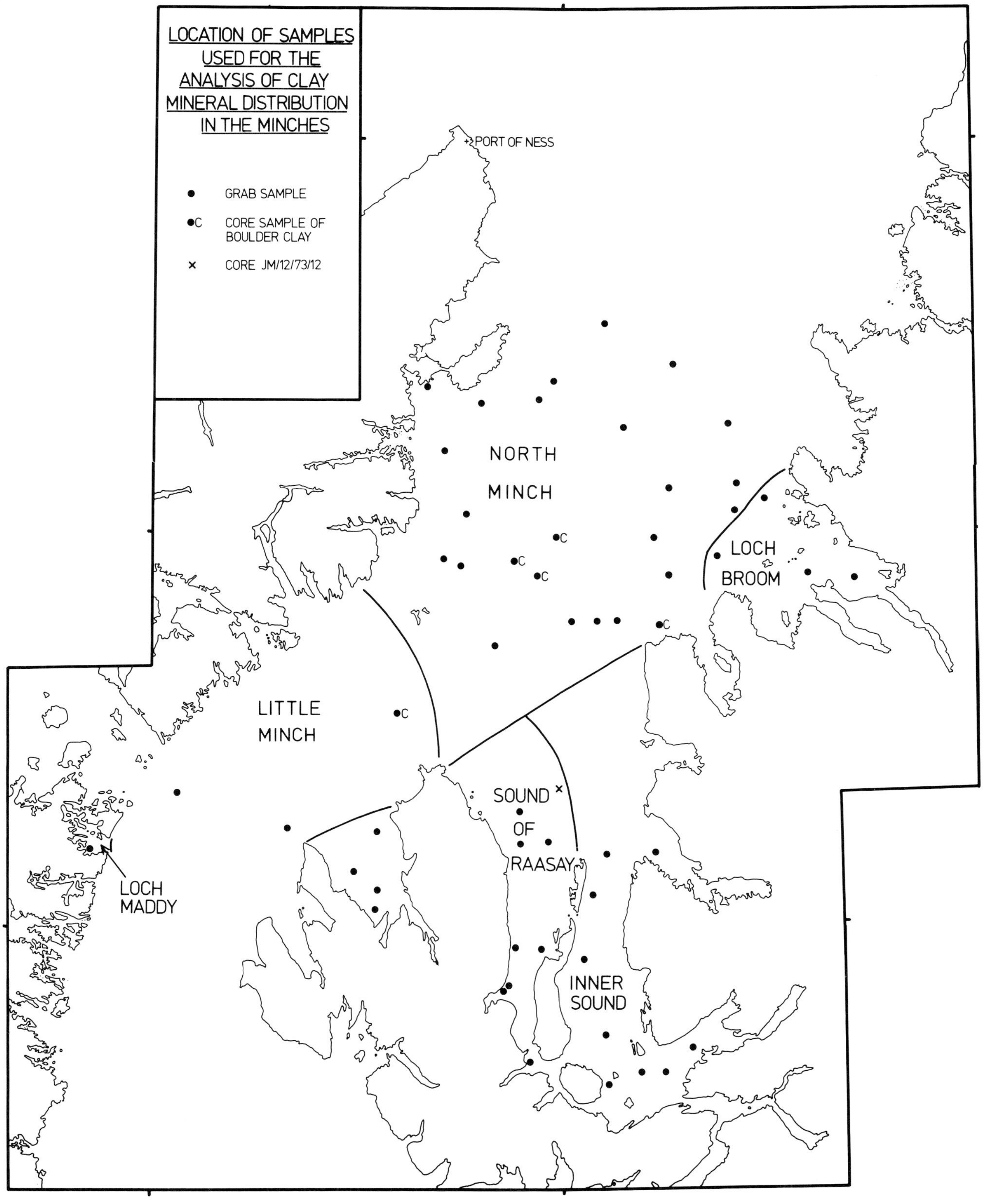
LOCATION OF SAMPLES USED FOR THE ANALYSIS OF CLAY MINERAL DISTRIBUTION IN THE MINCHES
GRAB SAMPLE
CORE SAMPLE OF BOULDER CLAY
CORE JM/12/73/12
PORT OF NESS
NORTH MINCH
LOCH BROOM
LITTLE MINCH
SOUND OF RAASAY
LOCH MADDY
INNER SOUND
C

Fig. 6.58.

presents results obtained from the analysis of boulder clays collected on land at the Port of Ness, Lewis, and from gravity cores on the Shiant East Bank and elsewhere. Fig. 6.59 shows the results of the analysis of five samples down the length of a single core from near the entrance to the Inner Sound. The location of all the samples is shown on Fig. 6.58, as well as the geographical divisions used in the discussion of the grab sample results. All the clays analysed have broadly comparable mineralogies, though significant variations are apparent.

The boulder clays analysed fall into two distinct groups. Those from core samples on the Shiant East Bank, Skye–Shiant Platform and Rubh'Re, have a higher chlorite content than all other samples, which seems to be in agreement with the previously noted association of this mineral with glacial sediments. However, a high chlorite content within a boulder clay ultimately depends on it having evolved from chlorite-rich metamorphic and igneous rocks. Mesozoic sediments, which were heavily eroded on the Minches floor, contain clay and shale deposits with contrasting clay mineral suites which could be reflected in the clay mineralogy of a boulder clay derived from them. This could well explain why the Port of Ness boulder clays from Lewis have much less chlorite and more illite than those from the Shiant East Bank, and elsewhere, for the ice that emplaced this deposit would have flowed over considerably more Mesozoic territory than that in any other part of the Minches. The Port of Ness boulder clays also lack calcite and have slightly more feldspar than any clays found elsewhere in the Minches.

The modern mud deposits have compositions occupying fields between the two boulder clay extremes and could, therefore, be explained by derivation from mineralogically variable boulder clay sources. However, the clays from the three most enclosed inshore areas, the Sound of Raasay, the Inner Sound and Loch Broom, have distinctly more illite than those from the open North Minch or Little Minch. Neither the chlorite/illite ratio nor the concentration of calcite and quartz bear any relationship with bathymetry. High illite values are usually taken to indicate nearby river influences, and although north-west Scottish rivers are not noted for large sediment loads, illite may be available from glacial deposits exposed on land and in the catchment areas. Chlorite concentration of detrital clays from glacial sources would be lowered by weathering in modern temperate conditions. Dilution of chlorite-rich clays from offshore sources in these areas by detrital illite could also explain the lower calcite and quartz values obtained. Nevertheless, such a clay-mineral distribution could equally well be explained by local sources of illite-rich boulder clay.

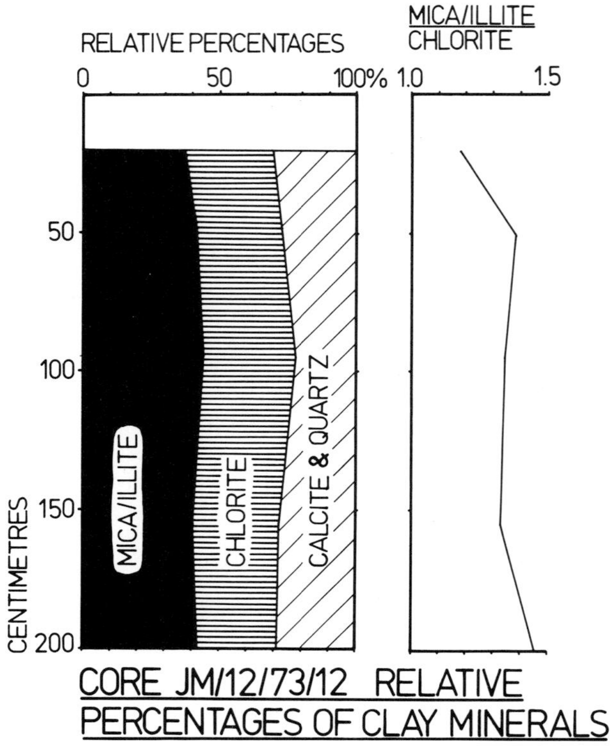

Fig. 6.59.

In summary, the observed clay mineral composition of modern Minches sediments nowhere offers incontrovertible evidence that anything but redistribution of glacial boulder clay forms the source of the fine material that is being deposited in the deep hollows, lochs and sounds. Finally, the analysis of five samples taken along Core JM 12/73/12 (Fig. 6.59) reveals little change in mineralogy within 200 cm of the surface, although the lower levels tend to have a slightly higher illite/chlorite ratio than at the surface, which may reflect the beginnings of diagenetic change.

Palaeomagnetic determination of sedimentation rate of muds in the Inner Sound

The declination of remanent magnetization of five sediment cores from the muds of the Inner Sound has been measured and discussed by Bishop (1975). Variations in declination have been tentatively correlated with declination curves from Lake Windermere, and reveal that 150 cm of sediment have been deposited during the last 2600 years in the north of the area, but only 60–80 cm in the south. This difference suggests that the recent sediment is being derived from the north and is probably carried in by the southward-flowing component of an estuarine circulation pattern in the Inner Sound.

Submersible observations

Dives by a manned submersible have provided descriptions and photographs of the sea floor in two parts of the Minches (Eden et al., 1973). Two dives in the Little Minch near the Sound of Harris visited a deep basin between rock ridges, where a dusting of fine sediment over crustacean shells on a flat and featureless seafloor suggested that sedimentation was in progress. Mud was also slowly covering rock exposures and glacial debris on slightly higher ground. On the southern end of the Shiant East Bank two further dives were made. The first at 70–76 m depth, encountered an ill-sorted, morainic drift sitting a metre or two above a plain of fine muddy sand, on which delicate bryozoan growth suggested stable conditions. A second dive on the bank, at a depth of 50 m, discovered an irregular gravelly sand cover, with sand and gravel ribbons, close to exposures of Precambrian rocks. The dive sites on the Shiant East Bank and the Little Minch are located on Fig. 6.2.

Side-scan sonar profiles

A considerable number of side-scan sonar and pinger recordings have been made in the Minches by both the I.G.S. and University College London (Fig. 6.4) and enough good data are available to make some pertinent contributions to our knowledge of the surface sediments. Especially when compared with pinger records, the side-can sonar data are generally good enough to delimit rock outcrops in areas of thin sediment cover. Line *SS7* (Fig. 6.60a) shows lighter patches of sand contrasting with darker areas of rock exposure, and includes the characteristic profile of a Tertiary sill through the Jurassic sediment floor. Another well displayed sill is shown on line *SS8*, this time east of Rubh' Hunish to the north of the Sound of Raasay. The intrusion here is surrounded by a weakly reflective and featureless mud-covered floor. In mud and sand-floored areas, in fact, it is common for the sea bed to be featureless on side-scan sonar records. Line *SS6* demonstrates this point, crossing a small mud-floored hollow in the Little Minch surrounded by rock exposure. Featureless records are typical of many kilometres of side-scan profiles from the Minches, the only common property being simply a slight darkening on higher ground, indicating the presence of coarser and more acoustically reflective sediment in the shallower water, as on line *SS4*. Sand waves or ribbons, or other evidence of sediment movement are almost totally lacking, except for a few localities near the mainland coast, where some poorly developed sand transport features may be present. Line *SS5* shows the most convincing of these – sand lobes north of Gruinard Bay that may indicate sand transport over a gravel or rocky substrate. Possible movement of sand along the mainland coast may result from the higher tidal currents experienced in the inshore waters, especially off the headlands (Fig. 6.22). However, no evidence of sand transport is revealed by the side-scan sonar records from the Skye/Shiant Platform, although the high tidal currents appear strong enough at the surface, as Belderson et al. (1971) have pointed out, for sand waves,

Sediment patches are well developed in restricted areas of the North Minch, particularly from the northern flanks of the Shiant East Bank northwards as far as the latitude of the Point of Stoer and eastwards to the Greenstone Ridge, but also in two smaller areas close to the Butt of Lewis and the Cape Wrath Bank. Records from these areas display a sea-floor consisting of alternating light and dark patches, often linear as in *SS1* and *SS3*, but also irregular in outline, as in *SS2*. The sediment patches are only found in association with seafloor that appears on pinger lines as strongly reflective and slightly uneven. Comparison of Pinger lines *1* and *SS1* shows that where the sea floor is flat and the surface sediment acoustically transparent – the reflective surface of the neighbouring areas passing below and fading from view – the sediment patches cease and are replaced by a characterless record. The area of sediment patches (Fig. 6.61) coincides closely with that region where the presence of glacial tills was indicated by textural analysis of the surface sediment (Fig. 6.38). Thus we may interpret this acoustic characteristic as indicative of an uneven sea-floor composed of relatively unsorted glacial sediment, which passes beneath a deposit of modern mud in the adjacent hollows. The area includes parts of the Shiant East Bank and the Greenstone Ridge, where continuous seismic profiles have revealed thick Formation 1 till sequences, but it terminates in the north, where Formation 3 fluvio-glacial deposits form the surface layer. Similar patterns on side-scan sonar records from till-covered areas in the Inner Sound and Little Minch have not been found.

Linear sediment patches in the North Minch have been noted on side-scan sonar records taken by Kenyon and Stride (1970), and have been further discussed by Belderson et al. (1971). Originally, it was suggested that the lineation of the patches represented the presence of a sand transport system responsible for moving sediment around the Butt of Lewis and Cape Wrath, but no supporting sedimentological evidence was offered. The study of the properties of the North Minch sediments

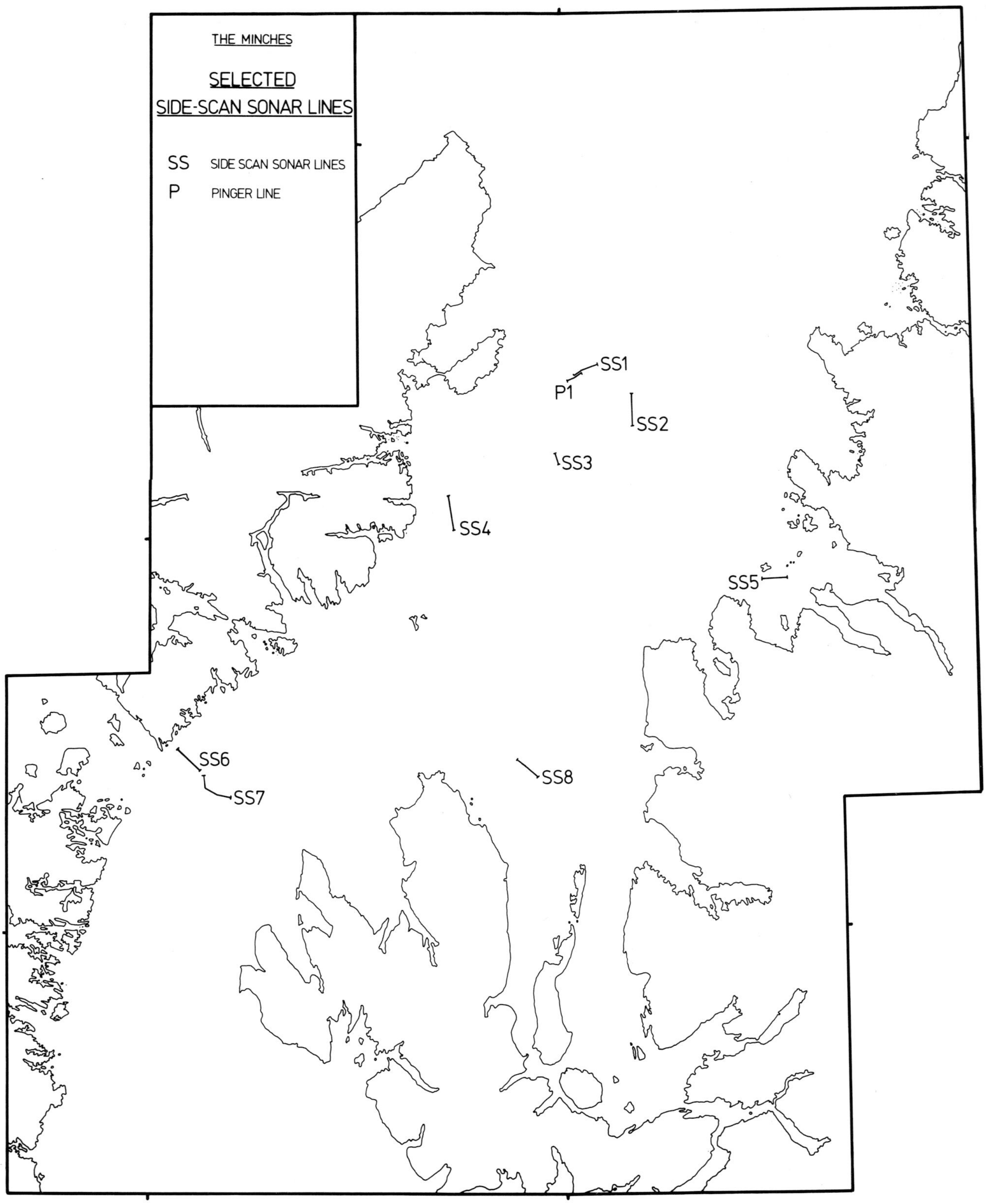
THE MINCHES
SELECTED
SIDE-SCAN SONAR LINES
SS SIDE SCAN SONAR LINES
P PINGER LINE
SS1
P1
SS2
SS3
SS4
SS5
SS6
SS7
SS8

Fig. 6.60a.

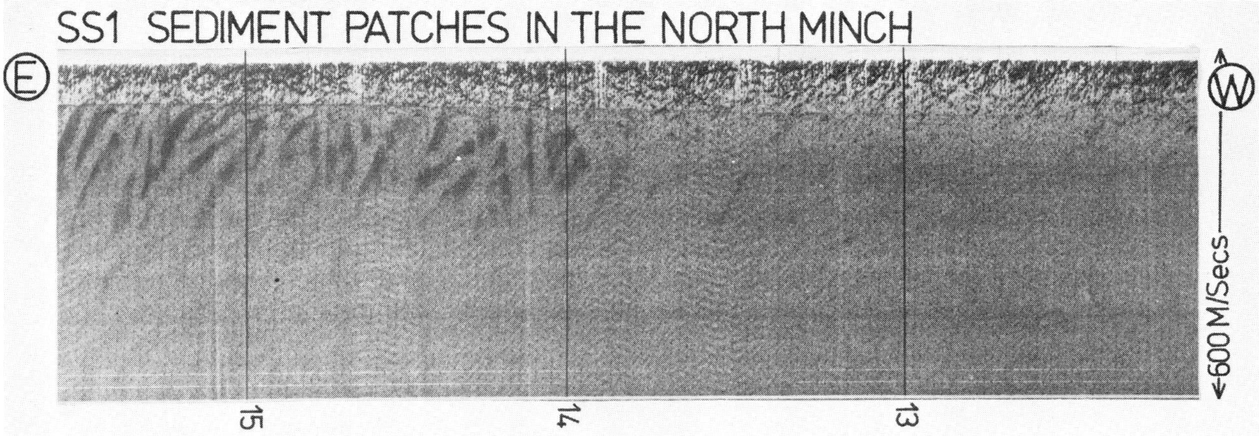

b

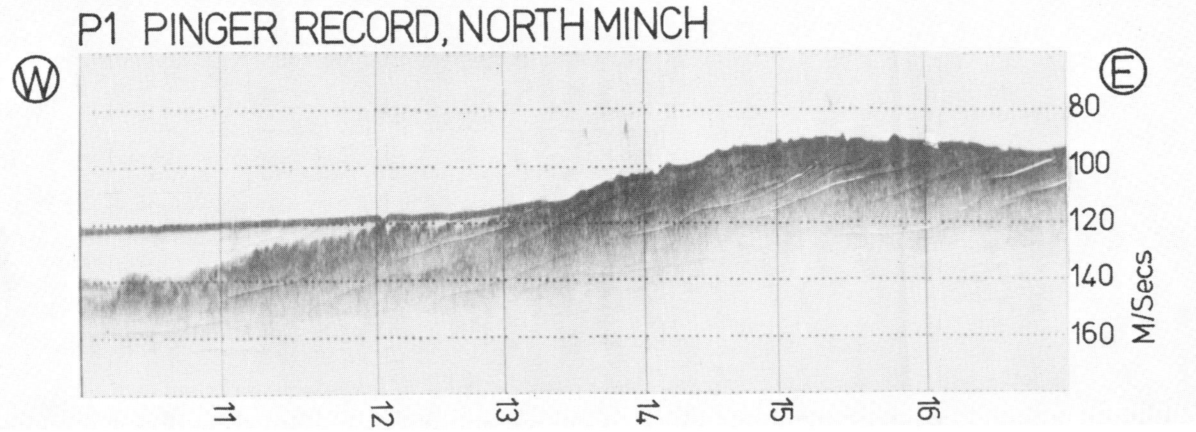

c

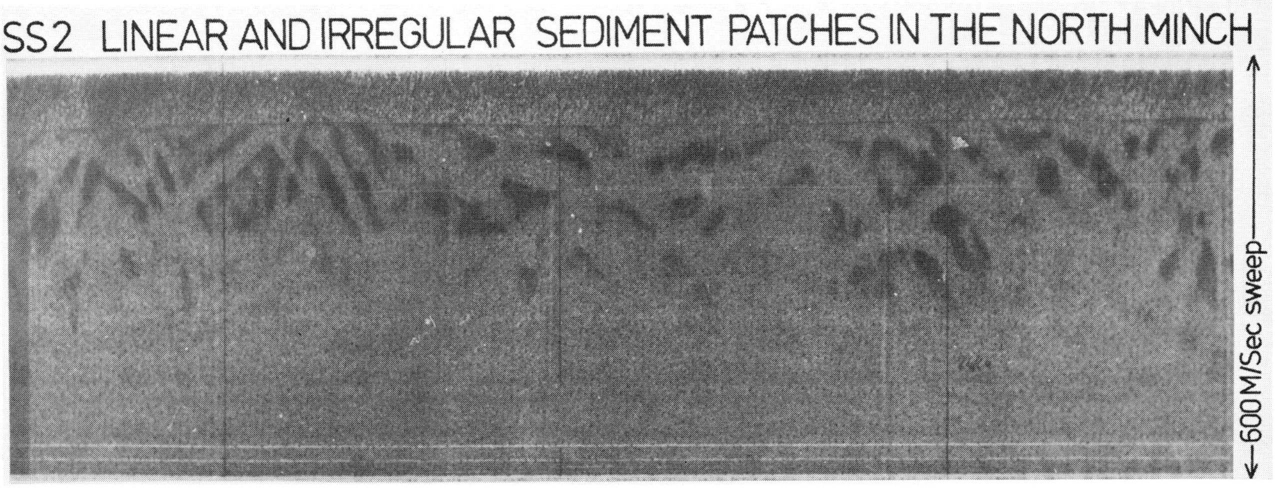

d

Fig. 6.60b–d. Selected side-scan sonar and pinger recordings.

SS3 LINEAR SEDIMENT PATCHES IN THE NORTH MINCH

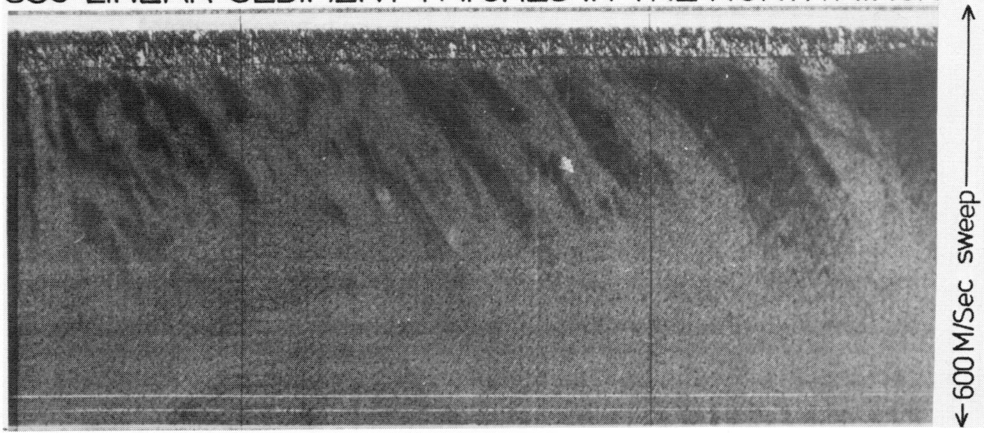

e

SS4 COARSER REFLECTIVE SEDIMENT ON HIGHER GROUND IN N. MINCH

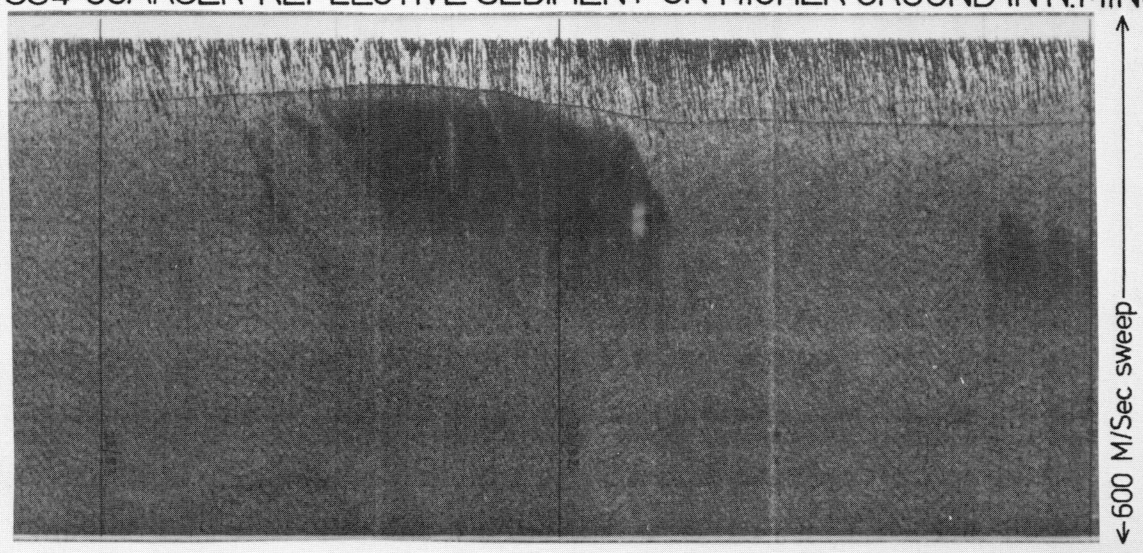

f

SS5 SAND LOBES NORTH OF GRUINARD BAY

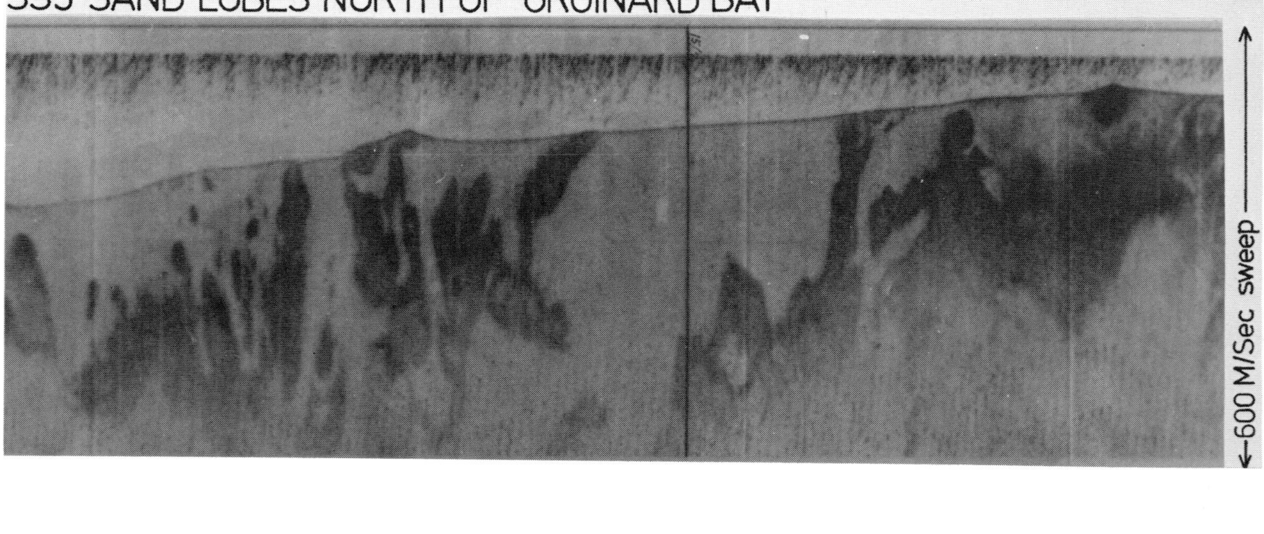

g

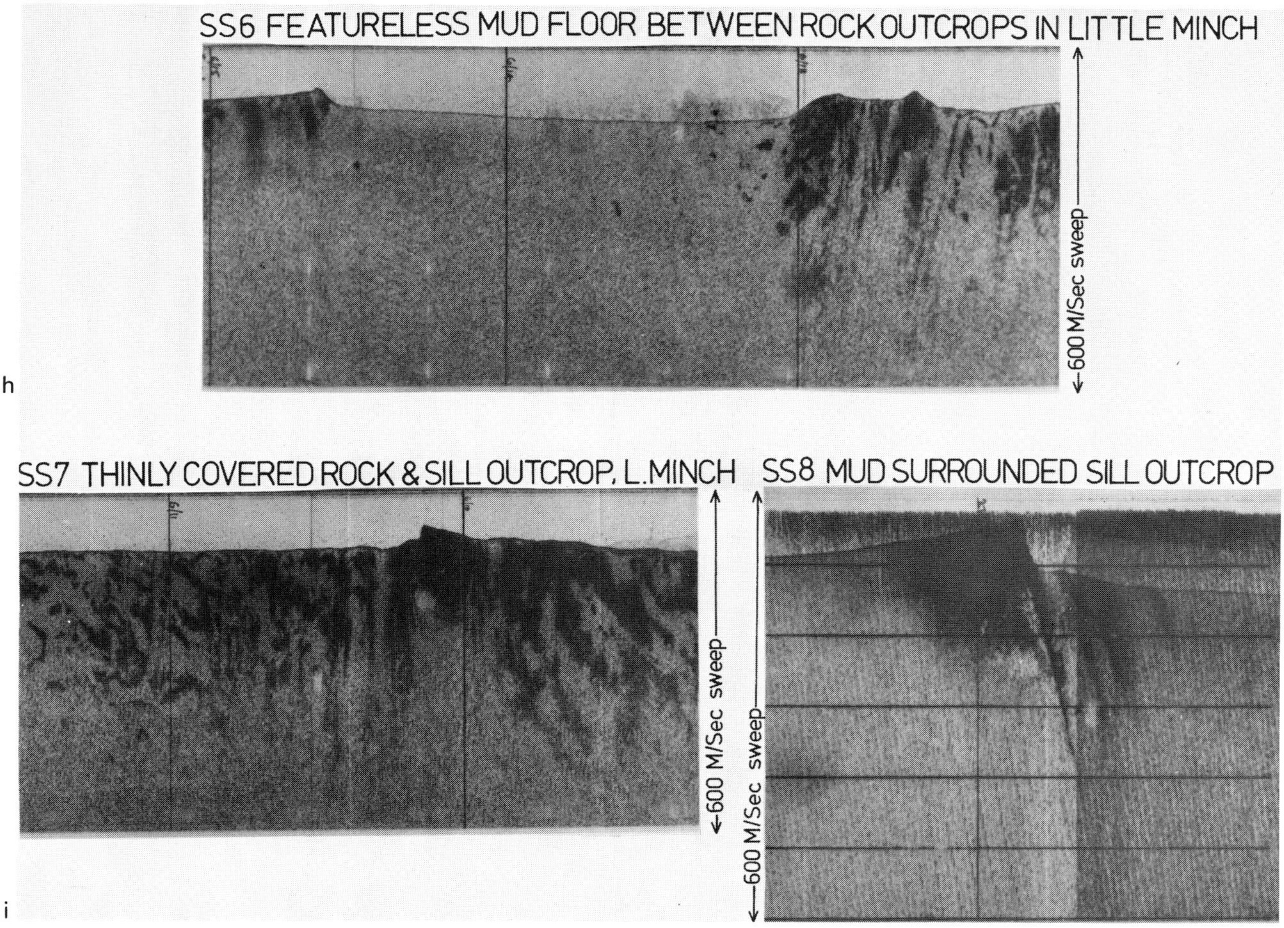

Fig. 60e–j. Selected side-scan sonar recordings

does not indicate large-scale sediment transport, except for the area close to the Lewis coast, yet the lineation of the sediment patches is hard to explain. Belderson et al. (1971) offered an explanation of their origin:

"Perhaps sand is gathered into them by a sweeping action or else the sand bodies themselves may migrate either back and forth to a limited extent in some places, while elsewhere a preferred direction of movement is suggested by their semi-crescentic plan view."

What energy is available at the sea floor to maintain a "sweeping action" cannot be very great, as mud is being laid down in the Stornoway Basin close to the sediment patches of line *SS1*, and at a very similar depth. In addition, the patches occur in up to 120 m of water, while on the southern margins of the Shiant East Bank, submersible dive records mentioned above indicate stable conditions and a muddy sand floor at only 70–76 m depth. Kenyon and Stride (1970) associated sediment patches with areas possessing tidal currents of less than 0.5 m/s at the sea surface, and this condition is certainly fulfilled in regard to their major area of distribution in the North Minch.

That the sediment patches, no more than a few centimetres thick in most cases (Belderson et al., 1971), are apparently limited to areas of glacial tills may only be a result of the acoustic contrast between the surface cover of sand and the till below, and they may indeed occur on the sandier floors of the rest of the North Minch but are not visible on side-scan records. Sediment sorting clearly increases away from the area of sediment patches in the North Minch, the best sorted material occurring in places where side-scan sonar records are featureless. Large scale sand transport is unlikely to explain the development of the sediment patches. However, it seems probable that local resorting of sand grade material results in the production of low and essentially linear sand bodies, although the tendency for the lineation to break down suggests that the preferred direction of movement is not always strongly determined.

Surface sediments: a summary

It is possible to classify the surface sediments of the

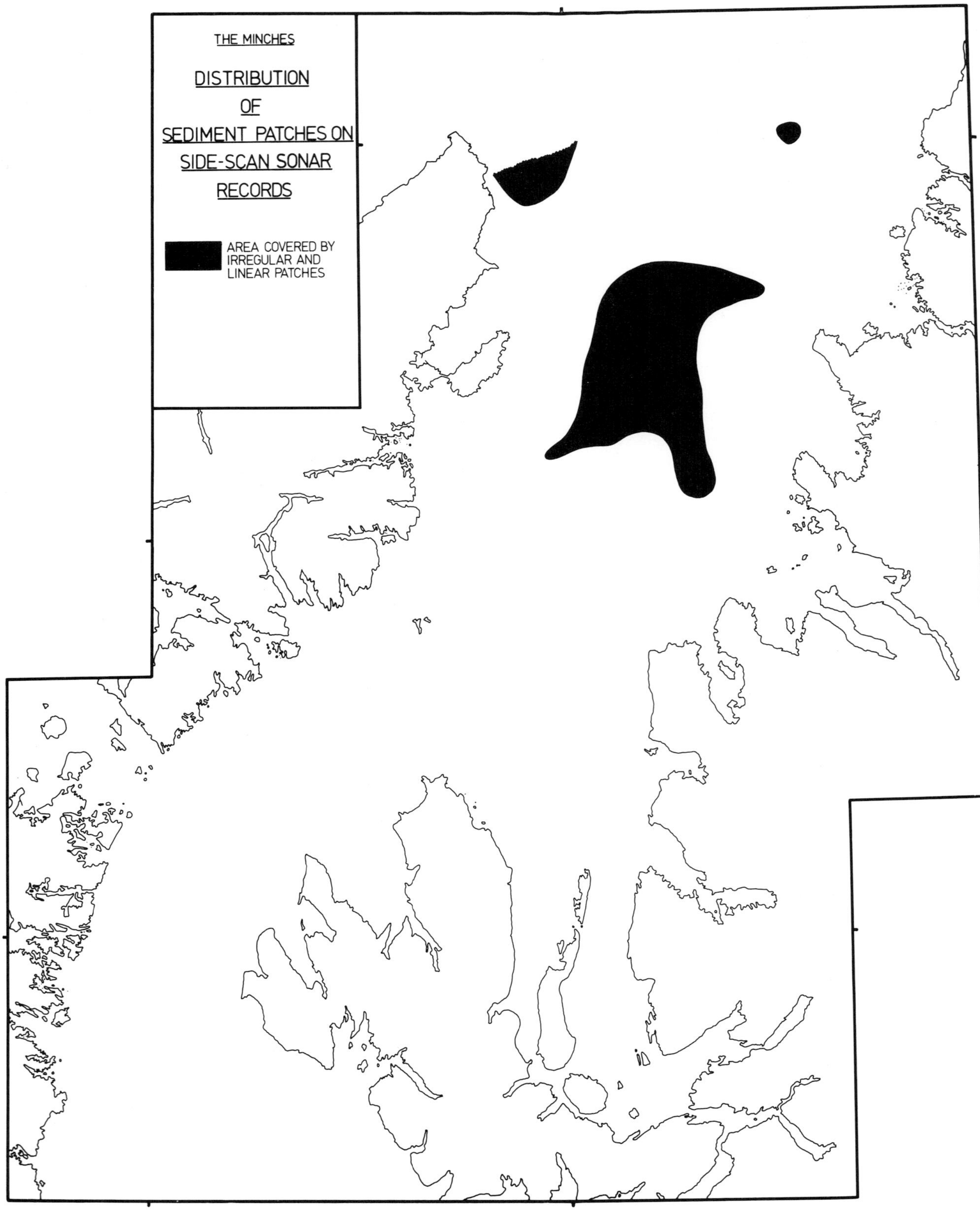
THE MINCHES
DISTRIBUTION
OF
SEDIMENT PATCHES ON
SIDE-SCAN SONAR
RECORDS
AREA COVERED BY
IRREGULAR AND
LINEAR PATCHES

Fig. 6.61.

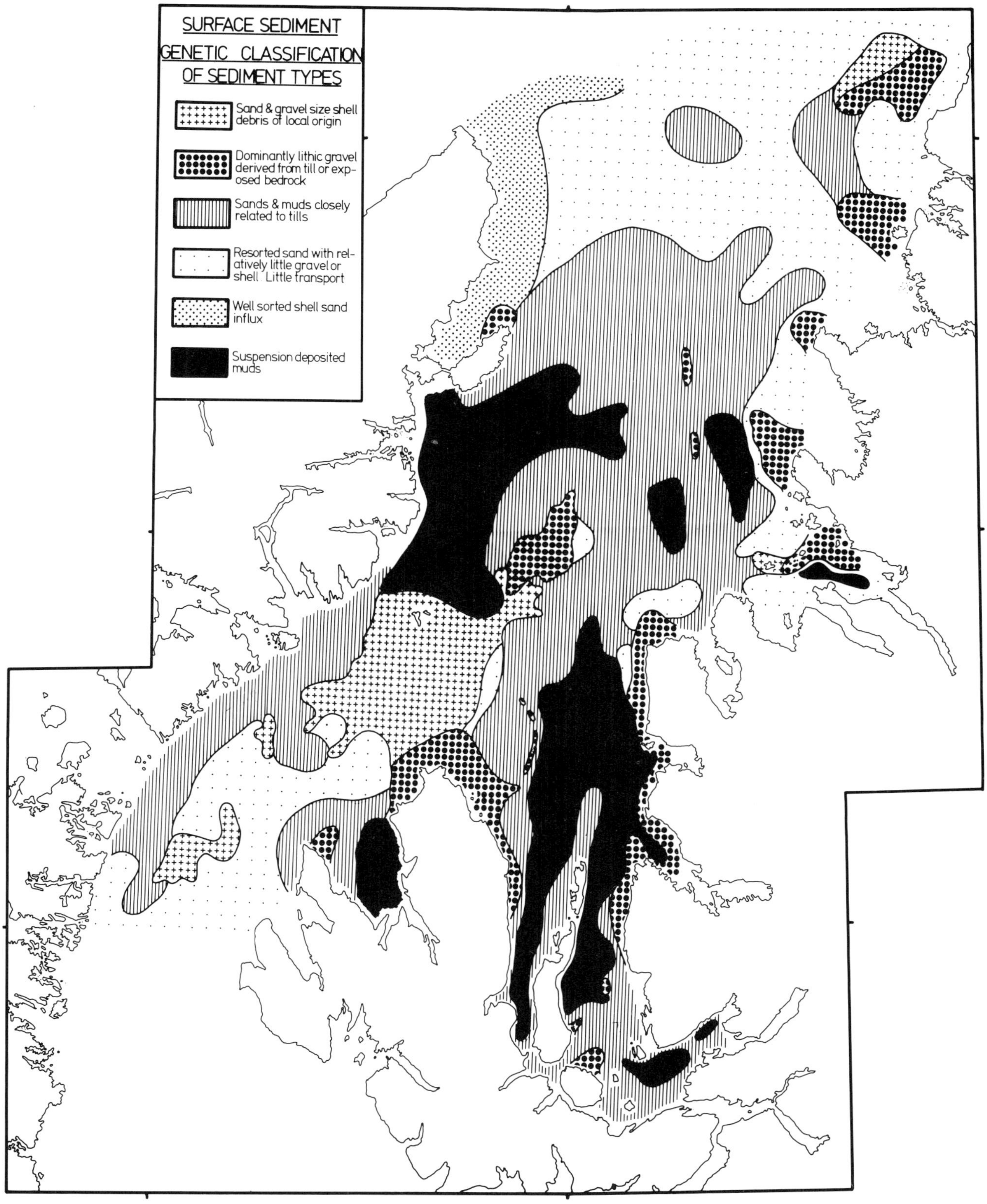
SURFACE SEDIMENT
GENETIC CLASSIFICATION OF SEDIMENT TYPES
Sand & gravel size shell debris of local origin
Dominantly lithic gravel derived from till or exposed bedrock
Sands & muds closely related to tills
Resorted sand with relatively little gravel or shell. Little transport
Well sorted shell sand influx
Suspension deposited muds

Fig. 6.62.

Minches into six genetic categories, the geographical distribution of which is shown on Fig. 6.62.

(1) The lithic gravels of the Shiant East Bank, the mainland and the Hebridean headlands display in their lithology a close relationship to the local solid geology. Although glacial transport probably also had a role to play in their distribution, to what extent is difficult to determine.

(2) The rugged bottom topography of the banks provides a suitable habitat for the organisms whose broken skeletal remains form a blanket masking the non-biogenic sediment of the northern Skye–Shiant Platform and southern Shiant East Bank. The topography developed over Tertiary igneous sills within the Jurassic seems to be particularly well suited to the development of this sediment type. Shell debris is also present within the lithic gravels, but not in such abundance. The biogenic sediment of the banks reflects its local origin in its variety of abraded and fresh shell fragments, although better sorting in some areas suggests a certain amount of reworking by currents and waves. There is no evidence, however, that biogenic sediment produced on the banks is dispersed beyond their margins into deeper water.

(3) It has been shown that many Minches sediment samples have textural properties similar to known tills, and sandy, gravelly and muddy sediments also occur which seem to be associated with a glacial parent material. Sediments closely related to tills cover wide areas of the Minches floor, especially in sheltered regions and at intermediate depths in more open water. They can be detected below superficial muds on pinger records, and can be mapped due to the acoustically contrasting sediment patches that partially cover them.

(4) North of the glacial sediments of the North Minch, and in the southern Little Minch, better sorted traction-laid sand deposits are found. Although possibly reflecting the more exposed environment of these extreme areas, these accumulations are probably derived from peri-glacial outwash deposits. Neither side-scan sonar records nor sedimentological properties indicate sediment transport presently in progress.

(5) Close to the north-east coast of Lewis a distinctive body of shell sand is in transit from a source west of the Outer Hebrides, and is impounded in Broad Bay. Transport probably occurs very close to the coast by wave-generated currents and longshore drift, but tidal currents within a zone a few kilometres from the coast probably also contribute to sediment movement.

(6) In the deep hollows in open water, and in the sheltered lochs and sounds, mud is being laid down along with minor quantities of fine quartz sand, mica and a rich foraminiferal fauna. The source of the muddy sediment is reworked and winnowed glacial deposits in the shallower parts of the Minches, and there are no indications that other sources contribute to the clay fraction of these deposits.

Conclusions

Both in detail and on the broadest scale, the bathymetry of the Minches can be related to fundamental geological controls. This region of drowned continental crust between the Scottish mainland and the Outer Hebrides is occupied by two long established sedimentary basins, separated by the Rubh'Re–Kebock Head basement ridge. The latter has been exhumed at the Shiant East Bank. The irregular topography of the numerous shoals around the Isle of Skye reflects the outcrops of resistant Tertiary sills within softer Jurassic sediments, and the absence of banks in the North Minch can be explained by its location outside the limits of shallow Tertiary igneous activity, the northward growth of intrusions apparently being halted by the Rubh'Re–Kebock Head Ridge. A system of pre-glacial drainage channels, possibly of Tertiary age, did much to control the forces of glacial erosion and subsequent deposition, the main basal ice-streams having followed the most important fluvial valleys as discordant under-currents of ice at maximum glaciation or as independent valley glaciers at a later stage. A drowned glacial breach through an important pre-glacial watershed has been recognised in the Sound of Shiant. Maximum glacial erosion occurred in the softer Mesozoic sediments, especially where faulted, or where ice was constricted or deflected by topographic barriers. A number of tenuously connected, deep, inshore basins was created, extending into lochs and sounds, and linking with a single over-deepened channel that carried ice through the Little Minch and Sea of the Hebrides to the outer shelf. Glacial and post-glacial sediments are thickest in the basins, and have partially obscured them, although they still form abnormally deep parts of the British continental shelf.

Although changes of sea level are of great complexity in northern Scotland, a sketched reconstruction of the late-glacial coastline in the Minches area has been made (Fig. 6.15). The deeply eroded troughs formed a ramiform system of sea lochs, while much of the North Minch floor was very close to sea level. Towards the south and east, the present coastline has resulted from emergence since late-glacial times, while there is substantial evidence for the submergence of the Outer Hebrides and extreme north of the mainland. The glacial sediments of the Minches have been deposited in a way concordant with the late-glacial palaeogeography and processes of sedimentation observed today in arctic areas (Carey and Ahmad, 1961). The fundamental control on sedimentation in late-glacial times was the position of the "buoyancy-line" of the wet-base glaciers that supplied detritus. On the grounded shelf in shallower areas, till and fluvio-glacial sediments were deposited in complex relationships. North of Rubh'Re, the flat base-level of glacial deposition and the foreset slope of the till below the "buoyancy-line" has been remarkably well preserved. Near the Cape Wrath Bank, fluvio-glacial deposits are better developed in less elevated areas where the ice attained a greater degree of "buoyancy" and towards the top of the Quaternary sequence, stratified deposits are of a peri-glacial origin, passing upwards into a thin modern sand. Similar shallow water glacial-marine and peri-glacial sediments exist in the Sea of the Hebrides, but data from the Minches allow more revealing interpretations of the sedimentary processes involved in forming these deposits. The sediments laid down beyond the "buoyancy-line" in the deeper water of the eroded troughs appear to be similar in the Minches and the Sea of the Hebrides, where they have been drilled and sampled, and have been interpreted as the result of till-flows and ice-rafting.

Post-glacial sediments attributable to the modern marine regime form only a thin layer above the glacial succession. In the deep basins as much as 10 m of post-glacial mud may have been deposited. In most other areas, sediment cores encounter substantially different lithologies within one metre of the sea floor, and textural and other properties of the surface layer indicate that the sediment exposed has strong affinities with glacial deposits. Grain size is closely related to water depth suggesting that the fundamental process responsible for modern sediment distribution is the winnowing of ill-sorted glacial debris in shallow areas by waves and tidal currents and the transport of finer fractions towards resting places in the sheltered deep hollows and adjoining lochs and sounds. Evidence for this pattern of redistribution comes from study of the mean and modal grain sizes, sorting and skewness and C-M characteristics of the sediment, as well as the lithological and mineralogical properties of the gravel, sand and mud fractions.

There is no evidence, except along the north east coast of Lewis, for the contemporary transport of coarser sediment in the Minches. Pebble lithology can be explained in terms of rock outcrop and glacial transport, with only a limited amount of direct contribution from the land in the vicinity of the coasts. Within the detrital component of the sand fraction, the distribution of rock fragments, iron-stained quartz and mafic minerals reveals no patterns indicative of modern transport and even the more easily dispersed mica component has a distribution strongly related to source areas. Although considerable amounts of gravel and sand grade biogenic material are produced on the shallow banks, there are no signs of the dispersal of shell fragments into deeper water. Local resorting of shell debris, however, does occur to produce a better sorted shell sand in some areas. Only certain small Foraminifera and broken tests find their way from the shallows into the mud of the deep hollows, along with some very fine angular quartz grains and mica. Side-scan sonar records also reveal no evidence for modern transport of the coarser grades, apart from a few restricted occurrences of bed forms within one or two kilometres of the mainland coast. Acoustic contrasts on till-floored areas in the North Minch reveal the existence of sediment patches, which may be continued over wider areas, but the degree of transport associated with such features is unclear and is not thought to have more than local significance.

Analysis of the clay-mineral composition of modern muds and glacial clays shows that glacial deposits are capable of providing all of the finer material presently being deposited in the deep hollows, and no other sources are directly indicated. The rate of mud deposition and a probable offshore source is indicated by palaeomagnetic studies of sediment cores from the Inner Sound. In such bodies adjoining the more open water of the Little and North Minches, estuarine circulation may be responsible for the transport of finer sediment inshore, although the evidence from a long current meter station near the mouth of the Inner Sound suggests a complex circulation in the deeper waters. Shorter current meter stations in more shallow water do not contradict the supposition of a more rigorous tidal controlled environment in these areas.

In only one part of the Minches have coarser sediments been recovered which can be attributed to contemporary transport on a larger scale. Off the coast of Lewis, between the Butt and Broad Bay, well sorted shell sand has been brought from west of the Outer Hebrides by longshore drift and strong tidal currents. This distinctive sediment body, possibly representing the encroachment of an outer shelf sediment into the inner shelf zone, does not continue south of Broad Bay, where it is impounded, nor eastwards for any considerable distance into the North Minch.

The surface sediments of the Minches, although possessing mineralogical properties that clearly assign to them a glacial origin, have attained by marine reworking textural characteristics indicating a state of near equilibrium with the modern environment in the area and, thus, they may be classified as *palimpsest* (Swift et al., 1971). Although palaeomagnetic studies in the Inner Sound (Bishop, 1975) demonstrate continuing deposition of mud to the present day, offshore sources of winnowed fine sediment must be becoming increasingly protected by a coarser superficial cover. More research is needed to determine the rate at which sediment redistribution is slowing down. It can probably be safely assumed that the rate of reworking today is much less than in the past, when protective sediment caps had not formed on the banks, and many parts of the Minch floor stood closer to sea level and thus in a higher energy environment than they do now. Only the finer sediment grades appear to be undergoing transportation at present (with the exception of the Butt of Lewis shell sands) and it is likely that the distribution of areas of coarser sediment may have to be explained by processes that operated in the immediate post-glacial period.

Acknowledgements

We are indebted to the Institute of Geological Sciences' Continental Shelf Unit in Edinburgh for providing the bulk of the samples and seismic records used in the study. The assistance of many of the Unit's staff is gratefully acknowledged, particularly the generous help of Dr. R.A. Eden, Dr. N.G.T. Fannin, and Dr. D.A. Ardus. We also thank our colleagues in the Department of Geology, University College London, the Hydrographic Department (Ministry of Defence), the Captain, Officers and Crew of H.M.S. "Hecla" and R.R.S. "John Murray", the Institute of Oceanographic Sciences, the Scottish Marine Biological Association's laboratory in Oban and the staff of the NERC Research Vessel Base in Barry, South Wales.

References

Armstrong, F.A.J. and Butler, E.I., 1960. Chemical changes in sea water off Plymouth during 1959. J. Mar. Biol. Assoc. U.K., 39: 525–528.

Armstrong, F.A.J. and Butler, E.I., 1962. Chemical changes in sea water off Plymouth during 1960. J. Mar. Biol. Assoc. U.K., 42: 253–258.

Baden-Powell, D.F.W., 1938. On the glacial and inter-glacial marine beds of northern Lewis. Geol. Mag., 75: 395–409.

Belderson, R.H., Kenyon, N.H. and Stride, A.H., 1971. Holocene sediments on the continental shelf west of the British Isles. In: F.M. Delany (Editor), The Geology of the East Atlantic Continental Margin. 2. Europe. ICSU/SCOR Working Party 31 Symp., Cambridge, 1970. Inst. Geol. Sci. London, Rep., 70/14: 161–170.

Binns, P.E., Harland, R. and Hughes, M.J., 1974a. Glacial and post-glacial sedimentation in the Sea of the Hebrides. Nature, 248: 751–754.

Binns, P.E., McQuillin, R. and Kenolty, N., 1974b. The geology of the Sea of the Hebrides. Inst. Geol. Sci., London, Rep., 73/14: 43 pp.

Binns, P.E., McQuillin, R., Fannin, N.G.T., Kenolty, N. and Ardus, D.A., 1975. Structure and stratigraphy of sedimentary basins in the Sea of the Hebrides and the Minches. In: A.W. Woodland (Editor), Petroleum and the Continental Shelf of North West Europe. 1. Geology. Applied Science Publishers, London.

Biscaye, P.E., 1965. Mineralogy and sedimentation of recent deep sea clays in the Atlantic Ocean, and adjacent seas and oceans. Bull. Geol. Soc. Am., 76: 803–832.

Bishop, P., 1975. The palaeomagnetism of muddy sediment cores from the Inner Sound, North West Scotland, and the glacial and post-glacial history of sedimentation in the area. Earth Planet. Sci. Lett., 27: 51–56.

Bott, M.H.P., 1974. Subsidence of Atlantic continental margins. Tectonophysics, 22: 283–299.

Cary, S.W. and Ahmad, N., 1961. Glacial marine sedimentation. Proc. Int. Symp. Arctic Geology, 1st. University of Toronto Press, Toronto, Ont., 2: 865–894.

Carrol, D., 1970. Clay minerals: a guide to their X-ray identification. Geol. Soc. Am., Special Pap., 126: 80 pp.

Charlesworth, J.K., 1956. The late glacial history of the highlands and islands of Scotland. Trans. R. Soc. Edinburgh, 62: 769–928.

Chester, J.A., Deegan, C.E., Ardus, D.A., Binns, P.E. and Fannin, N.G.T., 1972. IGS marine drilling with M.V. Whitehorn in Scottish waters, 1970-1971. Inst. Geol. Sci. U.K., London, Rep., 72/10: 25 pp.

Craig, R.E., 1951. Hydrographic conditions in the Minch, September/October 1951. Ann. Biol., Cons. Perm. Int. Explor. Mer, 8: 20–21.

Craig, R.E., 1954. Hydrographic conditions in the Minch and the northern approach to the Irish Sea. Ann. Biol., Cons. Perm. Int. Explor. Mer, 11: 34–36.

Craig, R.E., 1959. Hydrography of Scottish coastal waters. Scott. Home Dep., Mar. Res. Ser., 2.

Craig, R.E. and Slinn, D.J., 1955. Hydrographic observations off the Scottish west coast and the northern Irish Sea, Autumn 1955. Ann. Biol., Cons. Perm. Int. Expl. Mer, 12: 59–61.

Doodson, A.T. and Warburg, H.D., 1941. Admiralty Manual of Tides. H.M.S.O., London, 270 pp.

Doyle, L.J., Cleary, W.J. and Pilkey, O.H., 1968. Mica – its use in determining depositional regime. Mar. Geol., 6: 381–389.

Draper, L., 1967. Wave activity at the sea bed around North West Europe. Mar. Geol., 5: 133–140.

Eden, R.A., Deegan, C.E., Rhys, G.H., Wright, J.E. and Dobson, M.R., 1973. Geological investigations with a manned submersible in the Irish Sea and off western Scotland, 1971. Inst. Geol. Sci., London, Rep., 73/2: 27 pp.

Edwards, P.G., 1975. Ecology and distribution of recent foraminifera in the North Minch Channel, North West Scotland. Thesis, Dep. Geol. Micropalaeontol., University College, London, 115 pp. (unpublished).

Emery, K.O., 1969. The continental shelves. In: The Ocean. Sci. Am. – Freeman, San Francisco, Calif., 140 pp.

Ferentinos, G.K., 1976. Sediment distribution and transport processes on the outer continental shelf of the Hebridean Sea. Mar. Geol., 20: 41–56.

Folk, R.L., 1965. Petrology of Sedimentary Rocks. Hemphill's, Austin, Texas, 159 pp.

Folk, R.L., 1968. Petrology of Sedimentary Rocks. Hemphill's Univ. of Texas, 170 pp.

Friedman, G.M., 1967. Dynamic processes and statistical parameters compared for size frequency distribution of beach and river sands. J. Sediment Petrol., 37: 327–354.

Geikie, J., 1878. On the glacial phenomena of the Long Island, or Outer Hebrides. Q. J. Geol. Soc. London, 29: 532–570.

Geikie, J., 1894. The Great Ice Age. Stanford, London, 3rd. ed., 850 pp.

Gibbs, R.J., 1965. Error due to segregation in quantitative clay mineral X-ray diffraction mounting techniques. Am. Mineral., 50: 741–751.

Godard, A., 1965. Recherches de géomorphologie en Ecosse du Nord-Ouest. Société d'Editions de Belles Lettres, Paris, 701 pp.

Godwin, H. and Willis, E.H., 1958. Radio-carbon dating of the eustatic rise in ocean level. Nature, 181: 1518–1519.

Griffin, G.M., 1962. Regional clay mineral facies – products of weathering intensity and current distribution in the north-eastern Gulf of Mexico. Bull. Geol. Soc. Am., 73: 737–768.

Griffin, J.J., Windom, H., Goldberg, E.D., 1968. The distribution of clay minerals in the world ocean. Deep-Sea Res., 15: 433–459.

Grim, R.E., 1939. Properties of clay. In: P.D. Trask (Editor), Recent Marine Sediments. Am. Assoc. Pet. Geol., Tulsa, Okla, pp. 466–495.

Grim, R.E., 1953. Clay Mineralogy. McGraw-Hill, New York, N.Y., 384 pp.

Hadley, H.L., 1964. Wave-induced bottom currents in the Celtic Sea. Mar. Geol., 2: 164–167.

Hough, J.L., 1942. Sediments of Cape Cod Bay, Massachusetts. J. Sediment Petrol., 12: 10–30.

James, N.P. and Stanley, J.J., 1968. Sable Island Bank off Nova Scotia: sediment dispersal and recent history. Bull. Am. Assoc. Pet. Geol., 52: 2208–2230.

Judd, J.B., Smith, W.E. and Pilkey, O.H., 1970. The environmental significance of iron-stained quartz grains on the south-eastern United States Atlantic shelf. Mar. Geol., 8: 355–362.

Kenyon, N.H. and Stride, A.H., 1970. The tide-swept continental shelf sediments between the Shetland Isles and France. Sedimentology, 14: 159–173.

King, C.A.M., 1975. Introduction to Physical and Biological Oceanography. Edward Arnold, London, 372 pp.

Klug, H.P. and Alexander, L.E., 1954. X-ray Diffraction Procedures. Wiley, New York, N.Y.

Krumbein, W.C. and Pettijohn, F.J., 1938. Manual of Sedimentary Petrography. Appleton-Century-Croft, New York, N.Y., 549 pp.

McCann, S.B. and Chorley, R.J., 1967. Trend surface mapping of raised shorelines. Nature, 215: 611–612.

McKenzie, P., 1958. Rip current systems. J. Geol. 66: 103–113.

McQuillin, R. and Binns, P.E., 1973. Geological structure in the Sea of the Hebrides. Nature, 241: 2–4.

Meade, R.H., 1972. Sources and sinks of suspended matter on continental shelves. In: D.J.P. Swift, D.B. Duane, O.H. Pilkey (Editors), Shelf Sediment Transport. Dowden, Hutchison and Ross, Stroudsburg, Penn., pp. 249–262.

Milliman, J.D., 1974. Marine Carbonates. Springer, Berlin, 375 pp.

Milne, P.H., 1972. Hydrography of Scottish west coast sea lochs. Dep. Agric. Fish. Scotland, Mar. Res. Ser., 3: 50 pp.

Ministry of Defence, Hydrographic Dep., published annually. Tide tables for major and secondary ports. H.M.S.O., London.

Ministry of Defence, Hydrographic Dep., 1958. West Coast of Scotland Pilot (Corrected to 1971). H.M.S.O., London.

Ministry of Defence, Hydrographic Dep., 1971. Oceanography sea-floor sampling. Prof. Pap., 20 H(MO) 744/68; 59 pp.

Ministry of Defence, Hydrographic Dep., 1973. Tidal stream atlas: west coast of Scotland and north coast of Ireland. H.M.S.O., London.

Naylor, D. and Mounteney, S.N., 1975. Geology of the North-West European Continental Shelf. Graham, Trogman Dudley, London, 162 pp.

Norrish, K. and Taylor, R.M., 1962. Quantitative analysis by X-ray diffraction. Clay Miner. Bull., 5: 98–109.

Passega, R., 1957. Texture as characteristic of clastic deposition. Bull. Am. Assoc. Pet. Geol., 41: 1952–1984.

Passega, R., 1964. Grain-size representation by C-M patterns as a geologic tool. J. Sediment. Petrol., 34: 830–847.

Pettijohn, F.J., 1975. Sedimentary Rocks. Harper and Row, New York, N.Y., 3rd. ed., 628 pp.

Pettijohn, F.J., Potter, P.E. and Siever, R., 1973. Sand and Sandstone. Springer, New York, N.Y., 618 pp.

Phemister, J., 1960. British Regional Geology. Scotland: The Northern Highlands. Department of Scientific and Industrial Research, Geological Survey and Museum, H.M.S.O., London, 3rd. ed., 104 pp.

Richey, J.E., 1961. British Regional Geology. Scotland: The Tertiary volcanic districts. Department of Scientific and Industrial Research, Geological Survey and Museum, H.M.S.O., London, 3rd. ed., 120 pp.

Ritchie, W., 1966. The post-glacial rise in sea level and coastal changes in the Uists. Trans. Inst. Brit. Geogr., 39 (Spec. No. on Vertical Displacements of Shorelines in Highland Britain): 79–86.

Saburo, A., Oinuma, K. and Sudo, T., 1974. The distribution of clay minerals in the recent sediments of the Japan Sea. Deep-Sea Res., 21: 299–310.

Schlee, J., 1973. Atlantic continental shelf and slope of the United States: sediment textures of the north-eastern part. U.S. Geol. Surv., Prof. Pap., 529L: 64 pp.

Shepard, F.P. and Inman, D.L, 1950. Nearshore circulation related to bottom topography and wave refraction. Trans. Am. Geophys. Union, 31: 196–212.

Sissons, J.B., 1967. The Evolution of Scotland's Scenery. Oliver and Boyd, Edinburgh, 259 pp.

Sissons, J.B., 1970. Quaternary. In: G.Y. Craig (Editor), The Geology of Scotland. Oliver and Boyd, Edinburgh, pp. 467–502.

Sissons, J.B., 1976. The Geomorphology of the British Isles: Scotland. Methuen, London, 150 pp.

Southard, J.B. and Cacchione, D.A., 1972. Experiments on bottom sediment movement by breaking internal waves. In: D.J.P. Swift, D.B. Duane and O.H. Polkey (Editors), Shelf Sediment Transport: Process and Pattern. Dowden, Hutchinson and Ross, Stroudsburg, Penn., pp. 83–98.

Spencer, D.W., 1963. The interpretation of grain size distribution curves of clastic sediments. J. Sediment Petrol., 3: 180–190.

Steers, J.A., 1973. The Coastline of Scotland. Cambridge University Press, New York, N.Y., 335 pp.

Stephens, N. and Synge, F.M., 1966. Pleistocene shorelines. In: G.H. Dury (Editor), Essays in Geomorphology. Heinemann, London, pp. 1–51.

Sverdrup, H.U., Johnson, M.W. and Fleming, R.H., 1942. The Oceans. Prentice-Hall, Englewood Cliffs, N.J., 1087 pp.

Swift, D.J.P., 1969. Mesozoic versus modern sedimentation of the Carolina coast and the problem of the graded shelf. Geol. Soc. Am., Spec. Pap., 121: 468 pp.

Swift, D.J.P., Stanley, J.J. and Curray, J.R., 1971. Relict sediments on continental shelves: a reconsideration. J. Geol., 79: 322–346.

Ting, S., 1937. The coastal configuration of western Scotland. Geog. Annal., 19: 62–83.

Trumbull, J.V.A., 1972. Atlantic continental shelf and slope of the United States: sand size fraction of bottom sediments, New Jersey to Nova Scotia. U.S. Geol. Surv., Prof. Pap., 529K: 45 pp.

Tulloch, D.S. and Tait, J.B., 1959. Hydrography of the north-western approaches to the British Isles. Scott. Home Dep., Mar. Res. Ser., 1: 32 pp.

Valentin, H., 1953. Present vertical movements of the British Isles. Geogr. J., 119: 299–305.

Van der Plas, L. and Tobi, A.C., 1965. A chart for judging the reliability of point counting results. Am. J. Sci., 263: 87–90.
Venkatarathnam, K. and Biscaye, P.E., 1973. Clay mineralogy and sedimentation in the eastern Indian Ocean. Deep-Sea Res., 20: 727–738.
Von Weymarn, J. and Edwards, K.J., 1973. Interstadial site on the Island of Lewis, Scotland. Nature, 246: 473–474.
Wright, P.L., 1974. The chemistry and mineralogy of the clay fraction of sediments from the southern Barents Sea. Chem. Geol., 13: 197–216.

Chapter 7

The Quaternary Sediments of the North Sea

V.N.D. CASTON

Introduction

The North Sea is an epicontinental sea lying on the continental shelf of north-west Europe. It is virtually surrounded on three sides by land, being open to the Atlantic Ocean only to the north where it continues into the Norwegian Basin, and to the Baltic Sea in the east and English Channel to the south through narrow but important straits. During the greater part of the Quaternary the area was directly affected by the Pleistocene glaciation, and deposits from successive glacial advances are well documented from bordering countries.

This Chapter reviews published information concerning the Quaternary sediments of the North Sea at what is an active time in the growth of knowledge in this area. Some new data, principally from boreholes drilled for engineering purposes, is also included. It is too early yet to attempt a complete regional synthesis or syntheses, and the level of the treatment, which depends wholly upon the detail of individual studies, varies greatly from area to area.

The superficial sediments on the floor of the southern North Sea have been studied in some detail since the 1950's. However, until very recently little work had been undertaken either upon the comparable sediments of the central and northern North Sea, or upon the very considerable thickness of Pleistocene deposits present throughout the North Sea Basin, which, as has been pointed out by McCave et al. (1977), are the thickest and most extensive in north-western Europe. This situation is now changing, assisted in part by the offshore search for hydrocarbons which has contributed by the release of relevant data obtained during the course of exploration and production activities (cf. Vol. II, Fig. 18.1).

Despite this increase in our knowledge of the Quaternary over the past decade or so, only the most sketchy information is yet available for much of the North Sea. This is insufficient to permit the identification and mapping of the deposits of even the principal Quaternary subdivisions. The most comprehensively-surveyed areas are the Dutch sector and U.K. waters between 56° and 58°N, supplemented by detailed studies of limited areal extent carried out in connection with oil and gas production operations. In these areas high-resolution reflection surveys and shallow boreholes have provided information concerning Quaternary lithologies to maximum penetrations of ca. 120 m sub-sea-bed.

On the basis of this information it is now apparent that lower and middle Pleistocene deposits outcrop on the sea floor or are present at a shallow depth over much of the North Sea to the south of the Dogger Bank. To the north of this latitude only sediments of Weichselian age have been recorded, where they form a disproportionate part of the total Quaternary thickness, suggesting either very much slower deposition rates throughout the earlier Quaternary or extensive erosion of pre-Weichselian deposits. Nature of Quaternary lithologies is very variable, both on a general scale and in detail. In part this variability is caused by extensive channelling as revealed by reflection surveys. The majority of boreholes have penetrated beds of fine silty sand and clay, with stony clays, identified as tills, of more local occurrence.

The Pleistocene beds are intermittently covered by a veneer of "recent" sediments, which include (a) reworked fluvio-glacial sands – much of which is still being moved around by present-day tidal currents, (b) clays and silts – in part derived from reworking of glacial deposits and in part introduced by rivers or carried into the area in suspension by sea currents, and (c) coarser-grained deposits which commonly form lag gravels overlying glacial tills. The distinction between a "relict" sediment (Emery, 1968) and a truly autochthonous sediment (Swift et al., 1971) is frequently difficult to make, and thus description of sea-floor sediments does not necessarily attempt to assign provenance or to discuss depositional history. An exception is made for the mobile bed forms of the southern North Sea, which are of significant extent and are well documented. The present distribution of sediments thus reflects the events, both depositional and erosional, of the Pleistocene and immediately post-Pleistocene periods, and the northern North Sea in particular must be a classic example of an area of "palimpsest" sediments, consisting of the "textural mixing of the products of deposition of different periods of time and different sources" (Swift et al., 1971).

Relatively few absolute age dates are available for Quaternary sediments in the North Sea, and few palaeontological analyses have been undertaken. In this review the term "Pleistocene" is therefore taken for convenience as being synonymous with glacial and glacial marine sediments. In order to maintain a consistent terminology the north European nomenclature is used (Van Eysinga, 1975), with reference to Mitchell et al. (1973) concerning correlation with British stages.

Bathymetry

The relief of the sea floor is important in that it reflects both the underlying geological structure and present day sedimentary processes, the latter being more significant to the south of 54°40′N. However, despite extensive surveying by the appropriate hydrographic authorities of the coastal states, the North Sea has never been comprehensively surveyed. The southern North Sea is understandably the better known, with a history of surveying which dates back to the early nautical charts constructed by sixteenth and seventeenth century Dutch cartographers (Lang, 1968). On the U.K. side large areas to the south of 56°N have been accurately surveyed by echosounder since the 1930's, but north of this latitude only a relatively narrow strip has been surveyed by recent techniques, and the great majority of depths shown on navigational charts are based upon isolated lead-line soundings dating from the eighteenth and nineteenth centuries. Our knowledge of this northern area is therefore inadequate.

Surveys for scientific purposes in the North Sea were first undertaken in connection with the various deeps or "holes", and date from the 1930's. Gregory (1931) reported the presence of the Devil's Hole, which was first surveyed by Beaugé (1937), who also published charts of the Long Forties and Swatchway areas to the east of Aberdeen, and a line of soundings across the North Sea between Copenhagen and Stornoway. Beauge's surveys were reproduced by Furnestin (1939–1943). In the southern North Sea detailed maps of the Outer Silver Pit were first published by Berthois and Brenot (1957), of the banks in the Southern Bight off the Flemish coast by Van Veen (1936), and the Silver Pit by Guilcher (1951) and Valentin (1957). The Norwegian Channel had long been recognised, but discussion of its morphology and origin again dates from the 1930's. A part of it first surveyed as part of a comprehensive study of Norwegian coastal waters by Holtedahl (1940), and additional detail was given by Pratje (1952) and, more recently, by Sellevol and Sundvor (1974). Work in the Skagerrak was initiated by Holtedahl (1964) and continued by Sellevol and Aalstad (1971) and Van Weering et al. (1973). Additional bathymetric data has been provided off the Yorkshire coast by Dingle (1970). A shortcoming of the majority of the earlier surveys was the inadequate navigation, so the true distribution and shape of many submarine features, especially in the northern North Sea, is imperfectly known.

The results of all significant surveys to the mid-twentieth century in the area south of 55°20′N have been summarised by Stocks (1955), who has produced a valuable contour map on a scale of 1 : 1 M. North of this latitude the only comprehensive bathymetric map is that of the area between 56° and 63°N constructed by Flinn (1973) from a variety of data, including a compilation of all navigational chart soundings, detailed Admiralty echosounder surveys in the inshore area, and a large number of individual echosounder lines run by ships of opportunity. Caston's (1974) map covering a slightly larger area of the northern North Sea was essentially based upon that of Flinn, with some additional information. Figure 7.1 is based upon contoured charts on a scale of 1 : 900,000 for the whole North Sea area published by the Deutsches Hydrographisches Institut (1953). This map shows the major relief features, which for convenience will be discussed in terms of four principal areas.

Shelf edge

The edge of the continental shelf, which arguably may be considered to be the northern limit of the North Sea, is poorly defined in the area between the Shetland Isles and the Norwegian coast. The 200 m isobath, which corresponds approximately to the shelf break off the west coast of Scotland, trends in a north-easterly direction past the Shetland Isles as far north as 61°30′N. Here it swings round to run approximately west–east, but beyond that the shelf continues to deepen towards a fairly subdued break in slope, the most northern point

of which, at approximately 62°N 1°E, lies at a depth of nearly 300 m on the western margin of the Norwegian Channel. The shelf edge continues in a north-easterly direction across the mouth of the Channel at a depth of some 450 m, until it reaches the eastern Channel margin at 62°30′N 2°50′E, some 150 km off the Norwegian coast, where it swings gradually round to run in a north-north-easterly direction up the west coast of Norway.

Norwegian Channel

Paralleling the coast of Norway along the eastern side of the North Sea runs a deep submarine trough or valley. This trough, the Norwegian Channel, which is the largest single feature on the north-west European continental shelf, has an average width of 90 km, and runs for some 700 km down the west coast of Norway before curving round the southern tip and ending abruptly in the Skagerrak. In plan view it is the shape of an elongate funnel, the width diminishing from over 180 km to the north of 62°N to about 45 km at 57°35′N. Over much of its length it is relatively flat-floored, although the greatest depths in cross-section occur on the eastern side. Much of the central part of the trough lies at a depth of between 250 and 300 m, but south of 58°30′N it deepens, lying at over 450 m off the southern coast of Norway, and with a maximum of 700 m in the Skagerrak. North of 60°N it also deepens steadily, reaching 360 m at 61°20′N and 420 m at 62°30′N. The eastern side of the Channel parallels the Norwegian coast south of about 61°30′N but at this latitude the coastline bears away to the north-east, leaving a triangular-shaped area of relatively-shallow water centred at 62°30′N between the Channel edge, the coast and the Norwegian Sea.

In general the Channel is relatively straight-sided but with minor indentations. The eastern side is the steeper, except to the north of 61°30′N. On the western side an apparently detached shallow bank with a minimum depth of about 180 m is shown centred at 61°45′N 1°50′E. This is based upon a number of soundings indicated on Admiralty charts, but is not shown on Flinn's (1973) map. Until more evidence is forthcoming its presence must be regarded as suspect.

Northern North Sea

For our purposes the northern North Sea will be defined as lying between the British coast and the western edge of the Norwegian Channel, and from 56°N northwards to the edge of the continental shelf.

Depths increase fairly rapidly to about 100 m away from the Scottish coast and islands, and the greater part of the northern North Sea lies at depths of between 100 and 140 m. The sea bed is, however, far from flat and there is an apparently haphazard pattern, as yet poorly defined, of shallower banks with least depths ranging from 60 m in the south to 100 m in the north, separated by intervening deeper areas in which depths vary from 120 m to 160 m. Despite the irregular nature of the sea bed certain generalisations may be made. In the first place it is apparent that average depths significantly increase to the north of a sinuous line running from 57°30′N 01°W in the west to 58°30′N 03°E in the east, the depth increase averaging about 35 m. Secondly, depths are greatest, averaging about 145 m, within a central zone lying roughly between 0° and 1°30′E, and to the north of 58°N, and then decrease farther eastward, much of the ground bordering the western side of the Norwegian Channel as far north as 61°N lying at a depth of less than 120 m. Finally attention is drawn to the very different nature of the shelf beyond the 160 m isobath north of approximately 61°20′N. In this area the bottom deepens steadily towards the north without the irregularities and undulations which so characterise the bed of the North Sea south of this latitude.

Certain prominent features, although of localised extent, deserve mention. These are the narrow linear deeps or "holes" which have maximum depths in excess of 200 m, and which are found principally within an arcuate belt with its apex to the north of Aberdeen. The formation of these deeps was originally attributed (Flinn, 1967) to sub-glacial erosion under the front of an ice sheet. This belt might thus mark the position of a late-glacial ice front. Flinn suggested that the ice-free area might have lain within the arc, although admitting that there was no evidence for this.

Later geophysical studies (discussed below) have shown that infilled linear deeps are abundant in the Quaternary throughout much of the central and northern North Sea. The present consensus of opinion supports Flinn's view that such features are probably sub-glacial tunnel valleys, although not necessarily formed at an ice front.

The sediment-free deeps, however, remain an enigma. Holmes (1977) notes that the Devil's Hole erodes into the Swatchway Beds and must therefore be younger than Late Weichselian. He suggests that the Hole may have been kept free of sediment during waning glaciation and periods of lowered sea level by tidal scour adjacent to an ice front. Applying this concept to the whole belt of linear deeps would provide support for Flinn's suggestion of an arcuate ice front.

Southern North Sea

In the southern North Sea, the German Bight, bounded to the east and south by the Danish and German coasts, to the north by the line of banks bordering the southern edge of the Norwegian Channel at about 55° 30′ N, and extending westwards as far as the Dogger Bank at approximately 3° E, is a relatively shallow and featureless area ranging from 30 to 50 m in water depth. The coastal waters are extremely shallow, and the sea bed deepens only gradually to the north and west, maximum depths occurring at about 4° 10′ E. At the same latitude but on the English side of the North Sea the relief is somewhat different. North of about 53° 50′ N depths increase fairly rapidly in a north and easterly direction away from the shore, reaching maximum depths in excess of 80 m between the coast north of 55° N and the Dogger Bank (Dingle, 1970) and to the south of the Dogger Bank in the Outer Silver Pit (Berthois and Brenot, 1957).

To the south of this area lies a belt of shallow water, less than 30 m in depth, which extends almost continuously from the English to the Dutch coasts between about 53° N to 53° 50′ N. On the Dutch side this is known as the Texel Spur (Houbolt, 1968). South of the shallow zone water depths in the Southern Bight increase to 50–70 m, with the greatest depths lying along an approximately north–south axis closer to the English coast.

A notable feature of the southern North Sea is the abundance of sand banks. These occur almost continuously off the Danish, German and Dutch coasts, and off the English coast from the mouth of the Thames as far north as 53° 40′ N. They are most prominent on the Dutch side south of about 52° 10′ N and on the English side between 52° 33′ N and 53° 35′ N, where they extend as a series of parallel linear features in 30 m to 40 m water depth up to 100 km offshore (Van Veen, 1936; Guilcher, 1951; Off, 1963; Houbolt, 1968; Caston, 1972).

Quaternary thickness

Considerable thicknesses of Cainozoic sediments are present in the North Sea. From published maps based primarily on seismic data (Heybroek et al., 1967; Dunn et al., 1973) it is apparent that over 3,500 m of combined Tertiary and Quaternary deposits infill a broad basin centred at about 57° N which runs the whole length of the North Sea, extending southwards into The Netherlands and North Germany, and northwards beyond 62° N.

Subdivision into Tertiary and Quaternary is not straightforward. The base of the Quaternary rarely forms an identifiable seismic marker except adjacent to land, where relatively unconsolidated Pleistocene deposits commonly overstep the Tertiary on to harder pre-Tertiary strata. Detailed sparker surveys, supplemented by sea-bed coring and shallow boreholes, have however enabled the base of the glacial deposits to be mapped off the east coast of Scotland and in the Moray Firth, but at present this has been undertaken only over a limited are (Holmes, 1977; Chesher and Lawson, in preparation). Similar surveys off the west coast of Norway have enabled the base of the Quaternary to be identified on sparker profiles extending across the northern Norwegian Channel (Sellevol and Sundvor, 1974), and in the Skagerrak (Sellevol and Aalstad, 1971), thus enabling isopachs of the total Quaternary thickness to be constructed in these two areas. In addition, the Sub-Pleistocene Geological map of the British Isles (Institute of Geological Sciences, 1972) shows contours on the base of the Pleistocene down to several hundred metres off the east coast of Britain, and Quaternary thicknesses in certain individual wells or in field areas have been published (e.g. Howitt, 1974). In The Netherlands and North Germany, the thickness, nature, and distribution of the Quaternary is well known from extensive borings (Keizer and Letsch, 1963; De Jong, 1967; Zagwijn, 1974) and this has provided a valuable insight into the sedimentary and tectonic patterns that might be anticipated further north within the North Sea Basin.

These varied data, together with some additional seismic information concerning the Quaternary thickness in 188 wells throughout the North Sea, has recently been compiled by Caston (1977a) to provide a first Quaternary isopach map of the whole North Sea area (Fig. 7.2).

Information concerning the thickness in these wells has been provided by the Operators concerned; no re-examination of the basic data has been undertaken. Identification of the base Quaternary is dependent upon a number of factors, including, in the southern North Sea, lithologic changes and their associated electric log responses. The principal criterion has, however, been palaeontological. In the southern North Sea identification of the Pleistocene and, where present, the Pliocene, has commonly been based upon assemblages of benthonic foraminifera and, to a lesser extent, ostracods, using established zonations defined primarily in The Netherlands. Extrapolation seaward of this palaeontological zonation has provided satisfactory results; in addition, lithological changes, especially in the south and west where Pleistocene sediments overstep the Tertiary and rest directly upon Mesozoic strata, enable positive determinations to be made.

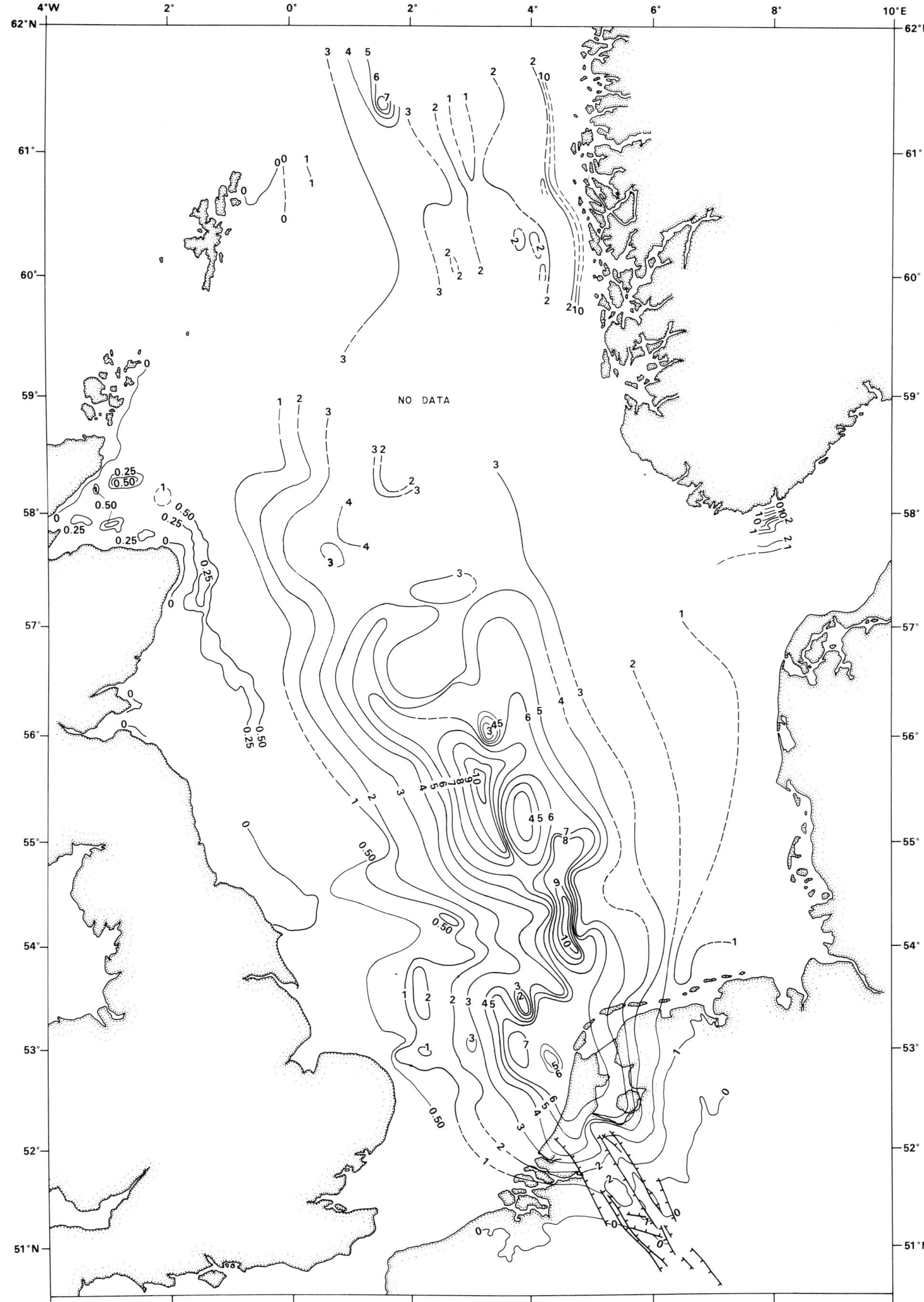

Fig. 7.2. Isopach map of total Quaternary thickness in North Sea based primarily upon data from 188 wells. Isopachs at intervals of 100 m, except locally off the U.K. coast where the 25 m and 50 m contours have been added. Sediment thicknesses within The Netherlands and location of faults active in the Quaternary taken from published sources given in text. (From Caston, 1977a).

Moving northwards into the deeper Tertiary basin, fine-grained sedimentation appears to have continued virtually unchanged through the Pliocene and across the Pliocene–Pleistocene boundary, providing few obvious lithological markers (Caston, 1977b). Distinct faunal assemblages are also less easy to identify, possibly because of the earlier influence of cold-water conditions with increasing latitude. As a consequence, many companies tend not to distinguish the Pleistocene from Pliocene. These problems are aggravated by the fact that, for operational reasons, samples are less frequently collected in those initial stages of drilling a well which commonly includes the Quaternary section, and also by the poor quality of cuttings in unconsolidated sediments. Evidence from the northern part of the North Sea, especially to the north of 58° N, is therefore scanty and insufficient for our purpose; the few isopachs based on well data in this area should thus be treated with considerable caution.

The resultant isopach map shows that maximum Quaternary thickness occurs within a linear trough trending north-north-west for over 750 km down the centre of the North Sea, the southern limit of which has been well documented in The Netherlands. This linear trough corresponds approximately in position to the underlying Central Graben, which has been identified as a major structural feature of Mesozoic and Tertiary age (Kent, 1975), and contains at least two closed basins in which maximum Quaternary thickness exceeds 1,000 m. Off the Dutch coast an area centred at 53° N 4° E containing in excess of 600 m of Quaternary sediments lies to the west of the principal depositional trough. This appears to be a direct extension of the zone of Quaternary graben and horsts recognised in The Netherlands, which may represent reactivation of older north-west trending faults shown on Kent's map. Further north, although there is inadequate data, there appears to be relatively little Quaternary immediately off the east coast of England, and virtually none at all north of about 54° 30′, which corresponds to the mid-North Sea High. What little evidence we have suggests that to the east of the Orkneys and Shetlands, Quaternary cover is also fairly thin, perhaps less than 100 m or so out at least as far as 0°, and this corresponds to the East Shetlands Platform. There is a small embayment, towards the Moray Firth, perhaps associated with the Moray Firth Basin. On the eastern side of the North Sea there is little information, but certainly off the Danish and German coasts there appears to be only a fairly gradual westward increase in Quaternary thickness as far as a line that corresponds roughly to the edge of the Central Graben. In the central northern North Sea the line of the Viking Graben is followed by a poorly-defined linear trough in which the Quaternary is at least 300 m thick, and in one well exceeds 700 m. Moving eastwards from this, there is a relatively narrow north–south high capped by less than 100 m of Quaternary, which may have some connection with the Vestland–Stavanger Ridge on Kent's map, and finally another linear Quaternary deep, which could correspond to the northern part of the Bergen Basin, although in this region sedimentary thicknesses undoubtedly reflect the morphology of the Norwegian Channel.

It is suggested that the relatively close correlation between the distribution of Quaternary sediments and the underlying Mesozoic–Tertiary tectonic features implies that Quaternary sedimentation in the North Sea has been strongly influenced by these older structural patterns.

Regional distribution and nature of the Quaternary sediments

Quaternary sediments of the North Sea will be described on a regional basis (Fig. 7.3). Selection of these regions is somewhat arbitrary and is dependent largely upon the coverage of data, which in certain cases is tied to the limits of territorial waters. In other instances, regional subdivisions are related to major relief features which reflect past erosional and depositional events and/or influence present depositional patterns.

Distribution of sea-floor sediments south of 55° 20′ N is best shown by Jarke's map on a scale of 1 : 1 M (1956), which incorporates earlier work, including that by Borley(1923), but is based primarily upon new samples, the positions of which are shown by Volpel (1959). This map has been simplified and used as the basis for surficial sediment maps covering parts of the U.K., Dutch and German sectors.

In the course of the following regional descriptions, reference will be made to various data points throughout the North Sea. These data points, principally oil and gas fields (see Vol. II, Fig. 18.1) and individual borehole locations are illustrated on Fig. 7.3 For convenience and ease of reference within this chapter, individual boreholes for which lithological sections are illustrated are lettered sequentially in the order in which they are discussed. The letters are unique to this chapter and should not be confused with the original borehole designations.

U.K. sector south of Dogger Bank

Despite its proximity to the United Kingdom, few

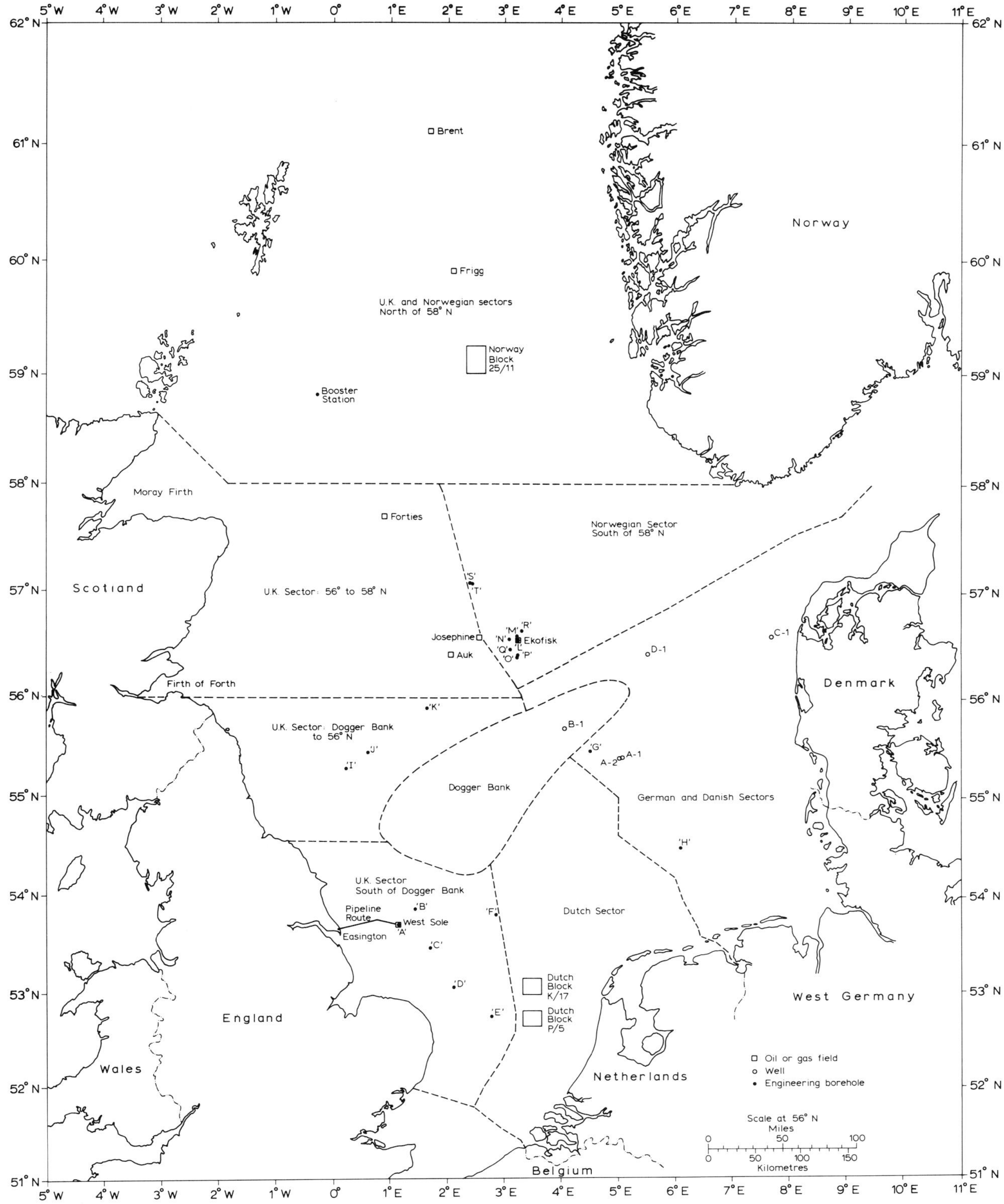

Fig. 7.3. Regional subdivisions of North Sea referred to in text, together with oil and gas fields, individual borehole locations, and other data points.

offshore geological studies have been published covering this area with the exception of those concerned with sediment transport. This chapter provides some additional data derived from offshore engineering studies, but there is still an outstanding requirement for a comprehensive survey comparable with that carried out in Dutch waters, or off the east coast of Scotland between 56° and 58° N.

Distribution of sediments on the sea bed is best shown by Jarke's (1956) map, upon which Fig. 7.4 has been based. The pattern consists essentially of a thin, locally discontinuous, superficial layer through which protrude the underlying Pleistocene deposits, the patchiness of the veneer being a function of the considerable transporting and distributive energy of tidal currents in this part of the southern North Sea. The

Fig. 7.4. Sea-floor sediments in U.K. sector south of 55°20′N, simplified from Jarke (1956). *Silt* category includes Jarke's "Mehlsand", "Mehlsand mit > 20% Feinsand", "Mehlsand mit > 20% Schluff und Ton", and "Schluff und Ton mit > 20% Mehlsand". *Fine sand* category includes Jarke's "Feinsand", "Feinsand mit > 20% Mittelsand" and "Feinsand mit > 20% Mehlsand". *Medium sand* category includes Jarke's "Mittelsand" and "Mittelsand mit > 20% Feinsand". *Coarse sand* includes Jarke's "Grobsand", "Grobsand mit > 20% Mittelsand", and "Mittelsand mit > 20% Grobsand". *Hardbottom and gravel* includes Jarke's "Felsboden" and "Kiesflachen".

greater part of the area is floored with sand. To the east of 2° E this decreases regularly in grain size towards the north from coarse sand at the southern end of the Southern Bight to very fine sand east of Indefatigable Bank. This forms part of a pattern of sediment distribution related to the diminishing power of tidal currents which is seen more extensively in the Dutch sector. The linear sandbanks of the Norfolk Banks and Outer Dowsing regions are clearly defined, being covered by sand of a finer grade than that on the surrounding sea floor. Further north, the Outer Silver Pit is partially infilled by clayey sediments of Holocene age which are up to 20 m in thickness (Zagwijn and Veenstra, 1966).

Off the east coast of the U.K., north of the Wash, a wide zone up to 55 km in width consists of hard ground covered in stones and gravel. Smaller outcrops of hard ground occur on the south-western flank of Dogger Bank, in the area of Indefatigable Bank centred at 2° E, and to the west of Haddock Bank at 1° 10′ E. Patches of gravel and isolated stones additionally lie on, or penetrate through, the sand cover throughout the greater part of the sector, and are especially common to the south of the Outer Silver Pit, and in part of the Southern Bight centred at 2° 30′ E. In this latter area the gravels consist almost exclusively of flint lying upon Tertiary clays, and are believed to represent the basal deposit of the Flandrian transgression (Veenstra, 1969). Gravels from the area of Inner Well Bank (south of Dogger Bank) and off the U.K. east coast betweeen 53° and 55° N are characterised by a sandstone–limestone–porphyry assemblage considered by Veenstra to represent a lag deposit overlying Weichselian age till of British origin.

The underlying Pleistocene deposits have been sampled and/or geophysically surveyed at various levels of detail in a number of areas. Between Dogger Bank and the Yorkshire coast, boomer surveys by Dingle (1965) revealed that Pleistocene deposits are virtually absent, with recent sands sitting directly upon Mesozoic strata. To the south of this area the sea floor, however, consists of a virtually unbroken Pleistocene mantle (Donovan and Dingle, 1965). Detailed work in the area of the Inner Silver Pit (Donovan, 1965; 1973) showed that the surrounding floor is composed of boulder clay, most samples being identical with the Purple Till of the east Yorkshire coast which dates from the Weichselian (Catt and Penny, 1966), although samples of Basement (Saalian) Till have also been recorded (Hull, in discussion of Donavan, 1973). Borings along a pipeline route survey from the coast at Easington to the West Sole Field (Fig. 7.3) in Block 48/6 have shown that stony boulder clay is continuous along the route, and is exposed on the sea floor along much of its length, being covered by superficial sands only to the east of 0° 50′ E. For the first four to five kilometres offshore the till consists of a blue-grey clay with chalk pebbles, identified as Drab Till (Weichselian) by Donovan (1973). "Grey clay", which might be Basement Till, has been reported from a depth of 4.7 m in a borehole 9 km from the coast. The greater part of the route is, however, floored by stony and locally sandy red-brown clay considered by Donovan to be Purple Till. The presence of pockets of coarser sediment (gravel, shingle or sand) within this upper layer has been proved by coring. There is some evidence that this Purple Till is no more than a few metres in thickness and overlies a more competent, less sandy, clay. Soil borings undertaken for platform foundations in the West Sole Field revealed a variable Quaternary thickness, averaging about 17 m, over Lias clay (Fox et al., 1970; Donovan, 1973). The section in Borehole "J" consists of 0.5 m superficial sand above stiff to hard boulder clay containing silt and sand seams, stones and cobbles. This has been interpreted by Donovan to consist of approximately 2.1 m Drab Till overlying Basement Till. At the base of the Quaternary succession is some 3–5 m of silty sand with gravel and clay seams. This is probably an outwash deposit (Borehole "A" in Fig. 7.5). To the east of the West Sole Field, samples of the Purple, Drab, and Basement Tills have been recorded from the vicinity of the Sole Pit, centred at 53° 39′ N 1° 33′ E (Donovan, 1973).

Further south, the underlying chalk is exposed in the walls and floor of the Inner Silver Pit, except at the southern end where a series of marine silts and clays dating from the Hoxnian (Holstein or Eemian) interglacial (Fisher et al., 1969) immediately underlies the till. Immediately to the south of the Pit is a small outcrop of Drab Till.

Under Smith's Knoll (52° 50′ N 02° 15′ E) the presence of a nearly flat surface of Eemian clay was reported by Brouwer (1964). This extends northwards under Well Bank (53° 05′ N 02° 15′) (Houbolt, 1968). Clays of a similar age (Brown Bank Bed) are present at equivalent latitudes in the Dutch sector, about 50 km away to the east (Oele, 1971b).

A number of shallow boreholes have been drilled in the south-western North Sea in connection with the search for hydrocarbons. Regrettably, few detailed sedimentological or palaeontological analyses have been undertaken on these data, which were obtained primarily for engineering purposes, and it must be stressed that descriptions are taken direct from contractors' reports, no re-examination of the samples having been undertaken.

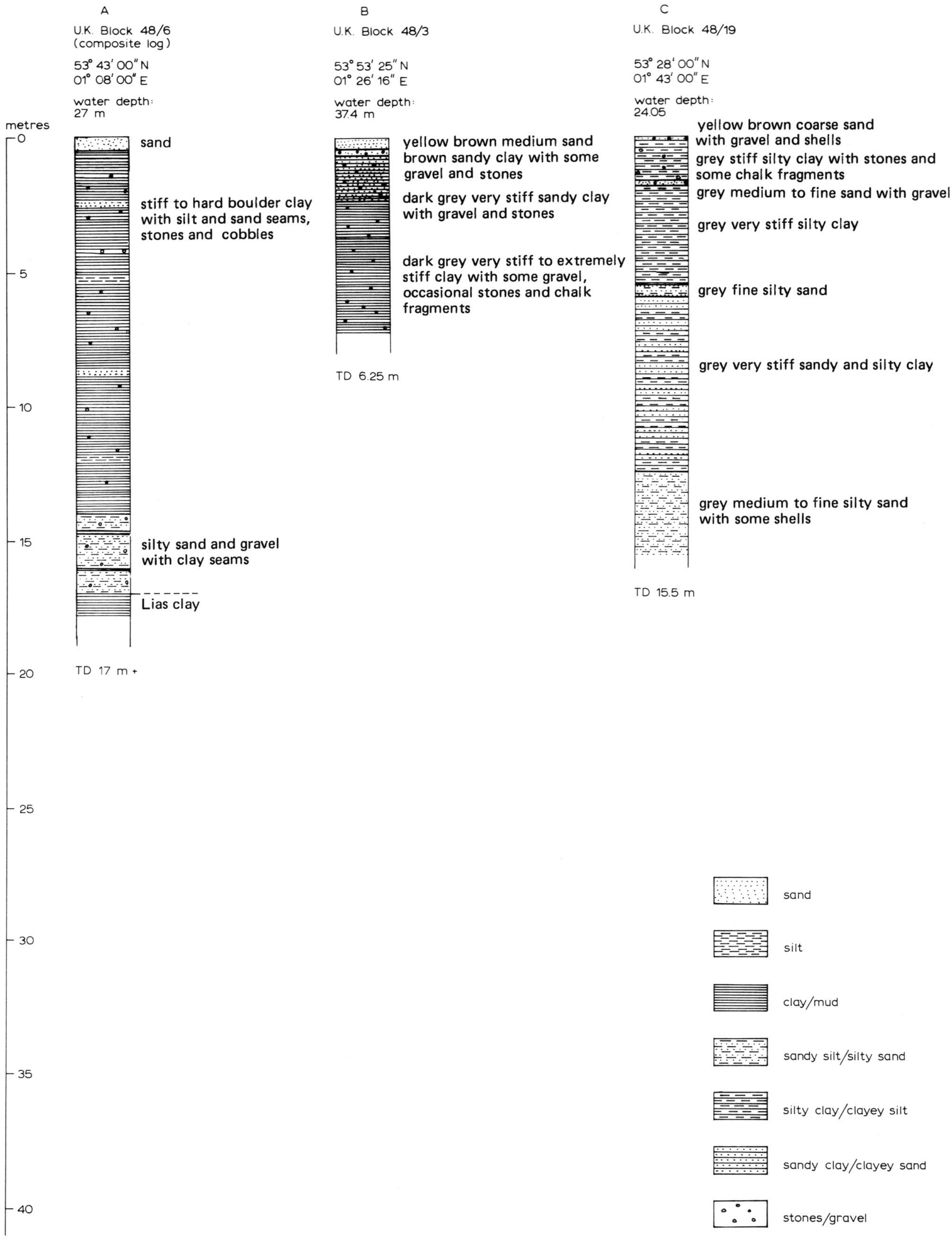

Fig. 7.5. Lithological logs of engineering boreholes "A"–"F" inclusive drilled in the U.K. sector, south of the Dogger Bank. For locations see Fig. 7.3. Borehole "A" is composite log of a number of boreholes drilled in Block 48/6. Borehole "D" is similarly a composite log of three boreholes drilled in Block 49/26.

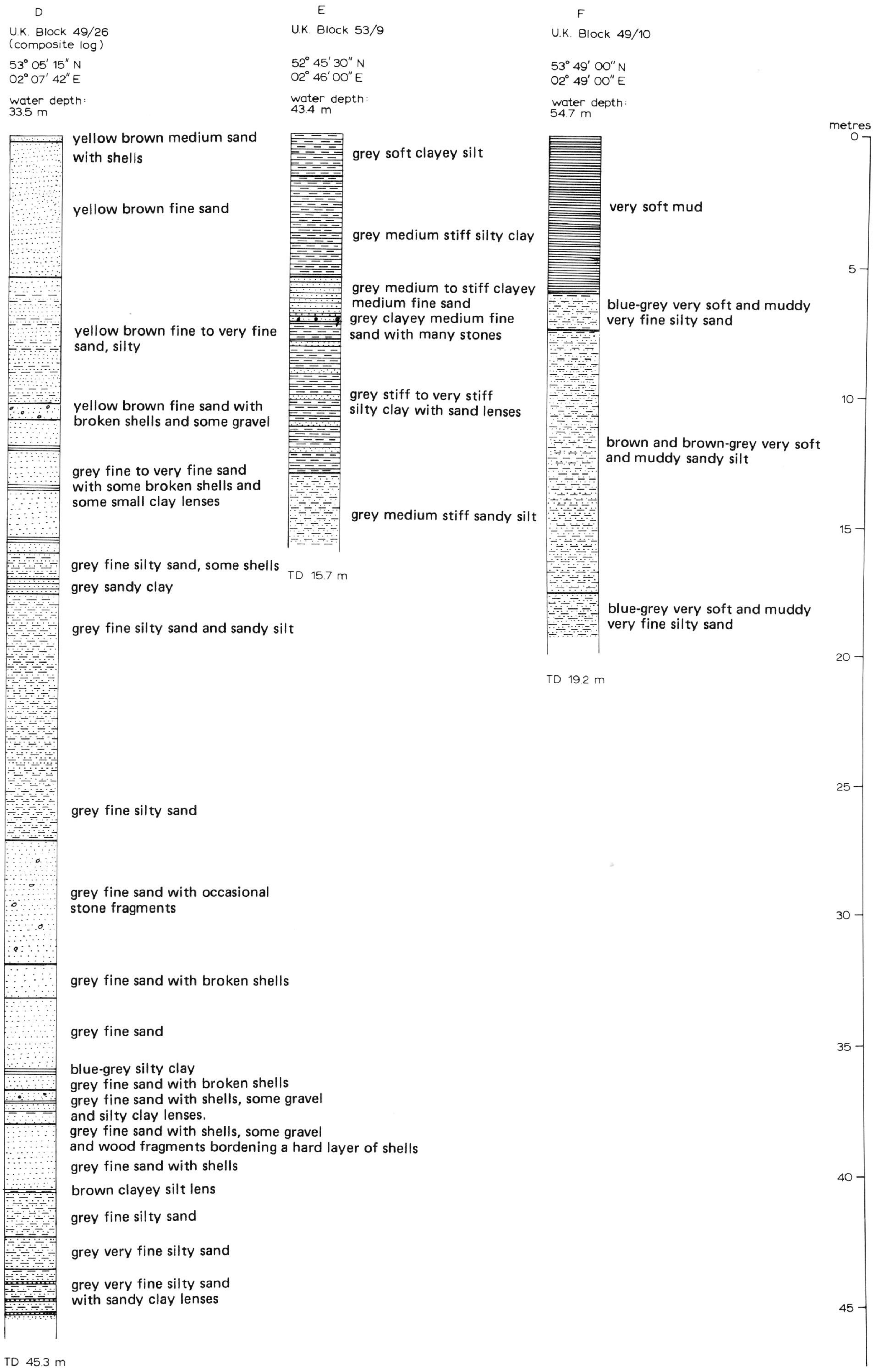
D
U.K. Block 49/26
(composite log)
53° 05′ 15″ N
02° 07′ 42″ E
water depth:
33.5 m
yellow brown medium sand
with shells
yellow brown fine sand
yellow brown fine to very fine
sand, silty
yellow brown fine sand with
broken shells and some gravel
grey fine to very fine sand
with some broken shells and
some small clay lenses
grey fine silty sand, some shells
grey sandy clay
grey fine silty sand and sandy silt
grey fine silty sand
grey fine sand with occasional
stone fragments
grey fine sand with broken shells
grey fine sand
blue-grey silty clay
grey fine sand with broken shells
grey fine sand with shells, some gravel
and silty clay lenses.
grey fine sand with shells, some gravel
and wood fragments bordening a hard layer of shells
grey fine sand with shells
brown clayey silt lens
grey fine silty sand
grey very fine silty sand
grey very fine silty sand
with sandy clay lenses
TD 45.3 m
E
U.K. Block 53/9
52° 45′ 30″ N
02° 46′ 00″ E
water depth:
43.4 m
grey soft clayey silt
grey medium stiff silty clay
grey medium to stiff clayey
medium fine sand
grey clayey medium fine
sand with many stones
grey stiff to very stiff
silty clay with sand lenses
grey medium stiff sandy silt
TD 15.7 m
F
U.K. Block 49/10
53° 49′ 00″ N
02° 49′ 00″ E
water depth:
54.7 m
very soft mud
blue-grey very soft and muddy
very fine silty sand
brown and brown-grey very soft
and muddy sandy silt
blue-grey very soft and muddy
very fine silty sand
TD 19.2 m
metres
0
5
10
15
20
25
30
35
40
45

Information from a series of holes provides a valuable north-west–south-east transect across this area (Fig. 7.5). Borehole "B" was drilled to a penetration of 6.2 m in Block 48/3 (53°53′25″N 01°26′16″E), some 23 km to the north-east of West Sole. This passed through 0.4 m of recent sand, 0.2 m of brown sandy clay with gravel and stones, and then into a dark grey chalky boulder clay. Moving to the south-east, another borehole ("C") was drilled in Block 48/19 (53°28′N 01°43′E). This recorded 12.4 m of the grey chalky boulder clay overlying a grey medium to fine silty sand, which continued to the bottom of the hole at 15.5 m. A composite log of three holes drilled in the immediate vicinity of a location (53°05′15″N 02°07′42″E) in Block 49/26 (Borehole "D") provided a total of 45.3 m penetration which consisted almost wholly of sand. The uppermost 10.8 m is composed of yellow-brown medium to fine sand with a layer of gravel at the base, which solely on lithology is assumed by the present author to be of Holocene age. This is underlain by grey fine sand or silty sand with rare grey clay or sandy clay lenses which persist to the bottom of the hole. Occasional stone fragments or gravel occur below 29 m, and at a 37.8 m "wood fragments bordering a hard layer of shells" were recorded. A final borehole ("E") in this transect was drilled at 52°45′30″N, 02°46′E in Block 53/9. This penetrated only 15.7 m, but was more variable in composition than the others. A layer of soft clayey silt, 1.6 m in thickness and probably of Holocene age, overlies grey medium-stiff clay, which has intercalations up to 1.7 m thick of grey medium-fine sand, locally stony at a depth of 7.2 m.

Another borehole was drilled in Block 49/10 at 53°49′N, 02°49′E. This was located in a water depth of 54.7 m adjacent to the Botney Cut, one of the deeper troughs extending south from the eastern end of the Outer Silver Pit (Stocks, 1955). The log (Borehole "F" in Fig. 7.5) recorded a total penetration of 19.2 m, of which the uppermost 6 m consists of very soft mud, overlying blue-grey soft and muddy very fine silty sand, and brown to brown-grey very soft muddy sand and silt.

It is desirable that these discontinuous pieces of evidence should be linked together to form a coherent picture. In view of the lack of palaeontological control and informative seismic lines, this is at present not possible. Some deductions may however be made concerning the possible relationship and ages of the deposits observed.

According to Donovan (1973) the greater part of the till in borehole "A" is of Saalian age, only the uppermost 2.1 m dating from the Weichselian. On the basis of lithological similarity, a Saalian age may also be proposed for the grey chalky boulder clay seen in boreholes "B" and "C". On this assumption, the fine-grained grey silty sands with clay lenses which underlie the boulder clay in the latter hole, and which comprise the total pre-Holocene section in borehole "D", must be of pre-Saalian, presumably Holsteinian age. However, borehole "D" is located within an area in which Eemian clays are exposed at the sea bed (Brouwer, 1964; Houbolt, 1968), and adjacent to that part of the Dutch sector where Eemian fresh-water clays overlie marine sands of the same age (Oele, 1971b). By analogy, the sands in at least the upper part of borehole "D" are also Eemian, and thus the boulder clay seen in boreholes "B" and "C" is Weichselian.

It is clear, both from this comparison with the Dutch sector and from the discussion after the presentation of Donovan's (1973) paper, that some doubt still exists concerning the age of the deposits in this south-western part of the North Sea.

Further to the north and east in Block 49/10, the sediments found in borehole "F" possess none of the superficial characteristics suggestive of a glacial origin. The hole is located in slightly deeper water, close to the southeastern end of the Outer Silver Pit. It is therefore reasonable to assume that at least the very soft mud found in the uppermost 6 m of the borehole is of Holocene age. By analogy with the adjacent part of the Dutch sector (Oele, 1969), the underlying sandy silt and silty sand could either date from the Boreal period of the early Holocene (c.f. the Elbow deposits), or possibly be of Weichselian age and comparable with the Twente Formation.

Dutch sector

The Netherlands State Geological Survey commenced offshore surveying in the Dutch sector during the late 1960's, and this area to the south and south-east of the Dogger Bank is now among the best documented of any in the North Sea (Oele, 1969, 1971a, 1971b). The geological history of at least the southern part of the area under review was similar to that of the western Netherlands during the late Tertiary to Saalian period, and the stratigraphy is summarised in Fig. 7.6, taken from Oele (1971b). Sedimentation of late Tertiary marine fine sands and clays continued through to the early Pleistocene, the North Sea Basin continuing to subside conttemporaneously with uplift of the northern flank of the Brabant Massif in the south. The marine deposits were later covered by fluviatile fine-grained sands and clays, probably during the late Tiglian, and their deposition continued until the end of the Lower Pleistocene.

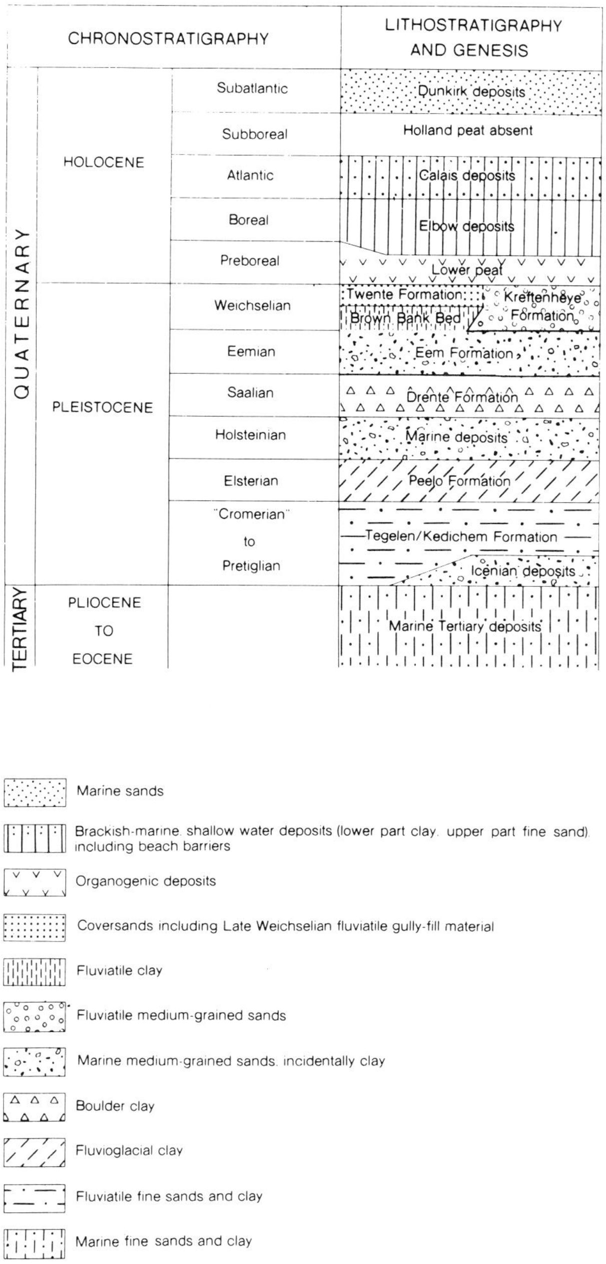

Fig. 7.6. Litho- and chronostratigraphy of Quaternary deposits in Dutch sector of the North Sea. (From Oele, 1971b.)

During the Elsterian glaciation, fluvioglacial clays up to 30 m thick (Peelo Formation) derived from Scandinavia were deposited over the greater part of the Dutch sector as far south as 53°10′N. Sediments dating from the subsequent Holsteinian interglacial have been recovered in borings at approximately 52°45′N and 53°05′N.

The renewed extension of the glaciers during the Saalian glaciation led to a land-ice cover over the greater part of the Dutch sector. The glaciers altered the topography, building ice-pushed and morainic ridges as far south as 52°30′N. The ice was, however, responsible for the deposition of only local accumulations of boulder clay, primarily on the south-east flank of the Dogger Bank (Fig. 7.7). Following the retreat of the Saalian ice, the transgressive Eemian Sea covered the whole area, depositing extensive marine sands which to a certain extent consist of reworked fluviatile deposits. The sea level then fell again, permitting the formation of fresh-water deposits in an extremely wide area around the Brown Bank (Brown Bank Bed) which represents the topmost regressional bed of the Eem Formation (Fig. 7.8). During the last (Weichselian) glaciation Eemian clays in the southern portion of the Dutch sector were sub-aerially exposed. Off the mouths of the Rhine, sands of the Kreftenheye Formation, representing fluviatile deposits of the Rhine, date from this time, and elsewhere, over much of the Dutch sector north of 52°30′N, wind-blown coversands of the Twente Formation were deposited under periglacial conditions.

Oele's maps do not show the distribution of superficial sediments, though included on his cross-sections. According to Jarke's (1956) map (Fig. 7.9) the sea bed of the Dutch sector south of 55°20′N is floored almost exclusively by sand and silt, and there is a gradual decrease in grain size from south to north along a presumed sediment transport path (Stride, 1963) towards an area of silt lying between 54° and 55°N in relatively deep water (> 50 m) to the south-east of the Dogger Bank. Oele refers to this recent cover as the Young Seasand, and notes that it ranges in thickness from over 10 m in the Southern Bight to less than 2 m to the north of the Hook of Holland. It is included within the Dunkirk Deposits of Late Holocene age, and has been dated in the south-west part of The Netherlands as commencing with a post-ninth century transgression, although it is probably older offshore. South of 53°20′ the sands are coarse, yellowish-brown in colour, and Rhenish in origin. Further north, their fine grain size, resemblance to underlying Pleistocene coversands, and specific heavy mineral assemblage (Baak, 1936) suggests that they are a separate assemblage and have not been transported northwards from the Southern Bight.

Gravel is exposed on banks, almost certainly all of glacial origin, around the periphery of the sector. These include the Borkum Riffgrund, Dogger Bank, Cleaver Bank and Texel Riff. Composition of the gravels on the latter two banks has been summarised by Veenstra (1969) (Fig. 7.10). Cleaver Bank consists of morainic boulder clay covered by a discontinuous cover of sand with patches of gravel, which probably represents a lag deposit. The gravels consist of a porphyry–granite–flint–sandstone–limestone assemblage originating primarily from Scandinavia, suggesting that the deposit is of Saalian age. This is supported by Oele's (1971a) report

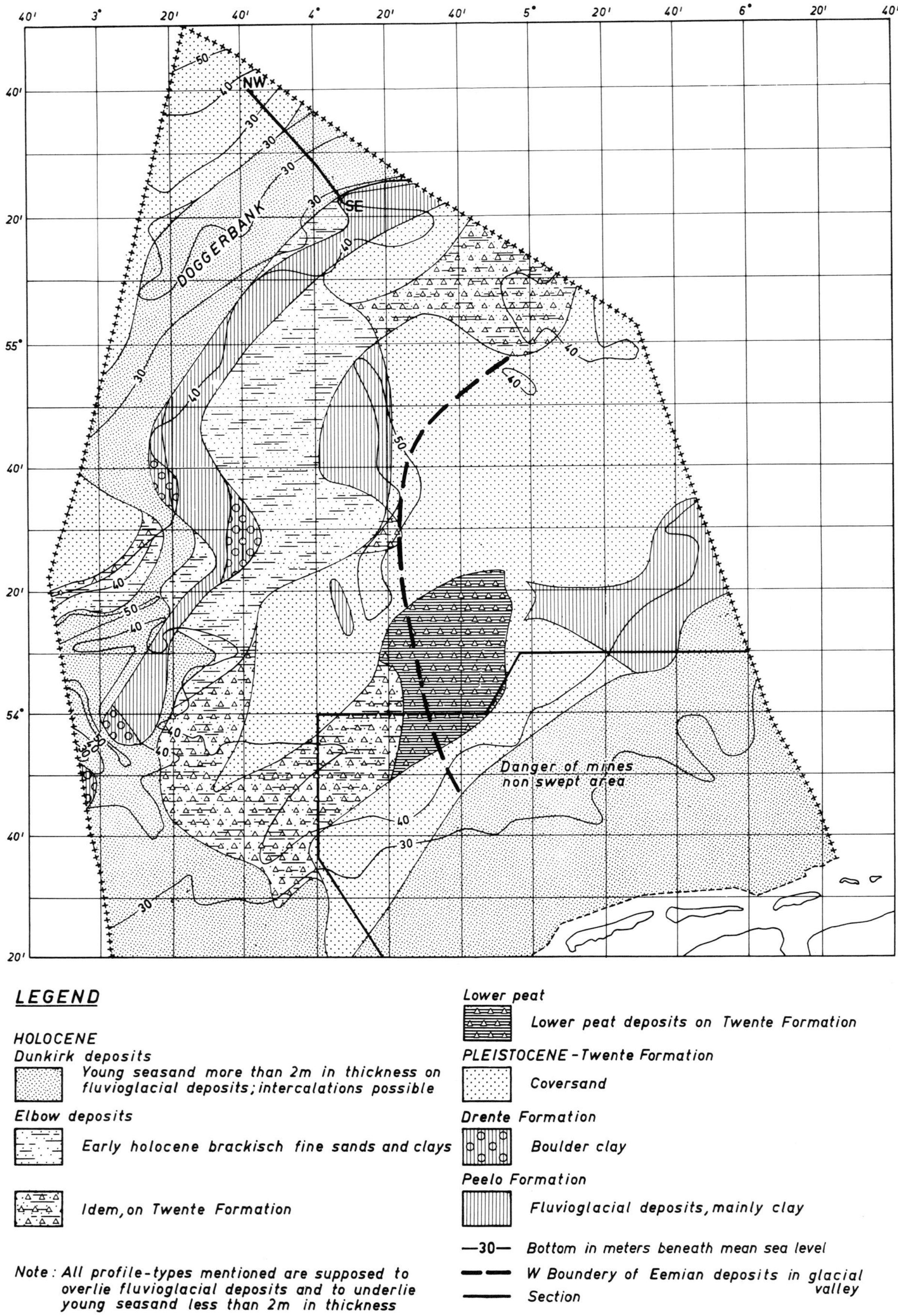

Fig. 7.7. Late Quaternary geology of the northern portion of the Dutch part of the North Sea. (From Oele, 1971a.)

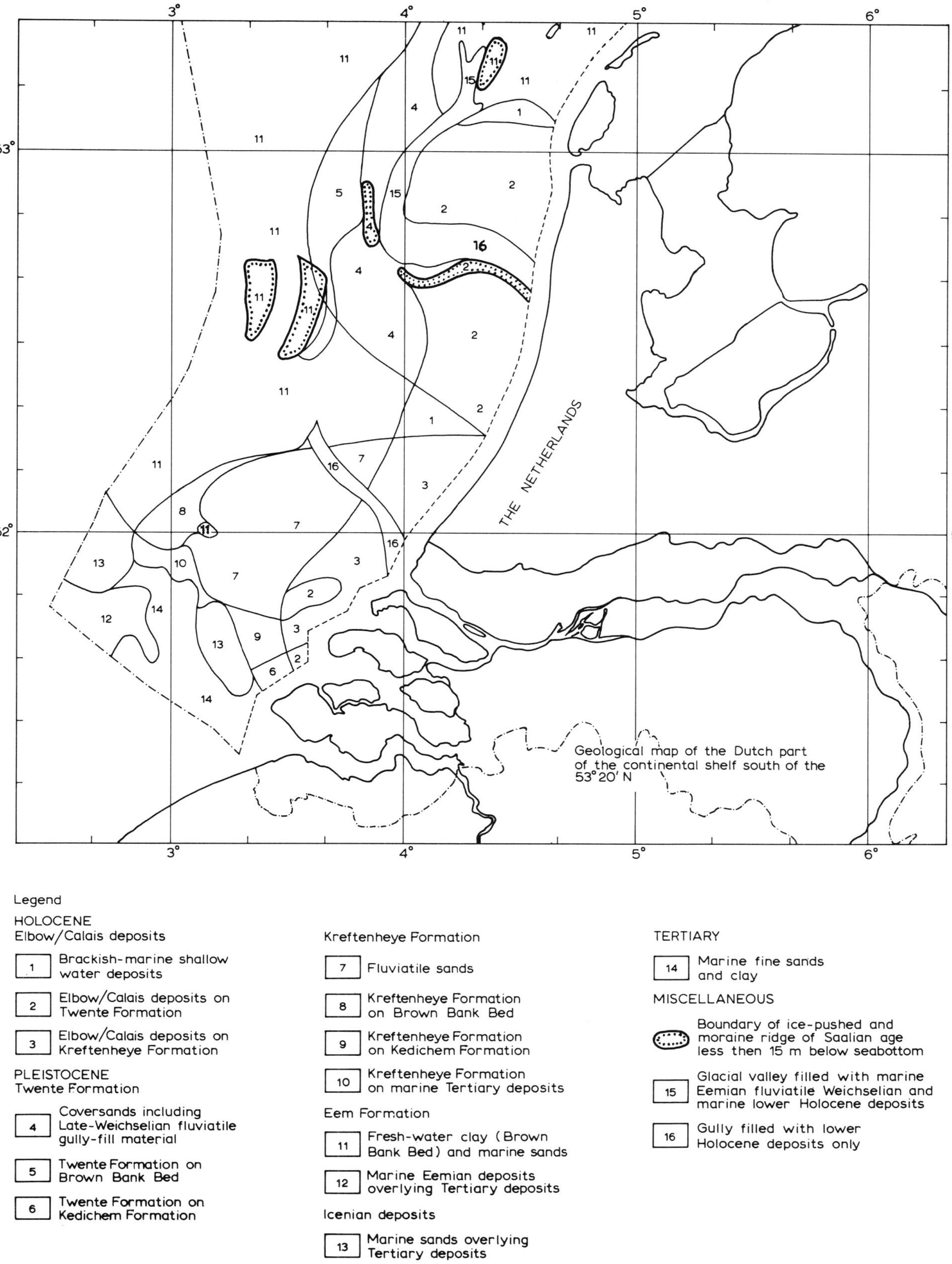

Fig. 7.8. Quaternary geology of the southern portion of the Dutch part of the North Sea. (Redrawn from Oele, 1971b.)

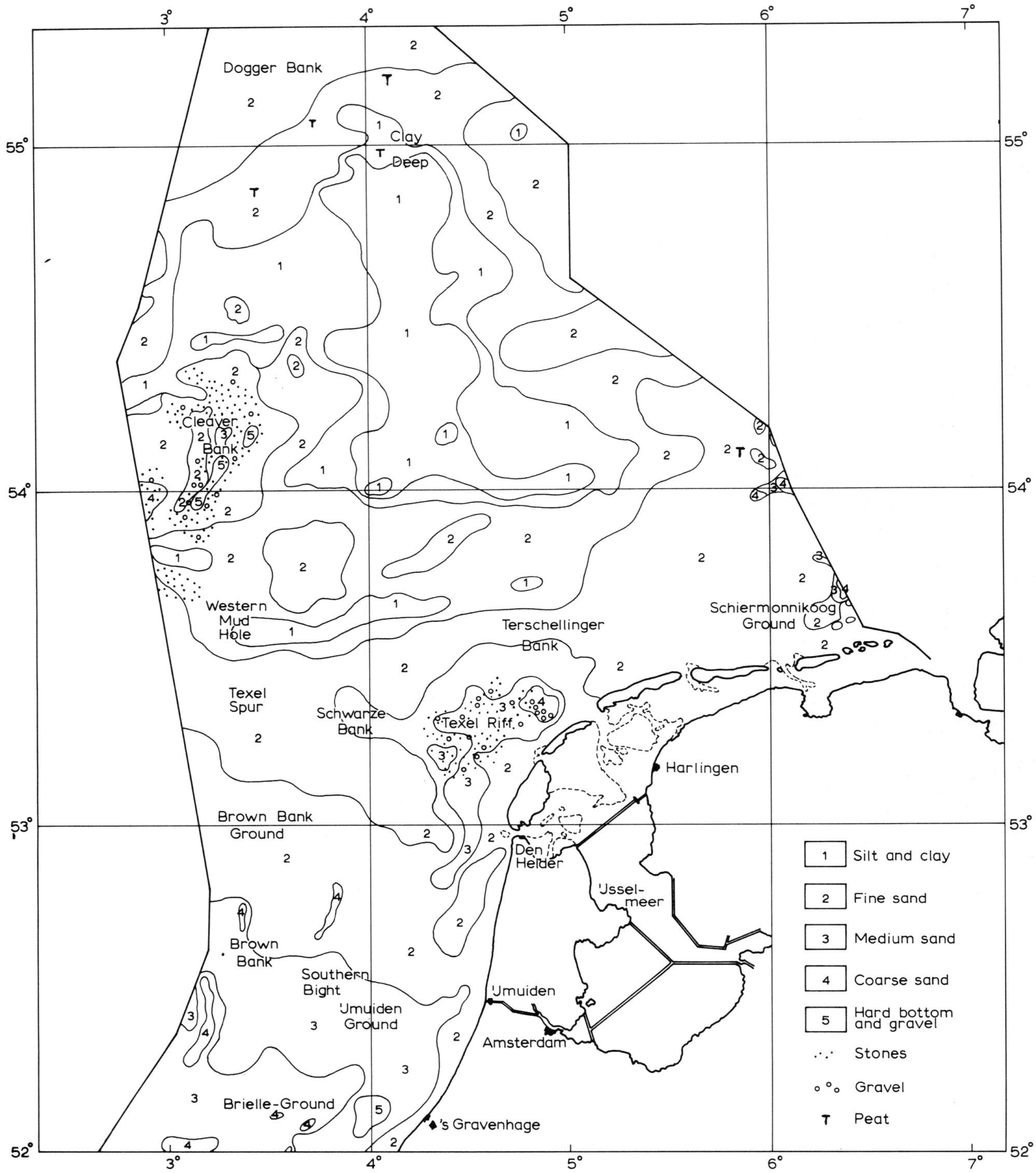

Fig. 7.9. Sea-floor sediments in Dutch sector south of 55°20′N, simplified from Jarke (1956). For legend, see caption for Fig. 7.4.

of Saalian boulder clay as far south as Botney Cut, just south-west of Cleaver Bank. Texel Riff similarly consists of boulder clay of Saalian age, and the gravels have a quartzite–flint–granite assemblage of Scandinavian derivation.

At several localities brackish-marine deposits (Elbow deposits) immediately underlie the Young Seasand. These consist of grey sands with occasional clay laminae and lumps of clay and peat. They overlie grey shelly sandy clays dated (on pollen analysis) as Boreal, which, on the basis of their ostracod content, are considered to have been deposited in a shallow marine environment or tidal flat. They are considered to be the lithological equivalent of the Velsen clay of western Holland, and

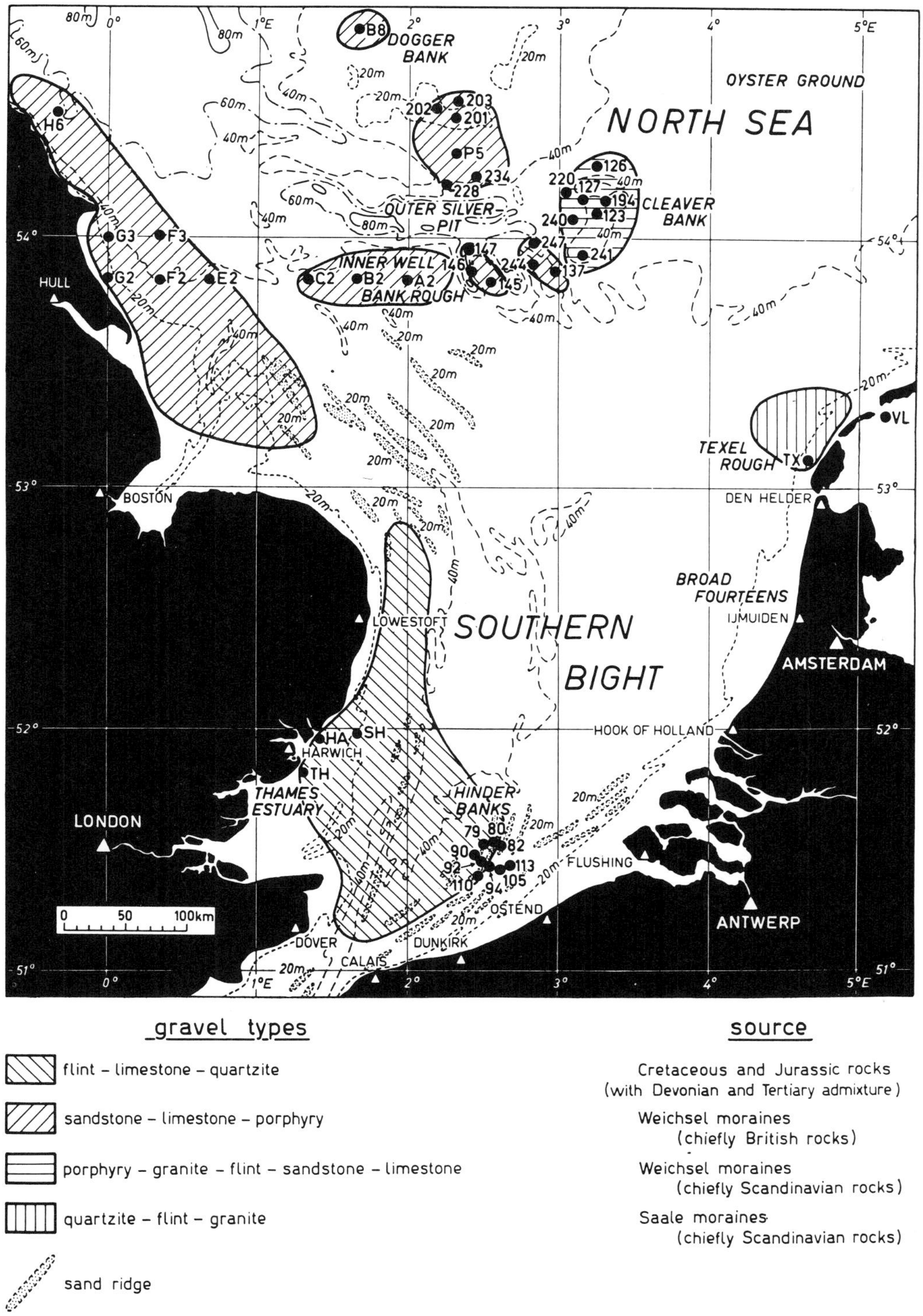

Fig. 7.10. Composition and extent of gravels on the floor of the southern North Sea. (From Veenstra, 1969.)

locally rest upon beds of peat, the Lower Peat, formed in Early Preboreal times by the rising sea level and the amelioration of the climate at the beginning of the Holocene. The peat deposit sampled by Oele north of 54°N lies at a depth of approximately 47 m below OD, which roughly fits Jelgersma's (1961) Early Holocene sea-level curve.

The shallow-water intertidal and immediately subtidal deposits of the Wadden Sea have been intensively studied, but are not considered here. The work is summarised by Van Straaten (1964).

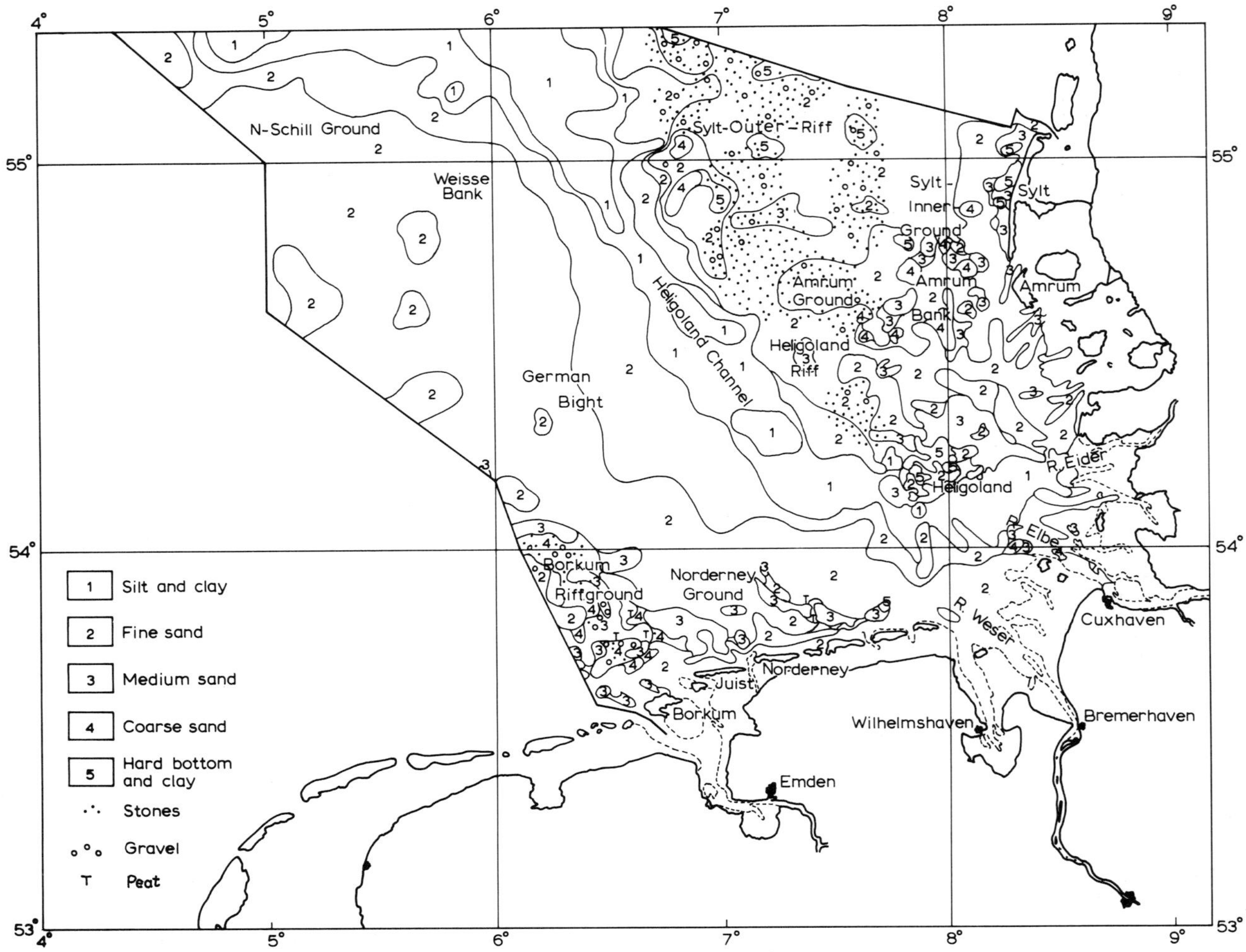

Fig. 7.11. Sea-floor sediments in German sector south of 55°20′N, simplified from Jarke (1956). For legend, see caption for Fig. 7.4.

German and Danish sectors

Little information is available concerning this area, which includes that part of the south-eastern North Sea known as the German Bight and its northern extension between the Danish coast and Dogger Bank. Jarke's (1956) map is still the principal source of information for the surface sediments to 55°20′N, together with his more detailed description of the area of the German Bight (Jarke, 1955). Further north, there are only the fisheries charts, and those maps constructed from them.

Jarke's work, which principally covers the German sector, shows that a clear distinction exists between the gravels and coarse to medium grained sands in the east, and fine sands and silts lying in relatively greater water depths further to the west, the abrupt dividing line trending in a north-westerly direction from Heligoland (Fig. 7.11). The finest sediments, of silt grade, in fact occur along a linear zone some 20 km wide corresponding to the Heligoland Channel, which lies immediately to the west of the gravel banks. North of 55°20′ the maps by Luders (1950) and Veenstra (1971) indicate that this area of mud widens to 50–70 km and trends more approximately north–south, terminating at approximately 56°45′N to the south of Little Fisher Bank, which corresponds approximately to the northern boundary of the Danish sector.

Jarke's map indicates considerable variability in the belt of coarse deposits to the east of Schleswig Holstein. Although most of the sea floor is covered by fine sand, many of the banks, including Amrum Bank, Heligoland Bank and Sylt Inner Ground are floored by hard ground, and gravels and stones cover the sea floor west of 8° E. There are also numerous patches of coarse and medium sand. The smaller scale maps by Luders (1950) and Veenstra (1971) show that these coarse, poorly sorted, deposits extend northwards in a belt 100–130 km wide immediately off the Danish coast, but are insufficiently detailed to show the extent of the small-scale variability. There is no reason however to doubt that the pattern seen off the German coast is not typical of that further north. In a well-known paper, Pratje (1951) proposed

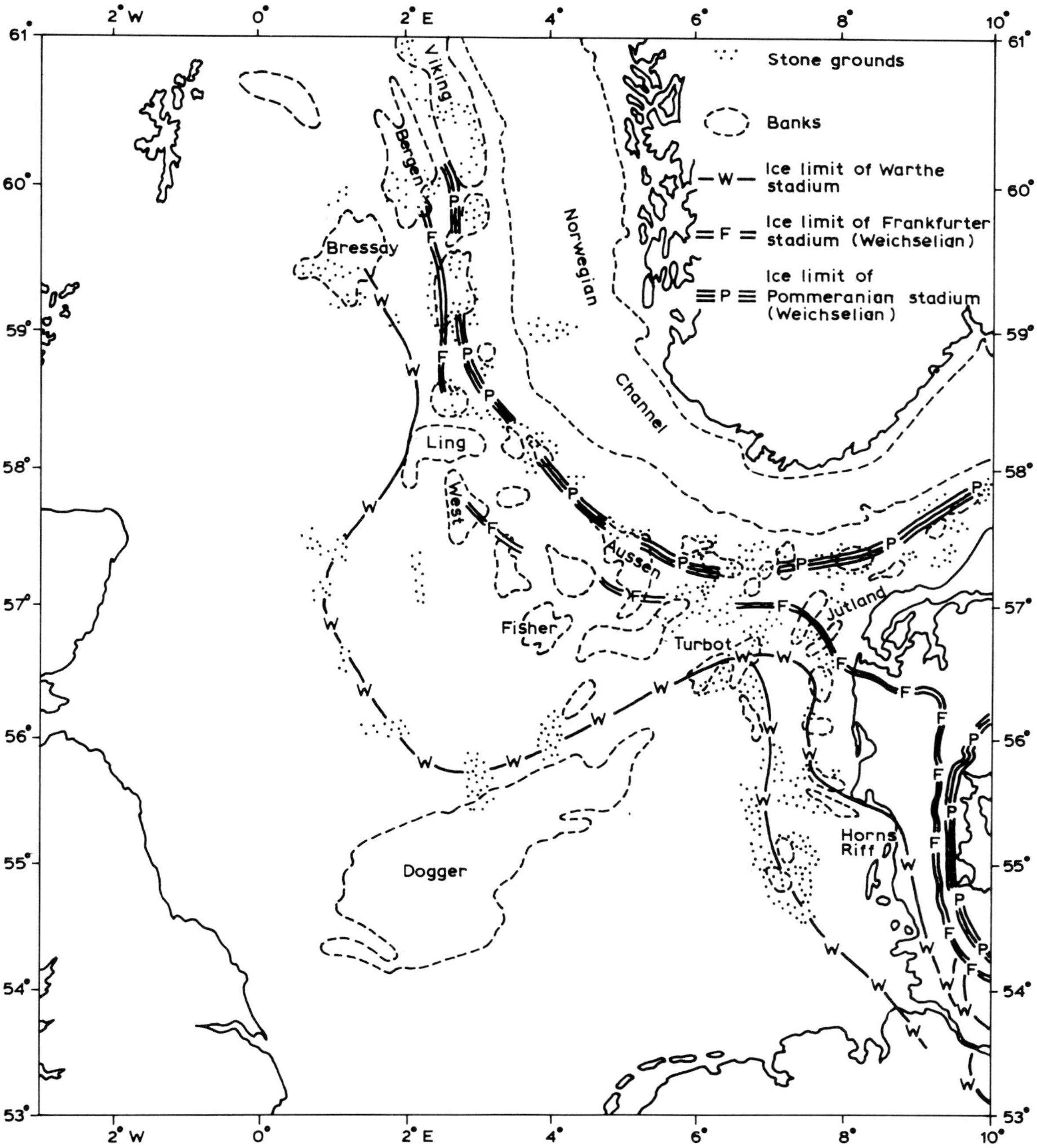

Fig. 7.12. Location of gravel-covered banks and proposed offshore extension of Warthe and Weichselian moraines. (From Pratje, 1951.)

that these sediments represented terminal moraines, marking the western, and in the north, the southern, extent of Warthe and Weichselian glacial advances. Pratje's map (Fig. 7.12) indicates that the deposits lying off the coast of Schleswig-Holstein form a moraine of Warthe (late-Saalian) age, the continuation of which has been identified on land between the Rivers Weser and Elbe. A second moraine of this approximate age has also been recognised in the north-west of Schleswig-Holstein. This is believed to extend into the sea at Horns Rift and to continue northward forming an inner belt of stony deposits off the Danish coast. On morphological evidence only, Pratje shows both moraines joining at Turbot Bank, at about 56°30′N, where they curve round to the west, lying approximately parallel to the northern boundary of the Danish sector. The terminal moraines of Weichselian age form the morphological axis of the Jutland peninsula, and are better preserved. To the south of Limfjord the moraines of the Frankfurt Stage curve westwards and extend offshore, where Pratje assumes that they form the core of Jutland Bank before extending westwards into the Norwegian sector.

Deposits of the Pomeranian stage, however, are believed to lie wholly offshore, forming the most northerly line of banks which comprise the southern edge of the Norwegian Channel.

In the southern part of the German Bight east of 6°30′E and south of 54°10′N, a more detailed map of the surface sediments has been constructed by Gadow and Schafer (1973) on the basis of samples taken at approximately 5 km intervals. This largely confirms Jarke's earlier map, showing that the area is dominantly covered by fine-grained sediments ranging from silt to fine sand. Two areas of medium to coarse sand with gravel, the Borkum Riffground and Norderney Ground, extend offshore from the north German coast in a north-westerly direction for 55 km and 35 km respectively. These may well represent other morainic accumulations. Gadow and Schafer's work was principally concerned with the heavy metal content of the clay fraction. Distribution of these metals, together with that of organic carbon, indicates that the principal source of fine grained sediments entering the area at present is provided by the major rivers.

The distribution of surface sediments up to and including sand-size on the floor of the southern North Sea has been related to the pattern of tidal current velocities. This general relationship undoubtedly exists, and there is clearly a relationship between changes in current velocity and sediment grade (see below). Results from engineering boreholes drilled, however, indicate that there is a considerable thickness of presumed Pleistocene-age sand in this area, and it is possible that much of the surficial sand represents only the reworked or partially reworked expression of the underlying deposits.

The most northerly hole, GP1 (Borehole "G"), was drilled at 55°29′41.73″N 04°29′56.10″E, some 18 km to the east of the Dutch sector at the north-eastern end of the Dogger Bank. It penetrated 131.7 m in a water depth of 31.7 m, passing almost exclusively through grey fine sand, locally silty or with shell fragments, clay seams and organic matter (Fig. 7.13). Two layers of very stiff grey clay were recorded between 9–31 m and 47–57.4 m. Within these layers occurred pockets of sand and organic material, and a varved structure was recorded within the lowest 2 m of the thicker layer. GP1A was drilled only a couple of kilometres away at 55°27′46.37″N 04°33′02.12″E in 32.6 m of water and terminated at a depth of 130.3 m. This also consisted almost wholly of grey fine sand with a single brown sandy and gravelly silt layer between 43.3 and 50.9 m. The sand is similar to that described from GP1 with the notable addition of some wood fragments from 31.8 to 37.2 m, and scattered gravel at 41.3 m and from 88.7 to 92.6 m. Boring GP2 (Borehole "H") was drilled further south at 54°30′34.6″N 06°02′07.31″E in 41.5 m of water. It passed through grey fine sand to total depth at 102.3 m, with the exception of a single thin stiff grey clay layer between 28.6 and 31 m. Locally the sand is medium to coarse with some gravel, as at 40 m, and pockets of organic matter and shell fragments are recorded (Fig. 7.13).

Geological results have been published from five of the first exploration wells to be drilled in the Danish sector (Rasmussen, 1974). Lithological descriptions of the Quaternary succession in these wells are, however, based upon ditch samples (cuttings), and are understandably limited, but in two wells the Quaternary has been subdivided on palynological evidence, and the results are thus of particular interest as they demonstrate the presence of sediments of Middle and Lower Pleistocene age. Of the five, wells C1 and D1 were sited on the morainic banks to the west of Jutland; wells A-1, A-2 and B-1 were drilled further to the south-west. Details are given in Table 7.1. Information concerning the thickness of Quaternary, though only approximate and limited to these five locations, suggests a gradual decrease from a maximum of 686 m at Borehole B1 in the west to a minimum of 91 m in the east.

Wells A-1 and A-2 were drilled within a couple of hundred metres of each other. The lithologies recorded were broadly similar: 0–4 m sand; 4–150 m clay and sand with lignite; 150–220 m grey silty clay and some sand with lignite and mollusc shells in places; 220–

TABLE 7.1

Details of the first five Danish exploration wells, taken from Rasmussen (1974)

Well No.	A-1	A-2	B-1	C-1	D-1
Latitude	55°24′17″N	55°24′23″N	55°42′32″N	56°36′41.5″N	56°25′29.7″N
Longitude	05°03′42″E	05°03′36.5″E	04°02′09″E	07°40.00″E	05°31′52.2″E
Water depth	44.5 m	44 m	41 m	27 m	49 m
Quaternary thickness	434 m	386 m	ca 686 m	140 m*	316 m**

* Revised figure of 91 m used by Caston (1977a).
**Revised figure of 207 m used by Caston (1977a).

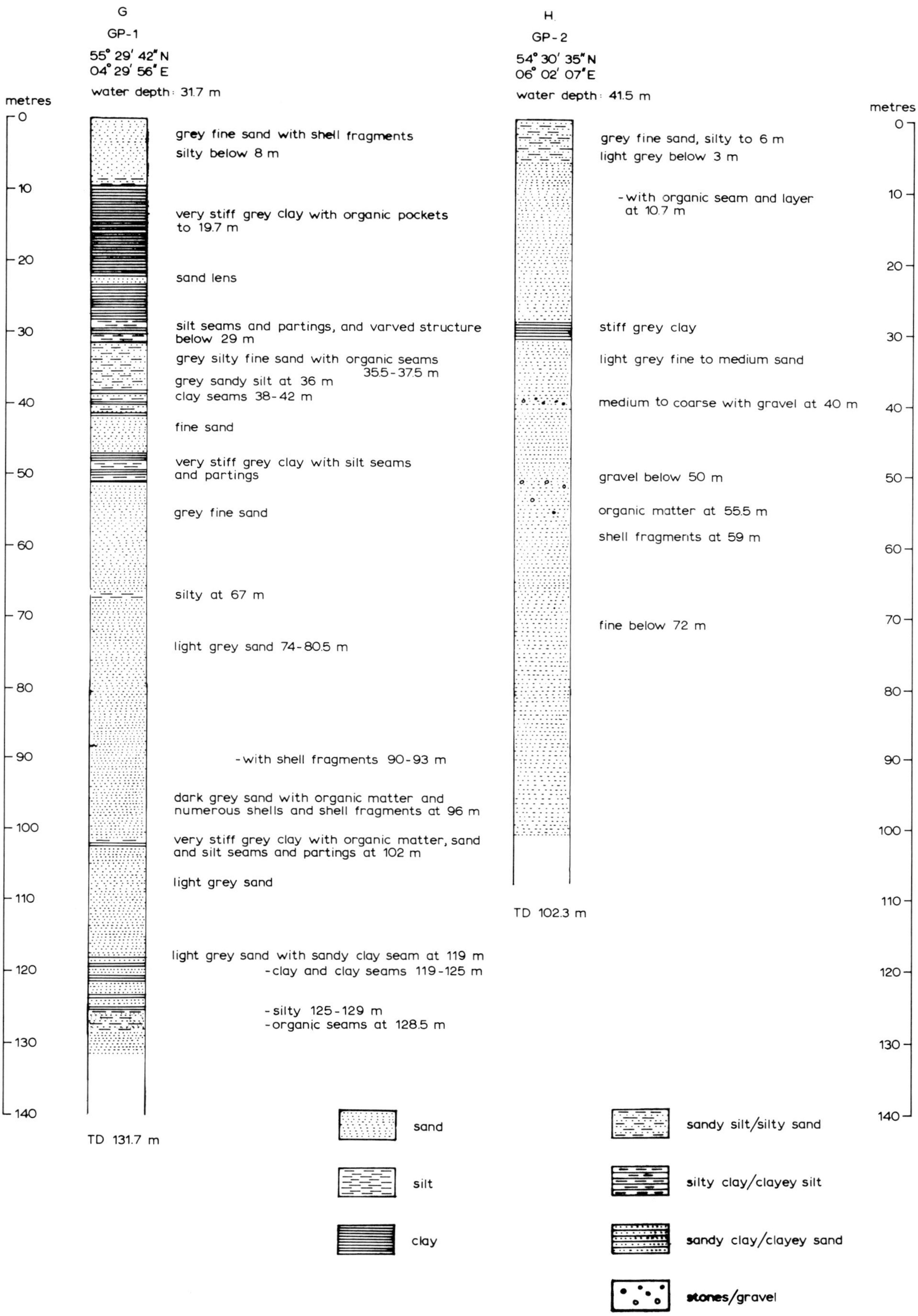

Fig. 7.13. Lithological logs of engineering boreholes "G" and "H" inclusive drilled in the German sector. For locations see Fig. 7.3.

330 m light grey clay with beds of sand, and from 330 to the base of the Quaternary, grey clay with beds of silt and sand and abundant molluscan shells. The greater part of the sequence apparently consists of marine strata which are often very shelly. One of the commonly occurring molluscs is the nuculid *Acila cobboldiae* well known from the Icenian of The Netherlands. Confirmation of an Icenian age for the section below ca. 150 m in A-2 is given by the foraminifera, for example *Elphidiella hannai* and *Elphidium clavatum*.

On the basis of studies of megaspores of two *Azolla* species (*A. tegeliensis* and *A. filiculoides*) the succession from sea bed to approximately 330 m in both wells has been subdivided by Bertelsen (1972), as shown in Table 7.2.

TABLE 7.2

Subdivision of Quaternary in Danish wells A1 and A2 on basis of megaspores (after Rasmussen, 1974).

	Depth sub-sea-bed (m)	
	A-1	A-2
Holocene and Weichselian	0– 2	0– 5
Elsterian and "Cromerian"	2–127	5–126
(?)Menarpian, Waalian and (?)Eburonian	127–225	126–227
Tiglian and Pretiglian	225–331	227–335

B-1 was drilled on the eastern "tail"of the Dogger Bank in what is now (since the change in the Danish–German median line) the German sector. The thickness of the Quaternary has not been precisely determined but is believed to be about 686 m; thick beds of marine Lower Pleistocene (Icenian) age are present. The lithology between sea bed and 197 m consists of grey sand, often silty, alternating with beds of clay and with lignite in places, below which greenish-grey sticky clay with beds of silt, the lowermost micaceous, continues to below the base of the Quaternary.

C-1 is the most easterly well for which we have information. It was drilled some 30 km off the coast of Jutland in a water depth of 27 m. Rasmussen quotes a preliminary thickness of 140 m for the Quaternary; a revised figure (used by Caston, 1977a) is 91 m. This consists of sand and gravel to a depth of 58 m underlain by olive grey clay to 82 m, and thence coarse-grained sand. D-1 recorded an estimated 316 m of Quaternary, later amended to 207 m (Caston, 1977a). The lithology consists of grey sticky clay, alternating at intervals with sand and in places gravel.

The Dogger Bank

The Dogger Bank is the largest positive relief feature in the North Sea, and therefore merits separate consideration, although in part it has been considered within the discussion of the Dutch sector (see this Chapter above). The bank stands more than 20 m higher than the surrounding sea-floor, and is 324 km long by up to 120 km wide at the 40 m isobath. It is highest and widest and has an abrupt termination at the west end (Stride, 1959a). Its origin has been variously attributed to rivers, tidal eddies, or to an extension of a scarp of Mesozoic rocks from the English coast. Glacial affinities were first suggested by Belt (1874), and from it Stather (1912) dredged shelley clay. The abrupt west and northern faces were considered by Valentin (1955) to mark the limits of the English–Scottish and Scandinavian ice sheets of the last glaciation respectively, but this was based upon virtually no field evidence.

Jarke's (1956) map indicates that as far north as 55°20′N the bank is covered by fine sand which according to Oele (1969) is from 50 to 200 cm thick in the Dutch sector. Patches of gravel and stones occur on the west and southwestern flanks, and boulder clay outcrops at ca. 54°37′N 1°20′E. The presence of peat is noted along the south-eastern margin. Further patches of gravel occur along the relatively steep northern edge (Veenstra, 1971), and the eastern end terminates at the edge of the linear zone of silt already described from the Danish sector. Petrographic work by Baak (1936) demonstrated that the surficial sands possess heavy mineral suites derived from northern England and Scandinavia.

A single reversed seismic refraction station, with a mid-point at 55°10′N 02°12′E, recorded an upper sedimentary layer some 380 m thick believed to be of Quaternary age (Stride, 1959a). Stride also noted the occurrence, in shallow cores, of stones and gravel derived from northern England, Scotland and Norway. He concluded that the bank was of Weichselian age, and in this was supported by Veenstra (1965, 1969) whose study of gravels obtained by grab sampling showed that they consisted of a sandstone–limestone–porphyry assemblage similar in composition to that of Weichsel boulder clay exposed on the adjacent U.K. east coast. On the basis of vibrocore samples, Oele (1969) reported that the southern margin of the Dogger Bank was underlain at a shallow depth by a fluvioglacial clay, which on palynological evidence was of Scandinavian origin. This he considered to be of Elsterian age (Peelo Formation) by comparison with similar deposits found in

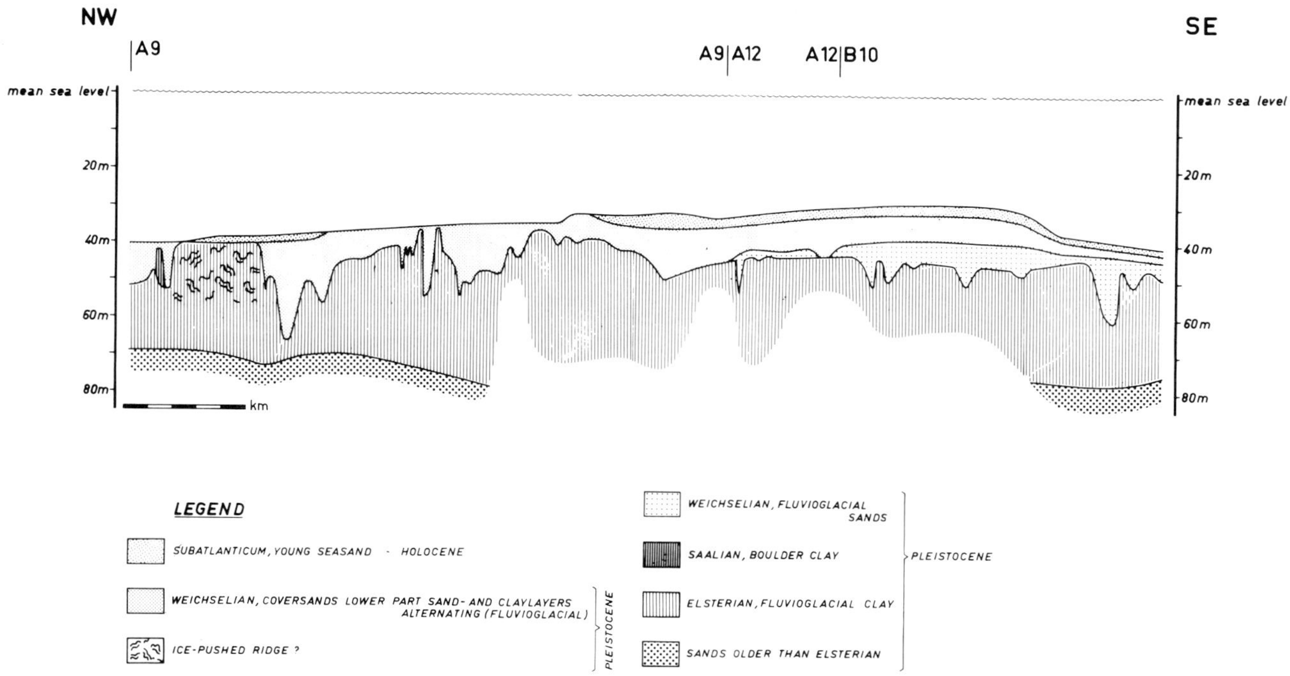

Fig. 7.14. Interpreted "Sonia" reflection profile over the Dogger Bank. For location see Fig. 7.7. Vertical exaggeration × 10. (From Oele, 1971a.)

the north of Holland. Later work (Oele, 1971a), using a Sonia reflection profiler, showed however that the surface of these fluvioglacial clays remained more or less horizontal under the bank (Fig. 7.14), and that the bulk of the feature appeared to be formed of proglacial sands deposited immediately in front of the southern limit of the Weichsel glaciers, thus confirming the earlier views of Valentin (1955) and Stride (1959a).

Reference to Fig. 7.2 shows that considerable thicknesses of Quaternary sediments underlie the Dogger Bank, which in fact crosses the principal depositional trough at one of its deepest points. Well 38/22-1, drilled some 5 km away from the refraction station reported by Stride (1959a) penetrated ca. 312 m of Quaternary, thus confirming his seismic results. It is, however, significant to note that according to Oele's (1971a) Sonia section (Fig. 7.14) the base of the Elsterian fluvioglacial clays occurs at a sub-sea-bed depth of only some 45 m. The great bulk of the Quaternary deposits appear therefore to date from pre-Elsterian or Lower to mid-Middle Pleistocene time. This is confirmed near the eastern end of the bank by well B-1 which recorded a Quaternary thickness of ca. 686 m and in which thick beds of marine Lower Pleistocene (Icenian) age are present (Rasmussen, 1974).

U.K. sector: Dogger Bank to 56° N

Relatively little is known about the deposits of this triangular-shaped zone, which is bounded on the west by the U.K. coast and to the south and south-east by Dogger Bank. In very general terms the bathymetry consists of the southern extremity of a broad basin which has many local variations in relief with depths ranging from 50 to 90 m. Information concerning the distribution of surface sediments is limited. The maps of Luders (1950) and Veenstra (1971) show that the greater part of the area is covered with sand, with patches of mud lying in deeper water between 1° and 2° E, at 55° 20′ N 01° 15′ E, and adjacent to the coast between 54° 45′ N and 55° 30′ N, where there is a linear deep with a maximum water depth greater than 100 m. Patches of gravel and/or coarse sand occur in a continuous belt along the coast, on the northern side of Dogger Bank centred at 3° E, and in association with the shallower banks. In the south-western corner of the area between 54° and 55° N, Quaternary deposits are absent along a zone some 40 to 50 km in width between the coast and Dogger Bank (Dingle, 1970). This absence is somewhat surprising in view of the thickness of Quaternary present on the adjacent land and further offshore.

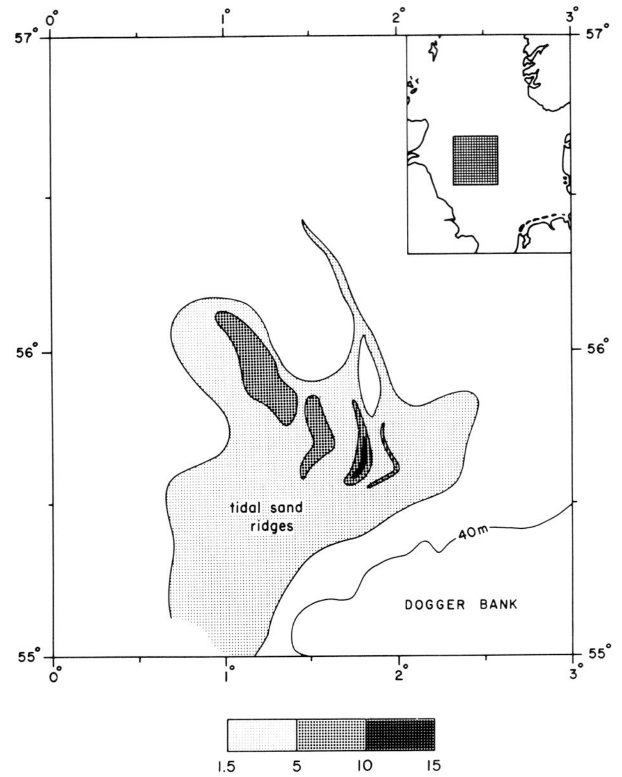

Fig. 7.15. Thickness in metres of the East Bank Deposit. In the area of tidal sand ridges no detailed contours are drawn (from Jansen, 1976).

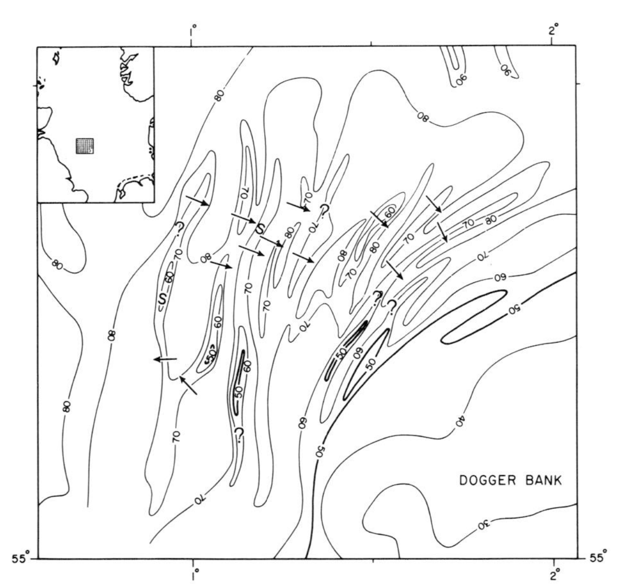

Fig. 7.16. Tidal sand ridges of the East Bank area, depths in metres below sea level. Arrows indicate observed slope direction of the steeper flank (From Jansen, 1976).

Sub-bottom deposits are known best from the vicinity of East Bank. Here the most recent sedimentary unit is a layer of rather uniform thickness, the East Bank Deposit, which varies in thickness up to 15 m, the greatest thickness occurring in the deeper parts of the area (Fig. 7.15). This unit is acoustically transparent, although with some weak horizontal internal reflectors (Jansen, 1976).

Further south, between 55° and 55°30′N, and 1° and 2°E, occur a group of linear banks (Dingle, 1971a; Jansen, 1976). These range in length from 17 to 60 km, are 1.5 to 5 km in width, and rise some 10 to 30 m above the surrounding sea floor. They are relatively straight but fan-like in plan view, the westerly bank trending approximately north–south, and the most easterly bank oriented north-east–south-west, parallel with the north-easterly margin of Dogger Bank. The majority of banks are rounded but asymmetric in cross-section with the steep flank facing east or south-east (Fig. 7.16). They lie on a relatively flat sub-bottom reflector forming the contact with the underlying Tertiary sediments, which are exposed between the banks. Jansen (1976) describes them as sand ridges, and ascribes their formation to tidal currents at a time of lowered sea-level about 40–45 m below the present, which he dates at about 12,000–9000 years B.P.

Underlying the East Bank deposits and to the east of ca. 0°, the presence of a complex pattern of buried channels up to 20 km long has been demonstrated by subbottom profiling (Dingle, 1971a). These valleys, which are up to 150 m in depth, are believed to be cut into strata of upper Tertiary age. Later work by Jansen (1976) has shown that these features form part of a generally north–south oriented series which he has mapped as far north as 57°30′N (see below). The age of the channels is unproven. Dingle considers that they probably date from the Weichselian, whereas Jansen favours a Saalian age or older. This will be discussed in the following section.

Three boreholes were drilled in this area in connection with a proposed pipeline from the Ekofisk Field to Teesside. Information is available concerning lithology but not the age of the deposits penetrated (Fig. 7.17).

Borehole "I" drilled in Block 36/22 at 55°17′26.5″N 00°12′11.6″E passed through some 6 m of tan fine sand overlying 4 m sandy silt, 18 m firm grey clay, and 8.2 m sandy clay and fine sand before passing into hard grey clay at 36.3 m sub-sea-bed. This hard clay continued without change to the end of the borehole at

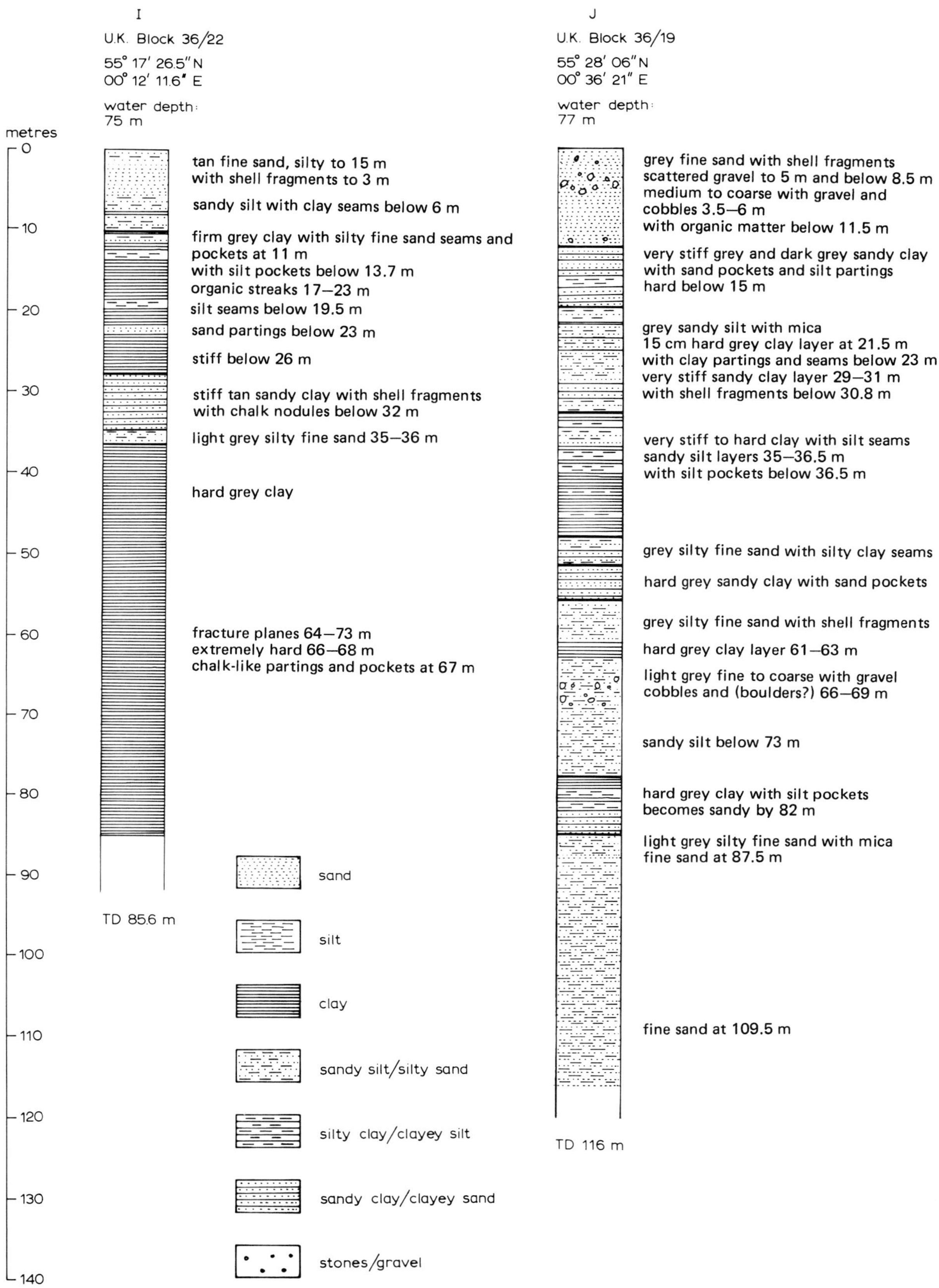

Fig. 7.17. Lithological logs of engineering boreholes "I"–"K" inclusive drilled in the U.K. sector north of the Dogger Bank. For locations see Fig. 7.3.

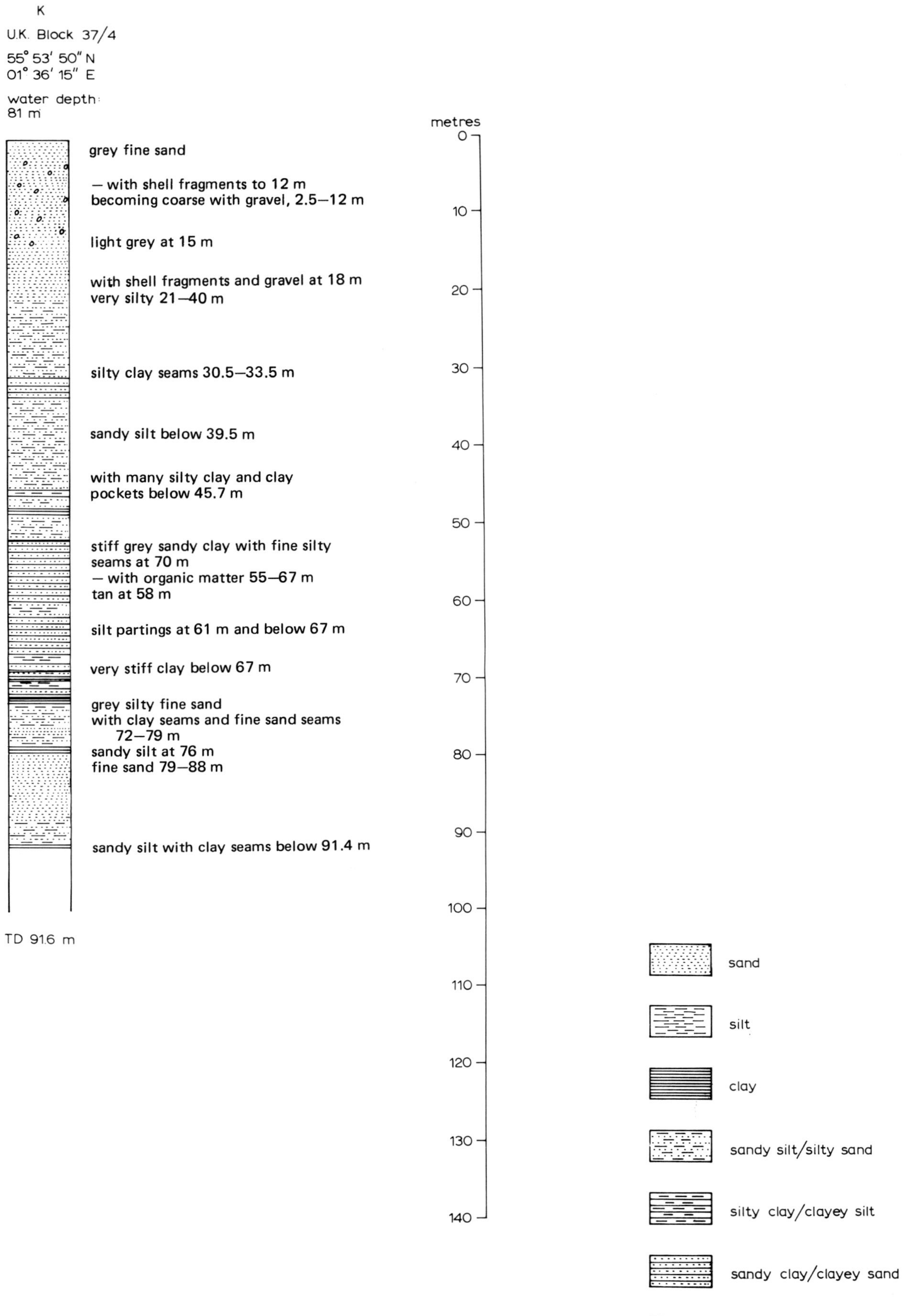

Fig. 7.17 (continued)

85.6 m, with shear strengths increasing steadily with depth, and is believed to represent the underlying pre-Quaternary strata which in this area is of Upper Tertiary age (Dingle, 1971b). The total Quaternary thickness of 36 m is comparable with a figure of 50 m recorded in an exploration well drilled some 18 km to the east. A second borehole at 55°28′06″N 00°36′21″E in Block 36/19 penetrated to a depth of 116 m and passed through alternate banks of locally sandy and silty hard grey clay, and grey silty fine sand (Borehole "J" in Fig. 7.17). The lowest 32 m consisted of apparently uniform fine sand with mica; it is not known whether this is of Tertiary or Quaternary age. Quaternary deposits are, however, believed to extend at least to 65 m at which depth a layer of gravel with cobbles was recorded. The hole was drilled within the East Bank area in which total Quaternary thicknesses are believed to range from zero to 150 m+ in the infilled erosion hollows. If located within such an infilled erosion hollow the whole of the borehole deposits could be of Quaternary age, and this explanation is preferred, as the lowermost grey fine sands are similar to other sands recorded further up the hole. 98 km to the northwest the third borehole ("K") was drilled in Block 37/4 at 55°53′50″N 01°36′15″E. This terminated at 91 m, having passed through a sequence consisting primarily of grey silty fine sand with a layer of stiff grey sandy clay or clay between 52 and 72 m. These deposits almost certainly all date from the Quaternary.

This borehole information is limited, but suggests that there is an increase in sand content in an easterly direction away from the stiff clays which lie closer to the coast. It is noteworthy that, apart from a thin layer of gravel in borehole "J", no sediments coarser than sand grade were observed. The stiff clays, whatever their origin, do not compare with the stony boulder clays of Weichselian age recorded to the south-west of Dogger Bank. The overall lithologies are, however, comparable with those seen further to the north-east in the Ekofisk and Cod Fields in the Norwegian sector (see below, this Chapter).

Norwegian sector south of 58° N

The Norwegian sector to the south of 58°N is approximately triangular in shape, and includes the southern portion of the Norwegian Channel, together with the wide area of shallow banks which surround Great Fisher Bank at a depth of less than 80 m. The southern limit of the sector lies just to the north of Dogger Bank. The various maps of sediment distribution already referred to show that the Skagerrak, like the rest of the Norwegian Channel, is floored by mud, and that other areas of mud occur in the southern and western parts of the sector in local deeps on the flank of Great Fisher Bank. The banks immediately to the south-west of the Norwegian Channel are covered by coarse sand, gravel and stones. They form a continuation of the morainic accumulations curving westward from the Danish sector (Pratje, 1951).

Norwegian Channel

As is common throughout the greater part of the North Sea, there is no consistent survey coverage in this area, and individual studies have been carried out for different objectives and at varying levels of detail. A number of geophysical surveys have been undertaken in the Skagerrak (e.g. Aric, 1968; Weigel et al., 1970). Air-gun profiles were shot by Sellevol and Aalstad (1971) south of Kristiansand. These show that sediments of presumed Quaternary age overlying an erosional unconformity have a maximum thickness of about 250 m below the southern margin of the Channel, and wedge out to zero against the northern edge adjacent to the Norwegian coast. No details are available from this survey concerning the internal structure or composition of the Quaternary, but in part these have been provided by a subsequent survey of the Norwegian Channel from the Skagerrak north to 60°N using a 3.5 kHz profiler (Van Weering et al., 1973). Four principal sedimentary units were distinguished, each with different internal characteristics and separated by well-defined reflectors. Greatest detail is available concerning the uppermost two layers, believed to be of post-glacial age. Unit 1 is acoustically transparent and coincides largely with the distribution of mud shown on the charts. It reaches a maximum thickness of 30 m in virtually the deepest part of the Channel south-east of Kristiansand. The underlying unit 2 is well-stratified, with layering near the base reflecting the topographic irregularities of the underlying glacial deposits. This is believed to consist of coarser material than unit 1, probably including sediment reworked from glacial deposits during the early Holocene, and is on the whole a much more massive deposit than the uppermost unit, reaching a thickness to the south-west of Kristiansand of greater than 90 m. Maximum thickness of post-glacial infill (units 1 and 2) within this linear trough in the Skagerrak is 127 m; to the north it rapidly decreases to an average of 20–30 m (Fig. 7.18). Unit 3 is interpreted as glacial drift and occurs below nearly the whole area of investigation. It has a typically irregular undulating surface that gives a strong reflection and usually prevents further penetration. No internal stratification has been observed, but

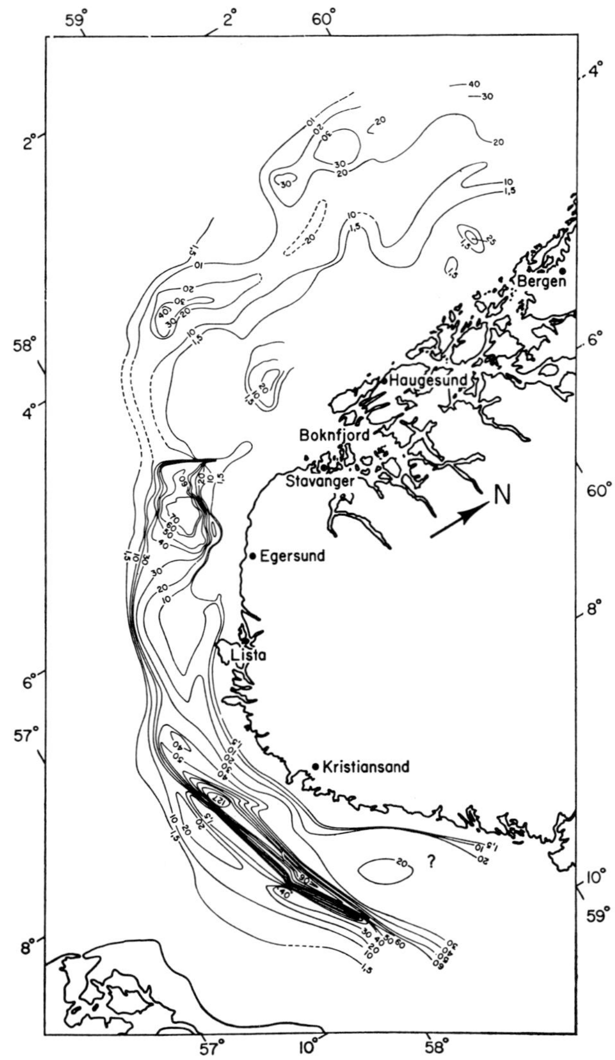

Fig. 7.18. Thickness (in metres) of post-glacial sedimentary infill (units 1 and 2) within Norwegian Channel. (From Van Weering et al., 1973.)

multiple hyperbolae occur on a large scale. The thickness (maximum 30–40 m) can only be determined in certain restricted areas where an underlying fourth unit (of similar appearance) was distinguished. Unit 3 outcrops on both sides of the Channel (to the north of 58° N), and continues to the west on the North Sea side where it presumably forms at least a part of the central moraine complex. The authors believe it to be of Weichselian age, although they have no positive evidence.

Ekofisk Field area

Moving westward onto the shallower banks, a number of borings have been undertaken within an area approximately 28 km north–south by 16 km east–west, in connection with the development of Ekofisk and adjacent Fields in Norwegian sector Blocks 2/4 and 2/7 (cf. Fig. 18.1). The maximum penetration was 122 m; a majority of holes terminated between 90 m and 120 m. Lithological logs for several of these boreholes are shown in Fig. 7.19; these are self-explanatory and will not be discussed in detail. The overall impression is one of extreme variability within the Quaternary sequence, despite the fact that water depths at all locations are virtually constant, ranging from 67 m in the north to 71 m in the south. At a majority of sites the uppermost 12–27 m consists of fine grey sand, locally silty at, and immediately below, the sea bed. Below this occur alternations of hard grey clay, occasionally slickensided, and grey fine to silty sand. The relative thickness of the layers, and of the proportions of sand and clay, varies considerably from hole to hole, borehole "M" penetrating almost wholly sand, and "N" predominantly clay. A layer of hard brown clay, in all but one case slickensided, was recorded at 76.5 m in "R", at 79 m in "L", at 94 m in "M", and 95 m in "N". The brown colour is distinctive, but from the descriptions appears to change to grey within 5–10 m, apparently without any other lithological change. It may represent a paleo-weathering phenomenon, and therefore could serve as a marker horizon. Gravel is very rare, thin streaks or pockets being noted only at 31 m in "L", at 40 m in "R", at 61 m and 100 m in "Q" and at 57 m in "O". None of the hard clay layers is pebbly and from the descriptions there is no way of judging whether any of them represent true glacial tills. Within the overall lithologies shown on Fig. 7.19 there is much small-scale variability, with frequent mention of "clay pockets", "silt seams and partings", "organic streaks", "sandy silt pockets" and "sand seams". Moreover, bearing in mind that the lithological descriptions are commonly based upon 0.76 m long samples taken at 3 m intervals, it is probable that the actual variability is even greater than indicated.

Cod Field

Two foundation boreholes were drilled in the Cod Field located in Block 7/11 in the Norwegian Sector (Fig. 18.1). This is situated on the western flank of the central moraine complex some 80 km to the north-west of the Ekofisk area. The two holes were drilled within a few hundred metres of each other, but the variation in lithologies recorded is sufficient to merit individual description.

Hole C2 (Borehole "S") was drilled in a water depth of 79.5 m at 57°04′04″N 02°26′05.9″E. A simplified lithological log is given in Fig. 7.20 and will not be described in detail. In brief, the succession consists of a

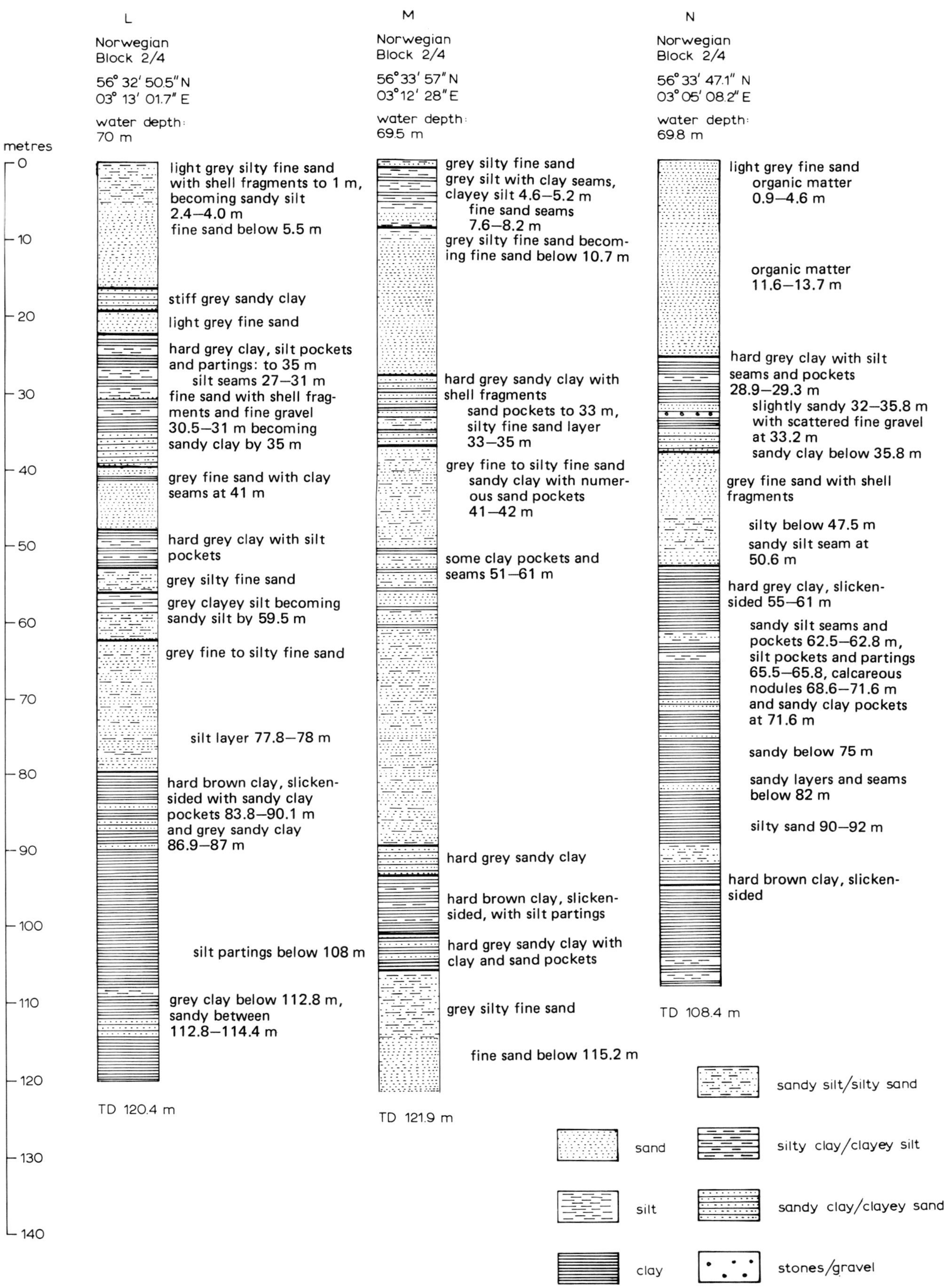

Fig. 7.19. Lithological logs of engineering boreholes "L"–"R" inclusive drilled in the Norwegian sector south of 58°N. For locations see Fig. 7.3.

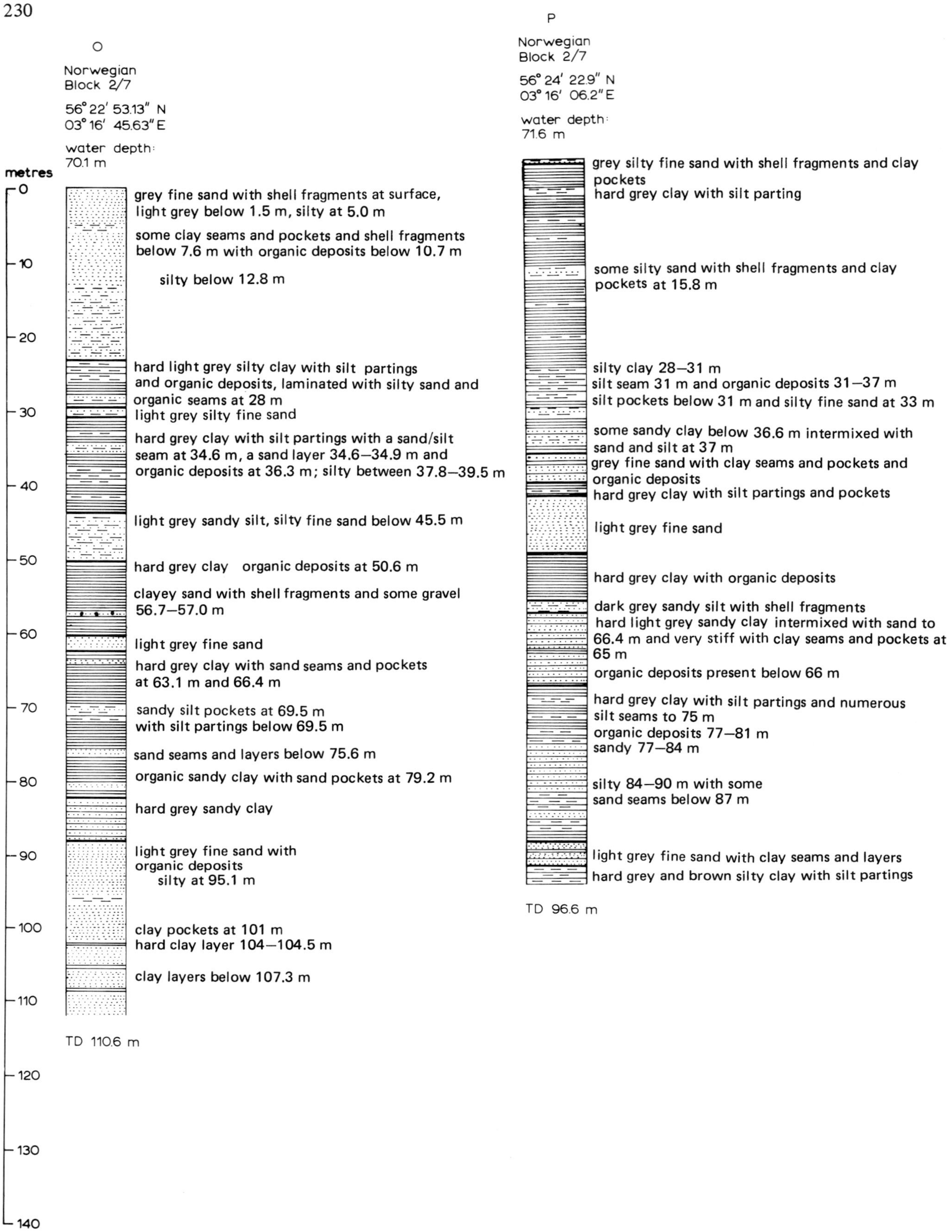

Fig. 7.19 (continued).

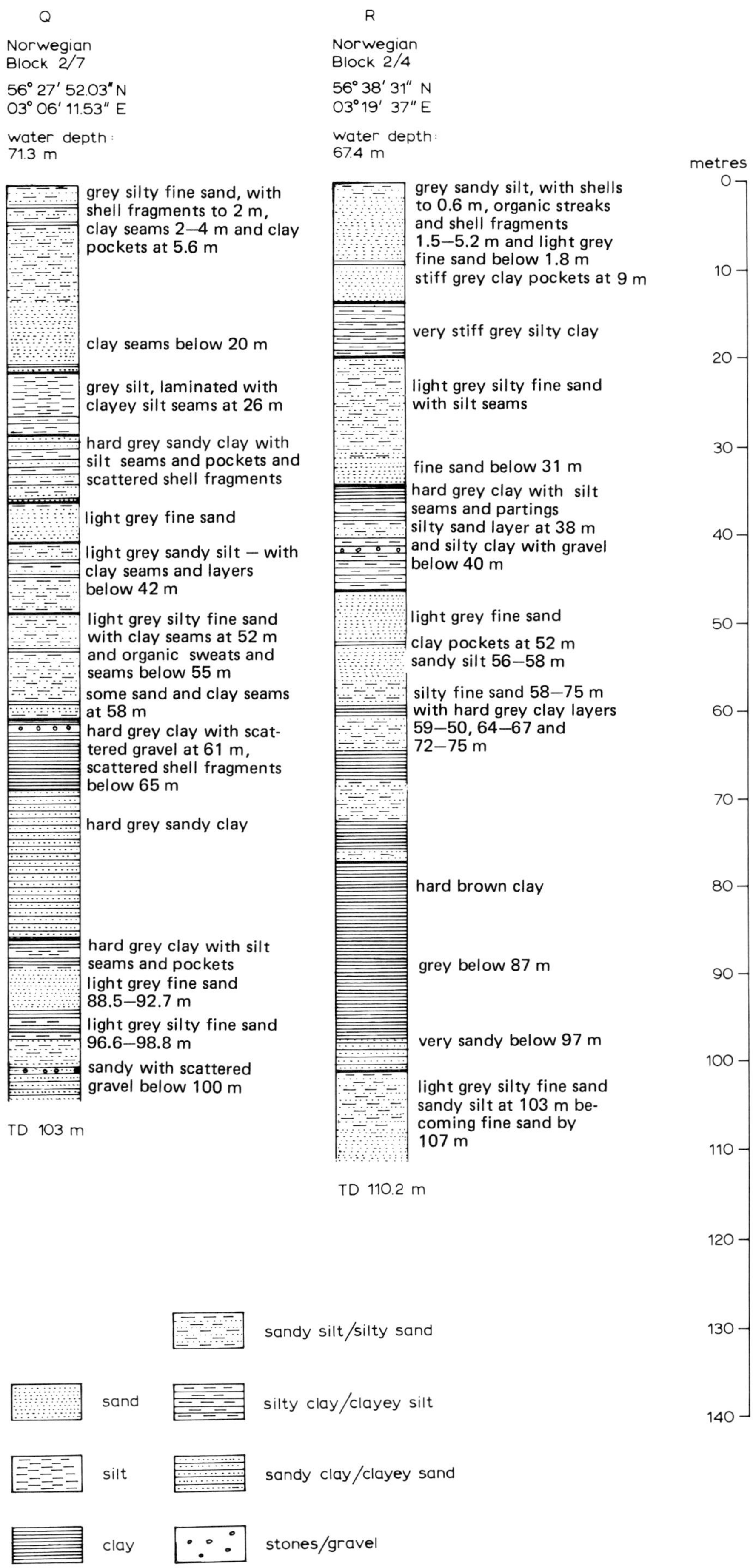
Q
Norwegian
Block 2/7
56° 27′ 52.03″ N
03° 06′ 11.53″ E
water depth:
71.3 m
grey silty fine sand, with shell fragments to 2 m, clay seams 2–4 m and clay pockets at 5.6 m
clay seams below 20 m
grey silt, laminated with clayey silt seams at 26 m
hard grey sandy clay with silt seams and pockets and scattered shell fragments
light grey fine sand
light grey sandy silt — with clay seams and layers below 42 m
light grey silty fine sand with clay seams at 52 m and organic sweats and seams below 55 m
some sand and clay seams at 58 m
hard grey clay with scattered gravel at 61 m, scattered shell fragments below 65 m
hard grey sandy clay
hard grey clay with silt seams and pockets
light grey fine sand 88.5–92.7 m
light grey silty fine sand 96.6–98.8 m
sandy with scattered gravel below 100 m
TD 103 m
R
Norwegian
Block 2/4
56° 38′ 31″ N
03° 19′ 37″ E
water depth:
67.4 m
grey sandy silt, with shells to 0.6 m, organic streaks and shell fragments 1.5–5.2 m and light grey fine sand below 1.8 m
stiff grey clay pockets at 9 m
very stiff grey silty clay
light grey silty fine sand with silt seams
fine sand below 31 m
hard grey clay with silt seams and partings
silty sand layer at 38 m
and silty clay with gravel below 40 m
light grey fine sand
clay pockets at 52 m
sandy silt 56–58 m
silty fine sand 58–75 m with hard grey clay layers 59–50, 64–67 and 72–75 m
hard brown clay
grey below 87 m
very sandy below 97 m
light grey silty fine sand
sandy silt at 103 m becoming fine sand by 107 m
TD 110.2 m
metres
0
10
20
30
40
50
60
70
80
90
100
110
120
130
140
sandy silt/silty sand
sand
silty clay/clayey silt
silt
sandy clay/clayey sand
clay
stones/gravel

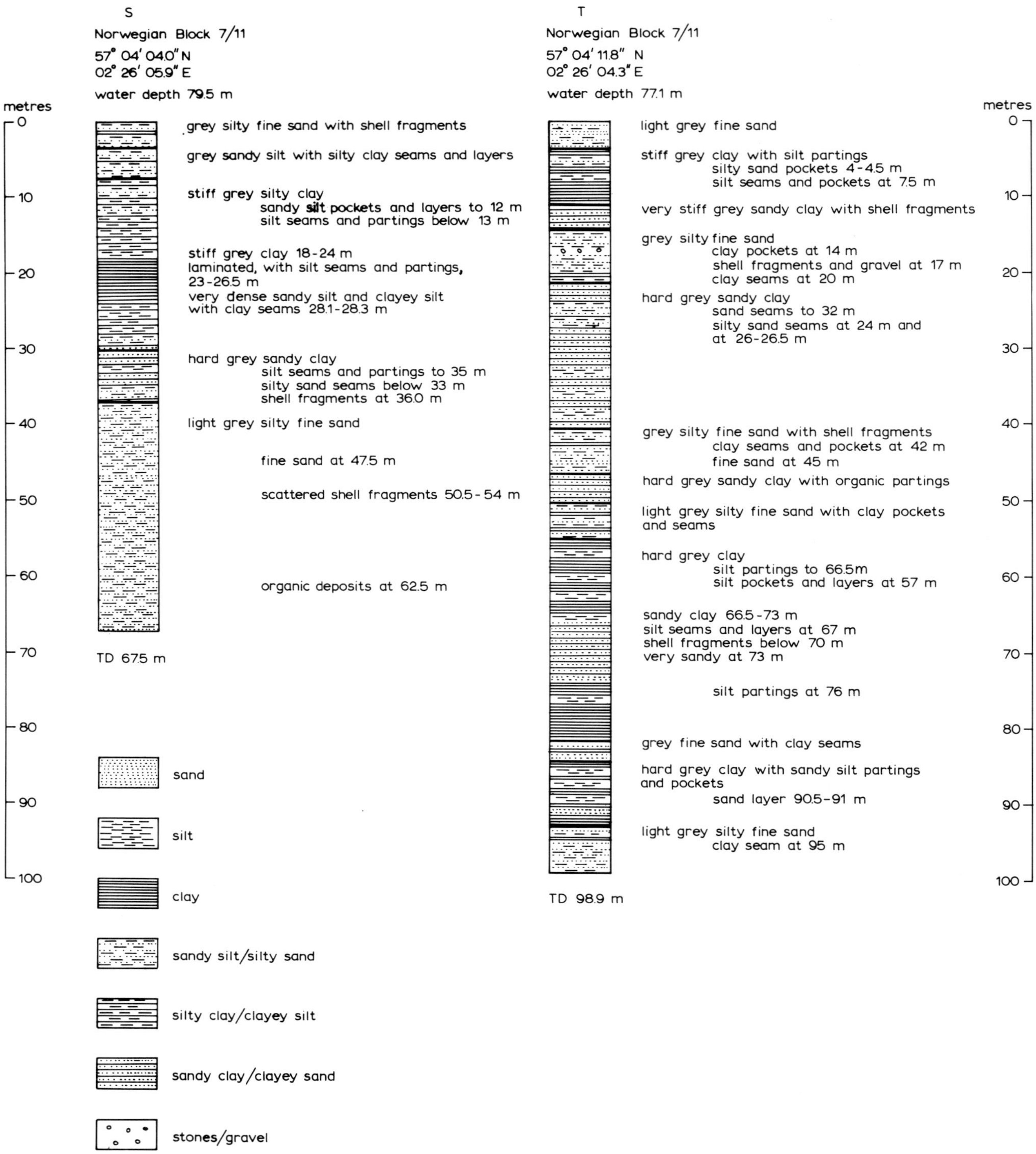

Fig. 7.20. Lithological logs of engineering boreholes "S" and "T" drilled in the Norwegian sector south of 58°N. For locations see Fig. 7.3.

superficial layer of silty sand to 3.5 m, sandy silt with clay seams to 7.6 m, and a predominantly stiff grey clay layer of somewhat variable composition to 30 m. This is underlain by a hard grey sandy clay from 30 m to 37.2 m in which cohesive shear strength increases abruptly from some 1–2 kg/cm^2 in the overlying silts to 3–6.75+ kg/cm^2. This is possibly an overconsolidated glacial till. It overlies at least 30.5 m of virtually featureless light grey silty fine sand.

Hole C3 (Borehole "T") was drilled to a penetration of 98.9 m at 57°04′11.8″N 02°26′04.3″E in 77 m of water. The succession consists predominantly of hard grey clay, locally sandy, interspersed with grey silty fine sand. Correlation with C2 is not feasible on present evidence. It may be that the top of the uppermost hard grey sandy clay at 21.3 m in C3 corresponds to the top of the only similar layer at 30 m in C2, but this is an association based solely upon a depth similarity, which may be entirely fortuitous. Certainly the 30.5 m thick layer of fine grey sand seen in C2 is non-existent in C3 despite the greater penetration of the latter.

U.K. sector 56°–58° N and the Moray Firth

The Quaternary sediments of this area are the most intensively investigated in the North Sea. Much of the region has been mapped by the I.G.S. on a reconnaissance basis and an informal stratigraphy established (Holmes, 1977). It is anticipated that this succession and identification of various characteristic features may be extrapolated further into much of the central and northern North Sea Basin.

Coastal zone

A series of boreholes put down by the I.G.S. in the Firth of Forth confirmed the presence of two sedimentary basins cut by glacial scour in Carboniferous bedrock (Chesher et al., 1972). Drilling results suggest at least four phases of fill within these troughs, which have a maximum thickness of 80 m. The first fill was a locally derived coaly boulder clay seen only at site 71/33 (Fig. 7.21). This was followed by a red boulder clay, probably derived from the Old Red Sandstone terrain to the north-west, which is mainly present in the northerly trough. Overlying the red boulder clay is a stiff reddish or brownish clay present in all the northerly boreholes and also in 71/30. This clay is similar to the brown plastic clays described from the St. Abbs area (Eden et al., 1970) and to the brown or pinkish clays in the Moray Firth (Chesher and Lawson, in preparation). The final phase of filling seems to have been the deposition

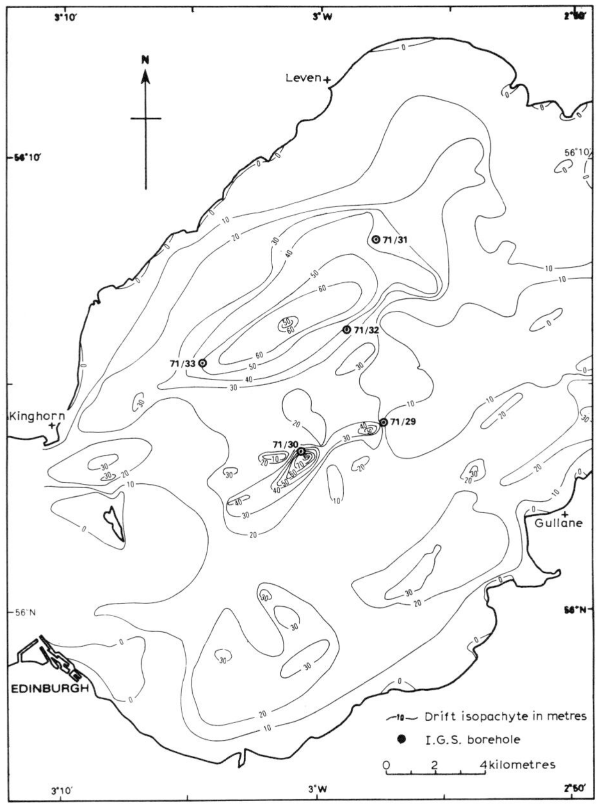

Fig. 7.21. I.G.S. borehole locations and thickness of drift deposits in the Firth of Forth. (From Chesher et al., 1972.)

of a sandy dark grey stiff clay. These are overlain by recent estuarine sandy muds.

30 km to the north of the Firth of Forth, investigations for a new road bridge over the Tay Estuary in the early 1960's revealed the presence of an almost symmetrical buried channel whose base reached at least 71 m below O.D. or 60 m+ below the bed of the estuary (McManus, 1970). The fill consisted of ablation tills, clays, estuarine sands and gravels, all considered by McManus to date from the late Weichselian (Fig. 7.22).

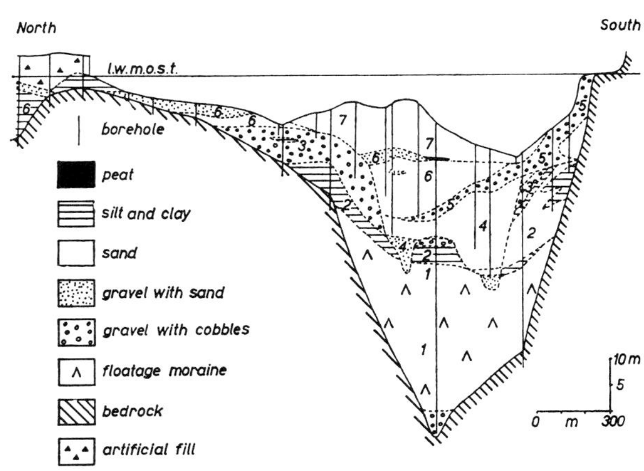

Fig. 7.22. Cross-section of Tay Estuary showing borehole locations and Quaternary stratigraphy. (From McManus, 1970).

The results of I.G.S. surveys, including a number of boreholes, within the offshore zone extending out to 1°W between 56° and 57°N have been reported by Thomson and Eden (1977). Within this area three provinces can be recognised: in the west an area of varied relief, much of it shallower than 60 m, is occupied by a series of north–south ridges and hollows made up largely of unstratified or poorly-stratified morainic drift lying on irregular rockhead. A maximum of 83 m of drift was proved in one borehole, but thicknesses more commonly range from 2.5 to 47 m. From its variable composition, it is inferred that this drift was deposited partly below grounded ice and partly in sea water in front of or below an ice sheet. In the east, beyond a low west-facing scarp, the sea bed changes abruptly to an area of very little relief lying at a depth of −60 to −70 OD. It is formed by a gentle eastward-dipping slope at the top of a series of homogeneous sands, reaching a maximum thickness of 40 m, referred to as the Marr Bank Beds. These sands were laid down on a gentle eastward-dipping plane. In the west of the area of Marr Bank Beds this uniform base coincides with rockhead, but further to the east rockhead deepens and the Beds lie unconformably on an older drift deposit, the Aberdeen Ground Beds. Contours on this planar surface show that it dips at about 1 in 200 from just under 60 m in the west to 100 m in the north-east. It is inferred to be a wave-cut platform cut by a transgressive sea which was subsequently responsible for depositing well-sorted fine sands with a basal gravel layer now comprising the Marr Bank Beds. Insofar as the area of the wave cut platform is not affected by glacial scouring, it would appear that neither the ice sheet nor its direct deposits extended into the area of the Marr Bank Beds. The escarpment marking the western extent of these beds thus appears to mark the approximate position of an ice front, considered by Thomson and Eden to be of Weichselian age, although they point out that further erosion and deposition occurred at a later date.

A third province is developed in the north of the survey area, where six northeast–southwest channels are present. The channels are trench-like features, nearly parallel-sided and much longer than broad, varying from 12 to 18 km in length and 1.5 to 3 km in width with a maximum depth of 130 m. The most prominent channels are cut into the low scarp face of the Marr Bank Beds, although they are also present in the irregular morainic area further west. Their formation is attributed to a combination of erosion by glacial melt-water and tidal scour, although it is noted that all channels occur over areas of collapse in underlying Zechstein anhydrites.

56°–58°N

To the north and east of this area, Holmes (1977) has described a reconnaissance-scale coverage of the Quaternary from 56° to 58° 10′N, and from the Scottish coast in the west to areas extending over the median line in the east. The work is based upon I.G.S. sparker surveys, including six limited areas surveyed in detail using processed sparker sections, together with commercial and I.G.S. borehole data. The Quaternary succession identified in this area is shown in Fig. 7.23; the names used to describe beds and groups of beds are used informally. Jansen (1976) has discussed in more detail the uppermost deposits of the Witch and Fladen Ground areas, north of 57° 30′N, together with the nature of the fill within buried channels extending southwards to the East Bank area (see above). The greater part of his work, however, refers to the region north of 58°N and will therefore be reviewed in the appropriate section.

(1) *Basal Beds.* The lowest Quaternary recognised, the Basal Beds, have only been identified on processed sparker data, principally in the vicinity of the South Fladen area centred at 58°N 0°35′E. They are seismically structureless. The upper surface consists of a strong crenulate reflector lying some 100–150 m above the top Pliocene. The crenulations appear as a series of low mounds and valleys rarely greater than 1 km in length and 50 m in amplitude. These are commonly emphasised by high-amplitude reflections or "bright spots" comparable with those described by Caston (1977b) in the Forties Field, which are associated with proven gas.

(2) *Aberdeen Ground Beds.* Above the crenulate reflector occur strong, laterally-persistent, reflectors correlated with the Aberdeen Ground Beds (recognised further east by Thomson and Eden, 1977), which, in those areas where detailed seismic and/or borehole information is available, have been sub-divided into the Upper and Lower Aberdeen Ground Beds. Together they form the thickest and most uniform sequence of sediments recorded from the Quaternary in the central North Sea.

The Lower Aberdeen Ground Beds are little known, but on seismic records consist of strong reflectors occurring at intervals less than 60 m apart and separated by intervening zones of weaker, laterally discontinuous, reflectors. The Upper Aberdeen Ground Beds consist of a dominantly structureless unit which in borings in the Auk and Josephine Fields and in Block 30/23 (56° 17′ 49″N 02° 34′ 42″E) consist of stiff to hard grey clays apparently similar to those at a comparable stratigraphic level (Layer "E") in the Forties Field (Caston, 1977b). Well preserved wood samples from between 240 m and 310 m depth below sea bed in the Josephine Field included Yew, Juniper, Birch, Alder, Poplar, Willow and

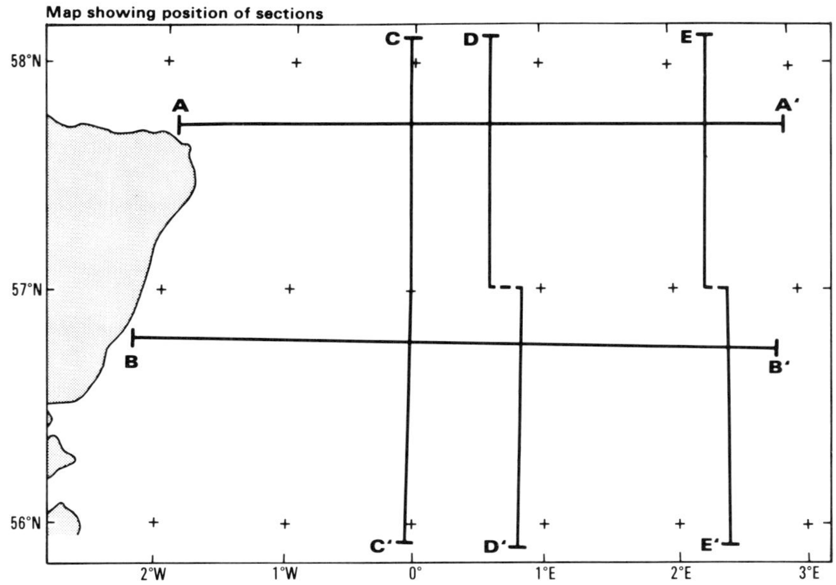

SYMBOL	SEISMIC TEXTURE	LITHOLOGICAL INTERPRETATION FROM SEISMIC AND BOREHOLE DATA	UNIT NAME
	Finely bedded acoustically transparent with numerous pockmarks, some of which are infilled	Very soft to firm, normally consolidated sandy and silty clays	Witch Ground Beds
	Finely bedded, passes downwards in deep channels to structureless, all acoustically transparent	Silty fine sands, and soft to firm clays deep structureless sediments probably fine sands	Upper Channel Deposits
	Structureless channel infill merging laterally into sheet structureless, all acoustically transparent	Fine sands	Upper Channel Deposits (lower series)
	Structureless to south and west, locally acoustically transparent, bedded, and acoustically opaque to north and east	Sands to south and west, bedded clays and sands to north and east	Marr Bank Beds
	Structureless or very poorly bedded	Stiff to very stiff clays	Wee Bankie Beds
	Strong reflectors, normally discontinuous, in an acoustically opaque medium	Very firm to hard clays with minor sand and gravel	Fisher Beds
	Weak discontinuous reflectors in a dominantly acoustically transparent medium	Dominantly firm to stiff clays, commonly normally consolidated	Upper Swatchway Beds
	Structureless, acoustically transparent	Fine sands *	Upper Swatchway Channel Deposits
	Structureless or very poorly bedded commonly with diffraction patterns, acoustically opaque and dense	Stiff to hard gravelly clays with boulders probably, in part, reworked Fisher Beds	Lower Swatchway Beds and Fisher Beds
	Poorly developed sub horizontal reflectors commonly separated by structureless sediment	Partially disturbed Fisher Beds or facies of Fisher Beds	Fisher Beds
	Strong reflectors and acoustically opaque infill to large channels, structureless acoustically transparent infill to most small channels	Gravel, sand and clay in large channels, dominantly massive sands in small channels	Lower Channel Deposits
	Structurless or very disturbed, acoustically opaque	Stiff to very stiff clays, probably distorted	Aberdeen Ground Beds
	Strong laterally continuous sub horizontal reflectors (often >20km length), acoustically opaque	Stiff to very stiff clays, rare gravel and sand	Aberdeen Ground Beds

* Not yet drilled

SUMMARY OF POSSIBLE CORRELATION

WEST	CENTRAL	EAST
Forth Beds ← ? →	Witch Ground Beds	
St Abb Beds ← ? →	Upper Channel Deposits	↔ Upper Channel Deposits (Upper Series)
Wee Bankie Beds = Marr Bank Beds ↔	Upper Swatchway Beds / Upper Swatchway Channel Deposits	↔ Upper Channel Deposits (Lower Series)
Fisher Beds ← ? →	Lower Swatchway Beds	← ? → Fisher Beds
Fisher Beds →	Fisher Beds	← Fisher Beds
Lower Channel Deposits ↔	Lower Channel Deposits	↔ Lower Channel Deposits
Aberdeen Ground Beds ↔	Aberdeen Ground Beds	↔ Aberdeen Ground Beds

Fig. 7.23. Quaternary succession in the U.K. sector between 56° and 58°N. (From Holmes, 1977.)

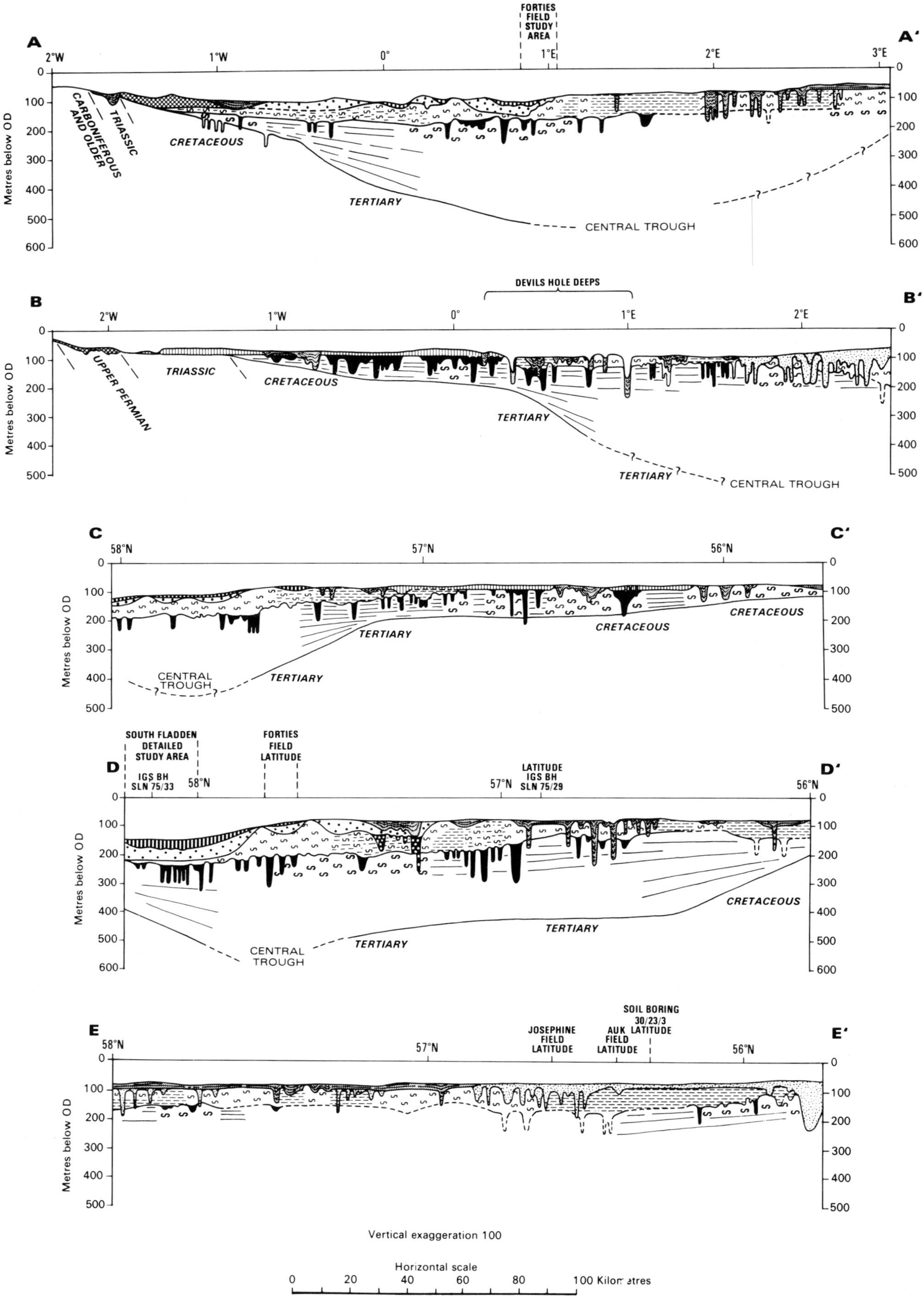

Fig. 7.24. East–west sections across Quaternary of central North Sea, U.K. sector, between 56° and 58°N. (From Holmes, 1977.)

Spruce, and yielded middle to late Weichselian ages. To date, micropalaeontologic analyses from the Upper and upper part of the Lower Aberdeen Ground Beds have proved an entirely marine, shallow-water sub-arctic to arctic succession.

(3) *Lower Channels*. The upper limit of the Aberdeen Ground Beds is defined by the Lower Channels. These extend over a wide area and, indeed, can be traced as far north as 60° N at 0° 30′ E (Fig. 7.24). They vary in depth from 20 to 200 m and may be over 20 km in length by 3 km in width. Because of the wide survey line spacing, the orientation of these features cannot be determined with confidence over much of the surveyed area, but in the Forties area they trend in a southwesterly–northeasterly direction. Infill may be seismically transparent, as at Forties where it has been proved by drilling to consist of fine sand, or well-bedded, which has been proved in an I.G.S. borehole to consist of rapid alternations of fine sand and clay. In the South Fladen area processed sparker data show that the Lower Channels and the reflector joining them are accentuated by "bright spots" which are more extensive than those occurrring on the crenulate reflector at the top of the Basal Beds. An I.G.S. borehole located at the edge of a "bright spot" recovered stiff clays honeycombed with cavities containing a gas of (as yet) unknown composition.

Holmes concludes that the nature and abundance of the Channels is consistent with a sub-glacial origin comparable with the tunnel valleys reported from East Anglia and Denmark (Woodland, 1970), although micropalaeontological evidence suggests that many channels had a mixed sub-glacial, glacio-fluvial and glacial-marine history which varied according to local conditions.

(4) *Swatchway Beds*. The Swatchway Beds comprise a complex unit which varies considerably from area to area, and may be divided into Upper and Lower, corresponding respectively to Caston's (1977b) Layer B and Layers C and D. The Beds overlie the Lower Channels, and in the west wedge out against the junction of the erosional base of the Marr Bank Beds (Thomson and Eden, 1977). In the Forties Field the Lower Swatchway Beds outcrop on the sea bed forming the Forties Bank, and similar stony clays forming resistant mounds or banks on the sea bed can be traced as the easterly limit of the Swatchway Beds between 1° and 2° E to the north of 57° 30′ N. In this region the Lower Swatchway Beds commonly change from sediments showing westerly-dipping reflectors to strongly folded strata (Fig. 7.24) and further east the folded strata merge to a series of sub-horizontal strong reflectors characteristic of the Fisher Beds. Thus it appears that the moraine seen in the Forties Field (Caston, 1977b) has some of the characteristics of a push moraine, which presumably marks the limit of a glacial advance during a period in the late Weichselian.

In the north of the area the Swatchway Beds reach their greatest thickness, and north of 57° 30′ N both Upper and Lower Swatchway Beds may be clearly distinguished on seismic records, the Lower Beds outcropping on the sea bed as a series of low mounds, with the Upper series lying against and between them, as softer and acoustically more transparent sediments. South of 57° 30′ N the Upper series is floored by a series of channels (the Upper Swatchway Channels) which are cut into the Lower Beds. In the Forties Field normally consolidated clays of the Upper Swatchway Beds contrast with heavily overconsolidated gravelly clays forming the moraine.

Deposition of Upper Swatchway Beds is believed to have occurred at approximately the same time as that of the Marr Bank Beds.

(5) *Fisher Beds*. The Fisher Beds occur east of the Swatchway Beds at the same topographic level above the Lower Channels, but unlike the Swatchway Beds appear on seismic records as a series of strong sub-horizontal reflectors. East of ca. 2° E and north of 57° N these strong reflectors are broken up into structureless or strongly distorted strata, perhaps deformed by the effects of glacial ice extending from the Norwegian North Sea. Between 57° 10′ N and 56° 50′ N detail in the Fisher Beds is lost in massive Upper Channels (see below). The Fisher Beds are heavily overconsolidated and contain a shallow-marine arctic to sub-arctic fauna which shows evidence for cooling conditions at the base.

(6) *Marr Bank Beds*. The occurrence of the Marr Bank Beds for the area immediately offshore has been described by Thomson and Eden (1977) and is referred to above. They may be traced as far east as 0° as a weaker reflector over the Aberdeen Ground Beds or Swatchway Beds before becoming confused with other reflectors of similar strength at the same level. In the area north of 57° N, and over rockhead and thin drift west of 1° W, the Marr Bank Beds appear to merge laterally with sediments which are acoustically opaque and structureless, and which have been tentatively mapped as the lateral equivalents of the Wee Bankie Beds also described by Thomson and Eden (1977). The sediments commonly consist of overconsolidated gravelly clays, although overall they are of very variable composition. Radiocarbon dates, structural relationships, and evidence for only one cold phase in the Marr Bank Beds, are indicative of a late Weichselian age.

(7) *Upper Channels*. Holmes' Upper Channels and their infill are the expression of a complex sequence of

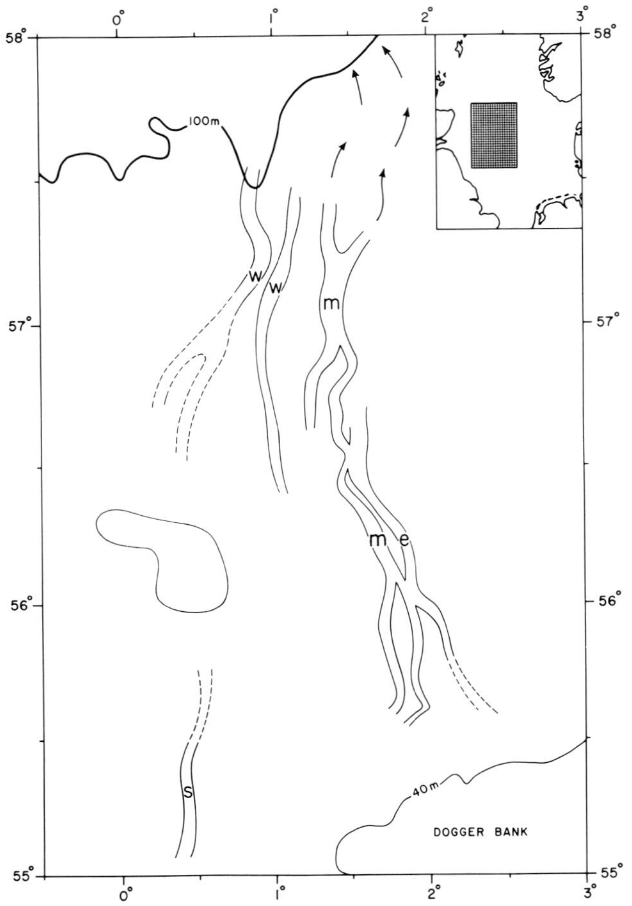

Fig. 7.25. Location of buried channels (From Jansen, 1976). Subdivision of channels into four groups: the e(astern), m(iddle), w(estern) and s(outhern) channels is based upon differences of infill.

erosional and depositional events post-dating the Fisher and Lower Swatchway Beds. In all probability they are equivalent to the buried channels reported by Jansen (1976), which trend in a general northerly direction from 55° to 57°30′N (Fig. 7.25), although Holmes traces them at least as far north as 58°N. In the east of Holmes' survey area the upper surface consists of a flat reflector which averages some 90 m below sea level between 56°N and 57°N and 100 m north of 57°30′N. This surface is overlain by acoustically transparent structureless sands in the south which appear to merge northwards to layered sediments. In contrast the Upper Channels further to the west are truncated by the sea floor. These features commonly erode down to depths in the range of 40–80 m in the west of the area and 80–90 m in the east. Although their detailed shape is not known, they are commonly broader than the underlying Lower Channels.

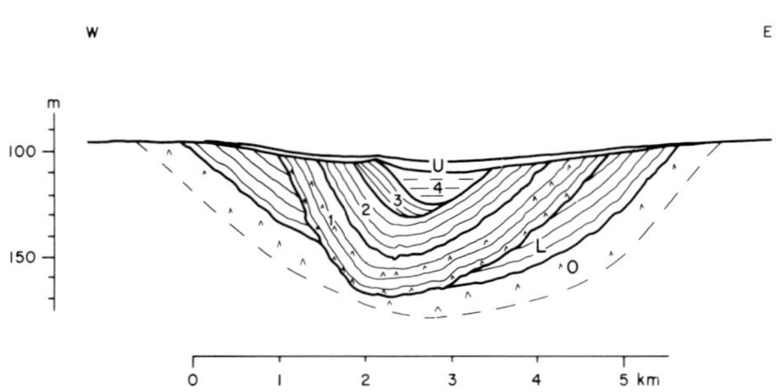

Fig. 7.26. Idealised profile of the buried channels: Older Channel Fill (*O*); Lower Channel Fill (*L*); M_1, M_2, M_3 and M_4 of Middle Channel Fill (*1, 2, 3, 4*); Upper Channel Fill (*U*). From Jansen, 1976.)

Three principal phases of fill are recognised by both Holmes (1977) and Jansen (1976). These have been described by the latter author as the Lower, Middle and Upper, of which the Middle is subdivided into four parts. Locally a fourth unit (the Older Channel Fill) occurs at the base of the sequence (Fig. 7.26). The Lower Channel Fill consists of a finely-layered sediment with a uniform thickness of 10–15 m which is ascribed to sedimentation from suspended load under submerged conditions. Character of the Middle Fill varies considerably, M_1 apparently containing large boulders, M_2 being finely layered without boulders, and M_3 and M_4, which are highly stratified. Jansen relates all these to deposition by currents flowing either longitudinally (M_1, M_2 and M_4) or transverse (M_3) to the channels. He considers that the Middle Channel Fill started with an incision related to a low sea level, and M_1 was subsequently deposited under glacial conditions. M_2 dates from a warmer period with high sea level, which in turn was followed by falling sea level and deposition of M_3 and finally deposition of M_4 in a subaerial or shallow water environment. Drilling results reported by Holmes (1977) have shown that the infill varies from silty fine dense sands to clays and gravel.

A late glacial origin is favoured for the Upper Channels, probably prior to 12,500 B.P. and the extinction of mainland glaciation prior to the Loch Lomond readvance. If they post date the maximum of the late Weichselian maximum at ca. 18,000 B.P., then the channelling may be tentatively correlated with events surrounding the Aberdeen–Lammermuir advance. In this case the asymmetric distribution of Upper Channels towards the eastern part of the Central North Sea may be attributed to a more significant glacial contribution originating from the larger Norwegian centre of ice accumulation than from mainland Scotland.

(8) *Witch Ground Beds.* The Witch Ground Beds are of Holocene and possibly late Weichselian age and are the time equivalent of the Forth Beds in the west of the area. They commonly infill hollows between push

moraines in the Swatchway Beds, and consist of a normally consolidated and multilayered succession of very soft to firm clays, silts and silty sands, which tend to become sandier towards the top. Holmes' unit of this name probably includes the Witch and Fladen deposits of Jansen (1976). Their most notable feature is the occurrence of pockmarks (King and Maclean, 1970) which are typically 2–10 m deep and 10–200 m across. Pockmarks are most common in water depths greater than 120 m deep, and are generally restricted to the areas of soft silts, sands, and clays of the Witch Ground Beds, where they occur in densities greater than 30/km², although there are scattered occurrences within the Swatchway Beds. Their mode of origin is still not proven, although the present consensus of opinion favours formation by gas. Holmes reports a possibly significant correlation of pockmark concentration with the presence of "bright spots" in the South Fladen area, and suggests a causal relationship. Distribution of pockmarks between 57°30′ and 58°N is illustrated in the following section.

Holmes concludes by pointing out that no pre-Weichselian age sediments have been recorded within this area of the North Sea, and that deposits of the Weichselian form a disproportionate portion of the total Quaternary thickness compared with the duration of the Weichselian within the Quaternary. It is clear that sedimentation has been extremely rapid within the middle to late Weichselian, with average rates of 7.5 m per 1,000 years in the Forties field, thus suggesting either very much slower deposition rates in the earlier Quaternary, or removal of huge sections of early Quaternary deposits prior to deposition of the Weichselian.

Forties Field

The general stratigraphy and nature of the Quaternary in the western central North Sea area is given above and is based primarily upon I.G.S. work (Holmes, 1977). A more specific study of the Forties Field area, centred at 57°45′N 0°55′E (Fig. 18.1), has been made in connection with the installation of oil-production installations (De Ruiter and Fox, 1975; Caston, 1977b). In this area the Quaternary sequence is approximately 370 m thick and may be broadly subdivided into two: a relatively featureless lower succession below 122 m sub-sea bed which has not been cored and has only been

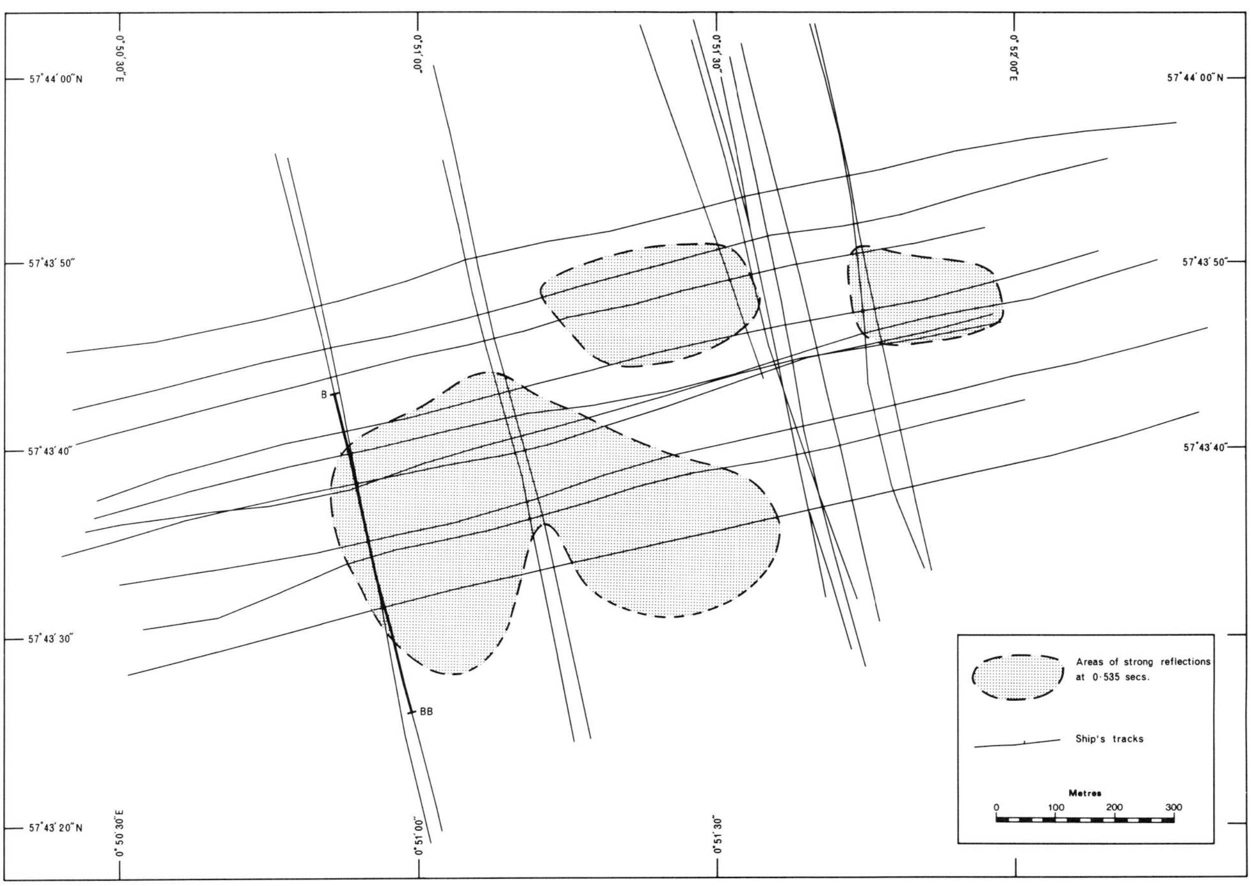

Fig. 7.27. Discontinuous lenses of low-velocity material, interpreted as gas-bearing sand, mapped in the area of the Forties Field, U.K. sector. (From Caston, 1977b.)

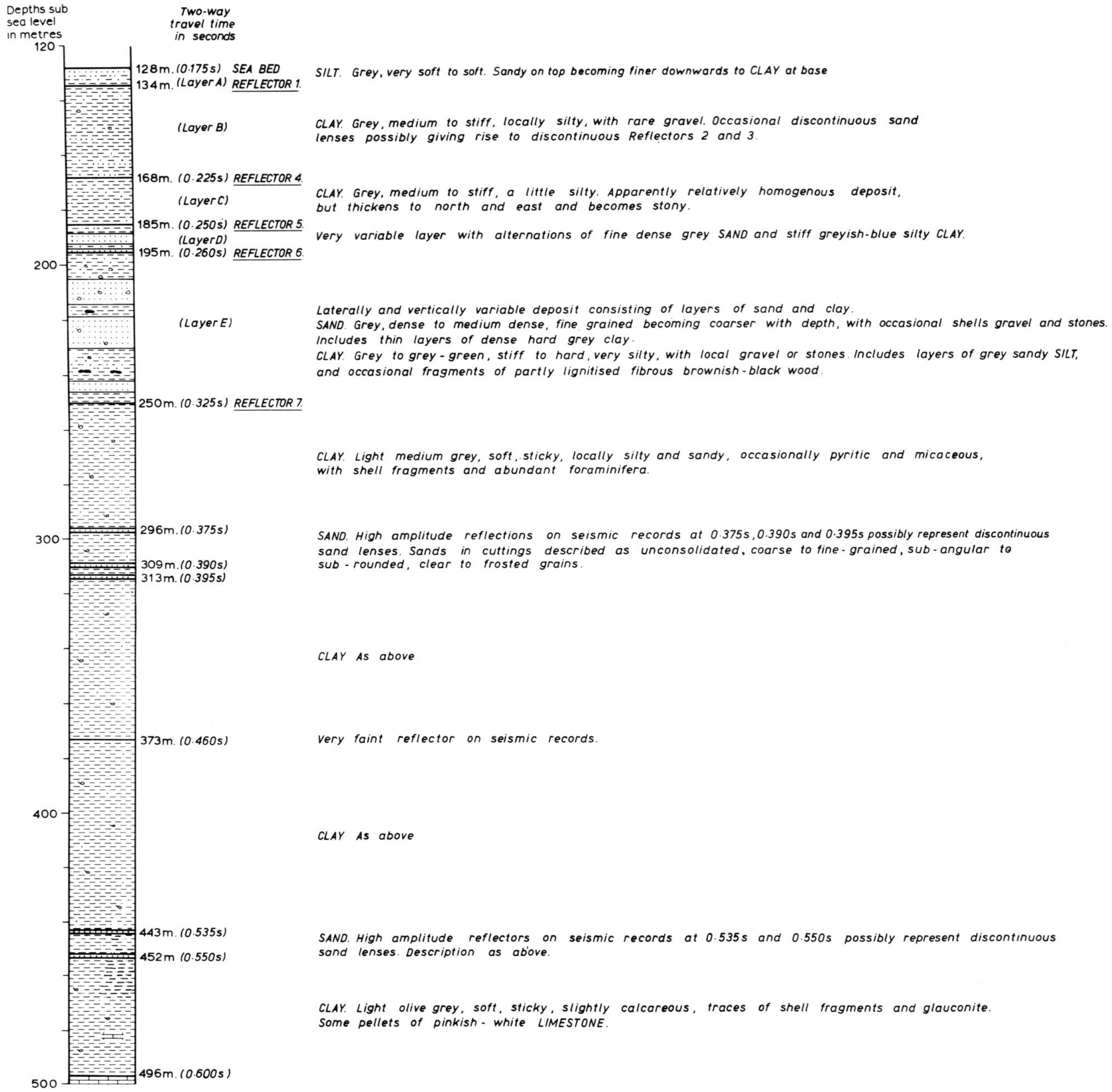

Fig. 7.28. Composite lithological log of Quaternary succession in Forties Field, U.K. sector. (From Caston, 1977b.)

described on the basis of seismic records and exploration well logs, and a more complex upper succession consisting of five distinctive layers composed of silts, clays and subordinate quantities of sand (Caston, 1977b). No dating has been undertaken from borehole samples, but the sedimentary layers identified on sparker records can be correlated directly with those described by Holmes.

A high resolution seismic survey has shown that the lower succession is apparently structureless with the exception of high amplitude but discontinuous reflections occurring at about 170 m and 320 m sub-sea-bed. These high amplitude reflections, giving a characteristic "bright spot" effect, have been interpreted (Lucas, 1974) as low velocity, low density material, and are believed to represent discontinuous lenses of gas-bearing sand. This interpretation is supported by the report of high-pressure gas recorded in well 21/10-3 at a depth of ca. 400 m sub-sea-bed. The lateral extent of some of these discontinuous lenses has been plotted (Fig. 7.27).

In the upper succession five distinct layers, A–E, have been identified on sparker records to a depth of ca. 0.325 s (ca. 122 m sub-sea-bed) and these have been correlated with lithological subdivisions recognised in engineering boreholes. The principal features of the suc-

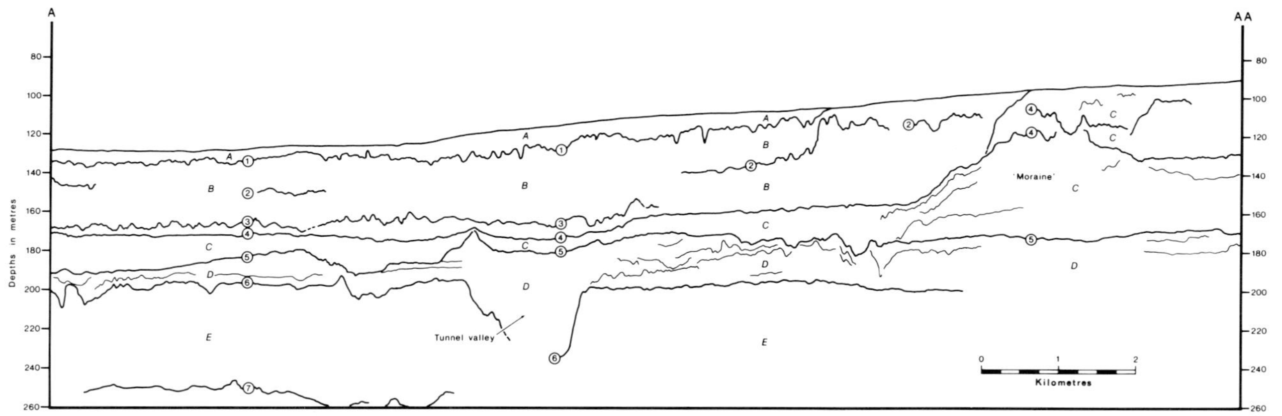

Fig. 7.29. Constant-scale section across Forties Field based upon sparker profile. (From Caston, 1977b.)

cession are shown in Fig. 7.28 which is a composite lithological log based upon the borehole data. The lateral distribution of the sediments across the field is shown in Fig. 7.29, an east–west constant-scale section based upon a sparker profile. Layer E, which corresponds to the Upper Aberdeen Ground Beds (Holmes, 1977) is composed of ill-sorted stony clays and gravels and would appear to be a poorly-stratified till. Into it has been cut closed and subsequently-infilled valleys typical of the Lower Channels. Within the area surveyed the Channels have a maximum length of 10 km, an average width of ca. 1 km, and a maximum depth of 0.078 s (ca. 65 m). The dominant trend is 045°/225°T. They are illustrated in Fig. 7.30, an isochron map of the upper surface of the Aberdeen Ground Beds, which here lies at an average depth of 200 m sub-sea-level. The sediments infilling the valleys, and present as a continuous layer above the till, consist primarily of fine silty sand believed to represent

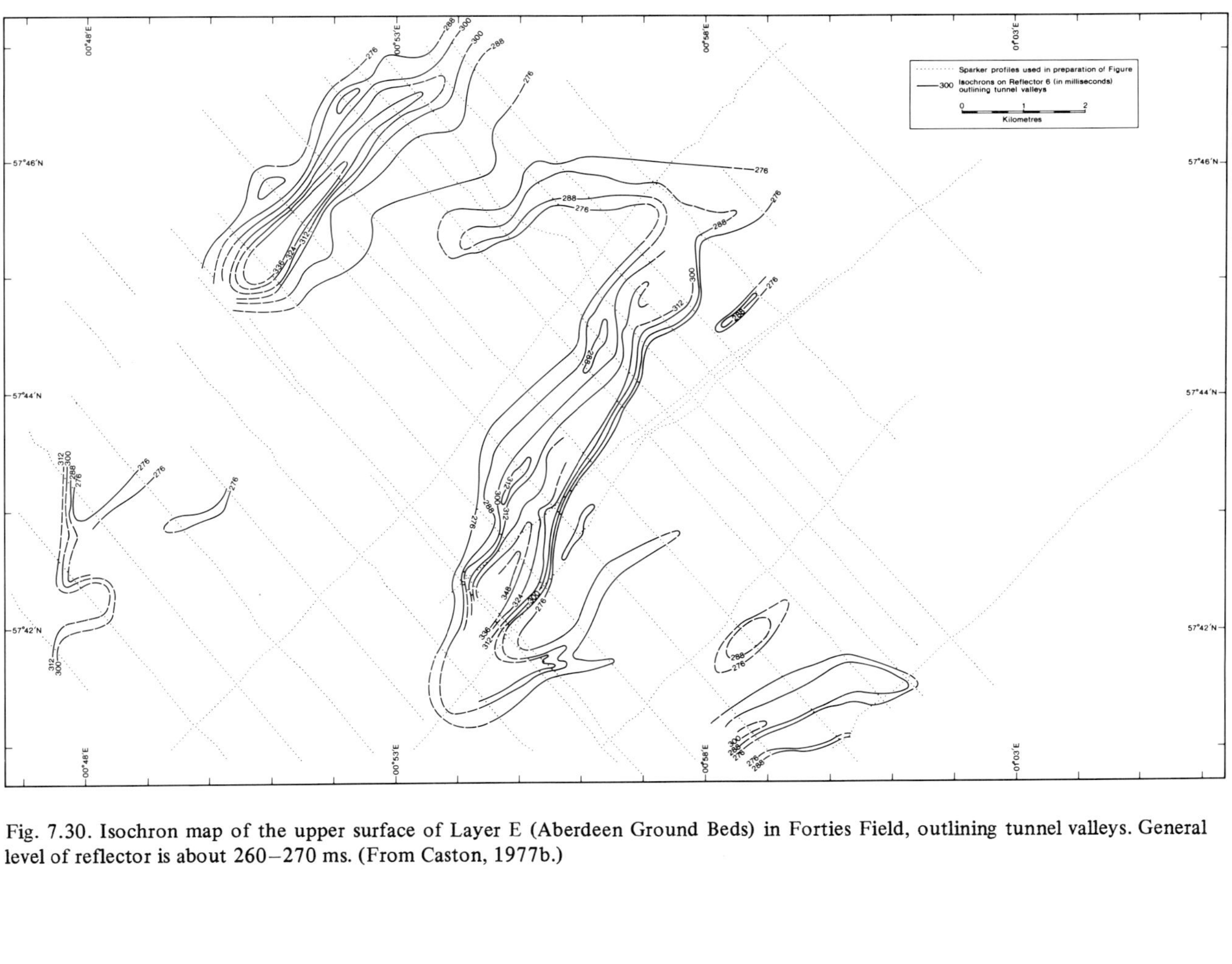

Fig. 7.30. Isochron map of the upper surface of Layer E (Aberdeen Ground Beds) in Forties Field, outlining tunnel valleys. General level of reflector is about 260–270 ms. (From Caston, 1977b.)

a fluvio-glacial deposit. Together with the overlying Layer C they comprise the Lower Swatchway Beds. In the south and west of the field area Layer C is an homogeneous clay, but in the north and east it consists of a thick accumulation of stony clays, sands and gravels, which outcrop on the sea bed. This is believed to be a moraine, which must represent the limit of a re-advance of ice over the Forties area at some fairly late stage in the Quaternary history of the region. The apparent acoustic transparency of the Layer C sediments to the south and west, and the lack of included stones, suggests that it is a waterlain sediment; in view of its proximity to the moraine this could be either lacustrine or marine. Layer B, equivalent to the Upper Swatchway Beds, is also a virtually stone-free clay, and in normally consolidated in the engineering sense. It is probably a glacial-marine clay. The silty clays of the uppermost Layer A are typical of the Witch Ground Beds, and are immediately distinguishable from all underlying deposits by their much higher water content and lower shear strengths. They occur in a bathymetric depression to the west of Forties Bank, upon which stony clays of the Swatchway Beds outcrop at the sea bed (Fig. 7.29). Examination of the relationship between the Witch Beds in the Forties area and the underlying clays suggests that erosion of the banks was contemporaneous with the initial deposition of sediment in the intervening deeps, presumably at the time of the Holocene transgression, the one providing material for the other. Numerous pockmarks have been mapped within the Witch Ground deposits; their occurrence has been reviewed by Holmes (1977) and has been discussed above.

Moray Firth

In 1971 the I.G.S. carried out a drilling programme within the Moray Firth (Chesher et al., 1972). A majority of the seventeen boreholes drilled in the area bottomed in Mesozoic or Palaeozoic strata, thus penetrating the complete Quaternary succession, which ranges from 9 m to over 70 m in thickness. Details of the Quaternary sediments and their distribution from borehole results, together with sparker and pinger profiles, have been given by Chesher and Lawson (in preparation). In the northern portion of the Firth the drift sequences are varied and local areas of accumulation are not well defined, but in the southern half, where the drift is thickest, five elongated west–east trending basins have been mapped on the basis of the 20 m drift isopachyte. From west to east these have been named the Nairn, South Lossiemouth, North Lossiemouth, Banff, and Fraserburgh Basins (Fig. 7.31).

In general the boreholes show that a layer, or layers, of dark grey silty boulder clay with numerous pebbles, predominantly of metamorphic rocks, covers much of the area and in many boreholes lies directly upon bedrock. In many holes this is overlain by a thin layer of pink-brown clay or pebbly clay, similar to that recorded in the Firth of Forth, and this is followed by soft clay or silt, locally laminated or pebbly, which may be of late or immediately post-glacial age. Wood fragments in rhythmites associated with ice melt sequences give ^{14}C age dates of ca. 14,000 B.P. (Late Weichselian). A surface layer 1–2 m in thickness of sand or gravel was recorded at most sites.

U.K. and Norwegian sectors north of 58° N

Because of the concentration of oil exploration activities, this northern part of the North Sea promises to become geologically better-known than other areas further south. At the present time, however, detailed information on the Quaternary is only available from relatively few widely-separated studies and it is not yet possible to build up a regional picture comparable with that for the area between 56° and 58° N. Much of the available data refer to the central part of the northern North Sea, on either side of the median line, and deposits of both sectors will therefore be summarised together in this section.

This northern portion of the North Sea consists essentially of the Norwegian Channel in the east, and a broad basin of irregular relief in the west. These are separated by a line of shallower banks lying parallel to the western side of the Channel which gradually increase in water depth from south to north. Within this area occur many of the linear deeps forming the arcuate belt believed by Flinn (1967) to mark a late-glacial ice front (see Bathymetry, above).

Surficial sediments

The relationship between submarine relief and grain size of the surficial sediments described in detail between 56° and 58° N continues northward into this area, finer-grained deposits occurring in deeper water and the coarser and more poorly sorted material on the banks. The various fisheries charts and those maps largely constructed from them (Luders, 1950; Veenstra, 1971; McCave, 1971; Eisma, 1973; Flinn, 1973; and Caston, 1974) show that the Norwegian Channel is floored almost exclusively with mud. Detailed studies of this deposit to the south of 60° N have been published by Van Weering et al. (1973). In this area unit 1 occurs principally on the western side of the Channel where it locally exceeds 15 m in thickness off Egersund. In this

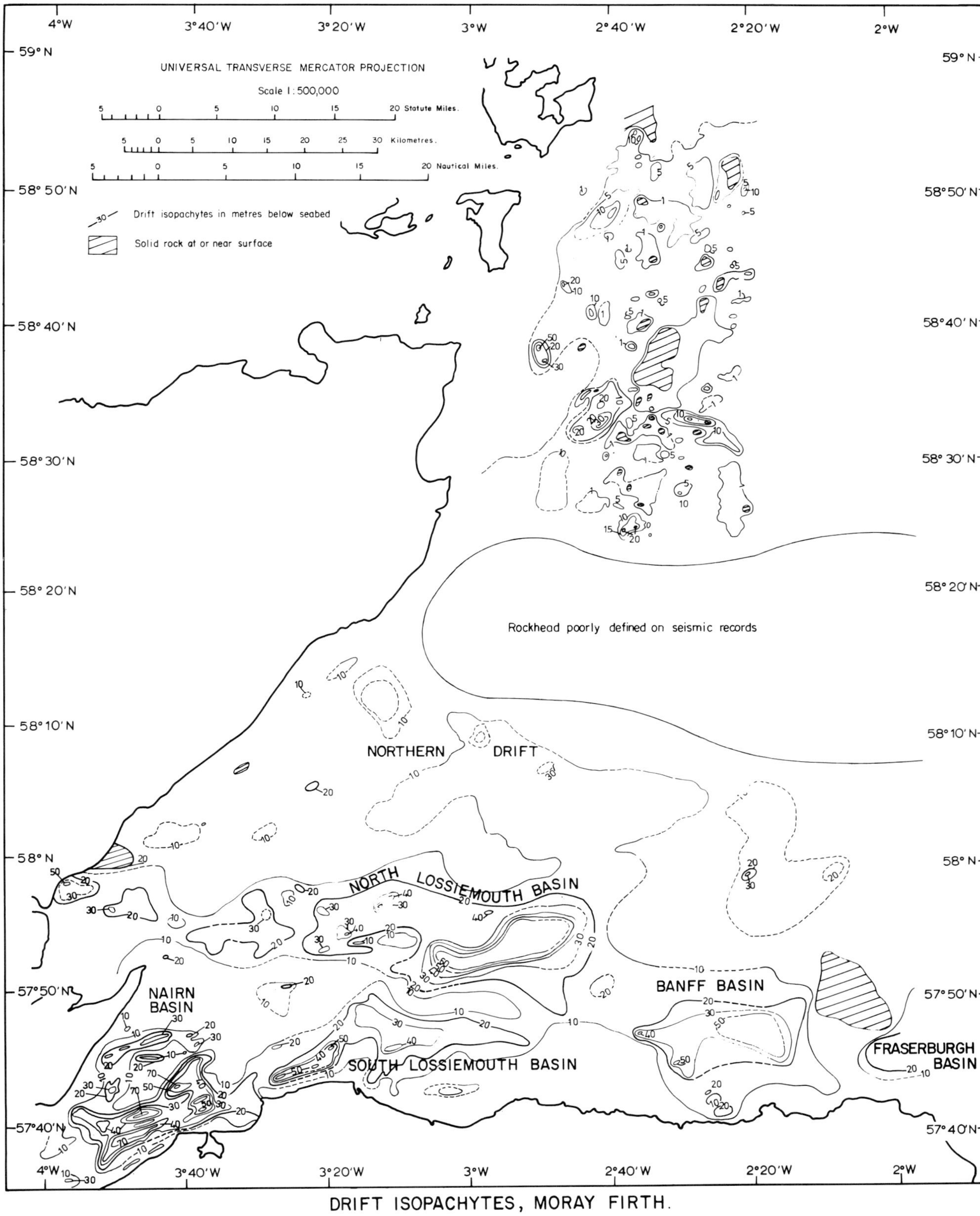

Fig. 7.31. Map of drift isopachytes in southern Moray Firth, showing elongate east–west oriented basins. (From Chesher and Lawson, in preparation.)

area unit 2 also achieves its maximum thickness of some 60 m. Total thickness of the post-glacial cover (units 1 and 2) to the north of 58°N is variable but generally less than 30 m; only in the Egersund basin does it exceed 70 m (Fig. 7.18) which is little more than half the maximum found in the Skagerrak (see above, Norwegian sector). On the western slopes of the Channel chiefly north of Egersund, abundant pockmarks occur in both unit 1 and unit 2 deposits (Fig. 7.32), but no new evidence is presented concerning their possible mode of formation.

The sediment distribution maps show that other patches of mud occur within local deeps in the U.K. sector, in particular in the central and southern Moray Firth; at 59°N 1°E; at 60°20′N 1°E; and in the Fladen Ground centred at 58°20′N 0°30′E (Flinn, 1967).

In this latter area, Jansen (1976) has shown that the Witch and Fladen deposits (equivalent to Holmes' Witch Ground Deposits) observed to the south of 58°N infill a shallow basin extending north to 59°20′N. The Witch deposits, which have a maximum thickness of 15 m (Fig. 7.33), are acoustically transparent with weak and intermittent horizontal reflectors. In the south-east of this basin their base is marked by an erosional reflector, which Jansen equates with the early Holocene (Flandrian) transgression.

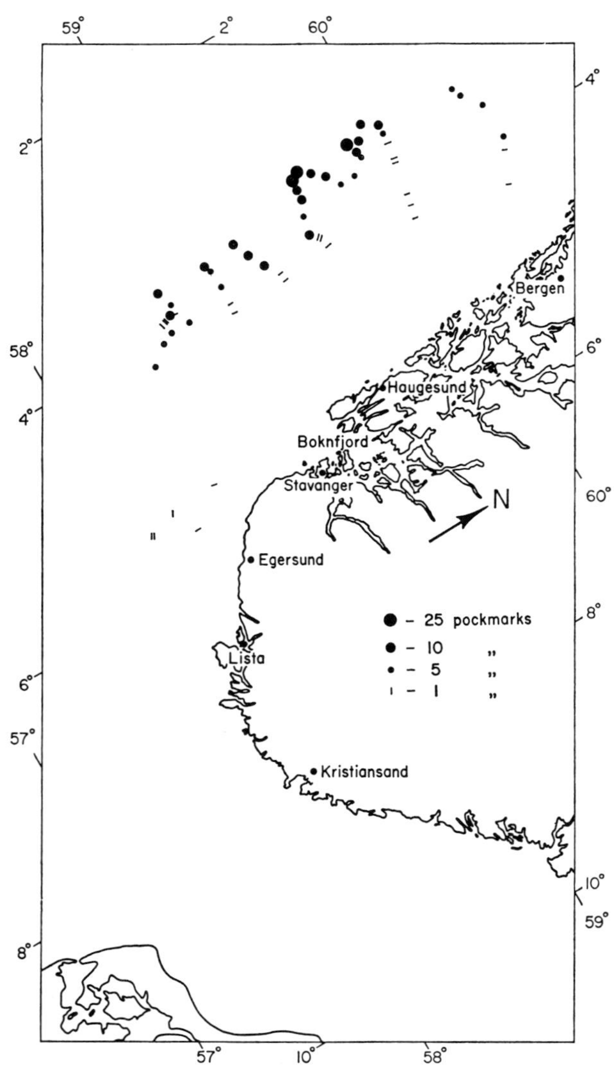

Fig. 7.32. Distribution of pockmarks in Norwegian Channel. (From Van Weering et al., 1973.)

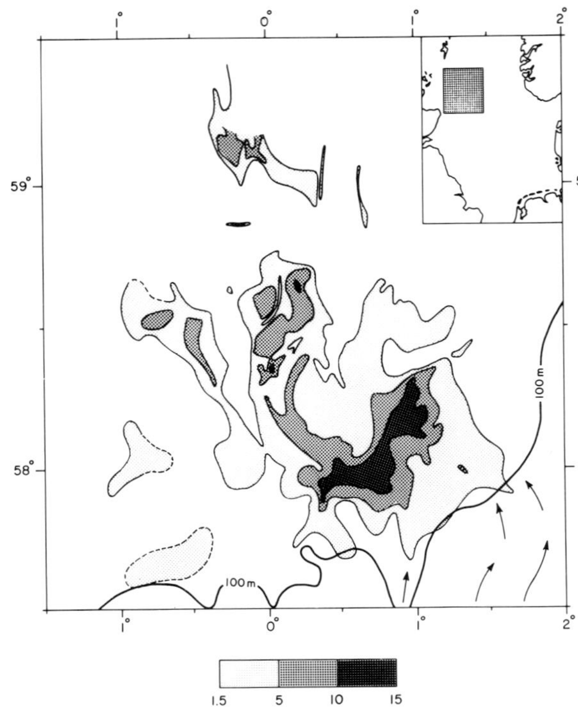

Fig. 7.33. Thickness in metres of (Jansen's) Witch Deposits. Arrows indicate the northern part of the courses of buried channels shown in Fig. 7.25 (From Jansen, 1976.)

The underlying weakly stratified Fladen deposits extend over much the same area within this basin and are also considered to be marine sediments, deposited during a late stage of the final Weichselian advance. Maximum thickness exceeds 15 m. In the south small areas of sound scattering (Beken effect) are recorded, believed to be caused by the presence of gas.

Numerous pockmarks up to 10 m in depth and 200 m in width occur within the Witch and Fladen Deposits. The greatest concentrations—up to 4 per nautical mile of survey track—are present in the centre of the basin (Fig. 7.34).

Patches of gravel and stones cap many of the banks forming the central "spine", and in the Brent area may include boulders up to 0.76 m in diameter (Milling, 1976). Smaller patches of gravel or coarse sand also occur on the shallower areas, such as Bressa Shoal at 59°30′N 0°30′E, lying to the west of the main belt of banks. This is a northward continuation of Pratje's

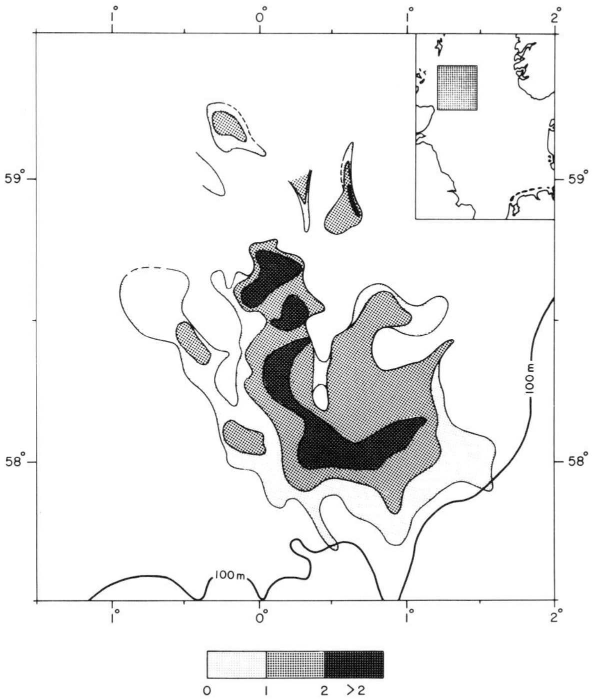

Fig. 7.34. Distribution of pockmarks in the Fladen Ground; density in numbers per nautical mile of echotrace (From Jansen, 1976.)

(1951) assumed westernmost Weichsel stage; the banks are undoubtedly of glacial origin but it is still not possible to establish geographical distribution of individual glacial advances in the western North Sea with even the remotest degree of certainty.

Apart from these instances of mud deposition in hollows and the continuing presence of glacial gravels on the banks, the greater part of the northern North Sea floor is covered by sand. According to Luders (1950) this decreases in grain size very roughly from west to east, the medium grained sands lying mainly to the west of 0° where they become coarse sand and even gravel in a narrow belt adjacent to the Scottish coast and islands. As recognised further south, the sands are of variable thickness and in many areas form only a veneer overlying both the older glacial clays, and, in places, soft early Holocene muds, which under this sandy cover are of greater extent than their present distribution on the sea floor would suggest. Carbonate sands, formed from comminuted shell debris, are common between the Orkney and Shetland islands and constitute the eastern limit of extensive areas of carbonate sand found round the northern and western coasts of Scotland and Ireland.

Information concerning the deeper Pleistocene deposits refers to widely-spaced localities, and thus will be discussed on the basis of individual studies.

Norwegian Channel

A framework for the Quaternary in the Norwegian sector has been provided by a series of air-gun profiles across and parallel to the Norwegian Channel between 59°30′N and 63°N (Fig. 7.35) (Sellevoll and Sundvor, 1974). These show Quaternary deposits unconformably overlying Tertiary, Mesozoic and older strata which dip at 3°–4° towards the west. The maximum thickness of Quaternary deposits within the Channel proper ranges from 220 to 245 m and lies beneath the eastern margin. Near the western edge they thin to a minimum of 45 m, and then increase again to 200–300 m below the shallow banks of the central morainal complex adjacent to the median line. Beneath the Quaternary cover the Channel is still present as a sub-bottom channel, although with a less pronounced topography. Profile J–K (Fig. 7.36) parallels the axis of the Channel, and shows that the underlying Tertiary erosional surface dips gently to the north, the Quaternary cover increasing from a thickness of ca. 200 m at the southern end to more than 500 m in the area where the Channel and the continental slope merge.

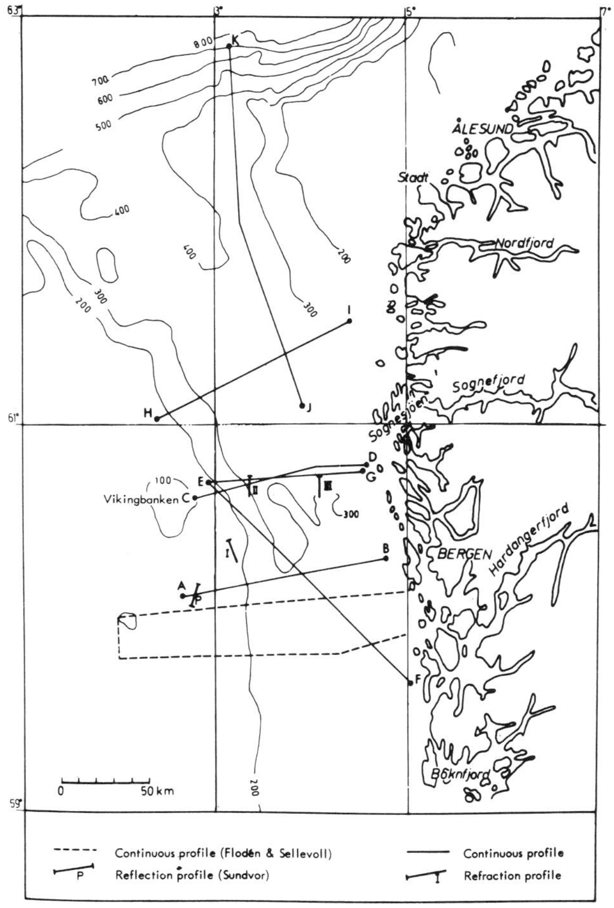

Fig. 7.35. Map showing locations of seismic profiles across Norwegian Channel. Water depths in metres. (From Sellevol and Sundvor, 1974.)

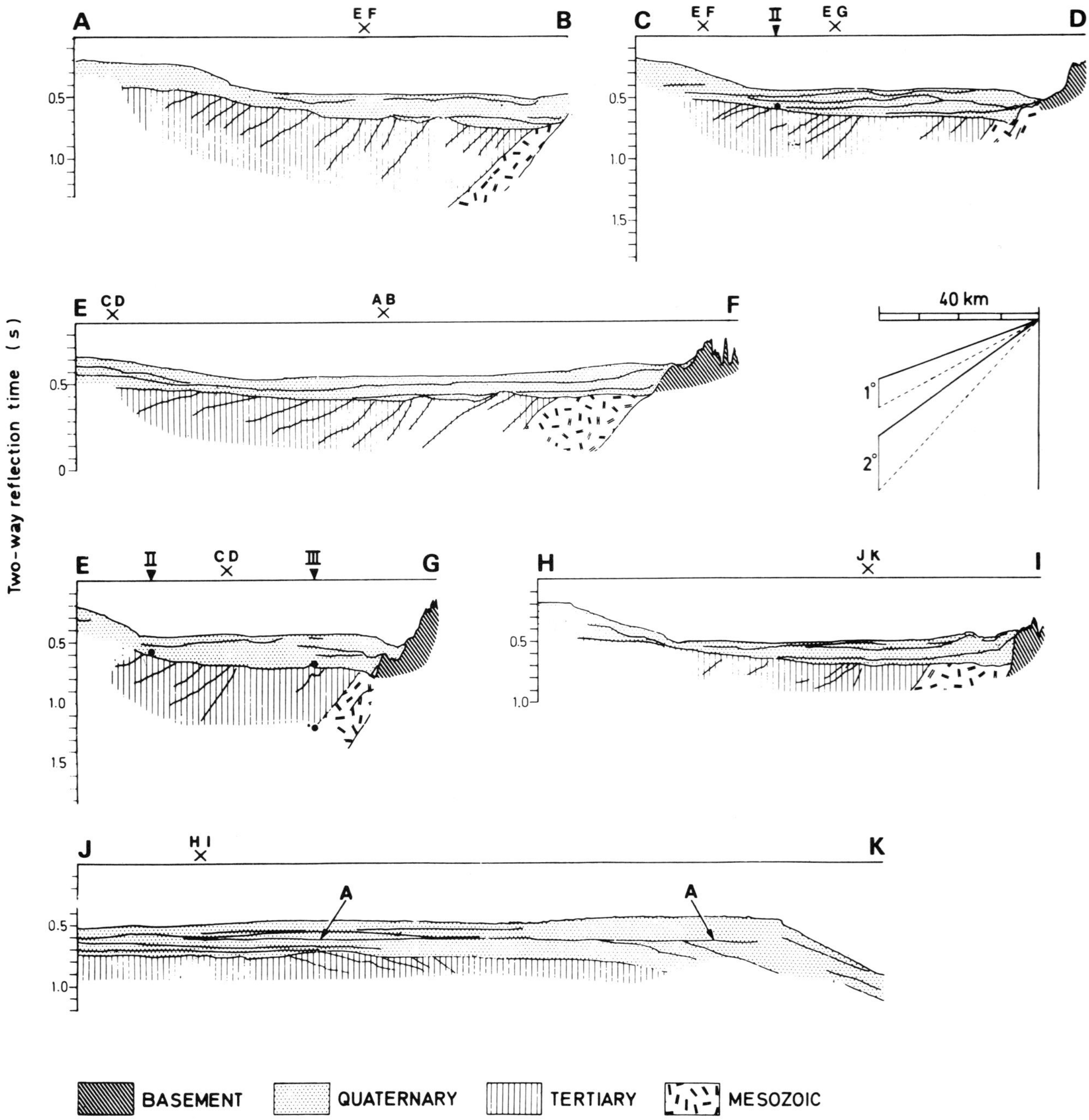

Fig. 7.36. Line drawings of seismic profiler records across Norwegian Channel. For location see Fig. 7.35. The scale shows the true slope for sediment velocities 2.0 and 3.0 km s^{-1}. Control from refraction profiles and crossover points of reflection lines are indicated. (From Sellevol and Sundvor, 1974.)

Stratification of the Quaternary sediments clearly indicates two main directions of transport, one across the Channel and another parallel to its axis. The transport across the Channel was glacial, since striae of the coast clearly indicate that the Quaternary glaciers moved in a westerly direction. Transport along the Channel was more likely fluvial or fluvio-glacial. The beds dip regularly towards the north and the northern part, at the mouth of the Channel, has a deltaic structure with topset beds resting on a foreset section (Profile J–K, Fig. 7.36).

Some results of a survey of the southern Norwegian Channel by Van Weering et al. (1973) have already been discussed. Because of the limited penetration of their profiling system, information concerning the uppermost glacial deposits of unit 3 between 58°N and 60°N has only been determined to any extent off Egersund in the south. In this area it locally exceeds 30 m in thickness.

The Fladen Ground area

Jansen (1976) has described this region which extends as far north as 59° N, and is the northward continuation of Holmes' South Fladen study area. The superficial sediments, the Witch and Fladen Ground Beds, have been described above (p. 244). Jansen also describes the underlying Quaternary deposits down to a depth of ca. 35 m sub-sea-bed. These consist principally of the Lower and Upper Hills Deposits which continue beneath the Fladen Deposits where present, but which outcrop on the sea bed to the north and north-east of these more superficial layers.

The Hills Deposits can be divided into a north-eastern and south-western part (Fig. 7.37). In the former, the deposits reach a maximum thickness of 25 m+ and locally have a hummocky relief, terminating to the south-west in a ridge some 15 m in height. There is an absence of stratification and strong point-source reflectors are apparent. Jansen interprets the bulk of this material as ground moraine, and the ridge as a terminal moraine. To the south-west the morainic material merges into horizontally layered sediments, the Upper Hills Deposits, which overlie the Lower Hills Deposits which are characterised by an absence of internal reflectors and occasional point-source reflectors. These sediments are believed to have been laid down under glacio-marine conditions ahead of the ice sheet. These observations and interpretations are virtually identical to those of Caston (1977b) in the Forties Field area 60 km to the south (see U.K. sector and Moray Firth, above), where similar relief forms and associated facies changes occur within his unit "C", which is equivalent to Holmes' Lower Swatchway Beds.

In the west of this Fladen Ground region occurs a series of roughly northwest–southeast oriented parallel ridges (Fig. 7.38), which appear to predate the Hills Deposits. Jansen notes that they have a "gently undulating surface suggesting erosional action", and proposes that they may be of Saalian age. Thickness of proven Weichselian deposits in this general area of the North Sea (Holmes, 1977), however, suggests that although the ridges may be older than the Swatchway Deposits, they are related to an earlier Weichselian glacial advance. Similar ridges are recorded to the west of the Forties Field at 57°40′N 0°30′E and locally show steeply-dipping and contorted silts lying to the west of a less well-stratified core. There seems to be little doubt that this steeply dipping material (Fig. 7.39) is an "ice-push" moraine, which, in view of its position to the west of a morainic ridge, and its easterly dips, has probably been formed by pressure of ice moving from east to west. These ridges have also experienced extensive erosion, as shown in Fig. 7.39, probably during the course of the early Holocene transgression.

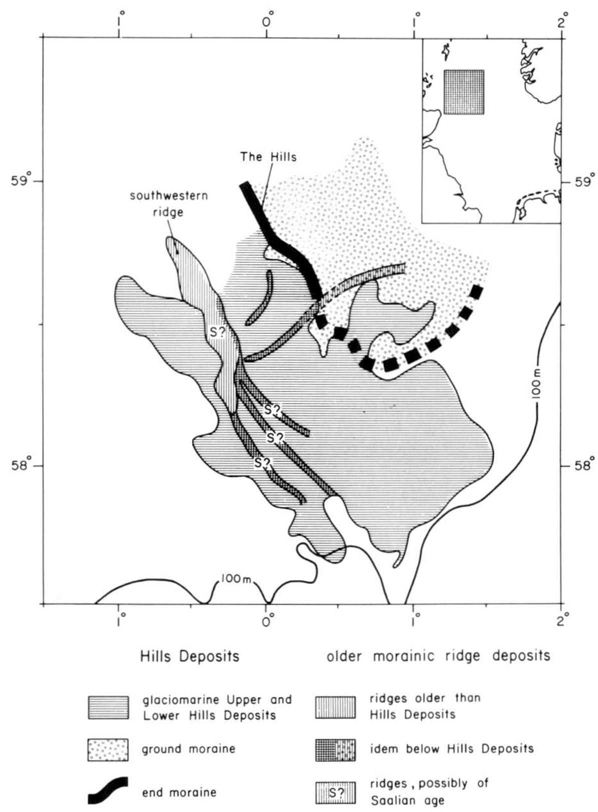

Fig. 7.38. Distribution of the various Hills Deposits and older morainic ridges in Fladen Ground. (From Jansen, 1976.)

Frigg Field

Løfaldli (1973) has reported primarily upon palaeontological investigations carried out on samples from two boreholes drilled in the Frigg Field (Norwegian

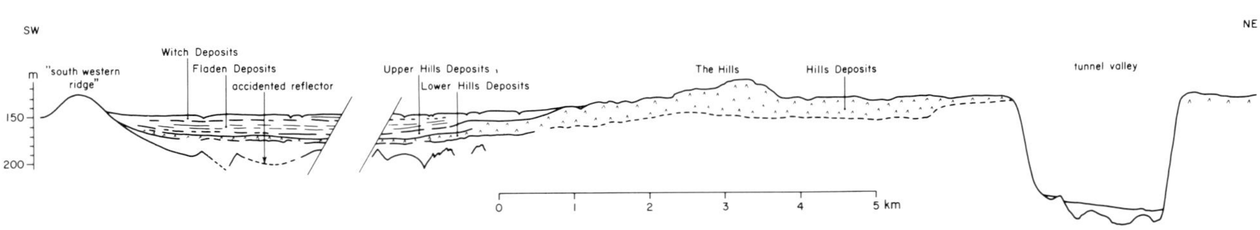

Fig. 7.37. Idealised profile across northern part of Fladen Ground shown in Fig. 7.38. (From Jansen, 1976.)

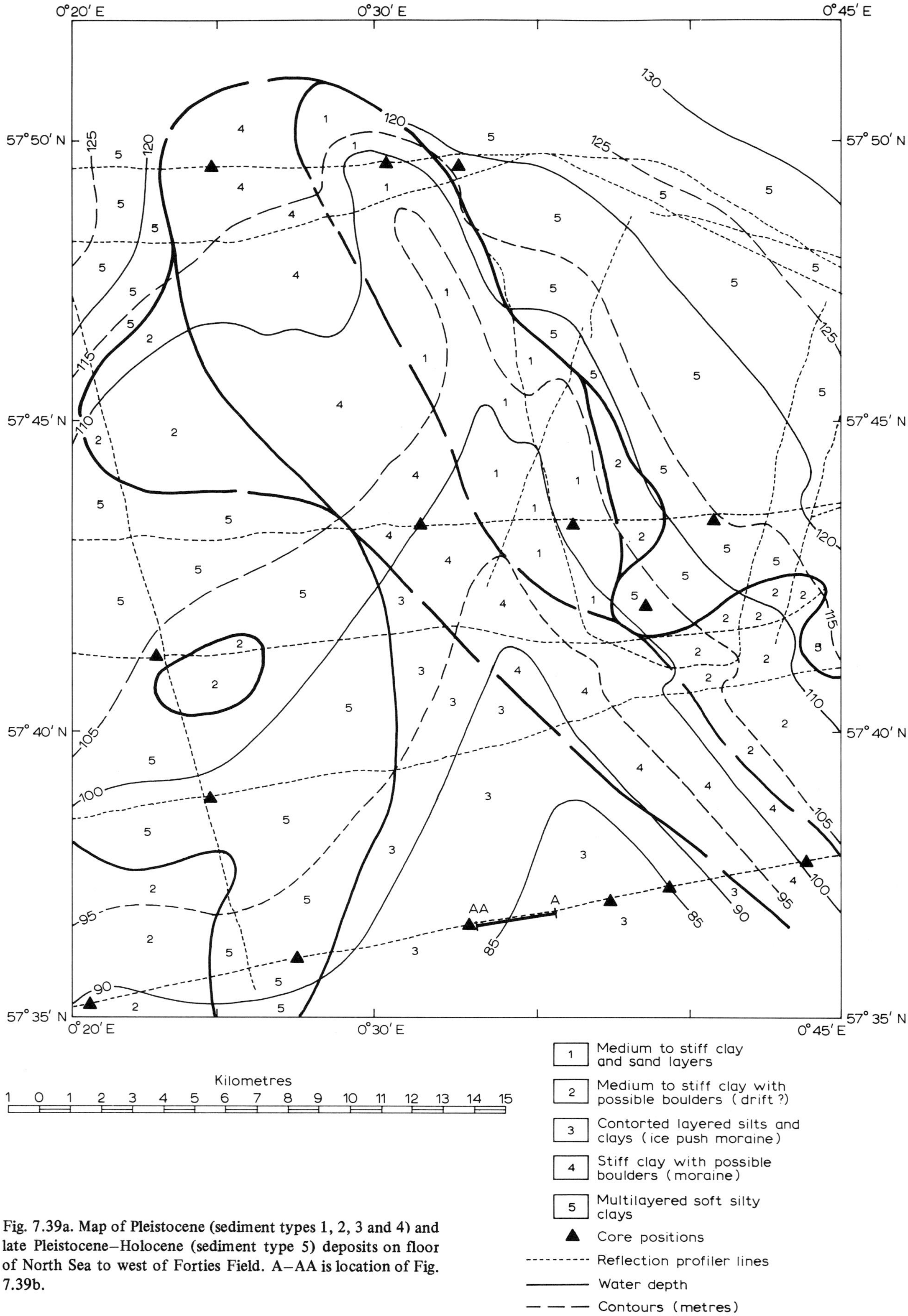

Fig. 7.39a. Map of Pleistocene (sediment types 1, 2, 3 and 4) and late Pleistocene–Holocene (sediment type 5) deposits on floor of North Sea to west of Forties Field. A–AA is location of Fig. 7.39b.

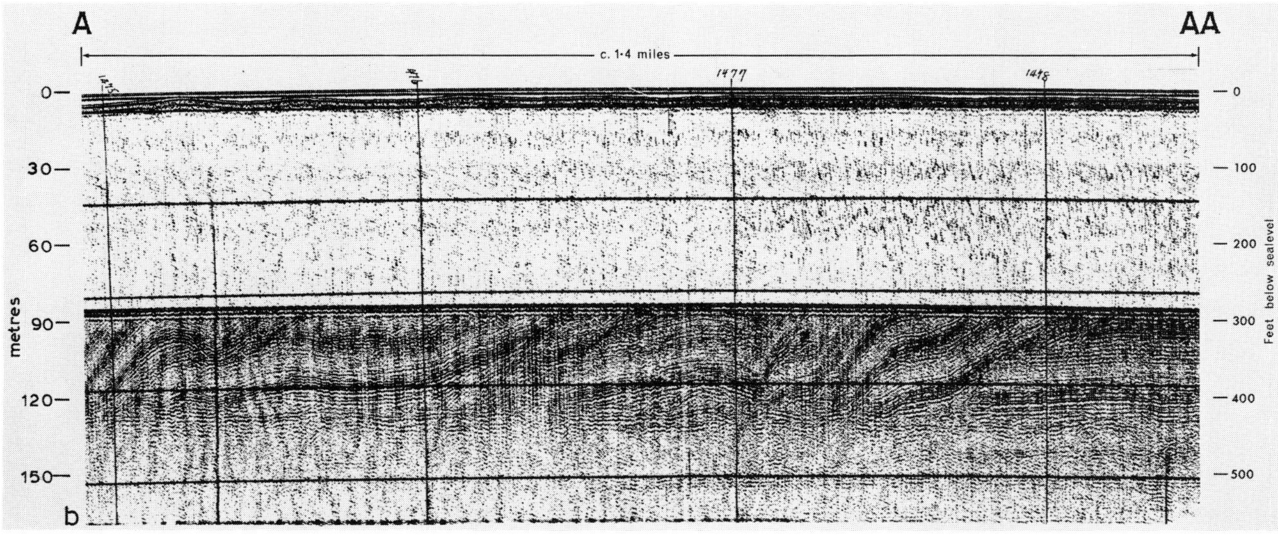

Fig. 7.39b. Portion of profiler record showing contorted silts and clays attributed to ice tectonics.

sector Block 25/1), but also includes a generalised lithological description.

The most complete succession was recovered in borehole B-3-A drilled at 59° 53′ 39.3″ N 02° 04′ 00.3″E in 95 m of water. This penetrated 100.7 m but samples from the upper 50 m only have been described. The following principal lithological units were recognised: 0–9 m fine gray sand with shell fragments; 9–15 m silty fine sand with gravel and shell fragments; 15–ca. 33 m layers of hard clay and fine sand with gravel; ca. 33–ca. 42 m sandy clay with some gravel; and ca. 42–50 m silty fine sand with some shell fragments and gravel. A five-fold palaeontological zonation was established based upon benthonic foraminifera. Zone I, at the sea bed, is considered to be of Holocene age. Zone II, from 1.8 m to 9.1 m, was deposited in from shallow to moderately deep waters during "the Postglacial time", although the oldest part of the zone is believed to date from immediately after the maximum extent of the Weichselian glaciers (about 15,000–18,000 years B.P.). Zone III, from 11.0 m to 20.2 m, contains a much richer foraminiferal assemblage than in Zones I and II, with a dominantly arctic population. The indications are that these sediments were deposited under shallow water conditions close to an ice front, probably before or during the latest advance of the Weichselian glaciation, i.e. about 18,000–25,000 years ago. Zones IV, from 21.9 m to 28 m, and V, from 31.3 m to 49.8 m, include an arctic fauna from a moderate water depth, and are believed to date from an interstadial before the last Weichselian ice advance, i.e. ca. 25,000–30,000 years and 30,000–35,000 years B.P., respectively. Average sedimentation rates in the investigated parts of the boreholes at both the Frigg Field and Booster Station range from 1.3 mm to 1.4 mm/year.

Frigg Field Booster Station

Further lithological and palaeontological information is given by Løfaldli (1973) for a borehole drilled on a shallow bank (water depth 95 m) adjacent to the Fladen Ground Deep at 58° 49′ 40.8″ N 00° 17′ 14.2″ W in U.K. Block 14/9. The hole was drilled to a penetration of 65.7 m, but the published information refers only to the upper 35 m, and lithological information is limited. It has the following succession: 0–11 m medium sand with gravel, stone and shell fragments; 11–20 m layers of fine sand, silt and clay with shell fragments; 20–30 m layers of sand and gravel with decomposed black organic material; and 30–35 m fine sand with layers of silty clay and silty sand. Three foraminiferal zones were identified; Zone I, in the uppermost sample only ; II from 2.1 m to 12.2 m; and III from 12.2 to 35 m, are as described for the Frigg Field and cover the Late Weichsel–Holocene period.

Norwegian Sector, Block 25/11

Two papers by Milling (1975, 1976) include some of the most comprehensive information available in the northern North Sea area, and are particularly valuable because they provide ^{14}C dates for the deposits at one site. The studies concern geological and engineering investigations undertaken in the vicinity of the Brent

Field (U.K. Block 211/29, centred at 61°05′N 01°42′ E), and in the northern half of Norwegian Block 25/11 centred at 59°7′30″N 02°30′E (M.E. Milling, personal communication, 1976).

The Norwegian block lies some 90 km south of the Frigg Field straddling the line of relatively shallow-water banks bordering the Norwegian Channel. Two boreholes and a high-resolution sparker profile (Figs. 7.40 and 7.41) show that up to 46 m of sandy silty clay, believed to consist essentially of reworked glaciomarine and glacio-fluvial deposits, overlie at least two separate stony till sheets separated by a layer of silty marine clay. The two tills are overconsolidated in an engineering sense, and have a massive non-laminated nature, poor sorting (pebbles up to 10 mm in diameter), low water content (< 20%) and high bulk density (> 1760 kg/m³). In contrast, the more normally consolidated units are well laminated, have a higher water content, and lower bulk density. The observed variations in geotechnical properties of the cohesive clay units are related to differences in their depositional regimes rather than to changes in their textural and/or clay mineral composition; grain-size fluctuates only slightly and clay mineral composition remains essentially constant. Milling considers that the tills were deposited by the Scandinavian ice sheet in successive westward advances across the Norwegian Channel during the Weichselian. The younger, well-laminated and sparsely-fossiliferous silty clayey sands and silty sands of Units B and C are believed to be reworked glacio-fluvial outwash and near-shore marine sequences deposited during the last transgression when the Late Weichsel (< 23,000 years B.P. – Morner, 1972) ice margin stood a short distance to the east.

High resolution seismic control reveals that the area is underlain by a series of deeply incised buried channels and inter-channel divide areas related to a Late Pleistocene erosion surface (Fig. 7.41). These channels are up to 3,000 m wide and greater than 75 m deep. They are relatively steep-sided and sharply truncate Units E–G, the upper surface lying at a depth of 45–60 m sub-seabed, which is somewhat less than that recorded in the Forties area (see above), but much the same as in the Brent Field (see below).

Brent Field

In the Brent Field (Fig. 18.1), which is the most northerly area in the North Sea for which information has been published, Milling has described Late Quaternary deposits in the immediate vicinity of the Late Pleis-

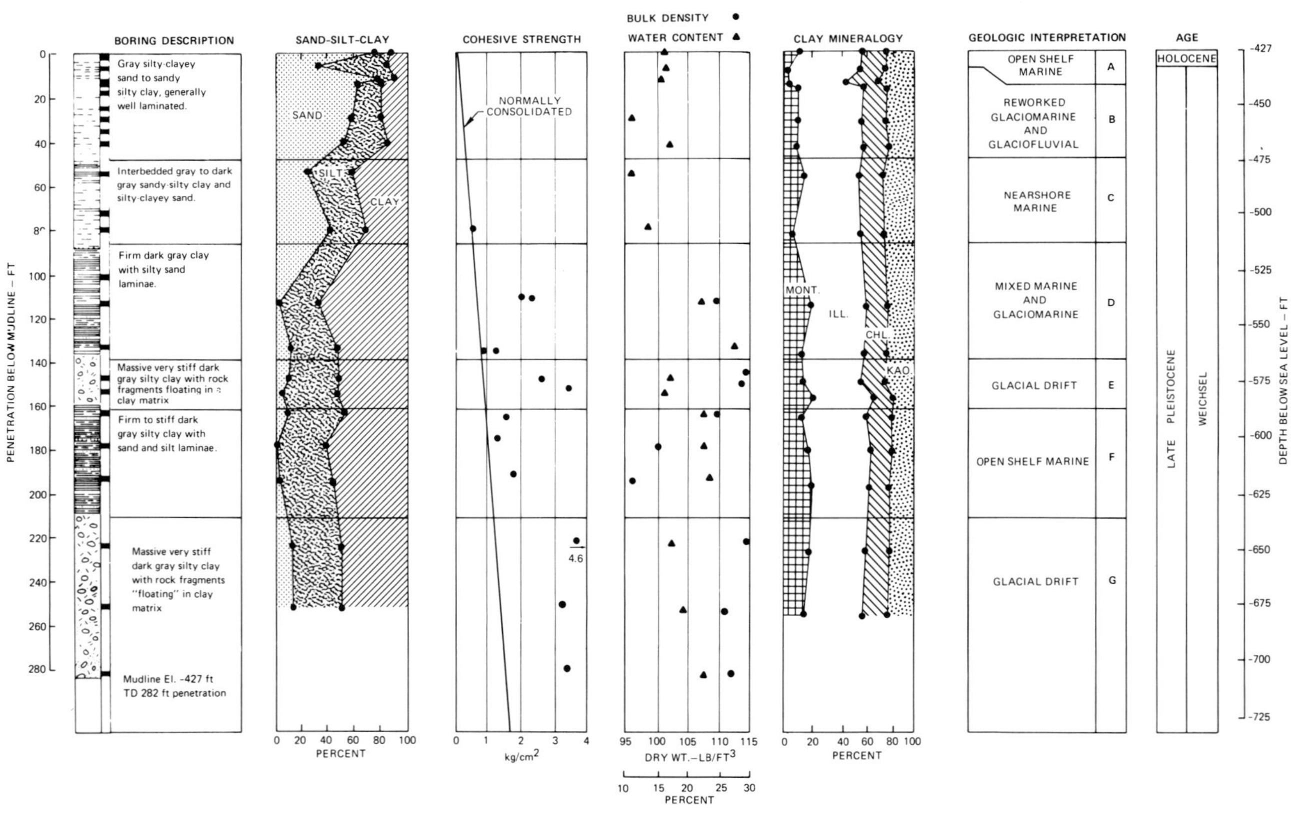

Fig. 7.40. Geologic description and interpretation of laboratory test results of cores from soil boring "A" in Block 25/11, Norwegian sector. (From Milling, 1975.)

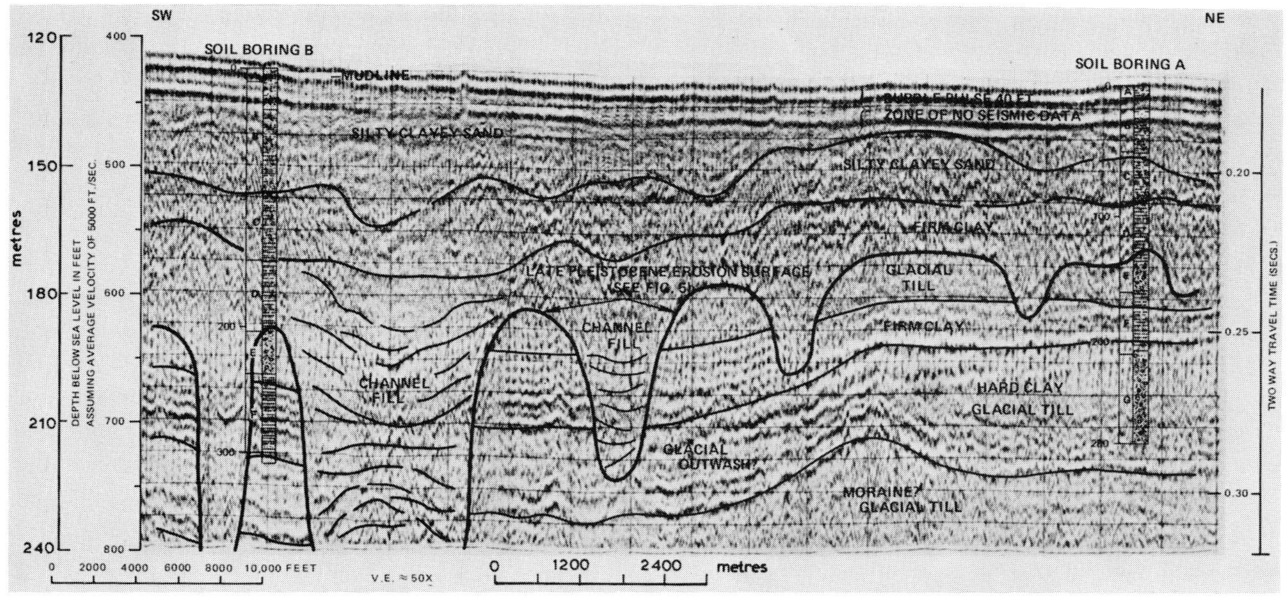

Fig. 7.41. Interpreted high-resolution sparker profile from Block 25/11, Norwegian sector, showing distribution of lithofacies units encountered in soil borings "A" and "B". The units of the lower foundation are sharply truncated by deeply-incised channels related to a late-Pleistocene erosion surface. (From Milling, 1975.)

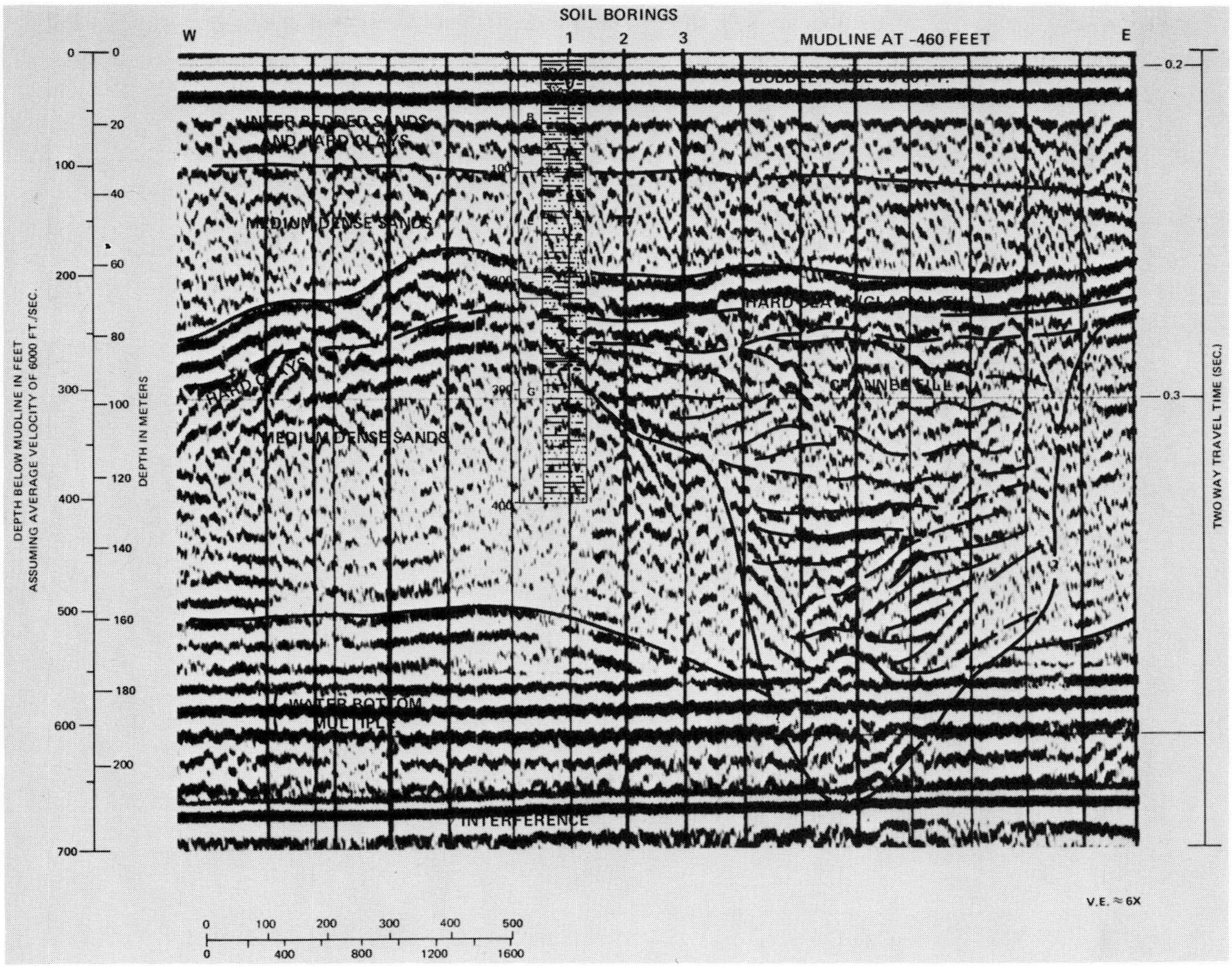

Fig. 7.42. Interpreted high-resolution sparker profile from Brent Field, U.K. sector, showing soil boring control and distribution of lithofacies units. (From Milling, 1975.)

tocene morainal complex along the west side of the Norwegian Channel. In this area hard overconsolidated glacial till occurs at a shallow depth below the surface. Otherwise foundation-zone sediments are grossly similar to those previously described from the Norwegian sector, and are characterised by interbedded stiff to hard clay and thick dense sand units truncated once again by deeply-incised buried channels. These channels are from 60 to 1,000 m wide, greater than 100 m deep, and occur at a depth of 60–75 m below sea bed (Fig. 7.42).

The nature of shallow near-surface sediment sequences within Unit A (Fig. 7.42) has been determined by shallow core and cone penetrometer soundings. Radiocarbon dates on shallow-water marine mollusc shells taken from cores places a Middle Weichsel (< 60,500 years B.P. – Morner, 1972) to Holocene age on this upper unit, which consists of interbedded sequences of dense silty sand and stiff to hard clay derived from the Middle and Late Weichsel ice sheets. These sediments were deposited in a morainal ridge complex adjacent to the outer edge of the ice front. The clays are glacial–glacio-marine tills and are characterised by their overconsolidated nature (e.g. shear strength > 4.5 kg/cm², Fig. 7.43). The dense sands are generally poorly sorted and are interpreted as glacial outwash fans and channel deposits. The surficial Holocene gravelly sands are derived from reworking of the Late Weichsel glacial deposits.

A detailed study of near-surface sediments has been based essentially upon stratigraphic correlation of closely spaced soil borings, shallow continuous cores and cone penetrometer soundings (Sanglerat, 1972), because acceptable equality high frequency sub-bottom profiles are not available. The resultant geological cross-sections show that the thicker dense silty sands have limited lateral extent and are confined to a series of buried shallow channel complexes (Fig. 7.44). The sand-filled channels are incised into the stiff glacio-marine clays. The interbedded sand–clay sequences can be subdivided into two intervals on the basis of their geometry and lateral extent. The upper interval, 0–3 m below the sea bed, is characterised by thinner widespread sheetlike sand bodies. Sand bodies in the lower interval (> 3 m below sea bed) are thicker but less widely distributed and appear to be clearly channelised. The lower channel sands are up to 10 m thick. Their gross three-dimensional distribution has been illustrated by the construction of a fence diagram (Fig. 7.45). This shows that the dominant trend of the sand filled channels is to the southwest, but overall the sand concentration in the shallow near-surface sequences is highly variable. In the northern and eastern margins of the area, net sand thickness decreases to less than 2 m, or 30% of the total section between 0 to 6 m below the sea bed. The highest concentration of sand occurs along a diagonal NE–SW trending line through the centre of the area where the net sand thickness is greater than 5.4 m, or 80% of the total section (Milling, 1976).

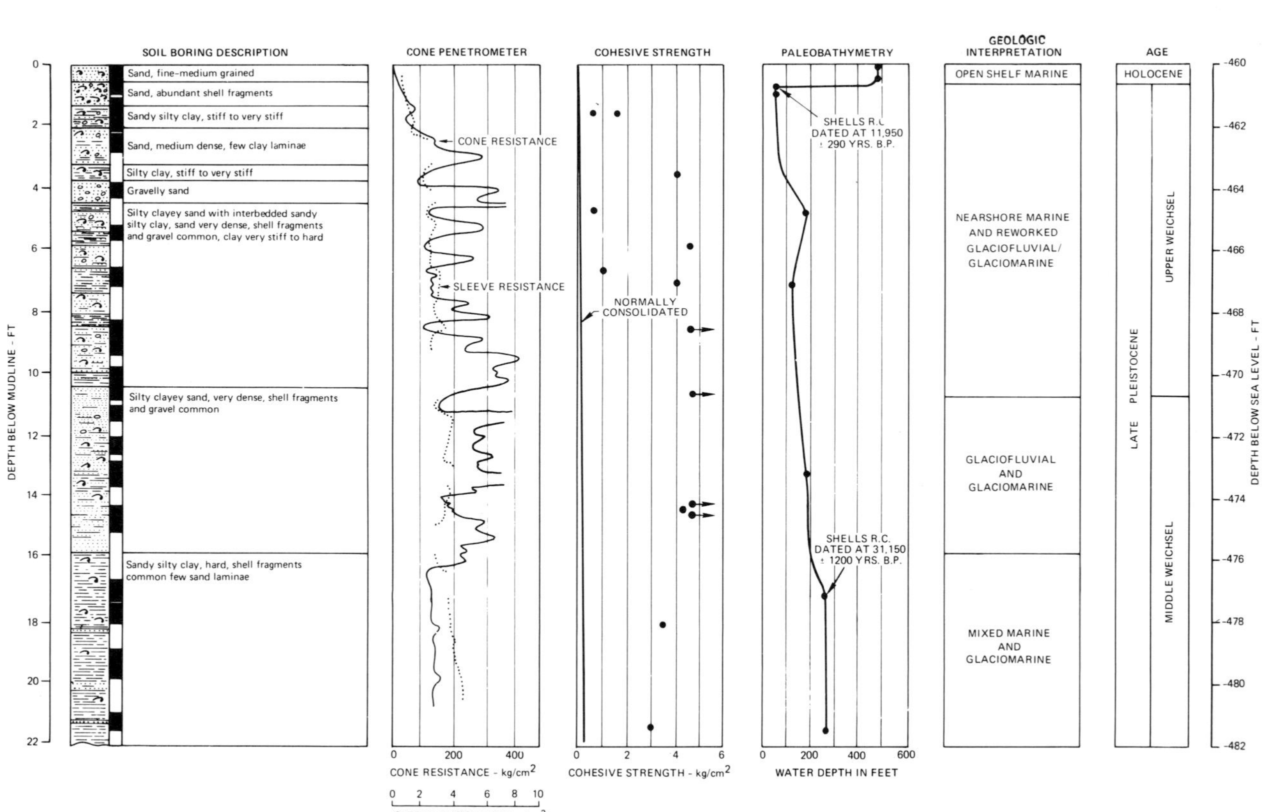

Fig. 7.43. Detailed description and laboratory test results of near-surface sediments encountered in soil boring 1 (see also Fig. 7.42 and nearby cone penetrometer sounding, Brent Field, U.K. sector. (From Milling, 1975.)

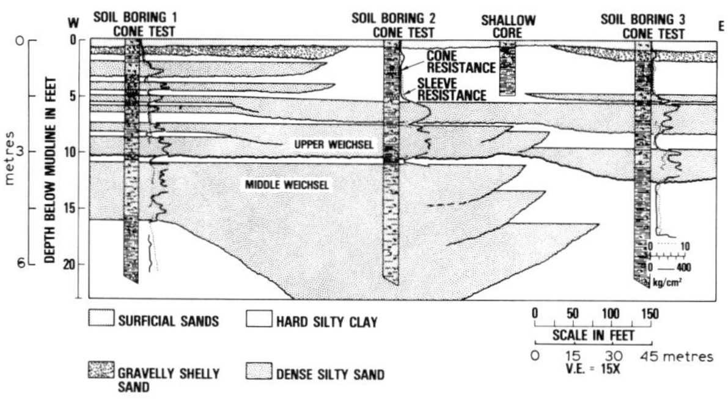

Fig. 7.44. Geologic cross-section based upon detailed correlation of near-surface soil boring cores, shallow sea-floor cores, and cone penetrometer soundings from Brent Field, U.K. sector. The dense silty sand bodies are interpreted as Late Middle Weichsel and Late Weichsel glacial outwash channel sands. (From Milling, 1976.)

Analysis of benthonic foraminifera and marine mollusc assemblages collected from shallow cores indicates that water depth in this area gradually decreased from a high of 100–200 m in late Middle Weichsel time, and then rapidly increased during the Holocene to more than 140 m (Fig. 7.43). Milling points out that a rise in sea level of 120 ±m during the last 12,000 years is not in agreement with published eustatic sea-level curves. He quotes Morner (1971) as saying that a generally accepted figure for the magnitude of sea-level rise during this time is 35–55 m, and suggests that the difference between the apparent 120 ±m and that predicted suggests that the sea floor in this area has undergone more than 60 m of subsidence during the past 12,000 years. Support for this concept is provided by a study of the coastal and submarine features surrounding the Shetland Islands (Flinn, 1964), which

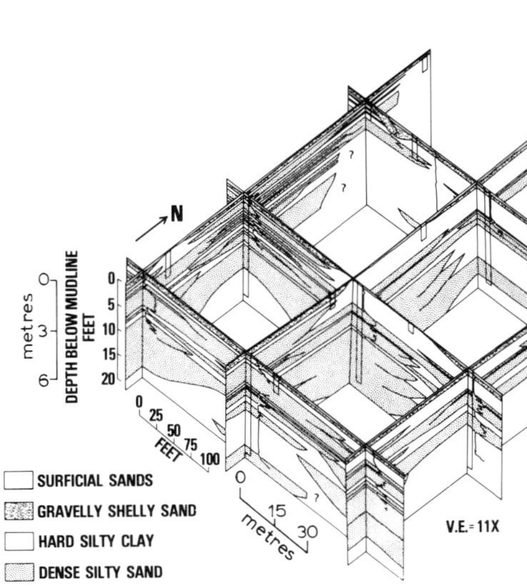

Fig. 7.45. Fence diagram illustrating three-dimensional distribution of near-surface sand-filled channels in Brent Field, U.K. sector. (From Milling, 1976.)

indicates that the Shetlands are an area of recent submergence which "may have continued unabated to the present time".

Recent sedimentary bed-forms

The present-day distribution of sediments on the floor of the North Sea has been described above. One aspect of this distribution concerns the characteristic bed forms which have developed in response to, and are indicative of, sediment transporting processes. These occur principally in the southern North Sea, and are sufficiently abundant to merit separate consideration. On this topic there is now a quantity of published information, as so much of our knowledge of sediment transport by tidal currents is derived from studies undertaken in the North Sea area. The bed forms concerned are composed essentially of sand, although there is some evidence that in high-energy environments particles up to gravel size are also involved. The question of sand transport and processes has recently been summarised for the North Sea by Stride (1973), and so will not be repeated in detail here. The present review concentrates rather upon the distribution and morphology of the principal features, together with implications that may be drawn concerning the pattern of transport (also see Chapter 8). Sediments and sedimentary processes in the intertidal and shallow coastal waters off the U.K., The Netherlands, and West Germany have been studied in detail by a number of workers (e.g. Evans, 1965; Van Straaten, 1964; Reineck and Singh, 1975), but are not considered in this Chapter.

Pioneering work by Van Veen in the 1930's (Van Veen, 1935, 1936, 1938) first drew attention to major sedimentary bed forms (sand banks and sand waves), and suggested that their formation was related to the flow of tidal currents. This was followed by the work of Cloet (1954) and then by Stride and his colleagues commencing in the late 1950's (Stride, 1959b, 1963, 1965, 1970, 1973; Stride and Cartwright, 1958; Johnson and Stride, 1969; Caston and Stride, 1970, 1973; Kenyon and Stride, 1968, 1970). Using side-scan sonar (asdic), the orientation, size and distribution of these features could be plotted, along with other newly-discovered bed forms such as sand ribbons (Stride, 1959b; Flemming and Stride, 1967), and their implications assessed (Belderson and Stride, 1966). By the early 1960's a correlation between the occurrence and orientation of these features and the direction and strength of flow of the dominant tidal currents was firmly established (Stride, 1963), stimulating more widespread interest in the nature of sedimentary processes and bed forms throughout the area of the southern North Sea (Dingle, 1965; Langeraar, 1966; Anonymous, 1967; Houbolt, 1968; McCave, 1971; Terwindt, 1971; Caston, 1972). Some attention has also been directed towards the transport of mud (McCave, 1970, 1973), although this material is primarily carried in suspension and does not form significant mobile bed forms (see Chapter 8). Increasing interest is being paid to the generation of non-tidal currents and their sedimentological importance (Johnson and Stride, 1969; McCave, 1971; Gienapp, 1973; Pitt et al., 1973; Stride and Chesterman, 1973; Koltermann and Lange, 1975; Caston, 1976), but little is yet known concerning their quantitative effects.

Sand waves

Van Veen (1935, 1936, 1938) and Cloet (1954) both recognised that sand waves were oriented at right-angles to the direction of flow of the tidal currents, but it remained for Stride and Cartwright (1958) and Stride (1959b) to develop this concept and to use sandwave orientation and asymmetry, together with the set of strongest tidal streams, to predict the directions of sediment transport over a wide area.

Distribution

The extent of the sand-wave fields in the southern North Sea steadily became apparent with increasing survey coverage through the 1960's (Stride, 1963, 1965, 1970; Dingle, 1965; Langeraar, 1966; Houbolt, 1968; Johnson and Stride, 1969; Kenyon and Stride, 1970; McCave, 1971; Terwindt, 1971; Caston and Stride, 1973 – and see Fig. 8.3). During a co-operative study in 1966–1968, eight oil companies and the National Institute of Oceanography pooled their data concerning sand-wave distribution and movement in the southern North Sea (Caston, 1969). On the basis of ca. 40,000 km of echosounder lines, a new map showing the distribution of sand waves between 52°30′and 55°00′N has been compiled (Caston and Stride, in preparation). This map (Fig. 7.46), which incorporates earlier data of Dingle (1965), Houbolt (1968), and McCave (1971), shows that there are two major sand-wave fields in the southern North Sea, one in the Southern Bight, which has been the object of considerable attention by English and Dutch workers, and the other off Flamborough Head, which has been studied in detail by Dingle (1965). The former has an area of ca. 15,000 km^2 according to McCave (1971), and the latter, on the basis of Fig. 7.46, an area of ca. 5,000 km^2. The extensive coverage available to the joint project enabled the limits of these fields to be determined with more precision than in the past; this is especially so of the northern limits of the Southern Bight field.

Between these two fields sand waves occur in patches ranging from single waves to areas in excess of 350 km^2. These patches are commonly, but not exclusively, found in association with shoal areas or banks.

In general terms the greatest concentration of sand-wave patches occurs off the coast of Norfolk, where trains of sand waves form intricate patterns in association with linear sand banks. A complex of patches is also found off the Lincolnshire coast in association with the Lincolnshire Banks (see below). A further zone of sand-wave patches extends north-north-west from the Norfolk coast towards the Flamborough Head field, although a lack of detailed survey data has prevented the full extent and shape of patches in this area from being determined.

Sand waves are infrequent and poorly-developed within an inshore zone extending for some 35–45 km off the Lincolnshire and Yorkshire coasts and north of 53°30′N. This scarcity reflects a general paucity of sand in this area, boulder clay and, in the north, Mesozoic rocks, being exposed on the sea bed over much of the zone.

The outer limit of sand-wave patches can be drawn quite clearly in a line from the Southern Bight around to about 54°30′N. North and east of this point sand waves recognised upon isolated survey lines are to be regarded with caution. This area corresponds to the south-western tip of the Dogger Bank and the sand waves shown are poorly formed. They may be true sand waves, or the features seen on the echosounder records may be related to some other form of sea-bed relief. It is possible that they are temporary sand waves, formed under extreme conditions. This outer limit cannot be simply explained in terms of a particular tidal current velocity, although in the Southern Bight the northern limit of waves has been ascribed by several authors to the known decrease in current velocity from south to north. The boundary is rather the result of a number of inter-related factors, including bottom current velocity, sediment grain-size, amount of sand, and water depth, which are of varying significance from location to location.

North of about 54°40′N sand waves are infrequent and are confined to a fairly narrow zone immediately offshore (Kenyon and Stride, 1970). Depending upon sand supply and current velocities they are, however, locally more abundant, as in the vicinity of the Orkney and Shetland Isles and off the east coast of Aberdeenshire where waves a maximum of 22 m in height (trough to crest) occur in a water depth of 70 m some 12 km offshore. On the eastern side of the North Sea lower

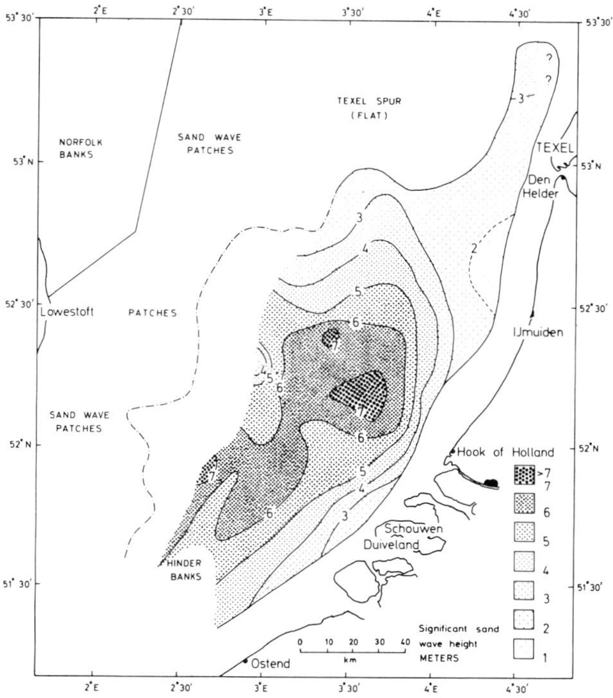

Fig. 7.47. Significant heights (in metres) of sand waves in Southern Bight of North Sea. (From McCave, 1971.)

tidal current velocities generally preclude the formation of true sand waves, although waves up to 7 m in height have been reported from off the north-western tip of Denmark, at ca. 57° N (Bruun and Vollen, 1972; Stride and Chesterman, 1973). In the same area the latter authors have recorded the presence of ill-defined linear sand patches oriented in a north-easterly direction; as mean spring tidal velocities in this area are little more than 12 cm/sec. (Von Sager, 1968), their formation is largely ascribed to the importance of wind-generated components.

Detailed figures for sand-wave heights and wavelengths are not available for the whole of the southern North Sea. The most complete coverage has been published for the large sand-wave field in the Southern Bight (McCave, 1971). His map (Fig. 7.47) of significant wave height H (the mean height of the highest one-third of sand waves) shows that the highest waves ($H > 6$ m) occur in the centre of the field south of 52°30′N with the principal peak of $H > 7$ m to the north-west of Hook of Holland. Sand waves decrease in height steadily to the north, east and south-east, and a pronounced tongue of waves extends northwards off the coast of the Island of Texel as far north as 53°25′N. A similar pattern, with the largest sand waves in the centre of the field, occurs off Flamborough Head (Dingle, 1965). Here the largest waves exceed 15 ft. (4.5 m) in height (Fig. 7.48), and there appears to be a relationship between wave height and wavelength, the longest wavelengths also occurring

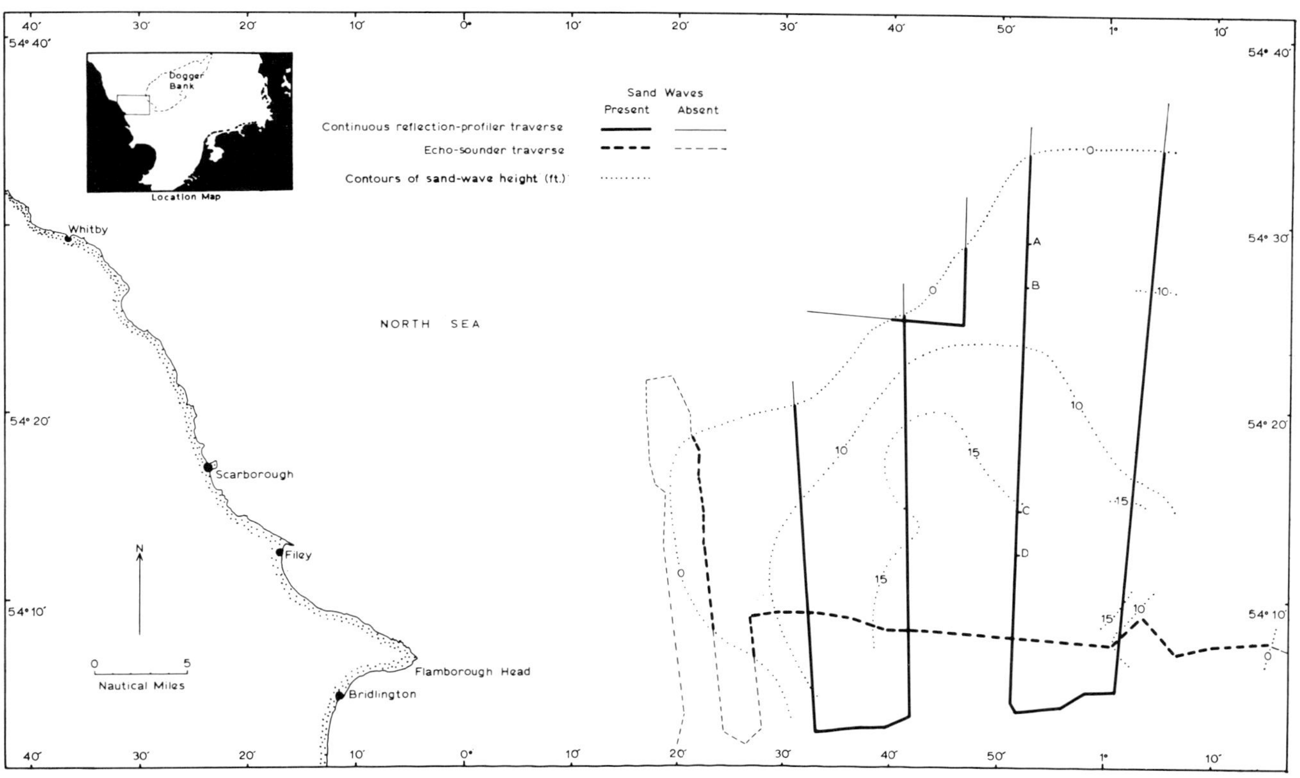

Fig. 7.48. Area of sand waves to east of Flamborough Head showing wave-height contours (in feet) as measured from wave trough to crest. (From Dingle, 1965.)

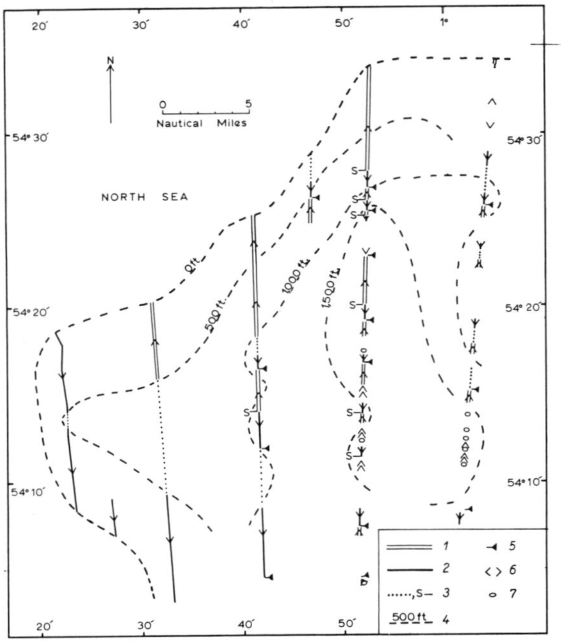

Fig. 7.49. Sand-wave area shown in Fig. 7.48 showing extent of areas of sand waves having similar wavelengths – "wavelength contours" – in feet (shown by broken line – no. 4 in Legend). (From Dingle, 1965.)

in the centre of the field (Fig. 7.49). In the Southern Bight, however, McCave (1971) noted no such relationship.

Inferred directions of sand transport

The relationship between sand-wave asymmetry and the direction of sand transport has been studied in the North Sea and elsewhere, and there is general agreement that transport proceeds in the direction faced by the steeper side, which corresponds to the direction of flow of the strongest tidal currents. On this basis the direction of sand transport in the western and southern North Sea has been plotted by Stride (1963, 1973) and Kenyon and Stride (1970). Fig. 7.50 (Stride, 1973) shows that sand is transported away from sites of erosion (bed-load partings) located in the Southern Bight off north-east Norfolk, off south-west Scotland, and in the Pentland Firth, towards bed-load convergences, found off Aberdeenshire, Flamborough Head, the Straits of Dover, and in the German Bight. In addition sand seems to be moving towards estuaries such as the inner part of the Moray Firth, Firth of Forth, the Wash and the Thames, as well as off Holland and in the Wadden Sea (see Fig. 8.2).

Superimposed upon these general transport directions however are smaller-scale movements which may run counter to the overall trend, the details of which are becoming apparent only with more comprehensive survey coverage. Such movements are significant in that they are believed to be responsible for the construction of sand patches and sand banks, as demonstrated by Caston and Stride (in preparation) for an area some 80 km long by 4 km wide lying parallel to the sand transport path in a north-north-westerly direction from just off the coast of Norfolk. In this zone sand-wave patches range from 2.7 to 6.5 km in length (parallel to the transport direction), and are consistently separated by from 8 to 10 km of sand-wave-free ground. The dominant tidal current direction is towards the north-north-west and a majority of the asymmetric sand waves face in this direction. There is however a small but persistent number of south-facing sand waves which appear to indicate movement in the opposite direction to that of the nett transport. The relative positions of these waves may be a clue to the significant processes. The northerly-facing waves are commonly found on the southerly or "upstream" side, symmetrical waves in the centre of a patch, and the south-facing waves on the northerly or "downstream" side.

To explain this, it is suggested that, although the north-north-westerly stream may be dominant, the (relatively) weaker south-south-easterly currents are responsible for the local movement of significant quantities of

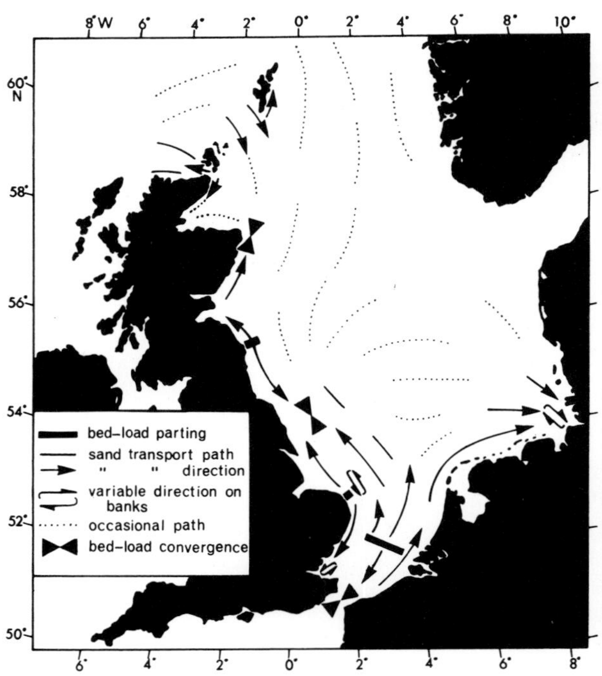

Fig. 7.50. Inferred sand transport paths (solid lines) and directions (arrows) for the sand wave zone, and the probable paths (dotted lines) for the regions of weak currents and correspondingly low transport rates. The symbol for sand movement around sand banks is schematic. (From Stride, 1973.)

sand (cf. Vol. II, Fig. 14.11). The sand wave patch, with wave heights frequently in excess of 4 m, affords a measure of protection on the northerly or "downstream" side thus allowing the south-south-easterly stream to become locally dominant. An extension of this hypothesis would provide a mechanism for the growth of a sand bank.

Further evidence for such construction processes will be discussed below.

Sand banks

Sand banks occur principally around the margins of the Southern and German Bights of the southern North Sea. They include prominent linear features aligned parallel or sub-parallel to the coastline together with other accumulations which are less well defined. The majority are constructional features apparently formed of sand by the action of tidal currents. Certain banks appear from profiler records to have a central "core", whilst other features, such as Brown Ridge (Van Eerde, 1964; Houbolt, 1968; Dujon, 1970; Oele, 1971b) probably consist almost entirely of erosional remnants now believed to be of early Holocene age (Oele, 1971b). Movement of certain of the inshore banks such as those off Lowestoft has been proven by repeated surveying (Cloet, 1963); further offshore, movements are less easy to detect and, in practice, probably proceed at a slower rate (Houbolt, 1968; Caston, 1972).

Van Veen (1936) first drew attention to the linear sandbanks off the Flemish coast (south of 52°N) and noted that they were oriented parallel to the direction of flow of the strongest tidal currents (Vol. II, Fig. 9.12). This relationship was also emphasised by Off (1963) who reviewed the nature and distribution of linear banks on a worldwide scale. The structure of both the Flemish and Norfolk Banks, off the East Anglian coast, has been described in detail by Houbolt (1968). He reported that all banks in the latter area were found to consist of sand and to rest upon an essentially flat, gravel-covered, sea floor, thus enabling an isopach map to be constructed of total sand thickness (Fig. 7.51). The banks all possess a strong asymmetry towards the north-east, and this, together with internal foreset bedding, suggests sand transport and thus eventual bank movement in that direction. Houbolt proposed a mechanism for the growth and development of these linear banks which involves the transport of sand by tidal currents obliquely along the gentle south-western flank towards the bank crest. From the crest the sand is deposited on the steep slope and in part transported obliquely downslope in a south-easterly direction (Fig. 7.52). There is thus a clockwise circula-

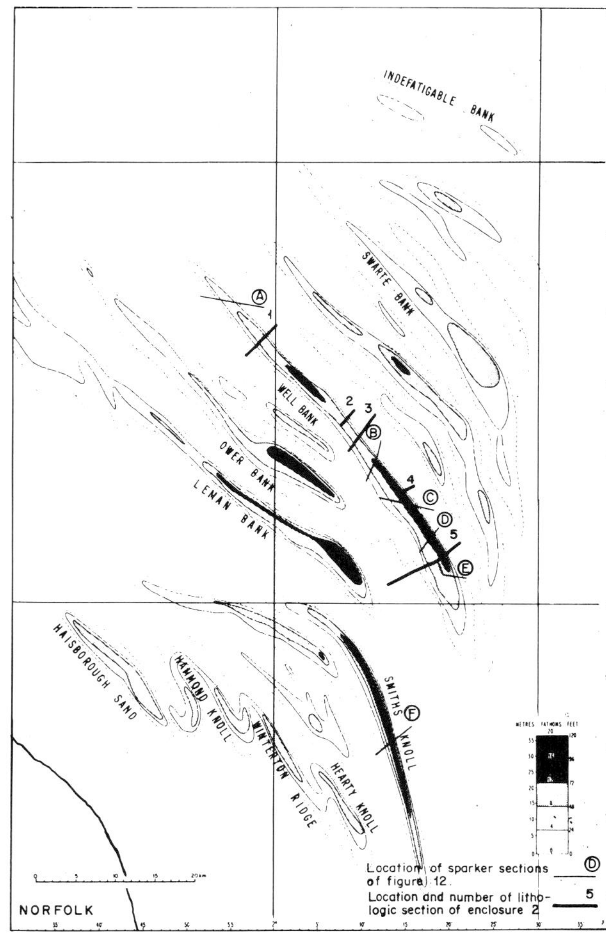

Fig. 7.51. Approximate isopachs of sand ridges in Norfolk Banks area. (From Houbolt, 1968.)

tion of sand around each bank. In order to explain the continued presence of the banks, which reach well into the zone of wave action, Houbolt further proposed that the tidal currents within the intervening channels or swales flow in elongate spirals in such a way that the streamlines are directed outward from the centre of the swale towards the adjacent banks, thus transporting sand from the swales to the bank crests (Fig. 7.53). Further studies in this area and that of the Flemish Banks, including measurements of tidal currents and sidescan sonar surveys, confirmed many of Houbolt's findings (Caston and Stride, 1970; Caston, 1972) but emphasised that on both sides of a bank sand streams are progessively deflected towards the crest, which is thus a line of convergence. This proposed convergence at the crestline would thus result in an accumulation of sand, providing material for the growth of the bank which, it is suggested, would proceed both vertically and laterally (parallel to the direction of flow of the tidal currents). Vertical growth would be limited by the depth of water, and the method of lengthwise growth is not proven, but

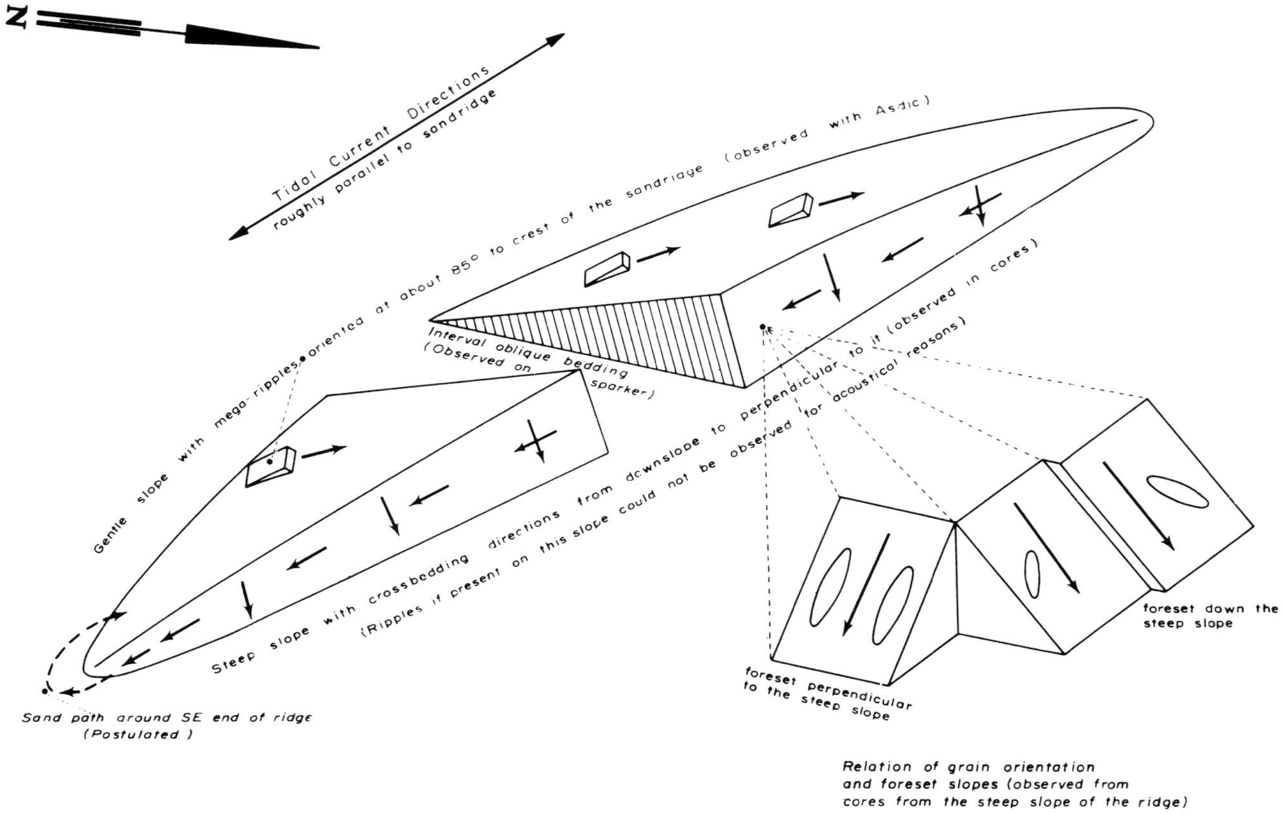

Fig. 7.52. Schematic presentation of sedimentary features of Well Bank, Norfolk Banks area. (From Houbolt, 1968.)

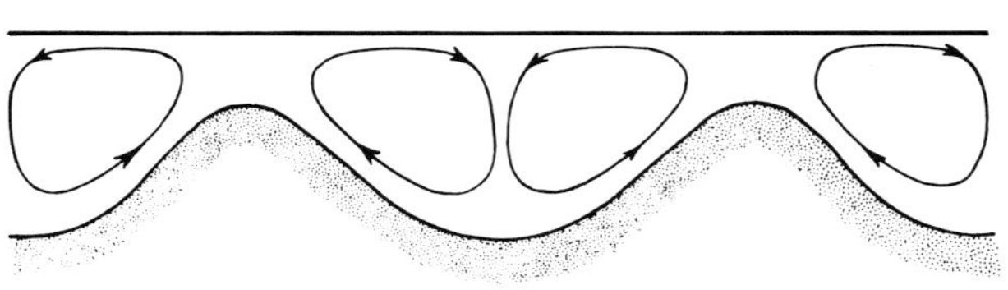

Fig. 7.53. Cross-section over a system of ridges and swales showing assumed spiral currents. (From Houbolt, 1968.)

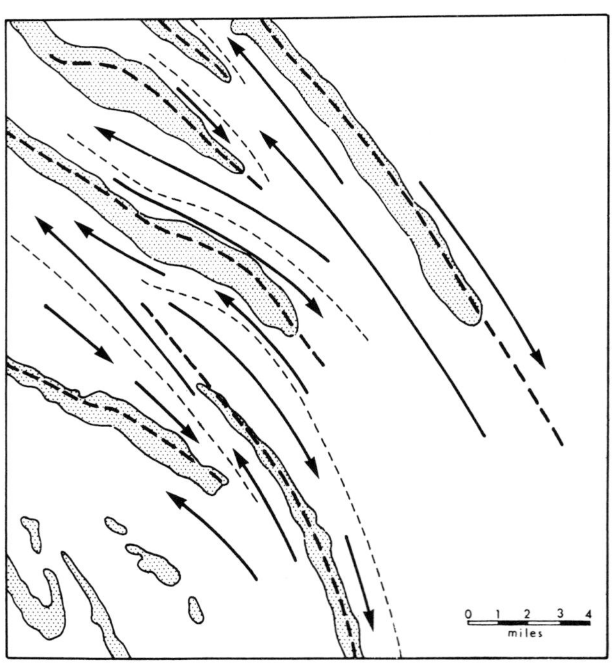

Fig. 7.54. Inferred sand-transport directions in the area of the Norfolk Banks. The approximate boundaries between the paths are shown by thin dashed lines extending along the channels and thicker dashed lines along the crest lines of the banks and at their terminations. (From Caston and Stride, 1970.)

is considered to be the resultant of two processes: (a) oblique movement of sand upslope under the influence of the near-bed tidal currents and (b) periodic down-slope dispersion of sand by wave action. The two processes would combine to produce a pattern of movement for sand particles along the flanks of a bank which would be analogous to that occurring along a beach under the influence of longshore drift, this leading to lengthwise growth which, in view of the dominance of the north-westerly-flowing tidal currents in the Norfolk Banks area, would be principally to the north-west.

The sand transport pattern postulated for the Norfolk Banks is summarised by Fig. 7.54, which shows how the north-westerly-flowing tidal currents, and thus sand streams, are dominant on the southwestern sides of the

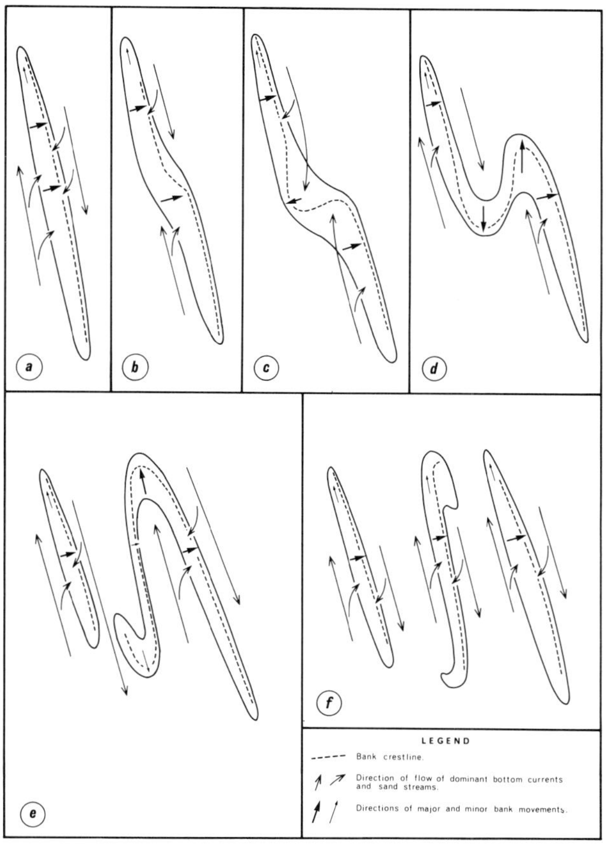

Fig. 7.55. Diagram showing stages in the growth and development of linear sand banks. For discussion, see text. (From Caston, 1972.)

banks, and how the southeasterly-flowing streams occur on the north-easterly-facing flanks.

On the basis of this and other evidence, Caston (1972) proposed a sequence in the movement and development of the banks which is associated with the growth of ebb and flood (tidal) channels (see Van Veen, 1936, 1950; Cloet, 1954; Robinson, 1960). The salient features are illustrated in Fig. 7.55, which shows how a "kink" in a linear sand bank may be emphasised by opposing currents on either side of the bank until it becomes a double curve (the Ower Bank may represent this stage at present) and then an incipient pair of ebb and flood channels. These new channels lengthen and eventually the parabolic terminal banks "blow-out" leaving gaps at the extremities of the channels. This latter stage is seen in the banks of the Haisborough Tail–Winterton Ridge complex.

Sand ribbons and patches

Sand ribbons, (Stride, 1959c; Flemming and Stride, 1967; Kenyon, 1970a) form in relatively high-energy environments (surface tidal current velocities approximately 100 cm/s, Vol. II, Fig. 9.12) where they parallel the flow line of the predominant currents (Fig. 8.3; Vol. II, Fig. 14.11). As such they do not provide a unique direction of sand transport. Detailed surveying, however, reveals that the sand surface is frequently wrought into asymmetric ripples or small sand waves, up to 1 m in height, which may indicate the direction of movement. Ribbons typically occur (or may be recognised) on a hard and/or coarser-grained substrate, typically gravel, which in a majority of cases in the North Sea is a lag deposit. They have little relief, and thus can rarely be recognised other than with a sidescan sonar. In the North Sea they have been reported only from relatively shallow waters off the U.K. east coast where they lie on the upper surface of the Pleistocene glacial clays (Kenyon and Stride, 1970), but also occur in association with sand waves off the east coast of Aberdeenshire.

Sand patches (Belderson and Stride, 1966; Kenyon, 1970b) also consist of medium to fine sand locally covering coarser material, but appear to be associated with relatively weak tidal currents reaching less than 50 cm/s near the sea surface (Vol. II, Fig. 9.12). In plan view they vary from long and straight, lying parallel to the transport patch, or sinuous to crescentic, and elongated normal to the transport path. Thicknesses vary from a few centimetres to about 2 m, and in the latter case can be identified on echosounder records. Patches have been mapped in the North Sea, in the Moray Firth and to the east of the Orkney and Shetland Isles (Kenyon and Stride, 1970); similar features have also been recognised in deeper waters on some of the glacial banks in the central North Sea. They are probably more abundant than known at present, and in certain areas near the outer limit of sand-wave formation (Figs. 7.46 and 8.3) may have been confused with true sand waves.

In terms of the overall distribution of sediments in the North Sea, however, these features are of relatively little significance.

Summary

Results from individual surveys undertaken to date in the North Sea are inadequate to provide anything like a comprehensive picture of the overall distribution and nature of the Quaternary succession. This applies in particular to the pre-Weichselian, concerning which there is virtually no information over the greater part of the region.

The principal and most significant aspect of the distribution is the very considerable thickness of Weichselian sediments present in the central and northern North Sea, where they form a disproportionately large percentage of the total Quaternary sequence, compared with their almost total absence in the south. The dividing line between these two provinces may be drawn very approximately from south-west to north-east along the axis of Dogger Bank to Turbot Bank, and thence eastwards to the Danish coast.

Pre-Weichselian

South of this dividing line, deposits of early Pleistocene age (Cromerian to Pretiglian) are present at or near the sea bed in the Dutch sector as far south as 52°40′N. A little further to the north, these are overlain by Elsterian fluvio-glacial clays which occur at a shallow depth over much of the northern portion of the Dutch sector (Oele, 1969, 1971b), and lie within a few metres of sea bed in Danish wells A-1 and A-2 (Rasmussen, 1974). Sediments dating from the Holstein interglacial have been recorded only locally within shallow borings in the southern portion of the Dutch sector. Other interglacial deposits sampled in the Inner Silver Pit have been dated as Hoxnian, formerly equated with the Holsteinian but now believed to be equivalent of the Eemian (Fisher et al., 1969). Some doubt exists concerning the age of the considerable thickness of silty sands seen in boreholes "C" and "D" in the southern U.K. sector (this Chapter). These could be of Holsteinian age but are more likely to date from the Eemian (see below).

Saalian deposits occur most extensively in U.K. waters where Basement Till of this age outcrops on the sea bed off the Yorkshire coast and in the vicinity of the Inner Silver and Sole Pits (Donovan, 1973). On the basis of lithological similarity, it is possible that the grey chalky boulder clay seen in boreholes "B" and "C" (this paper) is also Basement Till: on this assumption the underlying silty sands in borehole "C" and, by implication, in borehole "D" are of Holsteinian age, whereas by association with deposits in the Dutch sector they are more likely to date from the Eem interglacial. If the latter is the case, then the chalky till must be Weichselian.

Only local deposits of Saalian age have been identified in the Dutch sector, where extension of glaciers at this time was primarily responsible for remoulding earlier deposits, notably the construction of ice-push moraines. Still further east, Pratje's (1951) contention that the southernmost limit of the Warthe (Saal) stage is marked by the line of morainic banks extending seaward from Schleswig-Holstein (Fig. 7.12) is supported by the total absence of Saalian deposits in wells A-1 and A-2.

Much of the southern portion of the Dutch sector is covered by extensive marine sand overlain by fresh-water clays, both of Eemian age (Oele, 1971b). Immediately to the west in the U.K. sector, clays of the same age are exposed on the sea bed between some of the Norfolk Banks (Brouwer, 1964). On the basis of this association, it is proposed that the silty sands present in borehole "D" (drilled within the same area), which presumably underlie the dated clays, are also of Eemian age.

North of Dogger Bank no sediments of pre-Weichselian age have yet been positively identified. On empirical grounds, deposition of the Basal Beds in the central North Sea is considered to have commenced in the early Pleistocene, and the horizon marking their upper limit, the "Crenulate reflector", represents an erosive event estimated to be of upper Saalian age (Eden et al., 1977).

Weichselian

In Weichselian times the locus of sedimentation in the North Sea shifted north of 56°N, the only Weichselian deposits of any significance south of Dogger Bank being recorded in the south-west of the North Sea in U.K. waters off the Lincolnshire and north Norfolk coasts.

In this area, glacial till considered to be of Weichselian age has been reported from the region of the Inner Silver and Sole Pits (Donovan, 1973), and gravels of British origin sampled from the Inner Well Bank Rough and off the U.K. east coast between 53° and 55°N have been interpreted as a lag deposit overlying Weichselian till (Veenstra, 1969). Tills of this age are certainly present along virtually the complete route of the Easington to West Sole pipeline, where they are up to 13 m in thickness and notable for their high degree of overconsolidation. The age of the chalky boulder clay seen in boreholes "B", "C" and "D" has been discussed above; on lithological grounds it appears akin to the Saalian Basement Till, whereas assumption that the underlying sands are of Eemian age implies that the till must be Weichselian.

The Dutch sector appears to have been ice-free, but affected by periglacial conditions during this period. The resultant deposits consist principally of a thin cover of eolian sands, with fluviatile medium-grained sands derived from the Rhine south of about 52°15′N. Similar conditions probably prevailed further east in the Danish sector, where only 5 m of Holocene and Weichselian age sand was recorded in well A1.

The Dogger Bank itself consists largely of proglacial gravelly sands apparently deposited immediately in front of the southern limit of the Weichsel glaciers, which probably halted along a line corresponding to the present northern and western margins of the bank. Further eastwards, Jutland Bank and the associated line of banks trending west–east into the Danish coast undoubtedly include moraines of Weichselian age, as proposed in 1951 by Pratje, and these deposits may continue northwards under the southern Norwegian Channel, where a hummocky reflector seen on seismic records has been interpreted as marking the upper surface of Weichselian drift (Van Weering et al., 1973).

Quaternary deposits are absent in a wide zone between the U.K. east coast and the western edge of Dogger Bank (Dingle, 1970). This absence is surprising and there is no evidence yet to indicate whether it may be due to non-deposition or to deposition and subsequent erosion. Comparison with the Inner Silver Pit, believed to have been caused by marine erosion in the immediate post-glacial period (Donovan, 1965), suggests however that such erosion would have been concentrated into well-defined erosion hollows, leaving untouched much of the undoubtedly resistant stony boulder clay. This is supported by observations elsewhere in the south-western North Sea, where the tills have developed a lag gravel or cobble pavement which is extremely resistant to further erosion.

To the north of Dogger Bank, sediments of proven or probable Weichselian age have been recorded from maximum depths at such widely spaced locations as the Tay Estuary (60 m sub-sea bed), Moray Firth (70 m), I.G.S. borehole 74/9 off the Aberdeenshire coast (83 m), Forties Field (122 m), Josephine Field (310 m), Frigg Field (50 m), Frigg Field Booster Station (35 m), Norwegian Block 25/11 (91 m), and the Brent Field (122 m), thus indicating that the cover of Weichselian deposits is both thick and extensive.

In the U.K. sector, stony clays comparable with those recorded off the Lincolnshire coast; in boreholes "A", "B" and "C", have not been recorded in boreholes drilled to the north of Dogger Bank (boreholes "I", "J" and "K"), and the succession recorded in these holes consists essentially of layers of stiff clay and fine sand, the latter increasing in proportion from west to east. These deposits, however, have not been dated, nor have the sediments at the Ekofisk and Cod Fields described in this Chapter (boreholes "L" to "T" inclusive), but in view of their location within the moraine complex to the west of the Norwegian Channel, and the presence of proven Weichselian-age sediments to the north, north-west and south, it appears probable that a significant proportion of the deposits in the latter area are of Weichselian age. The most distinctive characteristic of these sediments, which penetrated to a maximum depth of 122 m, is their extreme variability, consisting of alternations of hard grey clay and grey silty fine sand. The relative thickness of the layers and of the proportion of sand and clay varies considerably from hole to hole. Gravel is very rare and none of the hard clay layers contains pebbles; shear strength values measured in one borehole in the Cod Field, however, suggest that at least one hard sandy clay layer is an overconsolidated glacial till.

To the north of 56°N, a wide range of lithologies have been recorded, including stiff stony clays, interpreted as tills, in the Brent and Forties areas, normally-consolidated glacio-marine clays, and fine silty sands, believed to represent fluvioglacial or outwash deposits. These variations in lithology reflect relatively localised depositional changes, extensive erosional action during at least two well-defined periods which formed complex patterns of subsequently-infilled channels, and sediment deformation caused by glacial tectonics. The complexity occurs both on a broad scale, as identified by the work undertaken by IGS between 56° and 58°N, and in detail at relatively shallow depths, as in the Brent Field. There is increasing evidence from such detailed studies, including those at Brent, Forties, and Ekofisk (assuming the deposits of the latter to be of Weichselian age) to suggest that small-scale variability may be anticipated over much of the northern North Sea, superimposed upon the large-scale variability recognised on reconnaissance-scale surveys.

Between 56° and 58°N, the erosional episodes already mentioned have cut widespread channels at two well-defined horizons. Some controversy exists concerning their mode of formation but a consensus of opinion favours excavation by sub-glacial streams. If this is so then the channels reflect two separate glacial readvances over the northern North Sea. Holmes (1977) quotes radiocarbon dates which suggest that formation of Lower Channels was associated with a major Weichselian glaciation at about 28,000 years B.P., and, more tentatively, correlates the Upper Channels with a late-Weichselian glacial maximum at ca. 15,000 years B.P.

Holocene

Distribution of sediments on the floor of the North Sea has been described in the body of this paper on a regional basis. The point has, however, been made that the North Sea is a region of "palimpsest" sediments, reflecting both the depositional and erosional events of the Pleistocene and post-Pleistocene periods. Identification therefore of sediments specifically of Holocene age, particularly those consisting of siliciclastics which may have experienced several episodes of reworking, is frequently impracticable. Concentration of sand into mobile bed forms has been considered separately and will not be summarised here.

The most notable aspect of Holocene sedimentation in the North Sea as a whole is the very limited amount of deposition which has occurred during this period. Sedimentation has been bathymetrically controlled, and significant quantities, principally of silts and muds, have accumulated mainly in the deeper water areas. The most extensive of these floors the whole length of the Norwegian Channel, reaching maximum thicknesses of 127 m in the Skagerrak and 70 m off Egersund (Van Weering et al., 1973). Others have been recorded from the Outer Silver Pit (ca. 20 m), the Fladen Ground (ca. 15 m), the Swatchway adjacent to Forties Bank (12 m), and other areas of deeper water present within a broad zone lying to the west of the line of shallower morainic banks that parallel the western edge of the Norwegian Channel. Within the best documented area of the central North Sea, between 56° and 58°N, these fine-grained deposits, the Witch Ground Beds, locally reach a maximum of 40 m in thickness, although this figure may include a lower, well-bedded unit of late glacial age (Holmes, 1977).

Deposition of fine-grained sediments has also taken place along a linear zone corresponding to the Heligoland Channel, which lies immediately to the west of the morainic banks that lie off the west coast of Denmark; thicknesses in this area have not been reported.

Away from these obvious hollows the greater part of the North Sea is floored by fine sand, although over many morainic banks this forms a veneer rarely more than a few metres in thickness which may not even conceal the underlying lag pavement. Locally this fine sand is found overlying the finer-grained Witch Ground and similar deposits in deeper water areas, although whether the sand forms a purely transgressive layer, originating on the shallower banks, or whether it represents a progressive upward coarsening within the depositional unit (as appears from some analyses) is not clear.

Distribution of sand in the southern North Sea shows a progressive decrease in grain size towards a zone of silt centred at about 4° E in the northern part of the Dutch sector. Although this distribution undoubtedly reflects the present energy pattern, the sediment layer so affected is of limited thickness (rarely more than 2 m) and within much of the area overlies considerable thicknesses of Pleistocene sands, which may have provided the source of the reworked superficial material.

In the higher energy zones of the south-western North Sea superficial sand is wrought into mobile bed forms and does not form a continuous layer, thus exposing Pleistocene tills and lag pavement over large areas off the U.K. east coast.

Deposits of the early Holocene have been recorded principally from the Dutch sector where they occur fairly extensively at a shallow depth. The sequence starts with a basal peat bed and is overlain successively by fresh water, brackish, and then marine clays of Preboreal and Boreal age, recording the progress of the basal Holocene transgression. An equivalent series of deposits has not been recognised elsewhere in the southern North Sea, although a little further to the north on Dogger Bank silty clays laid down under estuarine or tidal flat conditions and probably of an equivalent age have been cored a metre or so below sea bed below present water depths of ca. 30 m (Stride, 1965).

In the central and northern North Sea a majority of studies have been based primarily upon seismic profiles, and in other instances borehole deposits (this Chapter) have not been dated. Detailed subdivision of the late Weichselian and Holocene has therefore not been possible. From the profiles it is apparent that the sea bed on glacial banks throughout a wide area is an erosion surface, presumably dating from the early Holocene transgression (Fig. 7.39), assisted to a greater or lesser extent by wave action whilst the sea was still shallow during the early Holocene. Evidence from boreholes drilled through the Witch Deposits, however, suggests that deposition of fine-grained sediments commenced in the intervening hollows in the late Weichselian and continued unabated through the Holocene. Erosion of the shallower banks thus undoubtedly provided a source of sediment to infill the adjacent deeps, and it is apparent that these processes of erosion and contemporaneous deposition contributed throughout the early Holocene to the progressive reduction of late-glacial relief on the floor of the North Sea.

Sedimentation–subsidence rates

The rapid rate of subsidence in the North Sea was first noted by Clarke (1973), who showed that average subsidence rates have increased from 0.01 mm/year in the Cretaceous to a maximum of 0.5 mm/year in the Quaternary.

Any detailed consideration of sedimentation rates within the Quaternary is dependent upon reliable age/depth determinations, few of which are as yet available for the North Sea. No realistic estimates are possible for the pre-Weichselian. The extent of Weichselian sediments in the northern North Sea together with the availability of a few radiocarbon dates, however, permits some calculation of sedimentation rates for the Weichsel–Holocene period. Analyses from the Auk, Forties and Josephine Fields are consistent with very rapid average rates since 32,000 years B.P., including 7.5 mm/year at Forties and 7.3 mm/year at Josephine (Holmes, 1977). These rates are up to 25 times the overall average for the Quaternary quoted above, and suggest either that deposition during the Weichselian has proceeded at a uniquely high figure in comparison with the pre-Weichselian, and/or that there has been removal of considerable quantities of pre-Weichselian deposits (Holmes, 1977). The above high values, if extrapolated back throughout the Quaternary, would give a potential thickness of 13.5 km!

Contrasting with the well-documented post-glacial isostatic uplift of Scotland and Scandinavia, there is contemporary evidence, primarily from tide-gauge records, that the southern part of the U.K., southern Denmark, and the coastal strip bordering the North Sea of Germany, Netherlands, Belgium and France are subsiding at rates of up to 3 mm/year (West, 1968, quoting Fromm, 1953 and Valentin, 1954).

Further discussion of this topic by Eden et al. (1977) concluded that average subsidence rates for the late Pleistocene are too high to be credibly accounted for in terms of tectonic subsidence alone. They suggest that the Central Graben of the North Sea rose during glaciations in isostatic response to sinking of adjacent land areas, and subsequently sank during deglaciation as the land masses rose.

These effects were then superimposed on the accelerating tectonic subsidence inferred from Clarke's (1973) work. In this model, periods of elevation are represented by erosive episodes, and it is only the deposits of the periods of subsidence that are preserved. In the central North Sea apparent lack of subsidence of the Upper Channels horizon dated as ca. 22,000 years B.P. supports the concept that glaciation at this time was less severe than earlier glacial episodes, and substantial areas of open water remained in the central North Sea. The mechanism for large isostatic adjustments was therefore not triggered. Further north in the Brent Field, Milling (1975) has however reported a subsidence rate of 5 mm/year for borehole samples in comparison with published eustatic sea level curves. These differences between the central and northern areas may be explained by the greater intensity of glacial effects to the north.

Overall, therefore, there is evidence that subsidence of the North Sea, in particular of the Central and Viking Grabens, has continued through the Quaternary, and locally is active at the present day.

Acknowledgements

I am indebted to Phillips Petroleum, Shell (U.K.) and British Petroleum for permitting the publication of data from engineering boreholes and sea-bed surveys, and to M.E. Milling of Exxon for further information concerning those areas in the northern North Sea described in his recent publications. Figure 7.39A is based upon work undertaken by my colleague M.J. Davies, and originally prepared by him for an internal B.P. report. Lithological information for boreholes "A" and "G" to "T" inclusive was taken from reports prepared by McClelland Engineers of Houston, and for boreholes "B" to "F" inclusive from reports by Cesco N.V. of the Hague. My thanks also go to the many authorities (individually acknowledged), who have kindly allowed me to reproduce their published diagrams. Figures 7.2, 7.7, 7.14, 7.21, 7.23, 7.24, 7.27, 7.28, 7.29, and 7.30 of this chapter are NERC copyright and are reproduced by permission of the Director, Institute of Geological Sciences, London S.W. 7. Figures 7.35 and 7.36 are reproduced by permission of the National Research Council of Canada from the Canadian Journal of Earth Sciences. This review is published with the permission of the British Petroleum Co. Ltd.

References

Anonymous, 1967. The northern limit of megaripples in the North Sea. Hydrogr. Newslett., 1: 339–340.

Aric, K., 1968. Reflexionsseismische Messungen in Skagerrak. Z. Geophys, 34: 223–226.

Baak, J.A., 1936. Regional Petrology of the Southern North Sea. Thesis, State University, Leiden, Veenman, Wageningen, 128 pp.

Beaugé, L., 1937. Relevés hydrographiques exécutés au cours de quatre premières croisières: résultats de croisières scientifiques du navire "President Theodore Tissier". Rev. Trav. Off. Pêches Marit., 10: 117–232.

Belderson, R.H. and Stride, A.H., 1966. Tidal current fashioning of a basal bed. Mar. Geol., 4: 237–257.

Belt, T., 1874. An examination of the theories that have been proposed to account for the climate of the glacial period. Q. J. Sci., N.S., 4: 421–464.

Bertelsen, F., 1972. *Azolla* species from the Pleistocene of the central North Sea area. Grana, 12: 131–145.

Berthois, L. and Brenot, R., 1957. Topographie sous-marine de "l'Outer Silver Pit" et du "Markhams Hole" Lat. N. 54°, Long. E. Gr. 2°. Rev. Trav. Inst. Pêches Marit., 21: 431–434.

Borley, J.O., 1923. The marine deposits of the southern North Sea. Gt. Brit., Min. Agri. Fisheries, Fish Invest., Ser. II, 4(6): 1–73.

Brouwer, L.E.J., 1964. The North Sea. In: L.D. Stamp (Editor), The World Land Use Survey – Occasional Papers No. 5. Geographical Publications Ltd., Bude, 42 pp.

Brunn, P. and Vollen, Ø., 1972. Sand waves on the bottom of the sea, with special reference to conditions in the North Sea. Schiff Hafen, 24: 139–142.

Caston, V.N.D., 1969. Industrial and scientific co-operation in research – a practical example. Hydrospace, 2: 27–31.
Caston, V.N.D., 1972. Linear sand banks in the southern North Sea. Sedimentology, 18: 63–78.
Caston, V.N.D., 1974. Bathymetry of the northern North Sea – Knowledge is vital for offshore oil. Offshore, 34: 76–84.
Caston, V.N.D., 1976. A wind-driven near-bottom current in the southern North Sea. Estuarine Coastal Mar. Sci., 4: 23–32.
Caston, V.N.D., 1977a. A new isopachyte map of the Quaternary of the North Sea. In: Quaternary Sediments of the Central North Sea, 1 and 2. Rep. Inst. Geol. Sci. 77/11: 1–8.
Caston, V.N.D., 1977b. The Quaternary deposits of the Forties Field, northern North Sea. In: Quaternary sediments of the Central North Sea, 1 and 2. Rep. Inst. Geol. Sci. 77/11: 9–22.
Caston, V.N.D. and Stride, A.H., 1970. Tidal sand movement between some linear sand banks in the North Sea off northeast Norfolk. Mar. Geol., 9: M38–M42.
Caston, V.N.D. and Stride, A.H., 1973. Influence of older relief on the location of sand waves in a part of the southern North Sea. Estuarine Coastal Mar. Sci., 1: 379–386.
Catt, J.A. and Penny, L.F., 1966. The Pleistocene deposits of Holderness, East Yorkshire. Proc. Yorkshire Geol. Soc., 35: 375–420.
Chesher, J.A. and Lawson, D., 'in preparation'. Marine geology of the Moray Firth. Rep. Inst. Geol. Sci.
Chesher, J.A., Deegan, C.E., Ardus, D.A., Binns, P.E. and Fannin, N.G.T., 1972. IGS marine drilling with m.v. Whitethorn in Scottish waters 1970–71. Inst. Geol. Sci. Rep., 72/10, 25 pp.
Clarke, R.H., 1973. Cainozoic subsidence in the North Sea. Earth Planet. Sci. Lett., 18: 329–332.
Cloet, R.L., 1954. Sandwaves in the North Sea and Persian Gulf. J. Inst. Navig., 7: 272–279.
Cloet, R.L., 1963. Hydrographic analysis of the sandbanks in the approaches to Lowestoft Harbour. Hydrographic Dep., Admiralty – Admir. Mar. Sci. Publ., 6: 15 pp.
De Jong, J.D., 1967. The Quaternary of the Netherlands. In: K. Rankama (Editor), The Quaternary, Vol. 2. Interscience, New York, N.Y., pp. 301–426.
De Ruiter, J. and Fox., D.A., 1975. Site Investigations for North Sea Forties Field. Offshore Technol. Conf., Houston, Pap. no. 2246.
Deutsches Hydrographisches Institute, 1953. Fischereikarte der Nordsee (2 sheets), 1 : 900,000. Hamburg.
Dingle, R.V., 1965. Sand waves in the North Sea mapped by continuous reflection profiling. Mar. Geol., 3: 391–400.
Dingle, R.V., 1970. Quaternary sediments and erosional features off the north Yorkshire coast, western North Sea. Mar. Geol., 9: M17–M22.
Dingle, R.V., 1971a. Buried tunnel valleys off the Northumberland coast, western North Sea. Geol. Mijnbouw, 50: 679–686.
Donovan, D.T., 1965. Holocene origin of erosional hollows in the North Sea. Ann. Rep. Challenger Soc., 3: 43–44.
Donovan, D.T., 1973. The geology and origin of the Silver Pit and other closed basins in the North Sea. Proc. Yorkshire Geol. Soc., 39: 267–293.
Donovan, D.T. and Dingle, R.V., 1965. Geology of part of the southern North Sea. Nature (London), 207: 1186–1187.
Dujon, S.C., 1970. Sur la genèse des bancs de Mer du Nord meridionale. Rev. Geogr. Phys. Géol. Dyn., 11: 503–510.
Dunn, W.W. et al., 1973. North Sea is a tough theater for the oil-hungry industry to explore. Oil Gas J., 71(3): 127–128; (4): 90–93.
Eden, R.A., Small, Anne V.F. and McQuillin, R., 1970. Preliminary report on marine geological and geophysical work off the east coast of Scotland, 1966-68. Rep. Inst. Geol. Sci., 70/1: 30 pp.
Eden, R.A., Holmes, R. and Fannin, N.G.T., 1977. Depositional environment of offshore Quaternary deposits of the Continental Shelf around Scotland. Rep. Inst. Geol. Sci., 77/15.
Eisma, D., 1973. Sediment distribution in the North Sea in relation to marine pollution. In: E.D. Goldberg (Editor), North Sea Science. M.I.T. Press, Cambridge, Mass., 500 pp.
Emery, K.O., 1968. Relict sediments on continental shelves of world. Bull. Am. Assoc. Pet. Geol., 52: 445–464.
Evans, G., 1965. Intertidal flat sediments and their environments of deposition in the Wash. Q. J. Geol. Soc. London, 121: 209–245.
Fisher, M.J., Funnell, B.M. and West, R.G., 1969. Foraminifera and pollen from a marine interglacial deposit in the western North Sea. Proc. Yorkshire Geol. Soc., 37: 311–320.
Flemming, N.C. and Stride, A.H., 1967. Basal sand and gravel patches with separate indications of tidal current and storm-wave paths, near Plymouth. J. Mar. Biol. Assoc. U.K., 47: 433–444.
Flinn, D., 1964. Coastal and submarine features around the Shetland Islands. Proc. Geol. Assoc., 75: 321–341.
Flinn, D., 1967. Ice front in the North Sea. Nature (London), 215: 1151–1154.
Flinn, D., 1973. The topography of the sea floor around Orkney and Shetland and in the northern North Sea. J. Geol. Soc. London, 129: 39–59.
Fox, D.A., Parker, G.F. and Sutton, V.J.R., 1970. Pile driving into North Sea boulder clays. Offshore Technol. Conf., Houston, Pap. No. 1200.
Fromm, E., 1953. Map of glaciations and changes of level. In: Atlas over Sveridge. Generalstabens Lithografiska Anstalts Forlag, Stockholm.
Furnestin, J., 1939-1943. Remarques sur la géologie sous-marine du plateau continental de la Mer Celtique et de La Mer du Nord. Rev. Trav. Off. Pêches Marit., 13: 479–489.
Gadow, S. and Schafer, A., 1973. Die Sedimente der Deutschen Bucht: Korgrössen, Tonmineralien und Schwermetalle. Senckenbergiana Maritima, 5: 165–178.
Gienapp, H., 1973. Strömungen während der Sturmflut vom 2 November 1965 in der Deutschen Bucht und ihre Bedeutung fur den Sedimenttransport. Senckenbergiana Maritima, 5: 135–151.
Gregory, J.W., 1931. A deep trench on the floor of the North Sea. Geogr. J., 77: 548–551.
Guilcher, A., 1951. La formation de la Mer du Nord, du Pas de Calais et des plaines maritimes environnantes. Rev. Géogr. Lyon, 26: 311–329.
Heybroek, P., Haanstra, U. and Erdman, D.A., 1967. Observations on the geology of the North Sea area. Proc. World Pet. Congr., 7th, Mexico, 2: 905–916.
Holmes, R. 1977. The Quaternary geology of the UK sector of the North Sea between 56° and 58°N. Rep. Inst. Geol. Sci., 77/14.
Holtedahl, O., 1940. The Submarine Relief off the Norwegian Coast. Det Norske Videnskaps Akademi; Oslo, pp. 1–41.
Holtedahl, O., 1964. Echo-soundings in the Skagerrak. Norg. Geol. Unders., 223: 139–160.
Houbolt, J.J.H.C., 1968. Recent sediments in the Southern Bight of the North Sea. Geol. Mijnbouw, 47: 245–273.
Howitt, F., 1974. North Sea oil in a world context. Nature (London), 249: 700–703.
Institute of Geological Sciences, 1972. Map of the Sub-Pleistocene geology of the British Isles and the adjacent Continental Shelf. 1st Ed.
Jansen, J.H.F., 1976. Late Pleistocene and Holocene history of the northern North Sea, based on acoustic reflection records. Neth. J. Sea Res., 10: 1–43.
Jarke, J., 1955. Neue Ergebnisse zur Bodenbedeckung der Deutschen Bucht. Tag. Ber. Wiss. Abh. Dtsch. Geographentag Hamburg, pp. 369–375.

Jarke, J., 1956. Der Boden der südlichen Nordsee. 1. Beitrag: Eine neue Bodenkarte der südlichen Nordsee. Dtsch. Hydrogr. Z., 9: 1–9.

Jelgersma, S., 1961. Holocene sea-level changes in the Netherlands. Mededel. Geol. Sticht., Ser. C, 6: 101 pp.

Johnson, M.A. and Stride, A.H., 1969. Geological significance of North Sea sand transport rates. Nature (London), 244: 1016–1017.

Keizer, J. and Letsch, W.J., 1963. Geology of the Tertiary in the Netherlands. Verh. K. Ned. Geol. Mijnbouwk. Genoot. Geol. Ser., 21-2: 147–172.

Kent, P.E., 1975. Review of North Sea Basin development. J. Geol. Soc. London, 131: 435–468.

Kenyon, N.H., 1970a. Sand ribbons of European Tidal Seas. Mar. Geol., 2: 25–39.

Kenyon, N.H., 1970b. The origin of some transverse sand patches in the Celtic Sea. Geol. Mag., 107: 389–394.

Kenyon, N.H. and Stride, A.H., 1968. The crest length and sinuosity of some marine sand waves. J. Sediment. Petrol., 38: 255–259.

Kenyon, N.H. and Stride, A.H., 1970. The tide-swept continental shelf sediments between the Shetland Isles and France. Sedimentology, 14: 159–173.

King, L.H. and Maclean, B., 1970. Pockmarks on the Scotian shelf. Geol. Soc. Am. Bull., 18: 3141–3148.

Koltermann, K.P. and Lange, W., 1975. On the relation between residual currents and the wind field in the German Bight. Dtsch. Hydrogr. Z., 28: 193–206.

Lang, A.W., 1968. See Karten der südlichen Nord- und Ostsee. Deutsch Hydrographisches Institut, Hamburg, 105 pp.

Langeraar, W., 1966. Sand waves in the North Sea. Hydrogr. Newslett., 1: 243–246.

Løfaldli, M., 1973. Foraminiferal biostratigraphy of late Quaternary deposits from the Frigg field and Booster Station. NTNF Continental Shelf Div., Oslo, Rep., 18: 83 pp.

Lucas, A.L., 1974. A high resolution marine seismic survey. Geophys. Prospect., 22: 667–682.

Lüders, K., 1950. Die Sedimente der Nordsee. Neues Arch. Niedersachsen, 16: 213–226.

McCave, I.N., 1970. Deposition of fine grained suspended sediment from tidal currents. J. Geophys. Res., 75: 4151–4159.

McCave, I.N., 1971. Sand waves in the North Sea off the coast of Holland. Mar. Geol., 10: 199–225.

McCave, I.N., 1973. Mud in the North Sea. In: E.D. Goldberg (Editor), North Sea Science. M.I.T. Press, Cambridge, Mass., 500 pp.

McCave, I.N., Caston, V.N.D. and Fannin, N.G.T., 1977. The Quaternary of the North Sea. In: F.W. Shotton (Editor), British Quaternary Studies, Oxford University Press, London. pp. 187–204.

McManus, J., 1970. The geological setting of the bridges of the lower Tay Estuary with particular reference to the fill of the buried channel. Q. J. Eng. Geol. 3: 197–205.

Milling, M.E., 1975. Geological appraisal of foundation conditions, northern North Sea. Oceanology International 1975, Brighton, pp. 310–319.

Milling, M.E., 1976. Geologic factors control foundation design. Pet. Eng. Int., 48(3): 12–22.

Mitchell, G.F., Penny, L.F., Shotton, F.W. and West, R.G., 1973. A correlation of Quaternary deposits in the British Isles. Geol. Soc. London, Spec. Rep., 4: 99 pp.

Morner, N.A., 1971. Eustatic changes during the last 20,000 years and a method of separating the isostatic and eustatic factors in an uplifted area. Paleogeogr., Paleoclimatol., Paleoecol., 9: 153–181.

Morner, N.A., 1972. World climate during the last 130,000 years. Int. Geol. Cong., 24th, Sect. 12, pp. 72–79.

Oele, E., 1969. The Quaternary Geology of the Dutch part of the North Sea, north of the Frisian Isles. Geol. Mijnbouw, 48: 467–480.

Oele, 1971a. Late Quaternary geology of the North Sea southeast of the Dogger Bank. In: F.M. Delany (Editor), The Geology of the East Atlantic Continental Margin. Rep. Inst. Geol. Sci., 70/15: 26–34.

Oele, E., 1971b. The Quaternary geology of the southern area of the Dutch part of the North Sea. Geol. Mijnbouw, 50: 461–476.

Off, T., 1963. Rhythmic linear sand bodies caused by tidal currents. Bull. Am. Assoc. Pet. Geol., 47: 324–341.

Pitt, E.G., Carson, R.M. and Tucker, M.J., 1973. The current system around the British Isles as it relates to offshore structures. An assessment. National Insitute of Oceanography, NIO Intern. Rep., A.62.

Pratje, O., 1952. Zur Geologie der Norwegischen Rinne. Abhandl. Endmoranen. Dtsch. Hydrogr. Z., 4: 106–114.

Pratje, O., 1952. Zur geologie der Norwegischen Rinne. Abhandl. Naturw. Ver. Bremen, 33: 79–88.

Rasmussen, L.B., 1974. Some geological results from the first five Danish exploration wells in the North Sea. Dan. Geol. Unders., Afh., Raekke 3, 42: 46 pp.

Reineck, H.E. and Singh, I.B., 1975. Depositional Sedimentary Environments. Springer-Verlag, New York, Heidelberg, Berlin, 439 pp.

Robinson, A.H.W., 1960. Residual currents in relation to shoreline evolution of the East Anglian coast. Mar. Geol., 4: 57–84.

Sanglerat, G., 1972. The Penetrometer and Soil Exploration. Elsevier, Amsterdam, 464 pp.

Sellevol, M.A. and Aalstad, I., 1971. Magnetic measurements and seismic profiling in the Skagerrak. Mar. Geophys. Res., 1: 284–302.

Sellevol, M.A. and Sundvor, E., 1974. The origin of the Norwegian Channel - a discussion based on seismic measurements. Can. J. Earth Sci., 11: 224–231.

Stather, J.W., 1912. Shelley clay dredged from the Dogger Bank. Q. J. Geol. Soc. London, 68: 324–327.

Stocks, T., 1955. Der Boden der südlichen Nordsee. 2. Beitrag: Eine neue Tiefenkarte der südlichen Nordsee. Dtsch. Hydrogr. Z., 9: 265–280.

Stride, A.H., 1959a. On the origin of the Dogger Bank, in the North Sea. Geol. Mag., 96: 33–44.

Stride, A.H., 1959b. A pattern of sediment transport for sea floors around southern Britain. Dock Harbour Auth., 40: 145–147.

Stride, A.H., 1959c. A linear pattern on the sea floor and its interpretation. J. Mar. Biol. Assoc. U.K., 38: 313–318.

Stride, A.H., 1963. Current-swept sea floors near the southern half of Great Britain. Q. J. Geol. Soc. London, 119: 175–199.

Stride, A.H., 1965. Periodic and occasional sand transport in the North Sea. 1. 111, 4 pp. In: La Pétrole et la Mer – Commun. – Congr. Int., 1e, Monte Carlo. La Revue Pétrolière, 1072.

Stride, A.H., 1970. Shape and size trends for sand waves in a depositional zone of the North Sea. Geol. Mag., 107: 469–477.

Stride, A.H., 1973. Sediment transport by the North Sea. In: E.D. Goldberg (Editor), North Sea Science. M.I.T. Press, Cambridge, Mass., 500 pp.

Stride, A.H. and Cartwright, D.E., 1958. Sand transport at southern end of North Sea. Dock Harbour Auth., 38: 323–324.

Stride, A.H. and Chesterman, W.D., 1973. Sedimentation by non-tidal currents around northern Denmark. Mar. Geol., 15: M53–M58.

Swift, D.J.P., Stanley, D.J. and Curray, J.R., 1971. Relict sediments on continental shelves: a reconsideration. J. Geol., 79: 322–346.

Terwindt, J.H.J., 1971. Sand waves in the southern bight of the North Sea. Mar. Geol., 10: 51–67.

Thomson, M.E. and Eden, R.A., 1977. The Quaternary sequence in the west-central North Sea. Rep. Inst. Geol. Sci., 77/12.

Valentin, H., 1954. Gegenwärtige Niveauänderungen im Nordseeraum. Petermann's Geogr. Mitt., 98: 103–108.

Valentin, H., 1955. Die Grenze der letzten Vereisung im Nordseeraum. Verh. Dtsch. Geografentag, Hamburg, 30: 359–366.

Valentin, H., 1957. Glazialmorphologische untersuchungen in Ostengland. Abhandl. Geogr. Inst. Freien Univ. Berlin, 4: 86 pp.

Van Eerde, L.A., 1964. Geological investigations on the Brown Ridge. Hydrogr. Newslett., 1: 81–83.

Van Eysinga, F.W.B. (Compiler), 1975. Geological Time Table, 3rd ed. Elsevier, Amsterdam.

Van Straaten, L.M.J.U., 1964. De bodem der Waddenzee. In: Het Waddenboek. Netherlands Geol. Ver.

Van Veen, J., 1935. Sand waves in the North Sea. Hydrogr. Rev., 12: 21–28.

Van Veen, J., 1936. Onderzoekingen in de Hoofden. Algemeene Landsdrukkerij, The Hague., 252 pp.

Van Veen, J., 1938. Die unterseeische Sandwüste in der Nordsee. Geol. Meere Binnengewässer, 2: 62–86.

Van Veen, J., 1950. Eb- en vloedschaarsystemen in de Nederlandse getijwateren. Wadden-symposium. Tijdsch. K. Ned. Aardrijksk. Genoot., 67: 45–65.

Van Weering, T., Jansen, J.H.F. and Eisma, D., 1973. Acoustic reflection profiles of the Norwegian Channel between Oslo and Bergen. Neth. J. Sea Res., 6: 241–263.

Veenstra, H.J. 1965. Geology of the Dogger Bank area, North Sea. Mar. Geol., 3: 245–262.

Veenstra, H.J., 1969. Gravels of the southern North Sea. Mar. Geol., 7: 449–464.

Veenstra, H.J., 1971. Sediments of the southern North Sea. In: F.M. Delany (Editor), The Geology of the east Atlantic continental margin. 3. Europe. Rep. Inst. Geol. Sci., 70/15.

Volpel, F., 1959. Studie über das Verhalten weitwandernder Flachseesande in der südlichen Nordsee. Dtsch. Hydrog. Z., 12: 64–76.

Von Sager, G., 1968. Atlas der Elemente des Tidenhubs und der Gezeitenströme für die Nordsee, den Kanal und die Irische See. Deutsche Akademie für Wissenschaften, Berlin; Institut für Meereskunde, Rostock, 45 pp.

Weigel, W., Hjelme, J. and Sellevoll, M.A., 1970. A refraction profile through the Skagerrak from Northern Jutland to Southern Norway. Geodaet. Inst. Medd., 45: 1–28.

West, R.G., 1968. Pleistocene Geology and Biology with special reference to the British Isles. Longman, London, 377 pp.

Woodland, A.W., 1970. The buried tunnel-valleys of East Anglia. Proc. Yorkshire Geol. Soc., 37: 521–578.

Zagwijn, W.H., 1974. The palaeogeographic evolution of the Netherlands during the Quaternary. Geol. Mijnbouw, 53: 369–385.

Zagwijn, W.H. and Veenstra, H.J., 1966. A pollen-analytical study of cores from the Outer Silver Pit, North Sea. Mar. Geol., 4: 539–551.

Chapter 8

Sediments of the North-Western European Shelf

F.T. BANNER (with a contribution by S.J. CULVER)

Sea-bed macrotopography, geology and sedimentology (F.T.B)

It is evident from preceding chapters that the topography of the sea bed of the north-west European shelf seas is rarely relatable in detail and directly to the underlying structural geology of the shelf itself. It is a classic example of a mature "trailing edge" in the geotectonic sense of Inman and Nordstrom (1971) and many of the major deep structures of the shelf have little or no topographic expression. Over much of the shelf sea-bed, isobaths at 10 m intervals reveal a hummocky surface of very low relief (Fig. 7.1, p. 197); isobathymetry plotted at greater intervals can reveal little topographic variation at all. The average depth of the North Sea is about 90 m, and, south of the Norwegian Trough, deeps greater than 200 m (e.g. the Devils' Hole, which reaches more than 260 m) are very rare and confined to the northern sector. Much of the southern sector is less than 40 m deep. A comparison between small-scale bathymetric charts and maps depicting the major geological structural units (e.g. Kent, 1975; Naylor and Mounteney, 1975) shows few relationships between them. The shelf has undergone so many successive episodes, since the Miocene period, of emergence, marine transgression, submergence, regression and emergence again, and so much of it has been subjected to so many glaciations and eustatic-isostatic regressions and transgressions during the Quaternary, that only where the older, harder and more resistant rocks outcrop does any trace of the underlying geological structure reveal itself on the sea bed (e.g., in the Malin and Irish Seas, see Chapter 3).

Of course, the geomorphology and coastal outlines of Scotland and Northern Ireland are still primarily determined by the NE–SW Caledonide structural lineations, just as the Armorican–Hercynian E–W lineations underlie the form of the land masses of Southern Ireland, South Wales, Cornubia and Brittany (Chapter 4), and these deep structural trends find their expression in the topography of the sea bed in adjacent coastal waters where hard igneous and sedimentary rocks outcrop. The most obvious Caledonide topographic feature is the Outer Hebrides archipelago of basement and Palaeozoic islands, emergent above sea level on a NE–SW horst block (Watson, 1977), separated from Scotland by the similarly trending Mesozoic–Tertiary sediment-infilled troughs which underlie the Sea of the Hebrides and the Minch (see Chapter 6). Similar Caledonide features are displayed by the Shetland and Orkneys, the Scottish Islands and the Isle of Man, with the seas immediately around them characterised by their associated submerged and partly resistant rocks, notorious as navigational hazards; the more deeply submerged sea bed also has smaller topographic features where hard rock outcrops reveal something of the underlying geology (e.g. the subparallel sea-bed ridges in the North Channel of the Irish Sea, caused by igneous rock dykes, observed on sonographs by Caston, 1975). Similarly, the Armorican–Hercynian basement lineations of Cornubia have a seawards expression in the rocks of the Longships Light, south-west of the Lizard, and by the basement outcrops of and around the Scillies (e.g., the Bishop Rocks and the Gilstone Ledges, where the *Association, Eagle* and *Romney* foundered in 1707), and those of Brittany in, for example, l'Ile de Ouessant. These, like the "island fragment of continental basement situated in the German Bight" (Fairbridge et al., 1966) known as Heligoland, are not typical of the shelf as a whole. The soft and easily eroded Mesozoic and Cainozoic rocks which infill the structural basins of the North Sea, the English Channel, the Celtic, Irish and Hebridean Seas, have no positive effect on topography. Even the great North Sea and Viking Grabens (a failed arm of a triple plate junction, according to Naylor and Mounteney, 1975) are blanketed by Cainozoic sediments to the extent that their existence is unrecognisable in the morphology of the sea bed.

The low relief, both in general and in particular, must be explained not so much by geological structure as by the glacial and marine dynamic processes of erosion and sediment redistribution, which have operated during the episodic post-Palaeogene history of the shelf and which are operating today. Many of the major erosional topographic features of the shelf have long been explained in such terms (e.g., Godwin-Austen, 1850; King, 1954). There are sculpted linear features of such lateral persistence that their palaeogeographic ascription to past periods of low sea-level stand are in little or no dispute: among these must be included the linear depressions of the English Channel (Fig. 5.10; Hurd Deep, Fosse de la Hague, Fosse du Pluteus, Fosse de l'Ile Vierge), which are regarded as the partly buried relicts of a valley system excavated during early Quaternary low sea-level stands (Hamilton and Smith, 1968, 1972; Smith and Hamilton, 1970), and the drowned coastlines off South Devon and Cornwall (at 44 m, 54 m and 64 m below O.D.), which probably were cut during the late Tertiary (Donovan and Stride, 1975), maybe in the Late Miocene (see below). The Dover Straits probably were not finally breached until late in the Pleistocene (Prentice, 1972; post-Saalian, teste Gullentops, 1974) although the preliminary shaping of the English Channel itself, like the Bristol and St. George's Channels, must have occurred much earlier (North, 1964; Curry et al., 1971; Gullentops, 1974), probably in the late Miocene–Pliocene interval. The global, eustatic lowering of sea levels by 40–70 m which probably occurred during the Late Miocene (contemporaneously with the Mediterranean Messinian "salinity crisis", see Adams et al., 1977) would have enhanced the planation of the shelf by subaerial weathering and erosion by regressive and transgressive surf zones during that period. After the Late Tertiary and Early Pleistocene levelling of the shelf, it was extensively gullied, at least in the Celtic Sea and Western Approaches, during the low sea level (around 240 m below that of today) reached at that time, but many of these channels were filled with sediment during subsequent transgressions (Bouysse et al., 1976). Later Pleistocene regressions (to − 110 m or − 120 m during the Devensian period) left the outer part of the south western shelf submerged, beginning linear sand-bank accumulation there (Bouysse et al., 1976), while the inner north-western parts of the Celtic Sea shelf experienced tundra conditions and periglacial weathering (Culver and Banner, 1979). Around 18,000 y B.P., the Devensian ice sheets covered the previously-sculpted St. George's Channel and Irish Seas, and those areas were blanketed by thick deposits of glacial drift (Delanty and Whittington, 1977; Kidson and Tooley, 1977; Culver and Banner, 1979). Cooper (1977) has presented detailed arguments for the existence of almost stationary snow firnfields on the Celtic shelf and the English Channel bed, during Devensian time, with spilling spates of summer meltwater from marginal, seasonal lakes eroding spillways (now, sometimes, still navigable channels) and promontories (isolating modern offshore islands). Ridges on the sea bed of the northern Irish Sea, between Donegal and Scotland, large enough to be depicted on navigational charts, have been ascribed to moraines (Mitchell, 1963), possibly of this age.

Ice sheets had similarly repeatedly covered the North Sea (see Caston, Chapter 7, and Pratje, 1951; Veenstra, 1965, 1969, 1970, 1971; Flinn, 1967; Robinson, 1968; Oele, 1969, 1971a, b; Sindowski, 1970; Buch, 1972), with interglacial and interstadial periglacial, fluvial and lacustrine, and marine transgressive periods (Macar, 1974). Here, too, deep, steepsided, linear erosion channels were cut (by periglacial meltwater streams or by sub-ice tunnelling, during the Weichselian or Devensian glaciations, or, less probably, by tidal scour during periods of interglacial submergence); some were subsequently infilled (Dingle, 1971; d'Olier and Maddrell, 1970), but others still remain as trench-like depressions (e.g. the Silver Pit, see Donovan, 1973, and Zagwijn and Veenstra, 1966; for other linear deeps to the north, see Flinn, 1973). The great trench-like depression of the Norwegian Trough is an over-deepened glacial scour channel (Shepard, 1931) while the north–south alignments of the Silver Pit and Sole Pit may represent periglacial spillways or (as originally suggested by Valentin, 1957) subglacial drainage channels. The belt of tunnel valleys which runs from the Devil's Hole, off Aberdeen, to include the Fladen Ground Deep, indicates the past location of a major British ice sheet (Jansen, 1977); the erosion channels in the East Bank area, north-west of the Dogger Bank, were probably cut by meltwater streams from the edge of the Weichsel icesheet, which deposited the Dogger Bank terminal morain (Veenstra, 1965; Dingle, 1971). Former glacial landscapes lie buried beneath the later Pleistocene and Holocene deposits which have built up the major banks of glacial, periglacial, lacustrine, fluvial, intertidal-marsh and marine sediments since the Middle Pleistocene (Eemian) at least; most major North Sea banks (Dogger Bank, Fladen and Witch Grounds, Hinder Bank, Brown Ridge, etc.) originated in this way (Stride, 1959; Van Eerde, 1964; Veenstra, 1964, 1965; Jansen, 1977). The Fladen and Witch Grounds are made of horizontally stratified glaciomarine deposits (sandy clays, gravels) formed during the glacial maximum (Jansen, 1977). Even the ridge-like bedforms superimposed upon these banks (e.g., on the East Dogger Bank) and which are now drowned to at least 45 m below present sea-level, accumulated in the latest Pleistocene or earliest Holocene (12,000–9,000 years B.P.), during the lower sea-level of the late Devensian–Flandrian transgression (Jansen, 1976). The same transgression eroded and redistributed the sediments and cryoturbated rocks of the bed of the north-east Celtic Sea, the advancing surfzone transporting the fine-grained material up the Bristol Channel to form a succession of intertidal marshes there during the early Flandrian (ca. 10,000–9,000 years B.P. – Culver and Banner, 1979).

By the time of the establishment of the modern hydrodynamic regime, the sea bed of the north-west European shelf seas exposed, to erosion and transportation, muds, silts and sands of Late Cretacous, Palaeogene, Neogene, Pleistocene and early Holocene (Flandrian) ages (Culver and Banner, 1979). Cretaceous and Tertiary sediments still outcrop in the Celtic Sea (Hamilton, Chapter 5; Curry et al., 1967; Delanty and Whittington, 1977), in the western approaches to the English Channel (Curry et al., 1965, 1970, 1971) and in the central and eastern English Channel also (Donovan and Stride, 1961; Curry, 1962; Dingwall, 1969; Larsonneur, 1969; La Pierre et al., 1970; Lefort and Deunff, 1971; Robert, 1971; Donovan, 1972; Groupe NOROIS, 1972). Quaternary deposits, including early and late Flandrian peats and muddy sands, occur off the coast of Devon (Clarke, 1970; Hails, 1975; Kelland, 1975). Very little terrigenous sediment is nowadays contributed to the English Channel by rivers; for example, the Seine estuary is a sink for marine sands rather than a source, marine sediments being found up the estuary as far as Caudebec (Germaneau et al., 1972). The regression of the cliffs of the coasts of Hampshire, due to the erosion of soft Palaeogene sediments there, may be a source of supply of terrigenous sediment for coastal embayments (e.g. Christchurch Harbour) (Dyer, 1972) but not for the offshore sea-bed. The decrease in grain size of superficial sediments from the central area of the English Channel to the Baie de Seine suggests not only lack of addition of new terrigenous sediment but that a process of reworking and redistribution of pre-Flandrian deposits has been dominant for the last 7,000 years at least (Larsonneur, 1972). That erosion and redistribution of outcropping Palaeogene deposits (e.g., the Eocene rocks exposed around the Channel Islands, described by Curry et al., 1970, and by Curry, 1962) also occurs is shown by the presence, in thin, modern, mobile quartz and shell sands east of Sark (Culver and Banner, 1979), of abundant Middle Eocene microfossils (e.g. *Rotalia trochidiformis* Lamarck), evidently derived from limestone and marl outcrops nearby. Large areas of the Late Cretaceous chalk outcrop adjacent to the Hurd Deep are covered only in scattered flints, derived from the chalk; the thick unconsolidated sediments to the south, off Brittany, are probably no more than periglacial "head" (Ruellan et al., 1972). The chalk outcrops are otherwise covered mainly by modern biogenic, shell sands and gravels, consisting almost wholly of the debris of bryozoa (e.g. *Cellaria*), echinoids (e.g. *Echinocyamus*) and bivalves (e.g. *Glycimeris*) with little or no terrigenous, detrital component (see also Chapter 5). Organogenic gradients of composition and "grain size", for these and other biogenic deposits, have been discussed by Boillot (1965).

In summary, the English Channel sea bed is characterised by outcrops of soft, easily eroded Mesozoic, Tertiary and Quaternary sedimentary rocks and by mobile sediments largely derived from them or from biological production. The same is very largely true of the shelf sea bed in all of the other north-west European shelf seas to the north, except that they possess extensive covers of glacial and interglacial deposits which conceal most or all of the older sedimentary rocks. The sediments of the North Sea are described by Caston (Chapter 7) who points out the real difficulty often experienced, in sea-bed survey, is the determination of which sediment is still potentially mobile and is continually readjusting to the fluctuations of the modern hydrodynamic regime (to remain in equilibrium with it) and which is effectively immobile under present wave and current conditions (being out of equilibrium with present hydrodynamic forces), and is a product of environments long past. As is discussed further, later in this Chapter, not only the dynamic grading of the sediment itself, but also the bed forms into which it is built, must be the result of the maximum shear stresses experienced by the deposit, and so the one can be interpreted to indicate the nature of the other. Of course, some topographic features are unrelated to hydrodynamics – natural examples are the iceberg "ploughmarks" near the Norwegian Trough (Belderson and Wilson, 1973) and the "pockmarks", up to 300 m long and 137 m deep, which are probably excavated in silts and clays by escaping natural gas, in patches between the Forties Field (Vol. II, Fig. 18.1) and the Norwegian Trough (Platt, 1977, and Caston, Figs. 7.32, 7.34). Nevertheless, rhythmically developed bedforms in unconsolidated sediment are expected to be a product of hydrodynamic transportation. However, both the sediments and their bed forms often appear (not only in the North Sea, but in all the other areas of the north-western European shelf) to be the product of hydrodynamic environments which are not those of today (e.g. the "transverse sandpatches" of the Celtic Sea, crescentic bodies of fine to medium sand and coarse shell sand, often 500 m long and 2 m high, which are present in waterdepths of 70–80 m where surface current speeds do not exceed 50 cm/sec; cf. Figs. 8.3 and 9.12, and Kenyon, 1970a). Some major sedimentary bedforms are so clearly in discordance with the modern environment that their origins must be ascribed to the past (e.g. linear sand banks, up to 220 km long, postulated to have originated in a large Pleistocene estuary: Stride, 1963a). For other sediments, the origin is less clear.

In the North Sea, as in the Sea of the Minches (Chapter 6), an obvious problem is that the areas of mud and sandy mud which sporadically occur on the sea bed are sometimes associated with topographic depressions, but not at others. In his attempt to estimate a budget of supply, deposition and loss of mud in the North Sea, McCave (1973) has calculated a rate of deposition for mud in the Outer Silver Pit of 20 cm/100 years (i.e. 2 m since the beginning of the Flandrian if the supply has been constant) and he assumed that the Pit is still a trap for mud deposition. This may in part be so, although Donovan (1965) core-sampled the walls and floor of the Silver Pit and found no sediment within it which he believed to be younger than that of the last glaciation, other than sand (less than 30 cm thick in the northern part, thicker and developed as sand waves centrally); he believed that the Silver Pit was eroded by Holocene tidal currents and implied that no modern mud could accumulate there. It is virtually certain that most mud which now caps topographic highs owes its origin to pre-Holocene environments – or, at latest, to those associated with the Flandrian sea-level rise. Examples are the clay which caps irregular ridges, north of the sand-wave field, around the Sandettie and Fairy Banks, ascribed to deposition during the early Flandrian transgression (Kirby and Kelland, 1972) and the pre-Boreal peats and Boreal shelly fine sediments on the east and south-west of the Dogger Bank (Veenstra, 1965). Areas which possess such deposits at the sea bed can have experienced negligible net deposition since the establishment of the present hydrodynamic regime. Caston (Chapter 7) has discussed the problem of distinction between Flandrian and modern Holocene deposits in the North Sea, and in Chapter 11, Vol. II, we show the significance of such a distinction in the interpretation of dynamic sedimentological regimes in coastal environments, with particular reference to Swansea Bay. The distinction is rarely an easy one, and can usually be made with confidence only after analysis of internal evidence – radio-isotope dating (by ^{14}C, e.g. Veenstra, 1965; by ^{210}Pb, e.g. Bertine in McCave, 1973), palynology (e.g. Zagwijn and Veenstra, 1966), macrofaunas (e.g. Veenstra, 1965; Jansen, 1977), microfaunas (e.g. Culver and Banner, 1978, 1979) and artifacts (e.g. the first appearance of coal from steamships, used by Reineck, 1963). With coarser-grained sediments, the external evidence of the bedform, and its likely genesis and equilibrium with present day waves and currents, can be used, but much less conclusively (compare the opinions expressed by Valentin, 1957, and Donovan, 1965 and 1973, on the origin of the Silver Pit!).

Clearly, when the sediments of the sea bed are described, their implications for interpretation of the prevailing hydrodynamic regime, as well as for dynamic sedimentological studies, demand that discrimination be made between the deposits remaining from past times and those which are actively mobile today. Consequently, in any useful terminology, a subjective genetic term should be added to the objective descriptor.

The genesis, origin and description of unlithified sea-bed sediments (S.J.C. and F.T.B.)

Sediments deposited under environmental conditions now past have been termed "relict" by Emery (1952, 1968), and they have been distinguished from "modern" sediments in that the latter are "currently being supplied from the continent or other sources and transported to the area of deposition" (Curray, 1965, p. 725). Relict sediments, however, may be reworked and the resulting sediment has been termed "palimpsest" and defined as a sediment ". . . which exhibits petrographic attributes of an earlier depositional environment and, in addition, petrographic attributes of a later environment" (Swift et al., 1971).

Belderson et al. (1971) studied the shelf sediment around the British Isles and suggested that "much of the deposit (is) in equilibrium with present day water movements", so that the term "relict" is "largely superfluous" when considering, especially, the sediments on the continental shelf west of the British Isles.

Clearly, biogenic deposits which can be analysed qualitatively (and the evidence of the species-list supplemented, if needs be, by radio-isotope dating) can firmly be referred to modern or past origins, but inorganic deposits (or those with a large detrital, inorganic admixture) are not so readily determined. Every quartz grain in every sand is likely to have had a long history of sedimentation, reworking and redeposition before it has come to its present situation. The only valid distinctions which can be made sedimentologically are: (1) have the mineral grains been newly added to the estuarine or off-shore sediment from a terrigenous source? – i.e., are they new input from terrestrial sources (cliff erosion, fluvial sediment load from a drainage basin, wind blown terrestrial sands, etc.) or from submerged consolidated rock (colluvials)?; (2) have the grains, minerogenic or biogenic, been recently subject to hydrodynamic transportation? – i.e., are the present waves and currents capable of moving, sorting and redepositing them? The first question is important economically, in that all estimates of sediment budget depend on it; the second is also economically significant in that its answer leads to the recognition of sediment transport paths, determination of aggregate supplies, estimates of need for harbour dredging, the likelihood of scour around engineering constructions, and so on. Each question can be answered, and the answers can provide the basis of a practical and useful, as well as scientifically valid, terminology. The existing terminologies can lead to ambiguities (Fig. 8.1, p. 277) so we seek to revise them, by simplification or by elaboration, as is appropriate.

We have taken as the basis of this revision the sediments of the sea bed of Swansea Bay and the adjacent Bristol Channel, not merely because this area has been studied intensively by ourselves and our colleagues over the past ten years, but also because the sea bed of the area contains deposits of almost all size grades of more or less consolidated, organic and inorganic sediment (see Chapter 11,Vol. II) which have been thoroughly sampled and examined for age determination (Culver and Banner, 1978, 1979) and subjected to detailed continuous seismic reflection surveys. In the early stages of survey, very real problems of discrimination between sediments of past and present environments were experienced here. The sediments exemplify well the problems which could be encountered elsewhere. Tidal current and wave monitoring and analysis, and hydrographic and hydrological study have been undertaken. The input of material from cliff erosion is nil and that from rivers is negligible: all significant sources of sedimentary material are from within the neritic and sanidal zones of Swansea Bay and the adjacent Bristol Channel.

Below MHWS, six principal types of deposit are present (summarised on Fig. 8.1, p. 277; for areas of occurrence, see Vol. II, Fig. 11.3). These are:

(1) Consolidated greyish, clayey sands, sandy silts and clays (and associated peats). The benthonic foraminiferal fauna recovered from these deposits is dominated by euryhaline forms such as *Haynesina germanica* (Ehrenb.), *Elphidium williamsoni* Haynes and *Ammonia tepida* (Cushman). Both juvenile and adult specimens of these forms are present and they represent the indigenous foraminiferal fauna (see Culver and Banner, 1978). Some stenohaline forms are also present; their size-sorting and often poor preservation indicates that they were transported specimens washed into Swansea Bay from the central and outer Bristol Channel (Culver and Banner, 1979; cf. Murray and Hawkins, 1976). All foraminiferid tests are white and opaque due to slight etching of the test surface following burial in a slightly acidic environment (see Murray, 1967; Murray and Wright, 1970).

These greyish deposits (the distribution of their sea-bed exposure is shown on Fig. 11.3, Vol. II), were deposited under intertidal conditions as mud flats and sand flats. They are early Flandrian in age and were deposited during the period 9,500–2,500 years B.P. (Culver and Banner, 1978). Scanning electron microscopy of surface textures of sand grains obtained from these deposits show that the grains probably have undergone glacial transport followed by some marine reworking (Culver and Bull, unpublished manuscript).

(2) Poorly sorted, consolidated deposits containing clay, silt, sand, gravel, pebbles, cobbles and boulders. These are boulder clays deposited very close to the maximum southern extent of the last (Devensian–Weichselian–Wisconsin) glaciation at about 20,000–17,000 years B.P. (Charlesworth, 1929; Bowen, 1970). This deposit contains pebbles typical of "Welsh Drift" (Coal Measures, Millstone Grit, Old Red Sandstone) which were picked up as the ice moved southwards over the South Wales Coalfield. Some better sorted sands and gravels, encountered in boreholes, are possibly of fluvio-glacial origin. No microfauna was recovered from these drifts. They are exposed in places in the littoral zone of Swansea Bay, have an undulating surface and underlie the sub-recent (Flandrian) deposits. Offshore gravity cores (Culver, 1976) and continuous seismic profiling (Price, pers. comm., 1977) also show the presence offshore of glacial deposits, underlying Flandrian sediment and exposed, in places, at the surface of the sea bed (See Vol. II, Fig. 11.4).

(3) Consolidated gray-brown, well-laminated silty clay. These deposits, containing only 1–22% sand, yielded no microfauna although a few fragments of freshwater gastropods were present in some samples. Scanning electron microscopy of surface textures of quartz grains taken from these deposits indicates that the grains originally had a glacio-fluvial origin; subaqueous modification of the grain surfaces is very limited. These laminated deposits are considered to be of lacustrine origin and were laid down during the period of approximately 16,000–10,000 years B.P. (Culver and Bull, unpublished manuscript). This late Pleistocene sediment was encountered directly at the seabed in one gravity core only, but additional information concerning its characters was supplied by data from boreholes drilled by the Institute of Geological Sciences in Swansea Bay.

(4) Unconsolidated sand and mud. This material (its distribution is shown in Fig. 11.4, Vol. II) is available for transportation in both the littoral and sublittoral zones. It forms intertidal sand-flats, sand bars and the sand deposits of the berm. The mud and silt is often concentrated in the lee of intertidal sand bars. Offshore the sand forms large sand banks (e.g. Mixon Shoal) and smaller sand bodies. Quartz grains from these deposits examined under the scanning electron microscope, show surface features produced by glacial action over which the marked effect of marine action is superimposed (Culver and Bull, unpublished manuscript). The foraminiferal fauna consists of (probably local) stenohaline forms which originally lived on or attached to firm substrates such as rocks and sea weeds – e.g., *Cibicides lobatulus* (Walker and Jacob), *Rosalina globularis* d'Orbigny, *Planorbulina mediterranensis* d'Orbigny, *Quinqueloculina seminulum* (Linnaeus). After death the tests of these animals were transported as sedimentary particles away from their habitat and were incorporated in the unconsolidated sand bodies in Swansea Bay. Also present are modern benthonic forms which live sublittorally on sandy or muddy substrates and which are transported into

Swansea Bay from as far away as the central and outer Bristol Channel as pseudoplankton (Murray, 1965; Murray and Hawkins, 1976; Culver and Banner, 1978a). A few euryhaline, benthonic foraminifera derived by erosion of Flandrian intertidal mudflat deposits are also present. Due to their transportation as sedimentary particles, foraminiferal tests present in the unconsolidated sands show size-sorting to varying degrees and thus differ in their preservation, showing breakages due to transportation by water and also little evidence of the etching typical of Flandrian foraminiferid tests.

(5) Pebbles, cobbles and boulders, sometimes mixed with mud, sand and silt. Pebbles and cobbles were often recovered in the gravity core nose cone, thus precluding the recovery of sediment cores. Subangular pebbles, cobbles, and boulders of allochthonous (i.e. not colluvial) rocks have been recovered also by grab and dredge sampling. Most pebbles were encountered in the area of unconsolidated sand and mud but some were also recovered where Boulder Clay is exposed on the sea bed. The latter pebbles are considered to be lag deposits, formed by submarine winnowing of glacial deposits, the sand, silt and clay fractions being transported to other localities. The pebbles in the area of unconsolidated sand and mud are probably lag gravels to which sand and mud has been supplied and added later by current or wave action.

(6) A large area of Swansea Bay is covered by a layer of relatively fluid sandy mud, up to at least 1 m thick, which is probably the material continuously being dredged from the channels approaching the mouths of the Tawe and Neath rivers and Port Talbot Harbour (Vol. II, Fig. 11.4). This material is dumped mainly on the eastern side of the bay; later some may be transported to and redeposited in the dredged channels.

In summary, the bedrock of Swansea Bay is largely covered by deposits of the Devensian ice age and the succeeding Flandrian transgression; these deposits remain, sometimes, unaltered, but in other cases they have been eroded and finer-grained fractions removed, so that original or modified deposits of past times may be covered with relatively thin layers of unconsolidated sands, silts and muds; some of these sediments are mobile with periodic regularity (e.g. the fluid mud of the approach channels, see below and Fig. 8.7, and muds in suspension), others regularly achieve mobility under stress from springs, tidal currents or from approaching wind, waves or oceanic swell or all of these (e.g. the various sand and silt fractions), while others are transported only at times of storm and surge (e.g. cobbles and boulders). The extent of movement of the last of these categories can only be estimated from predictions of maximum waves and surge currents, and confirmed by the uniform development, all over a cobble or boulder, of annelid and other encrustations; the rarity of the necessary maxima of wave and current energy (and the practical difficulties of directly measuring such maxima) implies their absence in the hydrological, quantitative record. The rare event cannot be ignored, however, as the mobile sediment sampled from the sea bed is likely to have achieved its equilibrium sorting and bed form in response to the most energetic hydrodynamic event to which it has been subjected, even though that event may have occurred months (or years) before the sampling programme (see Introduction to Chapter 11, Vol. II).

The significance of the Swansea Bay deposits in the exemplification of a sedimentological terminology may be described and discussed as follows (Fig. 8.1).

Sediment deposit No. 1 (Flandrian, consolidated, greyish, clayey sands, sandy silts and clays) can be considered to be a relict deposit as it is "remnant from an earlier environment" (Emery, 1952, p. 1105) and the sediment was supplied in the past, and is not being distributed today (see McManus, 1975). However, sand grain surface features indicate that it is probable that the sands at least, and also probably the silts and clays, of these Flandrian deposits were largely derived by erosion of glacial deposits. That is, the deposit "exhibits petrographic attributes of an earlier depositional environment and, in addition, petrographic attributes of a later environment" (Swift et al., 1971, p. 343). Thus, the Flandrian deposits in Swansea Bay are an example of relict palimpsest sediments.

Sediment deposit No. 2 (Devensian, consolidated glacial deposits) is considered also to be a relict deposit as it was laid down in the past, mainly under sub-ice conditions. However, unlike the Flandrian deposits above, this deposit shows no "petrographic attributes of a later environment" (Swift et al., 1971) and so is a simple relict deposit composed of particles supplied before the present day.

Sediment deposit No. 3 (Late Devensian, consolidated, laminated clays and silts) is again considered to be a relict deposit as it was laid down in the past under lacustrine conditions. However, scanning electron microscopy of quartz grain surface features show that it is also a palimpsest deposit as the sand grains, at least, were probably derived from previously existing glacial deposits. Thus, the lacustrine clays and silts in Swansea Bay are a further example of a relict palimpsest deposit.

Although the above three deposits themselves are not being distributed at the present day, particles eroded from them, by current and wave action during periods of stormy weather or extreme tidal conditions, may be distributed and incorporated in modern deposits.

The various forms of sediment deposit No. 4 (unconsolidated sand and mud) are continually being modified by wave and current action. That is, the group of processes that distributes sediment particles on the shelf to form them into sedimentary deposits are operating at the present day. Thus, sand bars and most of the sand banks found in Swansea Bay are modern deposits by definition (McManus, 1975). Ferentinos and Collins (1979, in press) consider that some of the larger linear banks at the southern margin of

Swansea Bay can be related to present day tidal eddy systems while others may have originated during times of lower sea-levels. The sand grains of these deposits generally show evidence of earlier glacial effects. As the transgressing Flandrian sea reached the Swansea Bay area around 9,500 years B.P. (Culver and Banner, 1979) the sand grains would have been eroded from the glacial deposits flooring Swansea Bay over approximately the past 9,500 years. It is probable that such grains, and finer ones, are supplied at the present day to the, at times, mobile unconsolidated sediments by erosion of sea-bed exposures of boulder clays, lacustrine deposits and Flandrian clays, silts and sands. Thus, these modern deposits may be further described as amphoteric (McManus, 1975) as they contain some particles supplied at present and some that were supplied in the past. However, if these deposits of unconsolidated sand and mud are classified using the broader, original definition of "palimpsest" proposed by Swift et al. (1971), and followed by Slatt (1974), rather than the more limited definition proposed by McManus (1975), they must be placed in the category "palimpsest" as they exhibit attributes of an earlier environment.

Sediment deposit No. 5 (pebbles and cobbles) is present in two forms. In one, gravel particles of various sizes have been encountered over outcrops of boulder clay. These are lag gravels, resulting from submarine winnowing of the glacial deposits, the sand, silt and clay fractions having been transported to other localities. It is probable that such winnowing began when the Flandrian

Example of sedimentary deposit	Term from Emery, 1952, 1968	Term from Swift et al., 1971	Term from McManus, 1975	Term proposed here
SWANSEA BAY, NO. 1.				
Flandrian, consolidated, greyish, clayey sands, sandy silts and clays	relict	palimpsest	relict	relict palimpsest
SWANSEA BAY, NO. 2.				
Devensian consolidated boulder clays	relict	relict	relict	relict
SWANSEA BAY, NO. 3.				
Late Devensian, consolidated laminated clays and silts	relict	palimpsest	relict	relict palimpsest
SWANSEA BAY, NO. 4.				
Unconsolidated sand and mud	modern	palimpsest	proteric or amphoteric	modern palimpsest
SWANSEA BAY, NO. 5.				
Pebbles, cobbles and boulders.		palimpsest	relict	palimpsest lag gravel
Pebbles, cobbles and boulders with mud, silt and sand		palimpsest	relict or palimpsest	palimpsest, muddy, silty sandy gravel
SWANSEA BAY, NO. 6.				
Dredged and dumped sand and mud	modern	palimpsest	neoteric	modern palimpsest
HURD DEEP				
biogenic ("shell sands and gravels")	modern	modern	neoteric	modern
DOGGER BANK				
Hydrobia-Macoma shelly sands (Veenstra, 1965)	relict	relict	relict	relict
FRISIAN ISLANDS				
Littoral shelly sands (Luders, 1939)	modern	modern or palimpsest depending on Pleistocene proportion	amphoteric	modern palimpsest
DUTCH DELTA				
fluvial (high Mn) muds (Terwindt, 1967)	modern	modern	neoteric	modern
DUTCH DELTA				
deposited and recirculated muds (variable Mn) (Terwindt, 1967)	modern	palimpsest	amphoteric or palimpsest	modern palimpsest
THE WASH				
mobile muds (Shaw, 1973)	modern	palimpsest	palimpsest or amphoteric	modern palimpsest

Fig. 8.1. Descriptive terminology for sedimentary deposits reflecting their genesis and potential mobility.

sea first transgressed into Swansea Bay and has continued for the past 9,500 years with particles still being added to the lag deposit at the present time. This type of deposit does not fit into the classification of continental shelf sediments proposed by McManus (1975) as distributor processes, an inherent factor in that classification, are not involved in the in situ formation of lag gravels. McManus (cit., p. 1154) stated that if distributor processes are "operating at present, we may say the resulting deposits are modern; otherwise they are relict". Thus, according to this classification, lag gravels cannot be described as either modern or relict. Slatt (1974) considered lag gravels in Conception Bay, Newfoundland to be palimpsest, in the sense of Swift et al. (1971), as each particle in the lag deposit was originally part of an earlier deposit representing a different deposition environment (i.e. boulder clay). In Swansea Bay, lag deposits cannot be described merely as "relict" in the sense of Emery (1952, i.e. "remnant from an earlier environment") as the environment of formation of this deposit was the same 9,500 years ago as it is today. Thus, following Slatt (1974), the gravels in Swansea Bay are best described simply as palimpsest lag gravels, implying the qualifying statement that particles comprising the deposit have often been transported, and therefore supplied, both in the past and at the present day.

The second form of sediment deposit No. 5 is a poorly sorted deposit formed when mud, silt and sand, probably derived directly from sediment deposit No. 4, are deposited upon and within the interstices of lag gravels. In this case the gravel particles are likely to have been supplied only in the past as further winnowing of underlying glacial deposits would be curtailed after deposition of smaller particles upon and within the gravel particles already present. This poorly sorted deposit is, therefore, a mixture of a palimpsest lag gravel with amphoteric (McManus, 1975) sediment of deposit No. 4. However, as stated above, this mobile sediment may also be placed in the category "palimpsest" as proposed by Swift et al. (1971).

Sediment deposit No. 6 (dredged and dumped sand and mud) is composed of particles which are mobile, at times, under present hydrodynamic conditions. Thus, following McManus (1975), this deposit is a modern one. Particles are supplied to this deposit by dumping of dredged material and also probably by addition of sediment particles derived from sediment deposit No. 4. This deposit originated when dredging first began in this area (very recently in geological terms) and thus the deposit may be termed "neoteric" (McManus, 1975) as particles are being supplied only at present. However, as the majority of sediment particles were obtained by dredging of glacial and Flandrian intertidal deposits and still show evidence of their earlier depositional environment (e.g. surface features of quartz grains, typical Flandrian foraminiferal assemblages), and because other dredged muds have been repeatedly recirculated in the estuary, sediment deposit No. 6 may also be described as palimpsest.

The resulting terminology which can be applied to these sediments (Fig. 8.1) is applicable to all others that we know of elsewhere. For example, the shelly beach sands of the Frisian Islands, with their admixture of Pleistocene shells transported shorewards from eroding offshore relict deposits, are "modern palimpsest". The muds of marine origin which are accumulating in the Dutch Delta are subjected to modern dynamic processes but are being continually recirculated (Terwindt, 1967), so they, also, are "modern palimpsest": only the mud being newly added to the marine environment from the erosion by rivers of their subaerial hinterlands is truly "modern".

Far from agreeing with Belderson's suggestion (1971) that the term "relict" is "largely superfluous" when sediments on the western shelf are being considered, recognition of these genetic categories permits a far more realistic assessment of the modern sedimentological environment than would otherwise be the case. Application of the criteria to an analysis of the sediments sampled in Swansea Bay, for example, indicates the following. The Pleistocene boulder clays were probably deposited during the Devensian ice-maximum, 20,000–16,000 years B.P.; where they are now exposed on the sea bed, there can have been no *net* deposition over the ensuing period. Similarly, the Flandrian clays (lacustrine, estuarine, intertidal marsh and marine), indicate areas of nil net deposition since the times of their formation, varying from 9,500 to 2,500 years B.P. In the area of the Bay, sea level reached its present maximum about 2,500 years ago (Culver and Banner, 1979) and the shape of the Bay has not significantly changed since then; neither, probably, has the wave climate or tidal regime, nor the offshore or onshore sources of sediment supply (except by small interference by man, in his dredging and constructional activities). The Bay is one containing areas of high levels of hydrodynamic energy (Chapter 11, Vol. II) but also areas where encroaching wave trains are either of short fetch or are strongly refracted. Conditions occur which could be predicted, from what is known of critical transport velocities for noncohesive sediments (see later) and of the measured wave and current energies (skewed, as always, to periods of calmer weather), to permit sediment accumulation. However, much of the Bay has undergone no net deposition, and, everywhere, the layer of mobile sediment, where it exists, is thin. One is forced to conclude that the Bay, which has been in a time-averaged stationary state for at least 2,000 years, would have been full of sediment by now if net deposition were possible. Consequently, we must also conclude that the mobile sediment is always transitory, all of it recirculating (except for new inputs of material winnowed from the sea bed of the Bristol Channel and for loss of sand, through the intertidal zone, to the aeolian sand dunes which fringe the coast) and that all of it is, in different ways, palimpsest.

Noncohesive sediments and their transportation (F.T.B.)

The first serious scientific studies of marine sediment transport evolved on the north-western European continental shelf. It was from his first-hand knowledge of Swansea Bay that Sir Henry de la Beche postulated (1851) a causal relationship between the sediment grade on the sea bed and the strength and direction of tidal currents. Over a century later, Stride and his colleagues at the Institute of Oceanographic Sciences (then the National Institute of Oceanography) were to propose the

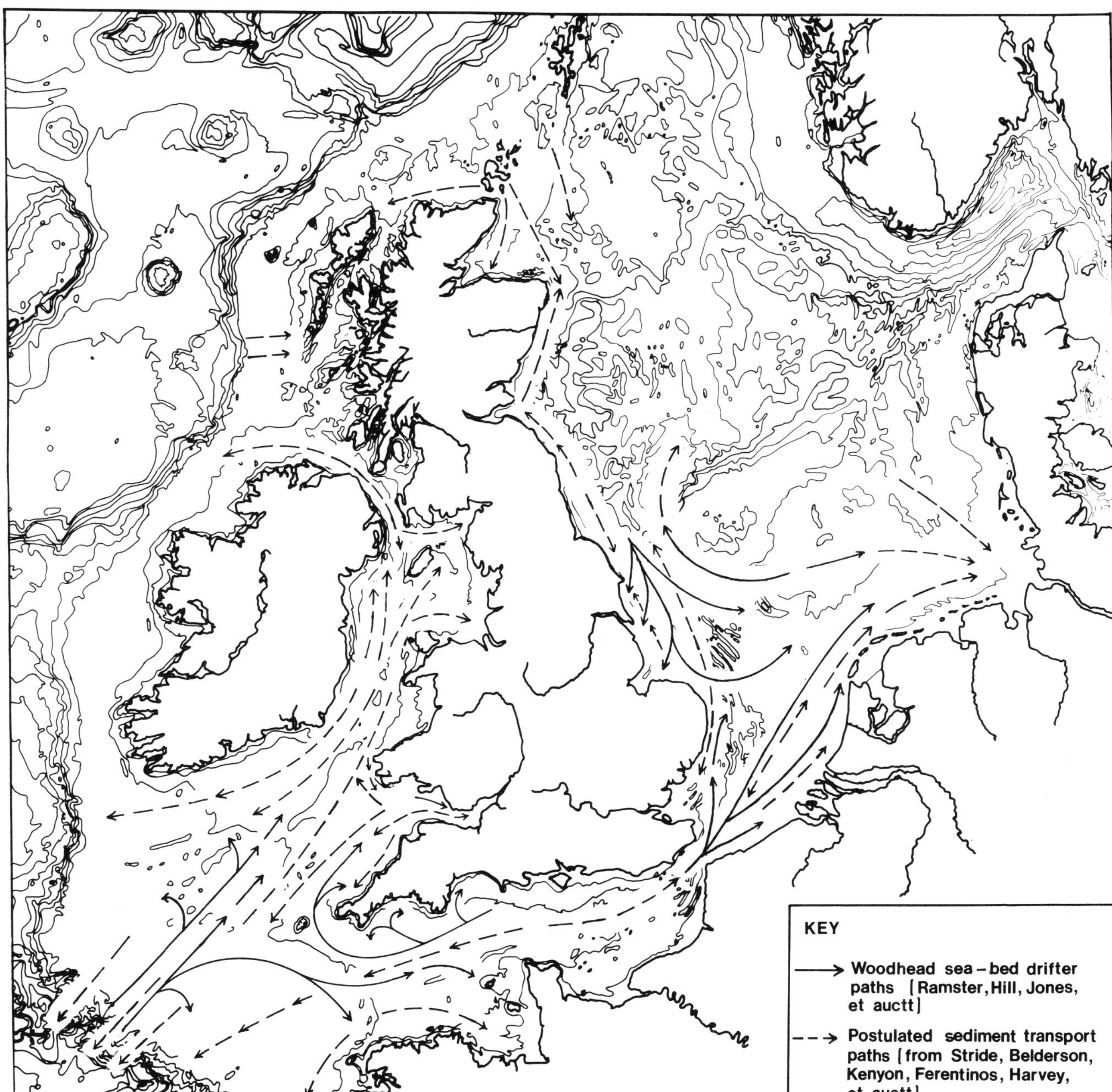

Fig. 8.2. Probable net, modern transport paths for non-cohesive mobile sediment in bed load.

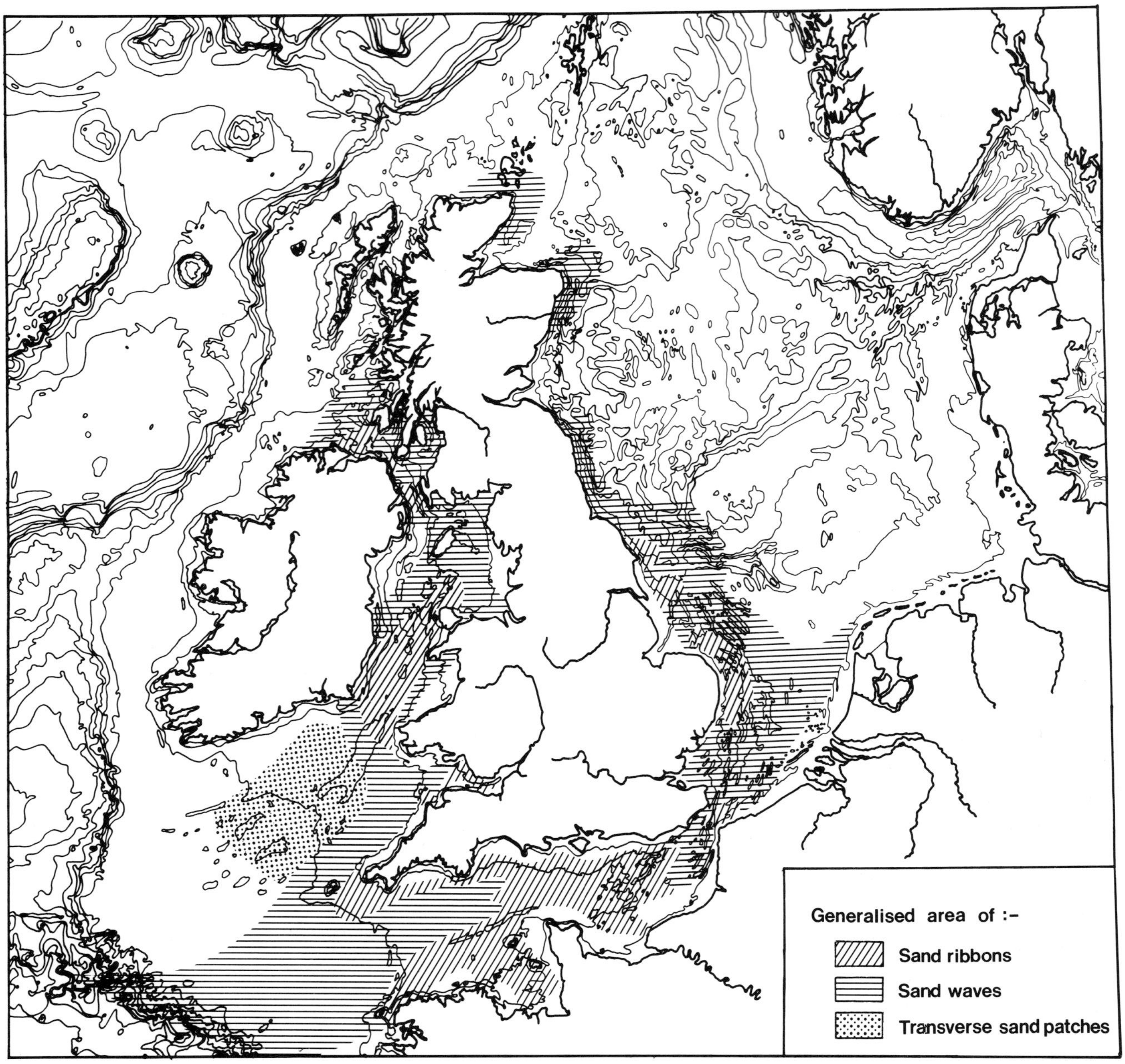

Fig. 8.3. Generalised distribution of rhythmic bedform classes, for unconsolidated, non-cohesive sediments.

hypothesis that tidal currents not only dominate the transportation and sorting of sediment on the shelf as a whole (Stride, 1963, 1965) but that they were responsible for the development of patterns of sedimentary bed-forms which, in turn, could be used to postulate corresponding regional patterns of sediment transport, sorting and deposition (Stride, 1973; Fig. 8.2 and 8.3).

The relationships between the form and distribution of sand waves, of some linear banks, and of sand ribbons and patches on the one hand, and the postulated transport paths of modern, mobile sediment, have been concisely described for the shelf as a whole (Belderson et al. 1971), for parts of the Irish Sea (Belderson, 1964; Caston, 1965; Jones et al., 1965; Harvey, 1966; Belderson and Stride, 1969), for the Celtic Sea (Belderson and Stride, 1966; Channon and Hamilton, 1976, and see Fig. 5.11), the western shelf (Kenyon and Stride, 1970), the English Channel (Stride et al., 1972; McCave, 1973a; Fig. 5.11), the North Sea as a whole (Stride, 1973; see also Caston, Chapter 7, Fig. 7.50) and for particular parts of it (Cloet, 1954; Dingle, 1965; Langeraar, 1966; Kenyon and Stride, 1970; Stride, 1970; Terwindt, 1970; McCave, 1971b, 1974; Caston and Stride, 1973). The first scientific description of sand waves as sedimentary bodies associated with hydrodynamic transport paths was probably that made on observations in the North Sea by Van Veen (1935).

The hydrodynamic theory postulated to explain,

quantitatively, the construction and mobility of sand waves and similar bed forms is beyond the scope of this book (but see, e.g., Kennedy, 1969; Allen, 1970; Smith, 1970), but the significance of rhythmic bed-form accumulations to interpretative dynamic sedimentology has been the subject of much relevant discussion (e.g. by Hamilton, Chapter 5; Caston Chapter 7; Belderson and Stride, 1966; Kenyon and Stride, 1968; Kenyon, 1970a; Stride, 1970, 1973; McCave, 1973a; in general terms, by, e.g., Off, 1963). In essence, the concept developed by Stride and his colleagues, for the north-western European shelf, is as follows.

There are certain areas of the shelf sea-bed where lithified sedimentary rocks outcrop on the sea bed or where they are only patchily covered by a thin veneer of coarse grained, unconsolidated sediment. Pleistocene and early Holocene deposits, if there ever were any, have been scoured from them, and no sediment has subsequently accumulated there. Such areas of scour, in the middle part of the Bristol Channel (where Jurassic strata are exposed) and middle and mid-western parts of the English Channel (where Cretaceous chalks outcrop) correspond to regions of highest tidal current surface speeds (in excess of 2.5 m/s, as mapped by Von Sager, 1963 and Von Sager and Sammler, 1968 (see Vol. II, Fig. 9.12), simplified by Belderson et al., 1971; note that the isolines of maximum current speeds, postulated for the North Sea by Kraav (1969), are a simplification derived from M_2 coheight and cotidal isopleths, and take no account of coastal, offshore bathymetric perturbations). There is no corresponding correlation between these scoured areas and maxima of wave energies (Vol. II, Fig. 10.10). Therefore, it was postulated that tidal scour was responsible for removal of all unconsolidated sediment, and that these "zones of erosion" (Belderson and Stride, 1966) are also "bed-load partings" (Stride, 1965) from which mobile sediment is being transported as bed-load towards areas of less hydrodynamic energy. Adjacent, shallower areas of lag gravel cover, where near surface tidal currents (Vol. II, Fig. 9.12) reach 1.5 m/sec or more at mean springs (e.g., adjacent to the scoured areas of the Bristol Channel and of the English Channel, see Fig. 13.7 and 13.8 in Vol. II), develop long, subparallel furrows about 1 m deep, 25 m apart and 9 km or more in length, in the loose gravels; the furrows lie parallel to the dominant tidal current direction and tend to converge in that direction (Stride et al., 1972); one can postulate that helical near-bed flow, with "herringbone" directions of bed-load transport, could create them, although current speeds in excess of 1 m/s within one metre of the bed would be needed (Fig. 8.4). The longitudinal furrows are likely to be hydrodynamically comparable to sand-ribbons, which have been morphologically systematised by Kenyon (1970a). Zones of subparallel "ribbons" of sand, elongated in the direction assumed to be that of sediment transport, occur immediately "downstream" from the areas of scour or lag-gravels (Fig. 8.3); they can be many kilometres in length and hundreds of metres in width but are only centimetres thick and their lateral boundaries are sharp (Belderson and Stride, 1966). Where surface current speeds reach 2.5 m/s (Vol. II, Fig. 9.12), the ribbons may have superimposed, shortwavelength bedforms developed with crests at right angles to the length of the "ribbon" (and to the dominant current direction); these "type A ribbons" lose their rhythmic bedforms in areas with surface currents around 2.0 m/s (to form "type B"), but lines of barchan-like dune forms develop along the ribbons ("type C") when surface speeds fall to about 1.75 m/s, to be succeeded by more rhythmically developed, asymmetric wave-forms at 1.5 m/s ("type D") (Kenyon, 1970a). Like true sand-waves, the waveforms of "type D" have steep slopes on their "downstream" sides and less steep ones on the sides facing the oncoming dominant current, believed to be the source of sediment supply. Zones of sand-ribbons are known from sonograph records adjacent to the scoured areas of the English Channel and Bristol Channel, and also in the Straits of Dover, the St. George's Channel and southern Irish Sea, in the North Channel of the Irish Sea and off northeast Norfolk (Belderson et al., 1971). The St. George's Channel – southern Irish Sea area has been taken by Stride (1965) to be a zone of bed-load parting; its sand ribbons appear to merge northwards and southwards with zones of true sand-waves (Belderson, 1964; Belderson et al., 1971) which indicate, by their asymmetry, advance of sand within them from the sand-ribbon zone. Surface currents in this area have maximum speeds ranging from 2 m/s off southeast Ireland to less than 0.5 m/s (in the northernmost parts of the sand-ribbon area and in the adjacent zones of sand waves) (Von Sager and Sammler, 1968). There is no significant source of new sediment supply to this bed-load parting except for the relict Quaternary deposits of the sea floor itself. The sand-ribbon zone off north-east Norfolk (Belderson et al., 1971), also rated as an area of bed-load parting (Stride, 1965), has a possible source of modern sediment in the eroding Pleistocene cliffs of the Norfolk coast, although much, at least, of this material is transported longshore (Robinson, 1966); mean maximum surface currents here are mapped as reaching 1.0–1.5 m/s (Von Sager and Sammler, 1968; Belderson et al., 1971), but bottom and nearbottom currents must be greatly influenced by the rapidly varying topography

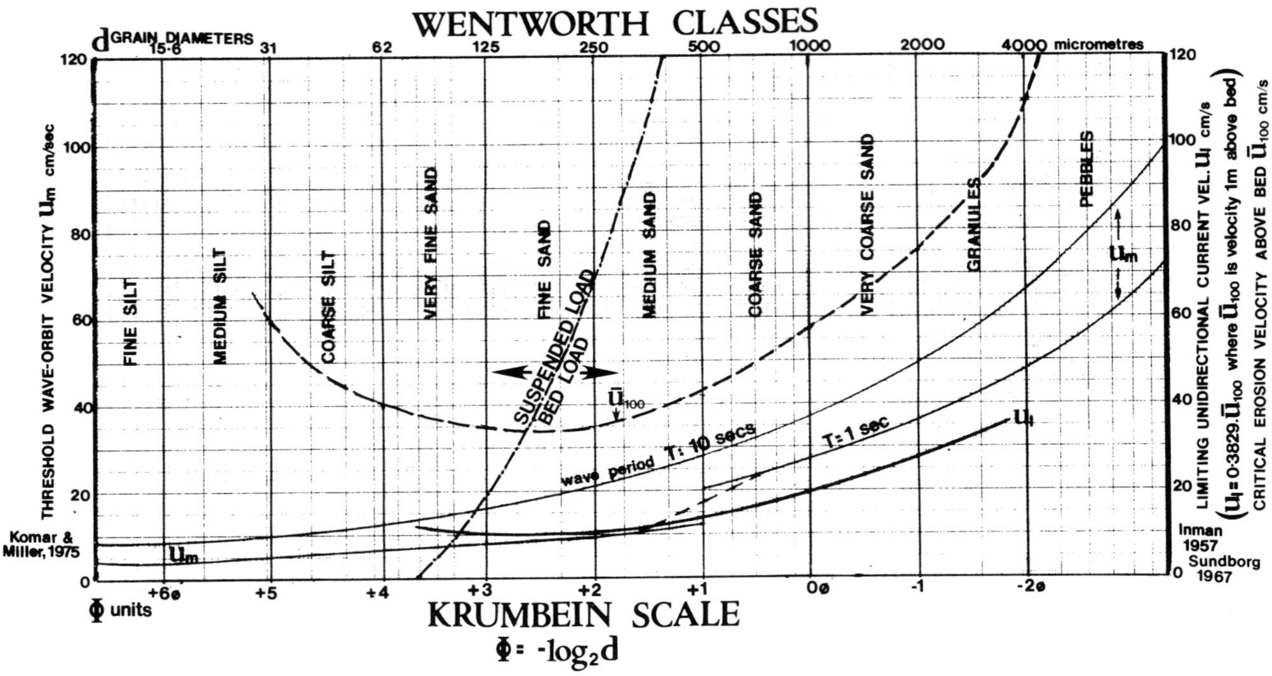

Fig. 8.4. Flow "velocities" (current speeds) needed for the initiation of movement in uncohesive, unconsolidated, well-sorted sediments composed of grains of specific gravity ca. 2.65 (i.e., quartz, cf. calcite 2.71, felspars 2.56–2.76). The wave-orbital, oscillatory, bottom currents of shallow-water Airy waves of 1 s and 10 s periods have speeds U_m, when:

$$U_m = \frac{\pi d_0}{T} = \frac{\pi H}{T \sinh (2\,h/L)}$$

where d_0 is the near-bottom orbital diameter for an Airy wave of height (i.e. surface orbital diameter) H and wavelength L in water depth h. U_m will be the threshold velocity of grains of diameter D and specific gravity ρ_s, in water of density ρ, when empirically:

$$\frac{\rho U_m^2}{(\rho_s - \rho)\, gD} = 0.21 \left(\frac{d_0}{D}\right)^{1/2} \text{ for } D < 0.5 \text{ mm} \quad \text{or:} \quad \frac{\rho U_m^2}{(\rho_s - \rho)\, gD} = 0.46\pi \left(\frac{d_0}{D}\right)^{1/4} \text{ for } D > 0.5 \text{ mm}$$

according to Komar and Miller (1975). The corresponding thresholds under unidirectional currents have been estimated for speeds at the bed (U_l, after Inman, 1957) and 100 cm above the bed ($\bar{U}_{100}$, after Sundborg, 1967); each assumes that the bed is initially plane and horizontal. The relationship $U_l = 0.3829 \cdot \bar{U}_{100}$, given by Inman (1957), should be compared to the equivalent $U_l = 5.47 \cdot 10^{-2}\, \bar{U}_{100}$ given by Sternberg (1972), who defined boundary shear-stress τ_0 as $3 \cdot 10^{-3}\, \bar{U}_{100}^2$ dynes/cm^2. The threshold ("limiting", "critical erosion") "velocities" are plotted here, lin-log, for the grain-size classes defined by Wentworth (1922) against the phi-scale proposed by Krumbein (1934, 1936). The limit between suspended-load transportation and bed-load transportation of these grains is from Sundborg (1967) and Sternberg (1972).

created by the complex of offshore linear sandbanks which is also there (Cloet, 1963; Robinson, 1966; Caston and Stride, 1970; Caston, 1972).

The sand-wave fields (Fig. 8.3) are visible on both echograms and sonographs, so their distribution is better known than that of sand ribbons (especially of "type B") and their asymmetry has, as noted above, been used to suggest the direction of their migration – or, at least, the direction of transport of sand grains within the field of rhythmic bed-forms. Little is certainly known yet regarding the possible migration of the waves themselves, principally because of the practical difficulty of reidentification of any particular wave crest or trough, but Langhorne (1973) has claimed that crests in the field at Long Sand Head, in the outer approaches to the Black Deep, Thames estuary, can move (in opposing directions in different zones of the field) at up to 25 m/year. The bathymetry of this area, with its closely spaced linear banks, is complex and so, too, must be its sea-bed hydrodynamics; the sediments of the channels and of the banks are emplaced by different mechanisms (Maddrell and Prentice, 1967). On the simpler topography of the Warts Bank, northern Irish Sea, the sand-wave fields on the two flanks of the bank are believed to travel in opposite directions (one being transported by the ebb, and the other by the flood tide), and waveforms up to 10 m high are stated to be transported up to 74 cm in a single tide (Jones et al., 1965). The large sand waves (3–6 m high) of the Southern Bight of the North Sea, believed to be moving northwards off the coast of The Netherlands from a bed-load parting at about 52°N (Stride, 1963a, 1965; Houbolt, 1968) have been estimated to be travelling at average rates of about 10 cm/day (Cartwright and Stride, 1958) or not more than 60 m/year (Langeraar, 1966). The mean tidal maximum surface currents in this region are about 75 cm/s, water depths are up to about 30 m, and the sand grains believed to be transported along the 107 km south-north path range from 0.4 mm in median diameter (medium sand, see Fig. 8.4) in the south to 0.2 mm (fine sand) in the North (Stride, 1970). Tidal currents, on the other hand, are frequently supplemented by wave-generated currents and storm surge: for example, during a gale, the transport rate of sand, on the sea bed west of The Netherlands, can be ten times greater than in a comparable tidal period of light winds (Johnson and Stride, 1969). The yearly total net sand transport along this 70 km wide path is estimated to be ten times greater than the $8 \cdot 10^5$ m^3 measured for The Netherlands littoral zone (Johnson and Stride, 1969).

Sand-wave forms are numerous: in addition to the fields of rhythmically repetitive, asymmetric waveforms from which a current transport direction can readily be inferred, Dingle (1965) described others from offshore east of Flamborough Head which were repetitive but symmetrical, sitting on "sand hills" up to 30 m above the surrounding sea floor (ascribed to accumulations resulting from the convergence of transport paths from opposite directions), and also large isolated waveforms in topographic hollows and on both irregular and apparently plane substrates. Some waveforms are almost certainly now immobile, possibly having cores of glacial material (e.g. the larger ridges in Tremadoc Bay, see Caston, 1965), and, as noted above, others are relict features of a previous, lower sea-level stand (e.g., those in 165 m water depth in the south-west of the Celtic Sea). Although there is no correlation between water depth and sand-wave height, those sand waves which probably form part of present-day sediment transport systems seem to be restricted to waters shallower than 70 m (usually shallower than 30 m), where surface currents flow at speeds between 65 cm/s and 125 cm/s, and where sufficient sand (from coastal or from sea-bed erosion) is available (Kenyon and Stride, 1970; Stride, 1970).

The known major sand-wave zones (Fig. 8.3), believed to be part of modern bed transport paths, extend across the Southern Bight of the North Sea (Stride, 1970, 1973), the central part of the eastern English Channel (just west of the Straits of Dover), from the central part of the western English Channel to the continental shelf edge and canyons of the south-western Celtic Sea, from the middle Bristol Channel to the northern Celtic Sea, and in belts across the northern Irish Sea and North Channel (Belderson et al., 1971). Further down the "velocity gradients", beyond the zones of sand transport, there are sheets of sand and muddy sand, often rippled but, on a large bed-form scale, featureless (Belderson et al., 1971); they have been considered to be due to deposition directly upon the basal conglomerate of the tidally scoured sea-bed (the "basal bed" of Belderson and Stride, 1966) but some of these heterogeneous sediments may be palimpsest, relict deposits formed after the passing of the transgressive surf zone of the late Devensian–Flandrian sea-level rise (Culver and Banner, 1979). The heterogeneous, muddy deposits of the central Celtic Sea, with their superimposed, anomalous sand patches, are in an area now occupied by rotatory tidal currents of no clearly defined directional dominance, where strong vertical density stratification can develop in the water column in summer (with the generation of internal waves at the pycnocline). Such an area cannot be expected to have developed modern, clearly directional, sediment grades. The heterogeneity of the sediments beyond the sand-wave fields of the North Sea is discussed by Caston (Chapter 7) who emphasises the problems presented by the relict sediments of past cycles of sedimentation. In such areas, the evidence for a simple regional pattern of transport and deposition is still far from clear.

There are other anomalies, too, which complicate the otherwise convincing story propounded by Stride et al. (1972). For example, the pattern of residual bottom currents in the southern North Sea (Fig. 8.2; Lee, Vol. II, Fig. 14.11), as deduced by Ramster (1965) and confirmed by Hill (1973), from direct observation of the transport of Woodhead sea-bed drifters, differs significantly from the pattern of sediment transport paths postulated by Stride (1965, 1973) from the sedimentological evidence. The area off northeast Norfolk, suggested to be one of bed-load parting by Stride (Fig. 7.50), was shown by drifter transport to be one of convergence; north-eastward postulated transport from this area conflicts with observed net sea-bed transport to the south-west. The postulated bed-load parting zone of the central Southern Bight was not confirmed, as drifters travelled, both in summer and winter, from the Straits of Dover north-eastwards (against the postulated path) with continuous net movement in that direction, to join the well-established north-eastward path lying offshore from The Netherlands. Similarly, the net transport of Woodhead sea-bed drifters in the south-western Celtic Sea has been found to be north-eastwards, towards the St. George's, Bristol and English Channels (Jones, 1974), not in the opposite direction (towards the shelf edge) postulated by Stride (1965) and by Belderson and Stride (1966). There are theoretical problems elsewhere, also, in the simplified Stride model: the scoured sea-bed in the central Bristol Channel may be a bed-load parting, but has no known modern, sediment supply, and upchannel (eastward) net sediment transport from it can hardly be significant, for, it if were, the confluence of that load with the fluvial input of modern sediment from the River Severn would have filled the Upper Bristol Channel and lower Severn Estuary by now. There must be subsidiary, narrower paths of sediment transport which bypass the others. Examples in the Bristol Channel are postulated by Collins and Banner (Vol. II, Chapter 11, Fig. 11.10, cf. Fig. 8.2); the same may be true for the Celtic Sea and the southern North Sea, at least (compare Fig. 7.54).

The importance of the recognition of true sediment transport paths is of great importance. Their role in the redistribution of modern mobile sediment is obvious, but the effect upon sea-bed ecology must not be underestimated. The distribution of benthic molluscs off The Netherlands, where *Gari fervensis, Dosinia exoleta* and other species form communities of distribution directly correlatable with that of the sandwave fields (Eisma, 1966), is not an isolated example. The often obliquely opposed ebb and flood currents in estuaries and estuarine approaches produce characteristically interfering patterns of bed forms (e.g. McManus et al., 1969) which reflect the circulation and recirculation of sediments which are potential navigational hazards (see also, e.g., Cloet, 1963, 1966, 1967; Maddrell and Prentice, 1967). The recreational beaches of most of the shores of north-west Europe depend upon offshore supplies of sand for their maintenance (Meyer, 1972), as do the commercial aggregate-dredging grounds (Vol. II, Chapter 18). Without clear understanding of the mechanism and routes of sediment transportation, none of these resources can be properly conserved or developed.

Cohesive sediments: modern muds and silts (F.T.B.)

The Flandrian, Pleistocene and older deposits, which once covered most at least of the continental shelf, have undergone erosion, transportation, "un-mixing" and redeposition since the beginnings of the late Devensian transgression. Not only has this resulted in a patchy distribution of relict, palimpsest and, to a much less extent, modern deposits exposed in the sea bed, it has also been responsible for much of the mud and silt which is in circulation both at the sea bed and in suspension. This is especially important in coastal and estuarine areas, where its effect upon the environment as a whole (and on navigational, dredged channels in particular) is of particular concern. Suspended mud increases turbidity and reduces photosynthesis in both phytoplankton and phytobenthos although it supplies the substratum for suspension and detritus feeding benthic organisms. It scavenges the sea water by adsorption for pollutant cations (e.g. mercury: see Lerman et al., 1974) and concentrates them on its deposition, and it provides suspended substrata for bacterial survival and persistence. It also provides the substratum for the development of intertidal mudflats, often capable of development for land reclamation, although it is also responsible for the siltation of harbours, dredged channels and estuaries and for the despoliation of pleasure beaches and even of fishing grounds (Vol. II, Chapter 18). Although the problems of mud erosion, entrainment, transport and deposition have been tackled experimentally and theoretically by chemists, civil engineers, physicists and geologists, as well as in laboratory, field and theoretical studies by sedimentologists, they are still not at all well understood. The literature has become vast (see, e.g. Swift et al., 1970; Graf, 1971; Gibbs, 1974, and relevant chapters in Ginsburg, 1975; Hails and Carr, 1975; McCave, 1976; Raudkivi, 1976; Stanley and Swift, 1976), but the conclusion of the 1969 Task Committee (Enger et al., 1968), that "the properties which control erosion and deposition of cohesive sediments have not been conclusively defined" still stands.

Although reasonably reliable estimates can be made of the speeds of current and the energies of waves which are needed to erode and transport unconsolidated, well-sorted quartz silts, sands and pebbles (Fig. 8.4), the same is not true for muds. Muds are, by definition, cohesive, due to their content of clays, with or without admixtures of various size grades of silt or even of larger detrital or bioclastic or organic particles. The complex aluminosilicate minerals which are classified together as "clays" have a well known mineralogy, physical chemistry and behaviour (under controlled, laboratory conditions, at least) which have been thoroughly described elsewhere (e.g. Velde, 1977). Although defined arbitrarily by civil engineers and field sedimentologists as merely comprising all inorganic (and, sometimes organic) particles smaller than 3.9 μm (on the Wentworth scale) or 2 μm (on the Atterberg scale), they have a sheeted crystalline structure which produces primarily flakey, not rounded, grains. The flat surfaces of the particles carry residual negative electrical charges, and the broken edges carry both positive and negative ones. It is because of the residual negative charges that cations can be adsorbed and retained in an exchangeable state, but the residual charges also produce effects of great sedimentological importance. In fresh water, suspensions of clay minerals can persist as colloids, as their gravitational settling is counteracted both by Brownian movement and by the repulsion between particles of like charges, but in salt water the electrolyte permits the particles to lose the repellant charges, so that they attract one another and form aggregates or "flocs". Flocculation of clays inevitably occurs in estuaries, where colloids suspended in river water meet the sea, with the result that the formation of estuarine mud-deposits, and even of broad intertidal mudflats or deltas, is enhanced. The inevitable cohesiveness of clays in sea water (which is affected by the mineralogy of the clay, the absolute and relative concentrations of dissolved cations, and the temperature and pH of the water) is emphasised when repeated interparticle collisions occur during fluid shear, a parameter particularly significant in the bottom boundary layer (McCave, 1976), in shoaling waters and in estuaries. Aggregates will inevitably form when the local shearing rate is not so great that it would break the interparticle bonds; even if such bonds are broken, the aggregates so dispersed can re-aggregate repeatedly. These aggregates (and reaggregates) become crudely rounded in form, unlike the parent flakes, but their density is, of course, much less than that of silica-quartz grains. According to Krone (1976), aggregates formed by collision during fluid shear (in the vertical gradient of current speeds) are denser and stronger than those formed by Brownian movement or differential settling alone, and the shear-strength, density and size of the aggregates determine the probability of formation of a cohesive deposit. They (and their subsequent, post-depositional compaction) also help to determine the sedimentological characteristics (and potential erodibility) of that deposit. Regardless of their mineralogy, the aggregates which form from a dispersed suspension ("zero order aggregates" of Krone, 1976) have high density (ca. 1.25 g/cm^3) and shear strength (21 dynes/cm^2), while at lower shear rates, these primary aggregates can themselves collide to form ("first order") aggregates of lower density (ca. 1.13 g/cm^3) and shear strength (9.4 dynes/cm^2). As shear rates decrease, aggregation of aggregates continues, producing "higher order" aggregates of decreasing shear strength (1.2 dynes/cm^2 for "third order") as each includes additional water in the new pore-volume, so decreasing the density of the final aggregate (1.07 g/cm^3 for "third order"). The increase of particle size by repeated aggregation must increase the void ratio (e, see Fig. 8.5b) and

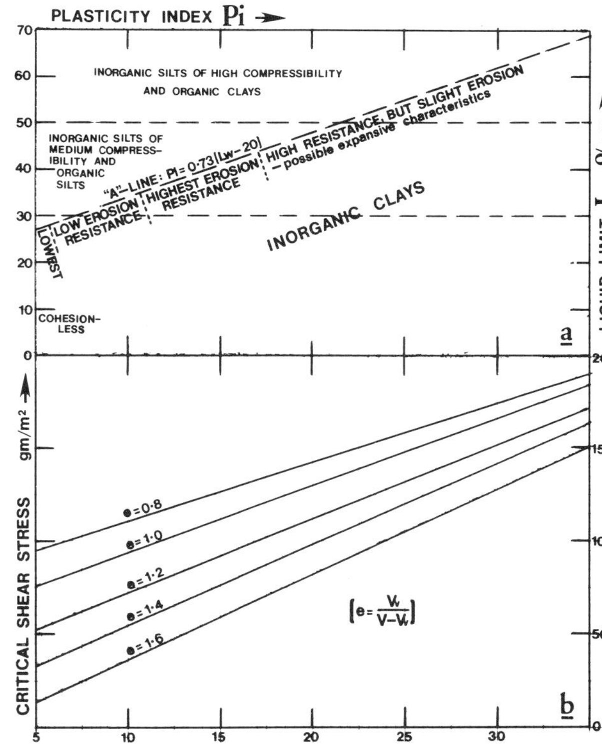

Fig. 8.5. The Plasticity Index of cohesive sediments related to their erodibility as "Casagrande diagrams" modified from (a) Gibbs (1962) and Terzaghi and Peck (1968), (b) Lyle and Smeardon (1965). The Atterberg terms are defined in British Stand. 1377: Liquid Limit (L_W) is the percentage (by weight) of moisture content at which the sediment starts to flow when jarred; the Plasticity Index (P_i) is the range of moisture content over which the sediment is plastic, and is defined as $P_i = L_W - P_W$ where P_W is the Plastic Limit, the lowest moisture content (wt. %) at which the sediment is still plastic (tested by rolling the mud or clay into 3 mm-diameter threads without breaking them, teste Raudkivi, 1976). The Void Ratio e is defined as shown, where V is the total sediment volume and V_V the total volume of its voids (Graf, 1971); the corresponding Porosity Ratio is defined as V_V/V. Figure (a), which indicates that inorganic clays are characterised by relatively low Liquid Limits compared to their Plasticity Indices, while inorganic silts, together with organic clays and silts have relatively high Liquid Limits (compared, again, to their Plasticity Indices), suggests that the greatest resistance to erosion is exhibited by cohesive sediments of intermediate to lower Plasticity Index and Liquid Limit values, erodibility increasing at still lower values of the P.I. (until the material becomes uncohesive) and at higher values where the material may assume expansive characteristics. Figure (b) is a best-fit graph based on samples of Texas soils, and shows how the Critical Shear Stress (τ_C) needed to initiate erosion, is less both at lower values of the Plasticity Index and at higher values of the Void Ratio. The actual Boundary Shear Stress (τ_0) has been related (Sternberg, 1972) to mean bed "velocity" $\bar{U}_Z$ of water (density ρ) by $\tau_0 = C_d \bar{U}_Z^2 \rho$, where C_d is the Drag coefficient, said to range from $0.89 \cdot 10^{-3}$ to $2.58 \cdot 10^{-3}$ in the Irish Sea (up to $4.56 \cdot 10^{-3}$ in the Menai Straits tidal channel) (Heathershaw, 1976), and from $3.0 \cdot 10^{-3}$ to $8.7 \cdot 10^{-3}$ in the North Sea (McCave, 1973a), the highest values of C_d being associated with rippled, sandy sea-beds (Heathershaw, 1976); ripples form in noncohesive sediments at relatively small values of shear-stress excess ($\tau_0 - \tau_C$) (Raudkivi, 1976, pp. 56–59).

decreasing boundary fluid shear stress will be needed to induce erosion of the deposited aggregates (Lyle and Smerdon, 1965; see Fig. 8.5b).

Although recently deposited material may become eroded at its surface, particle by particle, when the critical shear stress is reached there, this process is important only in specific localities (e.g., around emplaced structures, or other obstructions to the flow); generally, it is bulk erosion of the sediment which is important, and this occurs only when the shear stress is greater than the shear strength of the deposit. Although it is normal for the shear strength of a deposit to increase linearly with depth, "shear strength" is a parameter difficult to quantify in a reproducible form because of differences in instrumentation techniques which are employed; Krone (1977) has suggested that the "Cation Exchange Capacity" (CEC) of a cohesive deposit should, rather, be measured, as this has been found to be linearly proportional to "shear strength" as determined by any particular viscometer. This could introduce the appearance of reproducible precision to laboratory studies.

However, erosion resistance (and the boundary fluid shear stress needed to induce erosion and transportation) also varies with the Plasticity Index and Liquid Limits of the clay deposit (Fig. 8.5a); these properties, although crudely defined (principally by civil engineering practice), are realistic in the consideration of actual muds while more theoretical approaches do not lead to valid predictions. These properties vary not only with the aggregation and compaction history of the deposit but also with its mineralogy, which may often be composite. Pure kaolinites, often produced by advanced subaerial weathering of granite (e.g., in Cornwall, where the china-clay industry has flourished), range in Liquid Limit from about 25% to 75% while simultaneously varying in Plasticity Index from about 5 to about 35, respectively (Grim, 1962). Illites (often derived from geologically old argillites and a major component of many boulder clays and sediments produced from them) have Liquid Limit ranges from about 35% to 95% and comparable Plasticity Indices from about 25 to 45. In contrast, sodium–lithium montmorillonites (usually derived from volcanic rocks and relatively rare in European shelf sediments) have Plasticity Indices ranging from 250 to over 400, with a corresponding range of Liquid Limits from 350% to nearly 450% (Grim, 1962).

Differences in clay mineralogy and aggregation (flocculation) history, combined with differences of porosity index and void ratio, and of the content of water and particles of silt, sand, shell and organic material, produce natural deposits of mud for which the critical shear stresses or current speeds needed for their erosion are incalculable with precision – a situation made predictively worse by bioturbation and faecal production due to burrowing bivalves, annelids and other organisms. Therefore, the estimates published by, e.g., Postma (1967, see Fig. 8.6) can be no more than approximations for deposits of particular constitution and bed roughness. The problem has been studied intensively, especially by civil engineers (e.g. for soil mechanics, originally in land-based investigations) but it is inevitably intractable. As a result, it is of no surprise that a recent specialist working group was forced to report (Keller, 1976) that there is as yet no satisfactory practical way to predict the critical shear stress of cohesive bed surfaces in general, or their erosion rates as a function of excess stress. Material may be entrained, aggregate by aggregate, depending on the structure of the exposed surface, at low shear stresses, and local variations in shear stress occur because of changes in bed roughness due to organisms and included shell and other fragments; even over-consolidated sediment may fail along inhomogeneities, and the whole may be weakened by swelling (cf. Fig. 8.5a) and by biological and chemical processes. Even Drake's (1976) comment that "for the present, the most honest statement that can be made is that mud beds will be eroded at current velocities on the order of 10–30 cm/s (measured 100 cm above the bed) provided the water content exceeds 80%" may be true in some cases only (see Fig. 8.6).

Similarly, the settling rates of suspended fine particles are also very variable, as are the suspended particles themselves. Suspensions from the surface waters of Hydrographic Station England No. 1 (E.1), about 18 km south-west of the Eddystone lighthouse in the English Channel, were found to contain terrestrial wood fibres, iron oxides (from rusted steel), carbonaceous matter (probably from soot, pitch or tar) and fragments of shell, sponge spicules and mineral grains (up to 0.1 mm in diameter) as well as small organisms and clay (Atkins et al., 1954). Murray (1965) has recorded the frequent occurrence of benthic foraminifera (up to 0.2 mm in size) in English Channel plankton hauls after periods of storms. The classical modification of Stokes's formulae by Oseen (1927), Goldstein (1929) and others has been recalculated for particles of different, regular shapes (prolate and oblate spheroids, discs, cylinders) by Lerman et al. (1974, Fig. 1), but it has not yet been possible to propose realistic solutions to accommodate the extremely variable shapes and densities developed by natural clay aggregates and re-aggregates, even to their lower orders. Although it has been stated (emphatically)

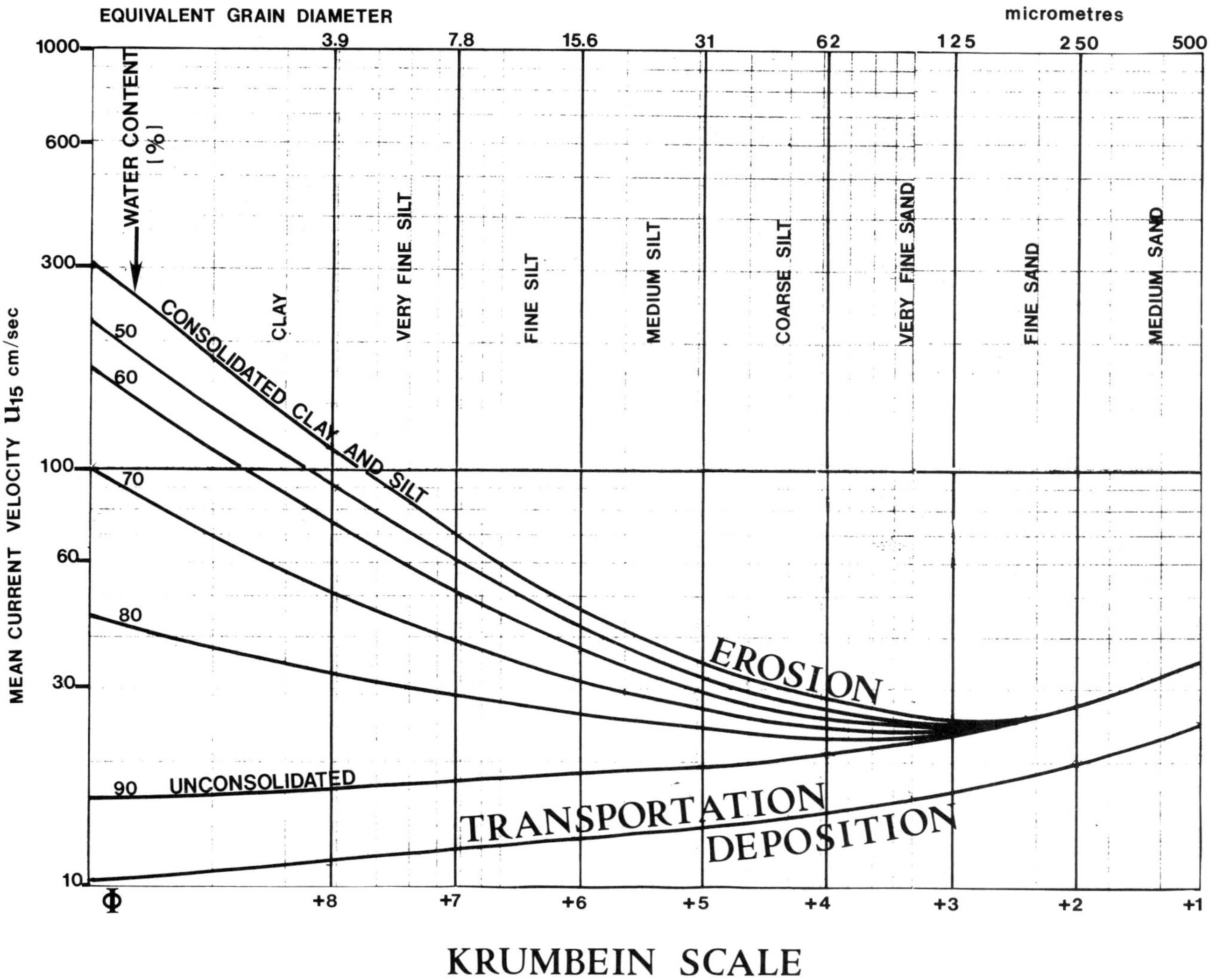

Fig. 8.6. Flow "velocities" (unidirectional current speeds) required for the erosion or to maintain transportation of well sorted, inorganic particulate materials (modified from Postma, 1967), referred to a current speed 15 cm above a plane bed (U_{15}). The graph indicates that the current speed needed to maintain sediment in movement is less than that needed to initiate that movement, and that the latter varies, for very fine sand, silts and clay, according to the consolidation and water content of the sediment. The "critical erosion velocity" will vary, therefore, according to the porosity of the fine-grained material and, for clay and "mud", with the liquid limit (see Fig. 8.5a). At very high water-content values, the clay will become "liquid mud" (see text). Note that the extrapolated values for the "critical erosion velocity" even for sands do not agree with those suggested by Inman and by Sundborg (Fig. 8.4).

that "les vases floculées ont des vitesses de chute de l'ordre de 0.5 mm/s en eau calme" (Migniot, 1977), turbulence and aggregate nature must greatly affect this in practice; for interpretation of settlement under natural conditions, it has been calculated that "mud" should settle from tidal currents (where U_{100} is less than 30 cm/s – in steady uniform flow, 25 cm/s is associated with a fluid shear stress at the bed of 0.9 dyne/cm^2, the limiting value for mud deposition, *teste* McCave, 1972, 1973a) at settling velocities around $5 \cdot 10^{-3}$ to 10^{-2} cm/s (McCave, 1969, 1970); it is inevitable that this too, will be far removed from most natural situations. Not only will wave action inhibit the deposition of mud from suspension (McCave, 1971a, has defined and discussed its effectiveness), but, in estuarine, coastal and inshore waters at least, eddy turbulence arising from all causes is potentially able to return deposited aggregates into re-suspension, adding to the range of physical properties possessed by the particles in suspension. Other factors affecting settling rates include variations in salinity, temperature and shear in the water column (especially in estuaries and their approaches) and eddy turbulence (see, e.g., Graf, 1971, section 4.2.4). The suspensions themselves cause changes in the total density and viscosity of the suspending sea-water, affecting, in turn, the settling rates of the suspended particles (Meade, 1972) as well as the "velocity profile" in the water column (Gust and Walger,

1976). Finely dispersed clay particles have been observed to settle at rates varying from $12 \cdot 10^{-3}$ cm/s (at 1000 mg/l concentration) to less than $1 \cdot 10^{-3}$ cm/s (at 120 mg/l) in water of more than 10 ppm chloride content; all settling rates reduce dramatically at very low salinities (Migniot, 1977), and this is especially true for kaolinite and illite, which (in brackish and saline waters) settle much more quickly than does montmorillonite (Meade, 1972). However, in the natural environment, not only will these rates be greatly altered by turbulence (McCave, 1971a), aggregation and so on, but also biological activity will radically alter them. Not only will particulate organic matter be present in suspension and in the aggregates, but heavily bioturbated muds will contain significant proportions of faecal pellets with settling velocities two orders of magnitude greater than that of the particles of untransformed mud (Haven and Morales-Alamo, 1972), just as the "critical erosion velocities" will be reduced to two orders of magnitude less (Rhoads and Young, 1970). Consequently, only field observations (ideally coupled with controlled laboratory experiments) can provide data valid for any particular situation.

Rates of sedimentation of consolidated mud in offshore natural "sediment traps" close to major mud inputs include from 40–60 cm/200 years (Bertine, in McCave, 1973a, for deposits off the Belgian coast using ^{210}Pb isotopes) to 15.5 cm/200 years (McCave, 1970, modified from Reineck, for deposits south of Heligoland) and 11–23 cm/200 years (for the Cleaver Bank, based on pollen analyses by Zagwijn and Veenstra, 1966); from an assessment of these, McCave (1973a) has calculated a total annual mud deposition in the natural "sediment traps" of the North Sea as about $1.5 - 6 \cdot 10^3$ ton/km^2.

Field observations of suspension concentrations must be treated with care, as the methods employed to determine the concentrations of suspended particles in natural sea-water are of very different kinds, and different methods necessitate different assumptions and interpretations. Water from samplers, or from continuous pumping, may be filtered and the amount of suspended material determined gravimetrically, or the attenuation of transmitted light may be measured and from the turbidity the concentration of suspended matter may be calculated. Estimates of the concentration, at and near the water surface, of suspended particulate matter (which must include phytoplankton, organic detritus and other particles as well as clays) vary from 0.796 mg/l (minimum, English Channel in March, according to Chester and Stoner, 1972, for waters pumped through a 0.45 μm membrane filter) to more than 200 mg/l (for Swansea Bay, by use of calibrated beam-transmittance meter, according to Davies, 1972) and even to 1400 mg/l in other parts of the Bristol Channel (Bassindale, 1943). Typical concentrations in the surface waters of the English Channel (centre of the Straits of Dover, McCave, 1973a) appear to be about 0.3 mg/l, which are comparable to those recorded for the North Atlantic inflow water into the North Sea (0.2 mg/l, Hagmeier, 1962); in offshore coastal waters, surface concentrations can increase by at least two orders of magnitude (e.g. 28 mg/l at the North Hinder lightship, about 54 km offshore from Zeebrugge; Terwindt, 1967), and, in inshore waters, concentrations can be two orders of magnitude greater again. Some of the differences in recorded values reflect real spatial or temporal changes in suspension concentrations, but others are undoubtedly due to differences in arbitrary definition of what shall be measured and how the measurements shall be made. Terwindt (1967) noted that Dutch researchers had variously defined "mud" as particles of diameter less than 16 μm or as "all matter in suspension" – the last could, in fact, incorporate up to 20% of sand, as well as silt and clay. British researchers have been similarly inconsistent. For example, Carr et al. (1976) estimated the "typical" concentrations of suspended sediment in the surface waters of Swansea Bay to be about 10 mg/l, considering the suspensions to consist of particles larger than 40 μm in diameter, although Davies (1972) had shown, by electron microscopy, that the dominant "floc" (aggregate) size in the Bay is about 30 μm. McCave has variously taken "mud" to comprise all particles smaller than 50 μm (1971, 1974) or 63 μm (1973a, c) – i.e. to comprise all clay and either most or all of the coarse silt of the Wentworth Scale. In other studies the division point between silt and clay has been taken variously, at 3.9 μm and 2 μm (Tanner, 1969) and twenty grain-diameter scales are available in the literature for adoption (Truesdell and Varnes, 1950); the Inter-Society Grain Size Study Committee of the Society of Economic Palaeontologists and Mineralogists recommended the arbitrary application of a 2 μm upper limit (i.e., the Atterberg value) for clay, and either 62.5 μm or 74 μm for silt (Tanner, 1969). We follow Folk (1966, 1968) and the Krumbein log scale (1934, 1936) (Figs. 8.4 and 8.6). McCave (1975) has suggested that "most of the particles suspended in sea water" have a diameter of less than 2 μm, but this ignores the aggregates and reaggregates formed in coastal and estuarine waters. As noted above, Chester and Stoner (1972), in their study of offshore suspensions, filtered out everything larger than 0.45 μm, apparently to relate their results to the lower size limit of "particulate" rather than "dissolved" matter – an arbitrary limit which depends "on the poresize and adsorptive properties of the filter used" (Strickland and Parsons, 1968) but which is customarily placed at 0.5 μm (Riley and Chester, 1971); Atkins et al. (1954) had used membrane filters of 1.09 μm average pore diameter. 0.1 μm is generally accepted as the upper size limit of the colloidal state of clay (Krumbein and Pettijohn, 1938). Thus, non-colloidal "mud" in the sense used by McCave (1971a, 1973a) comprises low- and high-order aggregates of clay and most silt grades as well, while that assumed by Carr et al. (1976) includes silt and the larger clay aggregates only. Similar variations apply to the records made by other authors. For example, all filtration techniques experience the hazard of filter clogging, which leads to the separation of increasingly large particles during the duration of the process.

Quantitative comparison between published records is inevitably complicated by the indirect methods which must be followed in the determination of the particle sizes of both suspended and settled "mud", as well as by the differences which exist in the mud constituents (different clay minerals, fine quartz and mica particles, fragments of biogenic skeletons, organic particles), which themselves will behave differently regardless of the measuring technique used. These problems may be well known to practising specialists, but must also be appreciated generally, when published records are to be assessed.

The classical "Pipette-method" (Andreasen et al., 1929) and "Hydrometer-method" (Casagrande, 1934) and their more elaborate developments (e.g. Fabricius and Muller, 1970), both entail interpretation of grain-size from measured velocities of settling from suspension, and are widely employed (Reineck, 1967) in studies by both sedimentologists and civil engineers. The method requires the dispersion of the settled mud by wetting, dispersing and oxidising agents, which must cause separation of both clay aggregates and faecal pellets; the settling is performed in fresh water, which excludes the electrolytic effects of the natural environment; the resulting "analysis" does not reflect the behaviour of the material in its original estuarine or offshore location (simple settling tubes could be said to produce equally valid results (Gibbs, 1972) and these can be used to give realistic values for mud content, even during field studies in the littoral zone). Other, more elaborate methods (electronic, microscopic, optical sedimentation) have been reviewed by Swift et al. (1972) and by many other authors (e.g. in Gibbs, 1974); similar comments, as to the relevance of the results to interpretation of natural processes, must, mutatis mutandis, apply to them.

Most published estimates of the concentration of "mud" in suspension are based upon the "turbidity" of the sea, determined by measurements of the degree of attenuation of light presumed to be due to the suspended particles. Such indirect methods, which have been used in the estuarine, coastal or offshore waters of the north-west European shelf seas, include those as diverse as the Secchi disk (e.g. Atkins, et al., 1954; Postma, 1961b; Otto, 1966; Visser, 1970), the Pulfrich photometer (Cooper, and Milne, 1938; Dietrich, 1963, pp. 79, 1950; Cooper, 1961), beam transmittance meters (e.g. Joseph, in Dietrich, 1963, p. 147; Otto, 1967; Heathershaw and Simpson, 1974) and air photographs (Moore, 1947; Cooper, 1961). These and other techniques, and the significance of the data which can be gained from them, have been summarised concisely by Drake (1976) and comprehensively by Jerlov (1976). Although there have been recent attempts precisely to define "turbidity" in terms of specific clay suspensions (e.g. McCarthy et al., 1974), light is variously attenuated at all wavelengths by organic and inorganic particles, the former including, of course, nanno- and microplankton. Field readings of extinction or attenuation values cannot yet satisfactorily discriminate the minerogenic component, even if only particular wavelengths of light are used (see, e.g. Otto, 1967, for red and blue light), although the volume-scattering functions of biological and minerogenic particles may broadly differ (Pak et al., 1970). However, the use of monochromatic or narrow spectral-band light does improve the experimental correlation between suspension concentrations estimated from "turbidity" and from filtration of in situ suspensions or resuspended deposited material (E. Allersma, Delft, pers. comm., 1977).

Typical values for the surface-water concentrations of suspended sediment off muddy coastlines throughout the world are between 10 mg/l and 100 mg/l (McCave, 1972), but these are frequently exceeded in north-west Europe. For example, in addition to the Bristol Channel records mentioned above, concentrations of 10 mg/l to 180 mg/l have been measured in the Ems estuary (Postma, 1960) and of 1 mg/l to 200 mg/l in the Thames estuary (Inglis and Allen, 1957); the former may (to some extent, indirectly) be associated with inputs of suspended material which lead to surfacewater concentrations of up to 620 mg/l in the Wadden Sea (Postma, 1961a), and the latter with a broad tongue of recognisably turbid, low salinity "English Coastal water" which stretches from the Thames estuary north-eastwards to at least 75 km north-east of the East Anglian coast, over the Winterton Twenties and the West Hole, beyond the Norfolk Grounds (see Joseph, 1953, Kalle,

1953 and Dietrich, 1953, in Dietrich, 1963, pp. 79 and 89). This turbid "English Coastal water" tongue is highly productive, as is shown by its high values for mean fluorescence, chlorophyll and protein (see Kalle, 1953 and Krey, 1953, in Dietrich, 1963, pp. 89 and 93), undoubtedly due to its enrichment in nutrients by the Thames discharge and other terrestrial contributions, and is believed to be carried north-eastwards by residual currents in a form said to be tantamount to an "advective mud stream" (McCave, 1972; Swift, 1976). The natural fluorescence of the English Coastal Water tongue, due to its dissolved organic matter (the yellow melanoidines of decayed cellulose of both higher and lower plants, including marine algae, rather than the non-fluorescent polyphenols and humic acids derived from the decomposed lignins of higher plants of terrestrial habitat), has been measured by Otto (1967) and its linear inverse variation with salinity was compared to a similar correlation found to occur between the salinity values and the degree of attenuation of monochromatic (red or blue) light. Otto (cit.) concluded that, as the English Coastal Water moved north and north-east from the Straits of Dover, its salinity decreased (by dilution due to mixing with English river plumes) and its suspended load at surface correspondingly increased (from less than 2 mg/l to more than 12 mg/l, as deduced by calibration with quantified suspensions of the degree of attenuation of red light in beam transmittance); the equally linear increase of mean fluorescence with decreasing salinity must indicate increased primary production, however, which suggests that some, at least, of the turbidity is due to an increasing concentration of particulate organic matter as well, probably, of mud.

The mud supply to the English Coastal water derives from both the rivers of the Thames estuary and eastern East Anglia and also the erosion of the poorly consolidated Quaternary sediments which form the bulk of the cliffs of the coasts of south-eastern England north of the Thames. The East Anglian cliffs alone are believed annually to contribute 300,000 metric tons ($300 \cdot 10^6$ kg) of mud to the sea, while the rivers of eastern Britain have been estimated to supply 1,470,000 metric tons/year (McCave, 1973a). In contrast, the discharge of suspended mud from rivers Rhine, Maas and Sheldt into the Dutch delta coast amounts to between 4,300,000 metric tons (Terwindt, 1967) and 5,100,000 metric tons (Veenstra, 1970) annually, to provide the biggest regional supply of terrestrial mud (without including contributions from coastal erosion) anywhere in the north-western European shelf seas. Terwindt (1967) (who defined "mud" as comprising "all matter of which the rate of fall in still water is less than that of a quartz grain 50 μm in diameter") has described how the muds discharged by the distributaries of the Rhine and Maas are transported northwards and north-eastwards, along the Dutch coast, to form a flocculated mud deposit covering the sea bed for about 2 km seawards from the shore, and how the turbid, coastal current carries concentrations of 125–150 mg/l of suspensions for at least 50 km from the delta source area. The Dutch inshore waters north of the delta are characterised, in consequence, by high levels of light attenuation (Joseph, 1953, in Dietrich, 1963) and of mean fluorescence (Otto, 1967). The sea bed, below the coastal "advective mud stream" (so called by McCave, 1972), ecologically reflects this turbid, muddy environment by supporting a characteristic benthic bivalve community (Eisma, 1966), with *Macoma balthica, Cerastoderma edule, Abra alba, Mysella bidentata,* etc.; while seawards of the "mud stream", where the sea-bed surface sediment has a lower mud content and where turbidity is much less, the community is characterised by *Abra prismatica, Arca lactea,* etc.; this is probably the clearest association, on such a scale in the north-western European shelf seas, of particular benthic associations with both modern sediment distribution and water mass characteristics and circulation, and it is intimately related to the transport and deposition of mud and its subsequent bioturbation.

The turbid, low salinity Dutch coastal water is the extreme expression of "Continental Coastal Water", recognisable (for some 50 km westwards and seawards of the coast of The Netherlands (Joseph, 1953, and Dietrich, 1953, in Dietrich, 1963, p. 79) and north-north-eastwards at least as far as the Frisian Islands) on its high mean fluorescence (Kalle, 1953; Otto, 1967) and chlorophyll content (Kalle, 1953) at surface. Otto (1967) has recognised, here also, a linear inverse relationship between the salinity and surface load of suspensions (as determined by in situ attenuation of red light); in the south, salinities of between 34 and 34‰ are accompanied by calculated inorganic suspension concentrations around 2 mg/l, but as the Continental Coastal Water moves northwards and is progressively diluted by river discharges, its salinity falls to less than 33‰ as its suspensions reach about 5 mg/l. As noted above, its natural fluorescence also linearly increases. On their progressive changes in mean fluorescence, turbidity and salinity, Otto (1967) believed that the characteristics of both the Continental and English Coastal Waters could be traced to the increasing coastal influences on inflowing "Straits of Dover Water" from the English Channel.

In contrast, the central North Sea possesses turbid waters, also high in chlorophyll, which owe their turbidity to no minerogenic particles in suspension. As described by Lee and Ramster (Vol. II, Chapter 14) the central and northern parts of the North Sea may be neither homohaline nor homothermal, so that surface water characteristics may not be the same as those at depth. North of the latitudes of The Netherlands and East Anglia, the central North Sea can develop a summer thermocline with a marked temperature and density gradient around 30–40 m depth. This pycnocline (which may be centred around the 10°C isotherm) can cover almost all of the central and northern North Sea, and in it can accumulate phytoplankton sufficient to create a "turbidity screen" (also high in chlorophyll) which reduces photosynthesis, in and below it, to the compensation point at which the phytoplankton may survive but no longer reproduce (Dietrich, 1963, pp. 287–289). This is an extreme example of turbid waters where the degree of light attenuation is *not* related to suspended sediment. Similar increases in turbidity are to be expected at tidal fronts, such as those described in the south-west approaches to the English Channel (Pingree et al., 1975) where, in summer, the frontal boundary is a zone of mixing of warm, surface water and the underlying colder water and becomes a site of phytoplankton bloom (and sometimes, even of "red tide", when reddish-brown streaks of *Gymnodinium* concentrations become visible even from aircraft at 3000 m altitude above sea level). Similarly, seasonal variations in water turbidity at surface are partly, at least, correlatable with primary biological production. "There is an inverse relation between the amount of phytoplankton, determined by spectrophotometric analysis of chlorophyll extracts, and the range of the 20 cm Secchi disk, though admixture of suspended inorganic matter may disturb the relation" (Atkins et al., 1954). Surface suspension concentrations of both biogenic and minerogenic origin may follow a seasonal cycle of turbidity involving the development and disruption of thermal stratification. The seasonal cycle of phytoplankton production in the surface offshore waters of the English Channel, which leads to chlorophyll maxima around March and again in September–October, was originally described by Harvey et al. (1935). Their work, which related the spring maximum (and a biological turbidity maximum) to the onset of high seasonal irradiance coupled with advective replenishment of surface nutrients during the preceding winter storms, and then related the summer phytoplankton minimum to the development of the summer thermocline (and the loss of advective nutrient replenishment from depth), showed how a winter turbidity maximum due to suspended particles from the sea bed could be followed by biogenically induced turbidity in the spring, and then by a turbidity minimum in the summer when neither inorganic particles nor phytoplankton could become abundant at the surface (for an elaboration of the classic generalisation of this seasonal pattern, see, e.g. Tait, 1968, Fig. 5.7). Of course, the phytoplankton themselves (at least those which have siliceous or calcareous skeletons) add their own contribution to the accumulation of fine particles which may be classed as "muds". Terwindt (1967) has stated that the "mud" content of coastal waters off the Dutch Delta is one and a half times greater in spring than in comparable weather in Autumn, due to biological production, and McCave (1973a) has calculated that the contribution of primary production to the sediment, in the North Sea as a whole, "lies somewhere between 1.15 and 3 million metric tons" annually.

However, most variations in the turbidity of coastal waters are due to changes in suspended matter due to "physical rather than biological temperature-dependent processes" (Newton and Gray, 1972); when turbidity values have a diurnal or semidiurnal periodicity, they are correlatable with tidal range and current rhythms, just as they are when the periodicity is monthly. For example, although concentrations of suspended matter in the Menai Straits, between the island of Anglesey and the mainland of north Wales, vary with the season due to bioproductivity, suspension concentrations are higher (around 50 mg/l) during spring tides, when tidal currents are at their maximum, than at neaps, when concentrations are about 5 mg/l (Buchan et al., 1967); this must be due to increased suspension, resuspension and vertical advection of particles during periods of increased eddy turbulence due to high inshore current speeds. Seasonally, also, the concentration of suspended material is influenced by the cycle of wave climate: for example, off the coast of North Yorkshire, statistical analyses of samples of suspended material, regularly collected between September 1968 and August 1970, showed significant differences in suspended solid values between the months and between inshore and offshore sampling stations (there was no significant difference between sampling points parallel to the coast), with low concentrations during the summer periods (high temperatures but low wave activity and low river flow rates) and high ones in the winters (Newton and Gray, 1972).

The advection of particles, both organic and inorganic, from sea-bed deposits to the surface by waves and swell, especially during winter storms, must be greatest in coastal waters where the waves shoal. The unidirectional transport of suspended particles by particular trains of waves may be enhanced by complementary tidal currents; the latter may maintain transport of the suspended material after movement has been initiated by wave action. Wave transport alone may be responsible for the shorewards transport of much suspended material, and it is probable that erosion of muds and silts of the offshore seabed has contributed largely to mud accumulations of the intertidal zone, even where there is also a large input of mud from adjacent fluvial sources. For example, Pleistocene muds and silts are patchily exposed in a 100 km-wide belt off much of the British, Dutch and Danish coasts and all is potentially available for transport in suspension (Veenstra, 1970). The difficulties of prediction of the physical and dynamic conditions which would permit such erosion have been outlined above, but there is little doubt that much of the mud which accumulates in the Dutch Delta is marine in origin (recognisable by its low manganese content, Terwindt, 1967) and has contributed extensively to the development of intertidal mud flats there. River muds, high in manganese, are transported seawards at ebbtide, partly to settle at slack low water in the outer estuarine reaches; fluid mud may accumulate then (see below). On the flood, it may be disturbed and returned inland, only to be retransported seawards at the next ebb, to meet the inflowing marine muds at the next flood. Although the total water flowing inland at flood will be less than that flowing out at ebb, the higher salinity and density of the flood water (in a "salt wedge") causes more effective movement at the bed, where mud content is highest, causing the recirculation with a relatively small net mud loss to the sea (Terwindt, 1967). The same process is responsible for the mud accumulations which have developed extensive intertidal flats in the Jade, Weser and Elbe estuaries (described so comprehensively by Rieneck, 1970) and even in British estuaries where there is very little contribution made at all by terrigenous, river-carried mud (including the Loughor estuary, where commercially important, cockle-bearing mudflats are developed on the shores of the Burry Inlet, South Wales: Nelson-Smith and Bridges, 1977). In the last of these, the offshore origin of the mud is proven by its content of marine biogenic particles such as echinoderm debris, stenohaline marine foraminifera, etc. (Banner and Collins, 1975). Similar evidence indicates that the shoreward transport of mud and its intertidal accumulation has been virtually continuous in many coastal areas since the beginning of the late Devensian transgression, 13,000 years ago (Murray and Hawkins, 1976; Culver and Banner, 1978), interrupted only by eustatic sealevel oscillations and by local changes in the available sediment supply. The flux of suspended sediment across intertidal flats can be estimated by sampling the vertical profile of suspensions at each tidal inundation (Collins, 1976); over the intertidal flats of the Wash, concentrations (of particles larger than 0.45 μm) from less than 200 mg/l to more than 1200 mg/l have been measured during high water, with sand varying from 3% to 96% by weight of the total suspended matter, indicating a longshore flux of at least 912 metric tons across a 2000 m transect at each inundation (Evans and Collins, 1976).

It has been suggested (Van Straaten and Kuenen, 1957) that the accumulation of muds in the Wadden Sea has primarily been caused by "scour lag" (difference between the current speed needed to bring a particle into suspension and that needed to keep it there), "settling lag" (time taken for a particle to settle after the current speed has fallen to that needed to permit settlement) and the bathymetry of the embayment: at rising tide, the inflowing water mass would spread over the shallower, coastal areas with current speeds reduced as the area occupied by the watermass would increase, allowing particles to settle. Postma (1961a) has described how the marked gradient of concentration of fine suspended matter, which is found from low values at the North Sea tidal inlets to high ones in the interior of the Wadden Sea, is upheld against strong tidal flushing by North Sea water and yet cannot be ascribed to a shorewards source of mud (e.g., by erosion of marshes or intertidal flats, or by fluvial discharges) or to an estuarine circulation like that of the Delta. He suggested (1961) that the gradient could be the result of the asymmetry between the tidal ebb and flood phases: the time from maximum ebb to maximum flood (i.e. over low water) is appreciably shorter than that from maximum flood to maximum ebb (i.e. over high water), producing stronger currents in the former (ebb to flood) phase. Groen (1967) followed this by a model which showed that the increased concentration of suspended matter from the North Sea towards the interior of the Wadden Sea could result from a purely alternating tidal current regime in which maximum current speeds were equal at both ebb and flood and which produced zero net (or residual) water transport, and that as the ebb current maximum is preceded, in the Wadden Sea, by a much longer period of ebb currents than the flood maximum is preceded by flood currents, there is more time for particles to settle during the ebb. Consequently, the ebb peak load is lower than that of flood peak, because it must be reached from a lower preceding load. Groen calculated (1967 cit.) that such a tidal regime could result in 38% more displacement of suspended matter in the flood direction than in the ebb direction, if there were an infinite supply of particles to be suspended at all times.

Such calculations, like all mathematical models of estuarine circulation (see, e.g. Odd and Owen, 1972; McDowell and O'Connor, 1977) are difficult to apply, not only because of the complex geometries of the embayments but also because of the time-variable stratification of the water columns in them. North-western Europe has provided spectacular evidence of the time-variability of stratifications of suspended sediment loads in large estuaries which may or may not have salinity stratification. Lenses of high concentrations of suspended sediment occur on the bed of the Thames Estuary (Inglis and Allen, 1957) and Parker and Kirby have reported (1977) the presence of highly turbid bottom layers in the waters of the Maas estuary; the bottom turbid layer, which can hold in suspension so much mud (up to 30,000 mg/l) that it may be termed "fluid mud" and is recognisable in echosounder records, is much less thick than the intrusive wedge of sea water which contains it and it may, itself, display, internally, separate maxima of density and velocity (i.e., the fluid mud layer may itself be stratified). As the Maas estuary has but low tidal energy, the only high energy events it experiences are storm waves; in consequence, the fluid mud once formed in it, is not tidally dispersed and must be dredged (Parker and Kirby, 1977). In contrast, the lower Severn Estuary and eastern Bristol Channel contains a vertically virtually homohaline body of water, strongly stirred at ebb and flood by tidal currents. Although no field studies have yet been undertaken in sea states higher than 4, the maximum observed wave effect homogenises only the uppermost part of the water column; the mixing is almost wholly tidal. Rapid, continuous transmissometer profiles on springs tides have revealed (Kirby and Parker, 1973; Parker and Kirby, 1977) that the lower 10 m or so of a water column 15 m deep may contain a virtually uniform distribution of suspended particles (at concentrations around 10,000 mg/l), but, as the tidal cycle moves towards neaps, stratification of the suspended material develops, with layers up to 5 m thick, containing 15–20,000 mg/l of material, appearing at successively deeper layers of the water column. At neaps tides, Kirby and Parker (1977) report the presence of fluid mud layers, 3–4 m thick and containing around 40,000 mg/l of "static suspensions", developed over areas of kilometer scale in breadth, below the water column with its normal "mobile suspension" load of about 10,000 mg/l of mud. Some 70% of all the mobile fine sediment of the immediate region is believed to go through a "static" fluid mud stage during neaps (Parker and Kirby, cit.). The "fluid mud" layers are, again, recognisable on echosounder records, and are there revealed to be, themselves, layered: recording transmissometer casts show the presence of short-lived fine structures, on decimeter scales, within them. At the return of springs tides, and a corresponding increase in the speeds of maximum flood and ebb currents, the static suspensions are dispersed to the mobile state and differentiation of the water column disappears.

The patches of "fluid mud" appear to have their greatest development in the area of Bridgewater Bay, south-eastern Bristol Channel (Kirby and Parker, 1977; Parker and Kirby, 1977), but similar fluid mud layers appear at slack water of neaps tides in narrow channels in the westward part of the Bristol Channel (Fig. 8.7). Joyce (1973) has been able directly to sample the "fluid mud" which occurs at slack water (at both high and low tides) in the dredged approach channel to Swansea Docks and the estuary of the river Tawe; his bottom-water samples not only contained concentrations of suspended material around 5,000 mg/l but were of lower salinity (27‰) than the overlying water (31‰); "as the temperature of both layers was the same, the stable vertical density gradient must have been due to the suspended load of sediment in the bottom layer" (Joyce, 1973, p. 743). The reduced electrical resistivity recorded by Parker and Kirby (1977) in the "fluid mud" at the bottom of the Maas estuary may, be due, partly at least, to the effect of the high suspension concentration on the conductivity sensor, but the existence of a salinity gradient within the layers of static suspensions cannot yet be excluded entirely; only direct sampling could resolve this. As only 10% of the sediment in the Tawe estuary comes from the river itself, 90% must be of marine origin; the low salinity of the bottom water containing the static suspension load was explained by Joyce (1973) by the entrapment, in the estuary approaches, of high levels of suspensions in low salinity surface water by wave activity at low tide, the resulting highly turbid, low salinity but high density water being available to "proceed as a bottom current up the estuary" during the tidal excursion. The reality of this process, and its implications, still needs study.

The processes of mud entrainment, transport and deposition – no matter how they may operate in particular coastal or estuarine situations – undoubtedly result in the entrapment of vast (and commercially expensive) quantities of mud in estuaries and harbours. McCave (1973a) has estimated an annual entrapment of 1,200,000 metric tons of mud in the estuaries of Rhine–Maas–Scheldt region (more than one quarter of the total estimated supply) and of 1,250,000 metric tons in the New Waterway to Rotterdam alone of all the dredged channels; he estimated that some 7 million

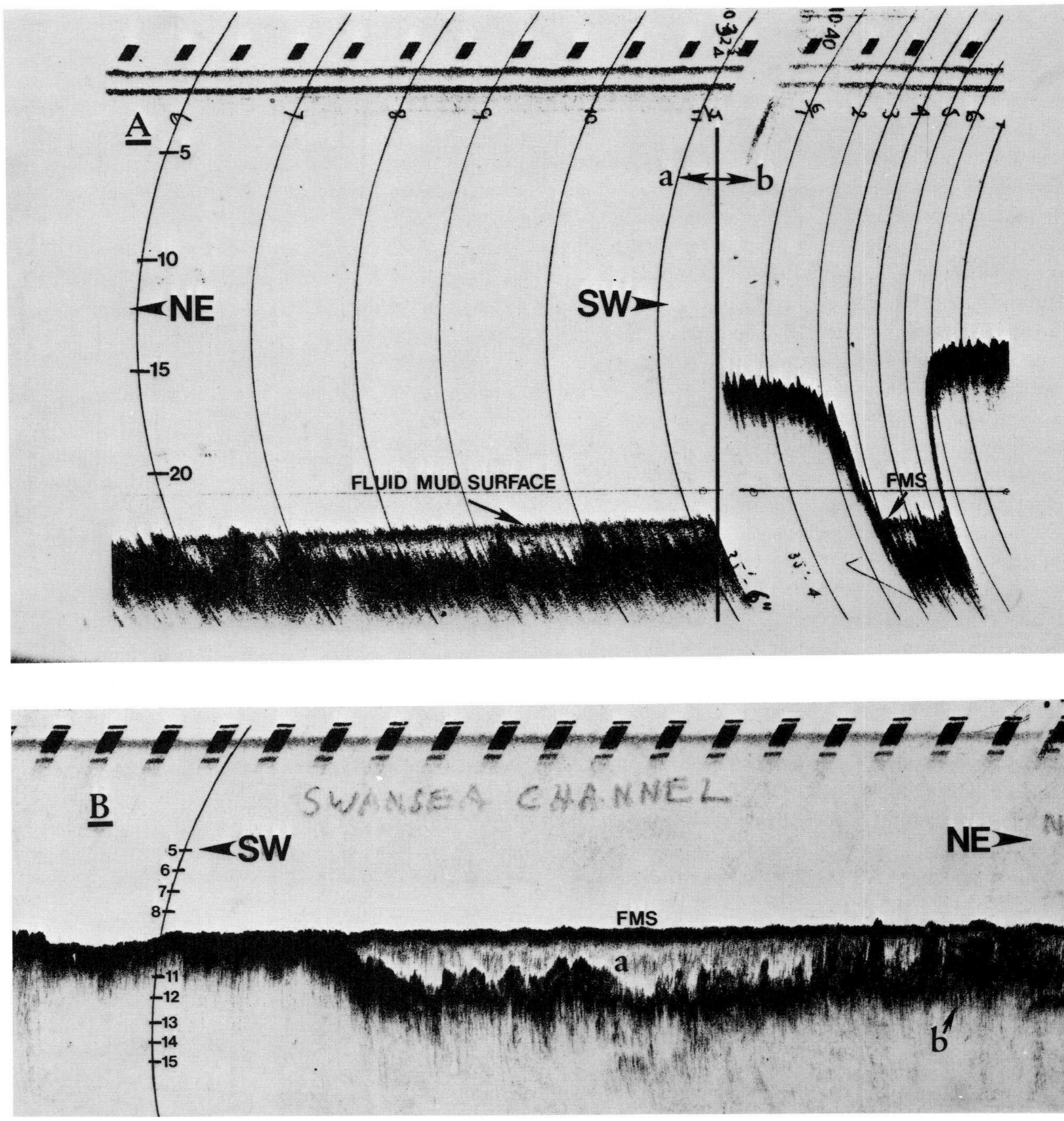

Fig. 8.7. A. Echosounder records (*a*) along and (*b*) across the dredged channel approaching the River Tawe estuary and Swansea Docks, showing stronger reflections from the consolidated muddy sands of the sea bed and above, in the dredged channel, weaker reflections from the upper boundary of the "fluid mud" static suspensions (*FMS*). Record taken in October, 1967, near High Water; depths in metres below hull-mounted transducers (uncorrected). Note the irregular topography of the consolidated bed below the "fluid mud" and the down-channel (to the *SW*) upward slope of the "fluid mud surface" (*FMS*). Minor wave-forms in each acoustic trace are largely artifacts introduced by ship movement in swell.

B. Echosounder record in the seaward part of the same channel, about 1½ hours after Low Water. The channel is full of "fluid mud" to down-channel sill depth. Note the weak, scattered reflections from within the midpart of the static suspension mass at *a*, and the stronger reflections of lenticular bodies enclosed within the mass at *b*. Depth in metres below transducers (uncorrected).

metric tons of mud were annually entrapped in the harbours of the North Sea, 1.8 million metric tons in the estuaries, and 800,000 metric tons in the Wadden Sea.

Following upon the publication of maps showing the known distribution of mud and other sediment grades on the sea bed (Luders, 1939; Pratje, 1949; U.S. Naval Oceanographic Office, 1965; Veenstra, 1971; McCave, 1973a; Lee and Ramster, 1976) attempts have been made (McCave, 1973a, c) to calculate a budget for the input, deposition and loss of muds into the North Sea. As McCave (1973a) has admitted, the figures for river supply are not well known or agreed; supply quantities from cliff erosion, oceanic inflow, aeolian supply, dumping and organic production can only broadly be estimated. The supply by erosion of consolidated, geologically sub-Recent, sea bed sediment is incalculable: "the net supply of 13 million metric tons. . . may, in fact, be completely overshadowed by a component derived from the winnowing of muddy sands, and this could possibly be order of magnitude greater" (McCave, 1973a, p. 94). The processes and patterns of sediment scour, transportation and redeposition by tides, waves, surges and all other hydrodynamic events on the northwestern European shelf still require much more investigation.

References

Adams, C.G., Benson, R.H., Kidd, R.B., Ryan, W.B.F. and Wright, R.C., 1977. The Messinian salinity crisis and evidence of Late Miocene eustatic changes in the world ocean. Nature, 269 (5627): 383–386.

Allen, J.R.L., 1970. Physical Processes of Sedimentation. Allen and Unwin, London, Earth Sci. Ser., 1: 248 pp.

Andreasen, A.H.M., Jensen, W. and Von Lundberg, J.J.V., 1929. Ein Apparat für die Dispersoid-Analyse und einige Untersuchungen damit. Kolloid Z., 49: 253–265.

Atkins, W.R.G., Jenkins, P.G. and Warren, F.J., 1954. The suspended matter in seawater and its seasonal changes as affecting the visual range of the Secchi disc. J. Mar. Biol. Assoc. U.K., 33: 497–509.

Banner, F.T. and Collins, M.B., 1975. Field meeting, 'Introduction to Oceanography'. Proc. Geol. Assoc., 86: 87–93.

Bassindale, R., 1943. Studies on the biology of the Bristol Channel. J. Ecol., 31: 1–29.

Belderson, R.H., 1964. Holocene sedimentation in the western half of the Irish Sea. Mar. Geol., 2: 147–163.

Belderson, R.H. and Stride, A.H., 1966. Tidal current fashioning of a basal bed. Mar. Geol., 4: 237–257.

Belderson, R.H. and Stride, A.H., 1969. Tidal currents and sandwave profiles in the northeast Irish Sea. Nature, 222: 74–75.

Belderson, R.H. and Wilson, J.B., 1973. Iceberg ploughmarks in the vicinity of the Norwegian Trough. Norsk. Geol. Tidsskr., 53: 323–328.

Belderson, R.H., Kenyon, N.H. and Stride, A.H., 1971. Holocene sediments on the continental shelf west of the British Isles. In: F.M. Delany (Editor), The Geology of the East Atlantic Continental Margin. Inst. Geol. Sci. Rep., 70/14: 157–170.

Boillot, G., 1965. Organogenic gradients in the study of neritic deposits of biological origin: the example of the western English Channel. Mar. Geol., 3: 359–367.

Boyse, P., Horn, R., Lapierre, F. and Le Lann, F., 1976. Etude des Grands Bancs de sable du sudest de la Mer Celtique. Mar. Geol., 20: 251–275.

Bowen, D.Q., 1970. South-east and central South Wales. In: C.A. Lewis (Editor), The Glaciations of Wales and Adjoining Regions. Longmans, London, pp. 197–228.

British Standards Institution, 1967. Methods of Testing Soils for Civil Engineering Purposes. BS. 13777. British Standards Institution, London, 234 pp.

Buch, A., 1972. Undersøgelser overdet marine Kvataer i Nordsoomvadet. Dansk. Geol. Foren. (1971 Års.): 86–90.

Buchan, S., Floodgate, G.D. and Crisp, D.J., 1967. Studies of the seasonal variation of suspended matter in the Menai Straits. Limnol. Oceanogr. 12 (3): 419–431.

Carr, A.P., Heathershaw, A.D. and Blackley, M.W.L., 1976. Swansea Bay (Sker) Project, Progress Report for the period August 1975 to July 1976. Inst. Oceanogr. Sci. Rep., 26.

Cartwright, D.E. and Stride, A.H., 1958. Sand transport at southern end of the North Sea. Dock Harbour Auth., 38: 323–324.

Casagrande, A. (Editor), 1934. Die Aerometer-Methode zur Bestimmung der Kornverteilung von Böden und anderen Mineralien. Springer, Berlin.

Caston, V.N.D., 1965. Localised sediment transport and submarine erosion in Tremadoc Bay, North Wales. Mar. Geol., 3: 401–410.

Caston, V.N.D., 1972. Linear sandbanks in the southern North Sea. Sedimentology, 18: 63–78.

Caston, G.F. 1975. Igneous dykes and associated scour hollows of the North Channel, Irish Sea. Mar. Geol., 18: M77–M88.

Caston, V.N.D. and Stride, A.H., 1970. Tidal sand movement between some linear banks in the North Sea off northeast Norfolk. Mar. Geol., 9: M38–M42.

Caston, V.N.D. and Stride, A.H., 1973. Influence of older relief on the location of sandwaves in part of the Southern North Sea. Estuarine Coastal Mar. Sci., 1: 379–386.

Channon, R.D. and Hamilton, D., 1976. Wave and tidal current sorting of shelf sediments southwest of England. Sedimentology, 23: 17–42.

Charlesworth, J.K., 1929. The South Wales end moraine. Q.J. Geol. Soc. London, 85: 335–358.

Chester, R. and Stoner, J., 1972. Concentration of suspended particulate matter in surface seawater. Nature, 240 (5383): 552–553.

Clarke, R.H., 1970. Quaternary sediments off southeast Devon. Q. J. Geol. Soc. London, 125: 277–318.

Cloet, R.L., 1954. Sandwaves in the southern North Sea and in the Persian Gulf. J. Inst. Navig., 7: 272–279.

Cloet, R.L., 1963. Hydrographic analysis of the sandbanks in the approaches to Lowestoft Harbour. Admir. Mar. Sci. Publ., London, 6: 15 pp. (multilith).

Cloet, R.L., 1966. First report on the Thames Estuary – development of the Edinburgh Channels environment. Thames Estuary Study Group, Tech. Pap., 18.

Cloet, R.L., 1967. Determining the dimensions of marine sediment circulation. Conference on the Technology of the Sea and the Seabed. H.M.S.O. London, pp. 546–563.

Collins, M.B., 1976. Suspended-sediment sampling towers as used on the intertidal flats of the Wash, Eastern England. Estuarine Coastal Mar. Sci., 5: 45–57.

Cooper, L.H.N., 1961. Comparison of three methods of measuring underwater illumination in coastal waters. J. Mar. Biol. Assoc. U.K., 41: 535–550.

Cooper, L.H.N., 1977. Studies by the American CLIMAP group considered as a foundation for understanding Quaternary events on the continental shelves peripheral to Great Britain and Ireland. In: M. Angel (Editor), A Voyage of Discovery. Pergamon, London, pp. 681–692.
Cooper, L.H.N. and Milne, A., 1938. The ecology of the Tamar estuary: II. Underwater illumination. J. Mar. Biol. Assoc. U.K., 22: 509–528.
Culver, S.J., 1976. A Study of the Post-Glacial Foraminiferida of Swansea Bay, South Wales. Unpubl. Ph.D. Thesis, University of Wales, 442 pp.
Culver, S.J. and Banner, F.T., 1978. Foraminiferal assemblages as Flandrian palaeoenvironmental indicators. Palaeogeogr., Palaeoclimatol., Palaeoecol., 24(1): 53–72.
Culver, S.J. and Banner, F.T., 1979. The significance of derived pre-Quaternary foraminifera in Holocene sediments of the north-central Bristol Channel. Mar. Geol., 29: 187–207.
Culver, S.J. and Bull, P.A., 1977. Late Pleistocene rock-basin lakes in South Wales. (Unpublished.)
Curray, J.R., 1965. Late Quaternary history, continental shelves of the United Sates. In: H.E. Wright and D.F. Frey (Editors) The Quaternary of the United States. Princeton University Press, Princeton, N.J., pp. 723–735.
Curry, D., 1962. A lower Tertiary outlier in the central English Channel with notes on the beds surrounding it. Q. J. Geol. Soc. London, 118: 177–205.
Curry, D., Murray, J.W. and Whittard, W.F., 1965. The geology of the western approaches of the English Channel, III. The *Globigerina* silts and associated rocks. In: W.F. Whittard and R. Bradshaw (Editors), Submarine Geology and Geophysics. Colston Pap. 17: 239–264.
Curry, D., Gray, F., Hamilton, D. and Smith, A.J., 1967. Chalk from the seabed south of Corke, Eire. Proc. Geol. Soc. London, 1640: 134–136.
Curry, D., Hamilton, D. and Smith, A.J., 1970. Geological and shallow subsurface geophysical investigations in the western approaches to the English Channel. Inst. Geol. Sci. NERC Rep. 70/3: 12 pp.
Curry, D., Hamilton, D. and Smith, A.J., 1971. Geological evolution of the western English Channel and its relation to the nearby continental margin. In: F.M. Delany (Editor), The Geology of the East Atlantic Continental Margin. Inst. Geol. Sci. Rep., 70/14: 129–142.
Davies, C.M., 1972. Aspects of Suspended Sediment Transport in Swansea Bay. Unpubl. Ph.D. Thesis, University of Wales, 241 pp.
De la Beche, H.T., 1851. Distribution and deposits of sediment in tidal seas. Geol. Obs., 7: 100–102.
Delanty, L.J. and Whittington, R.J., 1977. A reassessment of the "Neogene" deposits of the South Irish Sea and Nymphe Bank. Mar. Geol., 24: M23–M30.
Dietrich, G., 1963. General Oceanography. Wiley-Interscience, New York, N.Y.
Dingle, R.V., 1965. Sandwaves in the North Sea mapped by continuous reflection profiling. Marine Geol., 3: 391–400.
Dingle, R.V., 1971. Buried tunnel valleys off the Northumberland coast, western North Sea. Geol. Mijnbouw, 50: 679–686.
Dingwall, R.G., 1969. Geology of the central English Channel. Nature, 224: 358–359.
D'Olier, B. and Maddrell, R.J., 1970. Buried channels in the Thames estuary. Nature, 226: 347–348.
Donovan, D.T., 1965. Holocene origin of erosion hollows in the North Sea. Rep. Challenger Soc., 3 (17): 2 pp.
Donovan, D.T., 1972. Geology of the central English Channel. Mém. B.R.G.M., 79: 215–220.
Donovan, D.T., 1973. The geology and origins of the Silver Pit and other closed basins in the North Sea. Proc. Yorks. Geol. Soc., 39: 267–293.
Donovan, D.T. and Stride, A.H., 1961. An acoustic survey of the seafloor south of Dorset and its geological interpretation. Philos. Trans. R. Soc. London, Ser. B., 244 (712): 299–330.
Donovan, D.T. and Stride, A.H., 1975. Three drowned coastlines of probable Late Tertiary age around Devon and Cornwall. Mar. Geol., 19: 1935–1940.
Drake, D.E., 1976. Suspended sediment transport and mud deposition on continental shelves. In: D.J. Stanley and D.J.P. Swift (Editors), Marine Sediment Transport and Environmental Management. Wiley-Interscience, New York, N.Y., pp. 127–158.
Dyer, K.R., 1972. Recent sedimentation in the Solent area. Mém. B.R.G.M., 79: 271–280.
Eisma, D., 1966. The distribution of benthic marine molluscs off the main Dutch coast. Neth. J. Sea Res., 3: 107–163.
Emery, K.O., 1952. Continental shelf sediments of southern California. Geol. Soc. Am. Bull., 63: 1105–1108.
Emery, K.O., 1968. Relict sediments on continental shelves of the world. Am. Assoc. Pet. Geol. Bull., 52: 445–464.
Enger, P.F., Smerdon, E.T. and Masch, F.D., 1968. Erosion of cohesive sediments. Report of the Task Committee on Erosion of Cohesive Materials. Committee on Sedimentation. Proc. Am. Soc. Civ. Eng., J. Hydraul. Div., 94 (HY4): 1017–1049.
Evans, G. and Collins, M.B., 1976. The transportation and deposition of suspended sediment over the intertidal flats of the Wash. In: J.R. Hails and A.P. Carr (Editors), Nearshore Sediment Dynamics and Sedimentation. Wiley, New York, N.Y., pp. 45–57.
Fabricius, F. and Muller, S., 1970. A buret cylinder for grainsize analysis of silt and clay (with ALGOL program). Sedimentology, 14: 39–50.
Fairbridge, R.W., Ichiye, T. and Berry, J., 1966. The North Sea. In: R.W. Fairbridge (Editor), The Encyclopedia of Oceanography. Reinhold, New York, N.Y., pp. 543–552.
Ferentinos, G.K., 1976. Sediment distribution and transport processes on the outer continental shelf of the Hebridian Sea. Mar. Geol., 20: 41–56.
Ferentinos, G. and Collins, M., 1979. Effect of shoreline irregularities on a rectilinear tidal current and their significance in sedimentation processes. J. Sediment. Petrol. (in press).
Flinn, D., 1967. Ice front in the North Sea. Nature, 215: 1151–1159.
Flinn, D., 1973. The topography of the seafloor around Orkney and Shetland and in the northern North Sea. J. Geol. Soc. London, 129: 39–59.
Folk, R.L., 1966. A review of grain-size parameters. Sedimentology, 6: 73–94.
Folk, R.L., 1968. Petrology of Sedimentary Rocks. Hemphill's, University of Texas, 170 pp.
Folk, R.L., Andrews, P.B. and Lewis, D.W., 1970. Detrital sedimentary rock classification and nomenclature for use in New Zealand. N. Z. J. Geol. Geophys., 13: 937–968.
Germaneau, J. et al., 1972. Volcanic derivatives in the Seine and Central Channel. Mém. B.R.G.M., 79: 257–262.
Gibbs, H.J., 1962. A study of erosion and tractive force characteristics in relation to soil mechanics properties. U.S. Bur. Land Reclamation, EM-643.
Gibbs, R.J., 1972. The accuracy of particle-size analyses utilizing settling tubes. J. Sediment. Petrol., 42: 141–145.
Gibbs, R.J. (Editor), 1974. Suspended Solids in Water. Plenum Press, New York, N.Y., 320 pp.
Ginsburg, R.N. (Editor), 1975. Tidal Deposits. Springer, Berlin, 428 pp.

Godwin-Austen, R.A.C., 1850. On the valley of the English Channel. Q. J. Geol. Soc. London, 6: 69–97.
Goldstein, S., 1929. The steady flow of viscous fluid past a fixed spherical obstacle at small Reynolds Numbers. Proc. R. Soc. London, Ser. A., 123.
Graf, W.H., 1971. Hydraulics of Sediment Transport. McGraw-Hill, New York, N.Y., 513 pp.
Grim, R.E., 1962. Applied Clay Mineralogy. McGraw-Hill, New York, N.Y.
Groen, P., 1967. On the residual transport of suspended matter by an alternating tidal current. Neth. J. Sea Res., 3: 564–574.
Groupe NOROIS, 1972. Contribution a l'étude stratigraphique des terrains affleurant en marche centrale et orientale. Mém. B.R.G.M. 79: 123–190.
Gullentops, F., 1974. The southern North Sea during the Quaternary. In: P. Macar (Editor), L'Evolution Quaternaire des Basins Fluviaux de la Mer du Nord. Soc. Géol. Belge, Liège, pp. 273–280.
Gust, G. and Walger, E., 1976. The influence of suspended cohesive sediments on boundary layer structure and erosive activity of turbulent seawater flow. Mar. Geol., 22 (3): 189–206.
Hagmeier, E., 1962. Das Seston und seine Komponenten. Kieler Meeresforsch., 18: 189–197.
Hails, J.R., 1975. Submarine geology, sediment distribution and Quaternary history of Start Bay, Devon. Introduction. Sediment distribution and Quaternary history. J. Geol. Soc. London, 131: 1–6, 19–36.
Hails, J. and Carr, A. (Editors), 1975. Nearshore Sediment Dynamics and Sedimentation. Wiley, New York, N.Y., 316 pp.
Hamilton, D. and Smith, A.J., 1968. Sedimentary history of the Hurd Deep. Proc. Geol. Soc. London, 1650, 160.
Hamilton, D. and Smith, A.J., 1972. The origin and sedimentary history of the Hurd Deep, English Channel, with additional notes on other deeps in the western English Channel. Mem. B.R.G.M., 79: 59–78.
Harvey, J.G., 1966. Large sandwaves in the Irish Sea. Mar. Geol., 4: 49–55.
Harvey, H.W., Cooper, L.H.N., Lebour, M.V. and Russell, F.S., 1935. Plankton production and its control. J. Mar. Biol. Assoc. U.K., 20: 407–442.
Haven, D.S. and Morales-Alamo, R., 1966. Aspects of biodeposition by oysters and other invertebrate filter feeders. Limnol. Oceanogr., 11: 487–498.
Heathershaw, A.D., 1976. Measurements of turbulence in the Irish Sea benthonic boundary layer. In: N.I. McCave (Editor), The Benthic Boundary Layer. Plenum Press, New York, N.Y., pp. 11–32.
Heathershaw, D.C. and Simpson, J.H., 1974. Fine structure of light attenuation and its relation to temperature in the Irish Sea. Estuarine Coastal Mar. Sci., 2 (2): 91–104.
Hill, H.W. 1973. Currents and water masses. In: E.D. Goldberg (Editor), North Sea Science. M.I.T. Press, Cambridge, Mass., pp. 17–42.
Houbolt, J.J.H.C., 1968. Recent sediments in the Southern Bight of the North Sea. Geol. Mijnbouw, 47: 245–273.
Inglis, C.C. and Allen, F.H., 1957. The regimen of the Thames Estuary as affected by currents, salinities and river flow. Proc. Inst. Civ. Eng., 7: 827–878.
Inman, D.L., 1957. Wave generated ripples in nearshore sands. U.S. Army Corps Eng. Tech. Memo. 100: 1–41, A1–A23.
Inman, D.L. and Nordstrom, C.E., 1971. On the tectonic and morphological classification of coasts. J. Geol., 79: 1–21.
Jansen, J.H.F., 1976. Late Pleistocene and Holocene history of the northern North Sea, based on acoustic reflection records. Neth. J. Sea Res., 10: 1–43.
Jansen, J.H.F., 1977. Late Pleistocene and Holocene history of the North Sea north of the Dogger Bank. Abstr., Symp. Uppsala 7–12/3/1977, The Quaternary History of the Baltic, the North Sea and the Irish Sea, pp. 10–11.
Jerlov, N.G., 1976. Marine Optics. Elsevier Oceanogr. Ser., 14. Elsevier, Amsterdam, 232 pp.
Johnson, M.A. and Stride, A.H., 1969. Geological significance of North Sea transport rates. Nature, 224: 1016–1017.
Jones, N.S., Hain, J.M. and Stride, A.H., 1965. The movement of sandwaves on Warts Bank, Isle of Man. Mar. Geol., 3: 329–336.
Jones, S.R., 1974. Some results from Woodhead seabed drifter releases in the Western English Channel and Celtic Sea. MAFF, Fish. Lab., Tech. Rep. Ser., 10: 1–22.
Joyce, J.R., 1973. An improved bottom-water sampler. J. Mar. Biol. Assoc. U.K., 53: 741–744.
Kalle, K., 1953. Der Einfluss des englischen Küstenwassers auf dem Chemismus der Wasserkörper in den südlichen Nordsee. Ber. Dtsch. Wiss. Komm. Meeresforsch., 13.
Kelland, N.C., 1975. Submarine geology of Start Bay determined by continuous seismic profiling and core sampling. J. Geol. Soc. London, 131: 7–18.
Keller, G.H., 1976. Seafloor deposition, erosion and transportation. Report of the working group (leader: A.F. Richards). In: N.I. McCave, (Editor) The Benthic Boundary Layer. Plenum Press, New York, N.Y., pp. 247–260.
Kennedy, J.F., 1969. The formation of sediment ripples, dunes and antidunes. Annu. Rev. Fluid Mech., 1: 147–168.
Kent, P.E., 1975. The tectonic development of Great Britain and the surrounding seas. In: A.W. Woodland (Editor), Petroleum and the Continental Shelf of North-West Europe. 1. Geology, Applied Science Publishers, London, pp. 3–28.
Kenyon, N.A., 1970a. Sandribbons of European tidal seas. Mar. Geol., 9: 25–39.
Kenyon, N.H., 1970b. The origin of some transverse sand patches in the Celtic Sea. Geol. Mag, 107: 389–394.
Kenyon, N.H. and Stride, A.H., 1968. The crest length and sinuosity of some marine sand waves. J. Sediment. Petrol., 38: 255–259.
Kenyon, N.H. and Stride, A.H., 1970. The tideswept continental shelf sediments between the Shetland Islands and France. Sedimentology, 14: 159–173.
Kidson, C. and Tooley, M.J. (Editors), 1977. The Quaternary history of the Irish Sea. Geol. J., Spec. Iss., 7: 345 pp.
King, W.B.R., 1954. The geological history of the English Channel. Q. J. Geol. Soc. London, 110: 77–101.
Kirby, R. and Kelland, N.C., 1972. Adjacent stable and apparently mobile linear sediment ridges in the southern North Sea. Nature (Phys. Sci.), 238: 111–112.
Kirby, R. and Parker, W.R., 1973. Fluid mud in the Severn Estuary and Bristol Channel and its relevance to pollution studies. Inst. Chem. Eng., Grads. Stud. Sect., Ann. Symp., Est. Coastal Poll., Exeter, 27–28/9/73, Pap., A4 (Preprint),
Kirby, R. and Parker, W.R., 1977. Sediment dynamics in the Severn Estuary: background for studies of the effects of a barrage. In: T.L. Shaw (Editor), An Environmental Appraisal of the Severn Barrage. Dep. Civ. Engineering, University of Bristol, pp. 41–52.
Komar, P.D. and Miller, M.C., 1975. On the comparison of the threshold of sediment motion under waves and unidirectional currents, with a discussion of the practical evaluation of the threshold. J. Sediment Petrol., 45: 362–367.
Kraav, V.K., 1969. Computation of the semidiurnal tide and turbulence parameters in the North Sea. Oceanology, 9 (3): 332–341.
Krone, R.B., 1977. Application of rheological measurements to layer. In: N.I. McCave (Editor), The Benthic Boundary Layer. Plenum Press, New York, N.Y., pp. 143–156.
Krone, R.B., 1977. Application of rheological measurments to description of cohesive sediment transport. Euromech 95, Univ. College Swansea Symp. 13–15/9/77, Abstr. (et verb.), preprint.

Krumbein, W.C., 1934. Size frequency distribution of sediments. J. Sediment Petrol., 4: 65–77.
Krumbein, W.C., 1936. The application of logarithmic moments to size frequency distributions. J. Sediment. Petrol., 6: 35–47.
Krumbein, W.C. and Pettijohn, F.J., 1938. Manual of Sedimentary Petrography. Appleton-Century-Crofts, New York, N.Y., 549 pp.
Langeraar, W., 1966. Sandwaves in the North Sea. Hydrogr. Newslett., 1: 243–246.
Langhorne, D.N., 1973. A sandwave field in the outer Thames estuary. Mar. Geol., 14: 129–143.
La Pierre, F., Robert, J.P. and Ville, P., 1970. Esquisse géologique de la Manche orientale. C.R. Acad Sci. Paris, Ser. D.: Sci. Nat., 271 (4): 381–389.
Larsonneur, C., 1969. La géologie sous-marine aux abords du Cotentin, entre Cherbourg et Saint-Vaast-La Houge (Manche). C.R. Somm. Soc. Géol. Fr., 5: 155.
Larsonneur, C., 1972. The sedimentary model of the Baie de Seine/central Channel area in its geographical and historical frame. Mém. B.R.G.M., 79: 241–235.
Lee, A.J. and Ramster, J.W., 1976. Atlas of the seas around the British Isles. MAFF, Fish. Res. Tech. Rep., 20.
Lefort, J.P. and Deunff, J., 1971. Esquisse géologique de la partie meridionale du Golfe Normano-Breton (Manche). C.R. Hebd. Seanc., Acad. Sci. Paris, 272: 16–19.
Lerman, A., Lal, D. and Dacey, M.F., 1974. Stokes' settling and chemical reactivity of suspended particles in natural waters. In: R.F. Gibbs (Editor), Suspended Solids in Water. Plenum Press, New York, N.Y., pp. 17–48.
Luders, K., 1939. Sediments of the North Sea. In: P.D. Trask (Editor), Recent Marine Sediments. Am. Assoc. Pet. Geol. (Reprinted by Dover Publications, New York, N.Y., 1968), pp. 322–342.
Lyle, W.M. and Smeardon, E.T., 1965. Relation of compaction and other soil properties to erosion and resistance of soils. Trans. Am. Soc. Agric. Eng., 8 (3).
McCarthy, J.C., Pyle, T.E. and Griffin, G.M., 1974. Light transmissivity, suspended sediments and the legal definition of turbidity. Estuarine Coastal Mar. Sci., 2 (3): 291–299.
McCave, I.N., 1969. Deposition of fine-grained sediment from tidal currents. Nature, 224: 1288–1289.
McCave, I.N., 1970. Deposition of fine-grained suspended sediment from tidal currents. J. Geophys. Res., 75: 4151–4159.
McCave, I.N., 1971a. Wave effectiveness at the seabed and its relationship to bedforms and the deposition of mud. J. Sediment. Petrol., 41: 89–96.
McCave, I.N., 1971b. Sandwaves in the North Sea off the coast of Holland. Mar. Geol., 10: 199–225.
McCave, I.N., 1972. Transport and escape of fine-grained sediment from shelf areas. In: D.J.P. Swift, D.B. Duane and O.H. Pilkey (Editors), Shelf Sediment Transport. Douden, Hutchinson and Ross, Stroudsburg, Pa., pp. 225–248.
McCave, I.N., 1973a. Mud in the North Sea. In: E.D. Goldberg, (Editor), North Sea Science. M.I.T. Press, Cambridge, Mass., pp. 76–100.
McCave, I.N., 1973b. Some boundary-layer characteristics of tidal currents bearing sand in suspension. Mém. Soc. R. Sci. Liège, 6 (6): 187–206.
McCave, I.N., 1973c. A budget for mud in suspension and on the bed of the North Sea. Proc. Challenger Soc., 4 (5).
McCave, I.N., 1975. Vertical flux of particles in the ocean. Deep Sea Res., 22: 491–502.
McCave, I.N. (Editor), 1976. The Benthic Boundary Layer. Plenum Press, New York, N.Y., 323 pp.
McDowell, D.M. and O'Connor, B.A., 1977. Hydraulic Behaviour of Estuaries. MacMillan, New York, N.Y.
McManus, D.A., 1975. Modern versus relict sediment on the continental shelf. Geol. Soc. Am. Bull., 86: 1154–1160.
McManus, J., West, J.R. and Williams, D.J.A., 1969. Interfering bedforms produced by oblique ebb and flood currents on an estuarine sand bank. Geol. Soc. London, Circ., 155: 2–3.
Macar, P. (Editor), 1974. L'Evolution quaternaire des Basins fluviaux de la Mer du Nord. Soc. Géol. Belg., Liège, 318 pp.
Maddrell, R.J. and Prentice, J.E., 1967. The Black Deep – a critical area of the Thames estuary. Proc. Conf. Technol. Sea and the Seabed, H.M.S.O., London, 589–593.
Meade, R.H., 1972. Transport and deposition of sediments in estuaries. In: B. Nelson (Editor). Environmental framework of coastal plain estuaries. Geol. Soc. Am. Mem., 133: 91–120.
Meyer, R.E. (Editor), 1972. Waves on Beaches, and Resulting Sediment Transport. Academic Press, New York, N.Y.
Migniot, Cl., 1977. Action des courants, de la houle et du vent sur les sédiments. Houille Blanche, 1/1977: 9–47.
Mitchell, G.F., 1963. Morainic ridges on the floor of the Irish Sea. Irish Geogr., 4: 335–344.
Moore, J.G., 1947. The determination of depths and extinction coefficients of shallow water by airphotography using colour filters. Phil. Trans. R. Soc. London, Ser. B., 240: 163–217.
Murray, J.W., 1965. Significance of benthic foraminiferids in plankton samples. J. Paleontol., 39: 156–157.
Murray, J.W., 1967. Transparent and opaque Foraminiferid tests. J. Paleontol., 41: 791.
Murray, J.W. and Hawkins, A.B., 1976. Sediment transport in the Severn Estuary during the past 8,000–9,000 years. J. Geol. Soc. London, 132: 385–398.
Murray, J.W. and Wright, C.A., 1970. Surface textures of calcareous foraminiferids. Palaeontology, 13: 184–187.
Naylor, D. and Mounteney, S.N., 1975. Geology of the North-West European Continental shelf. Graham Trotman Dudley, London, Vols. 1 and 2.
Nelson-Smith, A. and Bridges, E.M. (Editors), 1977. Problems of a Small Estuary. Institute of Marine Studies, University College of Swansea, 381 pp.
Newton, A.J. and Gray, J.S., 1972. Seasonal variation of the suspended solid matter off the coast of North Yorkshire. J. Mar. Biol. Assoc. U.K. 52: 33–47.
North, F.J., 1964. The Evolution of the Bristol Channel. National Museum of Wales, Cardiff, 3rd ed.
Odd, N.V.M. and Owen, M.W., 1972. A two-layer model of mud transport in an estuary. Proc. Inst. Civ. Eng., Pap. 7517S.
Oele, E., 1969. The Quaternary geology of the Dutch part of the North Sea, north of the Frisian Isles. Geol. Mijnbouw, 48: 467–480.
Oele, E., 1971a. The Quaternary geology of the southern area of the Dutch part of the North Sea. Geol. Mijnbouw, 50: 461–474.
Oele, E., 1971b. Late Quaternary geology of the North Sea southeast of the Dogger Bank. In: F.M. Delany (Editor), The Geology of the East Atlantic Continental Margin, 3. ICSU/SCOR Working Party Symp., Cambridge, 1970; Rep. Inst. Geol. Sci., pp. 25–34.
Off, T., 1963. Rhythmic linear sandbodies caused by tidal currents Am. Assoc. Pet. Geol. Bull., 47: 324–341.
Oseen, C., 1927. Hydrodynamik (Chapter 10). Akademische Verlagsgesellschaft, Leipzig.
Otto, L., 1966. Light attenuation in the North Sea and Dutch Wadden Sea in relation to Secchi disk visibility and suspended matter. Neth. J. Sea Res., 3 (1): 28–51.
Otto, L., 1967. Investigations on optical properties and watermasses of the southern North Sea. Neth. J. Sea Res., 3 (4): 532–552.

Pak, H. Beardsley, G.F., Heath, G.R. and Curt, H., 1970. Light scattering vectors of some marine particles. Limnol. Oceanogr. 15: 683–657.

Parker, W.R. and Kirby, R., 1977. Vertical concentration profiles in estuarine fine sediment suspensions. Euromech 95, Symp. Univ. College Swansea, 13–15/9/1977, Preprint, Abstr., 14/9/77.

Pingree, R.D., Pugh, P.R., Holligan, P.M. and Forster, G.R. 1975. Summer phytoplankton blooms and red tides along tidal fronts in the approaches to the English Channel. Nature, 258 (5537): 672–677.

Platt, J., 1977. Significance of pock-marks for engineers. Offshore Eng., 8: 45.

Postma, H., 1960. Einige Bemerkungen über den Sinkstofftransport im Ems-Dollart Gebiet. Verh. K. Ned. Geol. Mijnbouwk. Genoot., Geol. Ser., 19: 103–110.

Postma, H., 1961a. Transport and accumulation of suspended matter in the Dutch Wadden Sea. Neth. J. Sea. Res., 1: 149–190.

Postma, H., 1961b. Suspended matter and secchi-disk visibility in coastal waters. Neth. J. Sea Res., 1: 359–390.

Postma, H., 1967. Sediment transport and sedimentation in the estuarine environment. In: G.H. Lauff (Editor), Estuaries. Am. Assoc. Adv. Sci., Washington, D.C., pp. 158–179.

Pratje, O., 1949. Die Bodenbedeckung der nordeuropäischen Meere. Handbuch der Seefischerei Nordeuropas, 1 (3). Schweizerbart, Stuttgart.

Pratje, O., 1951. Die Deutung der Steingrunde in der Nordsee als Endmoränen. Dtsch. Hydrogr. Z., 4: 106–114.

Prentice, J.E., 1972. Sedimentology of the North Sea and the English Channel. Mém. B.R.G.M., 79: 229–232.

Ramster, J.W., 1965. Studies with the Woodhead seabed drifter in the southern North Sea. Lab. Leafl., Fish. Lab. Lowestoft (N.S.), 6: 1–4.

Raudkivi, A.J., 1976. Loose Boundary hydraulics. Pergamon, London, 2nd ed., 397 pp.

Reineck, H.E., 1963. Sedimentgefüge in Bereich der Südlichen Nordsee. Abh. Senckenberg. Naturforsch. Ges., 505.

Reineck, H.E., 1967. Vergleich von Schlamm-Analysen mit der Pipette nach Kohn und dem Areometer nach Casagrande. Senckenbergiana Letheae, 48: 351–356.

Reineck, H.E. (Editor), 1970. Das Watt. Ablagerungs- und Lebensraum. Valdemar Kramer, Frankfurt-am-Main, 142 pp.

Rhoads, D.C. and Young, D.K., 1970. The influence of deposit-feeding organisms on sediment stability and community trophic structure. J. Mar. Res., 28: 150–178.

Riley, J.P. and Chester, R., 1971. Introduction to Marine Chemistry. Academic Press, New York, N.Y., 465 pp.

Robert, J., 1971. Results of a sparker reconnaissance survey of the eastern and central English Channel. In: F.M. Delany (Editor). The Geology of the East Atlantic Continental Margin, 3. ICSU/SCOR Working Party Symp., Cambridge, 1970; Rep. Inst. Geol. Sci. pp. 35–42.

Robinson, A.W.H., 1966. Residual currents in relation to shoreline evolution of the East Anglian coast. Mar. Geol., 4: 57–84.

Robinson, A.H.Q., 1968. The submerged glacial landscape off the Lincolnshire coast. Trans. Inst. Brit. Geogr., 44: 119–132.

Ruellan, F., Beigbeder, Y. and Dagorne, A., 1972. The Saint Malo and Saint Brieuc sheets of the sedimentological map of the sea floor. Mém. B.R.G.M., 79: 281–292.

Shaw, H.F., 1973. Clay mineralogy of Quaternary sediments in the Wash embayment, eastern England. Mar. Geol., 14: 29–45.

Shepard, F.P., 1931. Glacial troughs of the continental shelves. J. Geol., 39: 345–360.

Sindowski, K.H., 1970. Das Quartär im Untergrund der Deutschen Bucht (Nordsee). Eiszeitalter Ggw., 21: 33–46.

Slatt, R.M., 1974. Formation of palimpsest sediments, Conception Bay, southeastern Newfoundland. Geol. Soc. Am. Bull, 85: 821–826.

Smith, A.J. and Hamilton, D., 1970. Origin of the Hurd Deep, English Channel. Nature, 227: 828.

Smith, J.D., 1970. Stability of a sand bed subjected to a shear flow of low Froude number. J. Geophys. Res., 75: 5928–5940.

Stanley, D.J., 1968. Reworking of glacial sediments in the North West Arm, a fjord-like inlet on the southeast coast of Nova Scotia. J. Sediment. Petrol., 38: 1224–1241.

Stanley, D.J. and Swift, D.J.P. (Editor), 1976). Marine Sediment Transport and Environmental Management. Wiley, New York, N.Y.,

Sternberg, R.W., 1972. Predicting initial motion and bedload transport of sediment particles in the shallow marine environment. In: D.J.P. Swift, D.B. Duane and O.K. Pilkey (Editors), Shelf Sediment Transport: Process and Pattern. Dowden Hutchinson and Ross, Stroudsburg, Pa., pp. 61–82.

Strickland, J.D.H. and Parsons, T.R., 1968. A Practical handbook of seawater analysis. Fish. Res. Board Can., Bull, 167: 293 pp.

Stride, A.H., 1959. On the origin of the Dogger Bank, in the North Sea. Geol. Mag., 96: 33–44.

Stride, A.H., 1963a. Current swept sea floors near the southern half of Great Britain. Q. J. Geol. Soc. London, 119: 175–199.

Stride, A.H., 1963b. Northeast trending ridges of the Celtic Sea. Proc. Ussher Soc., 1: 62–63.

Stride, A.H., 1965. Marine geology at the National Institute of Oceanography. The Times Sci. Rev., Winter, 1965.

Stride, A.H., 1970. Shape and size trends for sandwaves in a depositional zone of the North Sea. Geol. Mag., 107: 469–477.

Stride, A.H., 1973. Sediment transport by the North Sea. In: E.D. Goldberg (Editor), North Sea Science M.I.T. Press, Cambridge, Mass., pp. 101–150.

Stride, A.H., Belderson, R.H and Kenyon, N.H., 1972. Longitudinal furrows and depositional sand bodies of the English Channel. Mem. B.R.G.M., 79: 235–240.

Sundborg, A., 1967. The River Klaralven, a study of fluvial processes. Geogr. Ann., 38: 127–316.

Swift, D.J.P., 1976. Coastal Sedimentation, In: D.J. Stanley and D.J.P. Swift (Editors), Marine Sediment Transport and Environmental Management. Wiley, New York. N.Y., pp. 255–310.

Swift, D.J.P., Duane, D.B. and Pilkey, O.H. (Editors), 1970. Shelf Sediment Transport: Process and Pattern. Dowden, Hutchinson and Ross, Stroudsburg, Pa.

Swift, D.J.P., Stanley, D.J. and Curray, J.R., 1971. Relict sediments on continental shelves: a reconsideration. J. Geol, 79: 322–346.

Swift, D.J.P., Schubel, J.R. and Sheldon, R.W., 1972. Size analysis of fine-grained suspended sediments: a review. J. Sediment. Petrol., 42: 122–134.

Tait, R.V., 1968. Elements of Marine Ecology. Butterworths, London, 272 pp.

Tanner, W.F., 1969. The particle size scale. J. Sediment. Petrol., 39: 809–812.

Terwindt, J.H.J., 1967. Mud transport in the Dutch Delta area and along the adjacent coastline. Neth. J. Sea Res., 3: 505–531.

Terwindt, J.H.J., 1970. Observations on submerged sand ripples with heights ranging from 30 to 200 cm, occurring in tidal channels of the S.W. Netherlands. Geol. Mijnbouw, 49: 489–501.

Terzaghi, K. and Peck, 1968. Soil Mechanics in Engineering Practice. Wiley, New York, N.Y.

Truesdell, P.E. and Varnes, D.J., 1950. Chart Correlating various Grain-Size Definitions of Sedimentary Materials. U.S. Geol. Survey, Washington, D.C.

U.S. Naval Oceanographic Office, 1965. Marine Geology. Oceanographic Atlas of the North Atlantic Ocean. Section 5.

Valentin, H., 1957. Die Grenze der Letzten Vereisung im Nordseeraum. Verh. 30 Dtsch. Geographentags. Hamburg, 1–5/8/ 1955, pp. 359–366.

Van Eerde, L.A.A., 1964. Geological investigations on the Brown Ridge. Hydrogr. Newslett., 1: 81–83.

Van Straaten, L.M.J.U. and Kuenen, P.H., 1957. Accumulation of fine-grained sediments in the Dutch Wadden Sea. Geol. Mijnbouw (N.S.), 19: 329–354.

Van Veen, J., 1935. Sandwaves in the North Sea. Hydrogr. Rev. 12: 21–29.

Veenstra, H.J., 1964. Geology of the Hinder Banks, south North Sea. Hydrogr. Newslett. 1: 72–80.

Veenstra, H.J., 1965. Geology of the Dogger Bank area, North Sea. Mar. Geol., 3: 245–262.

Veenstra, H.J., 1969. Gravels of the southern North Sea. Mar. Geol., 7: 449–464.

Veenstra, H.J., 1970. Quaternary North Sea coasts. Quaternaria, 12: 169–189.

Veenstra, H.J., 1971. Sediments of the southern North Sea. In: F.M. Delany (Editor), The Geology of the East Atlantic Continental Margin. 3. ICSU/SCOR Working Party Symp., Cambridge, 1970; Rep. Inst. Geol. Sci., pp. 9–24.

Velde, B., 1977. Clays and Clay Minerals in Natural and Synthetic Systems. Develop. Sedimentol., 21. Elsevier, Amsterdam, 218 pp.

Visser, M.P., 1970. The turbidity of the southern North Sea. Dtsch. Hydrograph. Z, 23 (2): 97–117.

Von Sager, G., 1963. Atlas der Elemente des Tidenhubs und der Gezeitenströme für die Nordsee, den Kanal und die Irische See. Deutsch. Akad. Wiss. Berlin, Inst. Meereskunde, Rostock 25 pp.

Von Sager, G. and Sammler, R., 1968. Atlas der Gezeitenströme für die Nordsee, den Kanal und die Irische See. Deutsch. Akad. Wiss. Berlin, Inst. Meereskunde, Rostock, 45 pp.

Watson, J., 1977. The Outer Hebrides: a geological perspective. Proc. Geol. Assoc., 88: 1–14.

Wentworth, C.K., 1922. Grade and class terms for clastic sediments. J. Geol., 30: 327–392.

Zagwijn, W.H. and Veenstra, H.J., 1966. A pollen-analytical study of cores from the Outer Silver Pit, North Sea. Mar. Geol., 4: 539–551.